D1559543

PRINCIPLES AND PRACTICE OF AVIATION PSYCHOLOGY

Human Factors in Transportation

A Series of Volumes Edited by
Barry H. Kantowitz

Barfield/Dingus • *Human Factors in Intelligent Transportation Systems*

Billings • *Aviation Automation: The Search for a Human-Centered Approach*

Garland/Wise/Hopkin • *Handbook of Aviation Human Factors*

Hancock/Desmond • *Stress, Workload, and Fatigue*

Noy • *Ergonomics and Safety of Intelligent Driver Interfaces*

O'Neil/Andrews • *Aircrew Training and Assessment*

Parasuraman/Mouloua • *Automation and Human Performance: Theory and Application*

Tsang/Vidulich • *Principles and Practice of Aviation Psychology*

Wise/Hopkin • *Human Factors in Certification*

For more information about LEA's Human Factors in Transportation series, please contact Lawrence Erlbaum Associates, Publishers, at *www.erlbaum.com.*

PRINCIPLES AND PRACTICE OF AVIATION PSYCHOLOGY

Edited by

Pamela S. Tsang
Wright State University

Michael A. Vidulich
Air Force Research Laboratory

LAWRENCE ERLBAUM ASSOCIATES, PUBLISHERS
2003 Mahwah, New Jersey London

Senior Acquisitions Editor:	Anne Duffy
Editorial Assistant:	Kristen Duch
Cover Design:	Kathryn Houghtaling Lacey
Textbook Production Manager:	Paul Smolenski
Full-Service Compositor:	Black Dot Group / An AGT Company
Text and Cover Printer:	Sheridan Books, Inc.

Cover photo courtesy of Daryl Chapman. Final approach to Hong Kong's Kai Tak airport during its final days of operation. Despite the exacting and least forgiving challenge of negotiating a wide body jet amidst precipitous terrain and hair-raisingly close highrises, pilots have routinely landed safely for many years.

This book was typeset in 10/12 pt. Times Regular, Italic, and Bold. The heads were typeset in Americana Bold and Times Italic.

Lawrence Erlbaum Associates, Inc., Publishers
10 Industrial Avenue
Mahwah, New Jersey 07430

Library of Congress Cataloging-in-Publication Data

Principles and practice of aviation psychology / edited by Pamela S. Tsang, Michael A. Vidulich.
p. cm. — (Human factors in transportation)
Includes bibliographical references and index.
ISBN 0-8058-3390-0
1. Aviation psychology. I. Tsang, Pamela S. II. Vidulich, Michael A. III. Series.
RC1085 .P75 2002
629.132′52′019—dc21 2002024377

Books published by Lawrence Erlbaum Associates are printed on acid-free paper, and their bindings are chosen for strength and durability.

Printed in the United States of America
10 9 8 7 6 5 4 3

To my parents, Michael and Pansy,
for their wisdom
and
to my grandmothers, Yeung Guk, Lee Sui Woo, Chu Hew Fa, and Lai Sui Kim
for their fortitude

—Pamela S. Tsang

To my parents, Joseph and Therese,
for their love and support

—Michael A. Vidulich

Contents

Preface

This book has two main objectives. First, to serve as a training tool for a graduate-level course in aviation psychology or more generally in engineering psychology. In aviation psychology, aviation is the domain of interest; in the engineering psychology, aviation serves as a context on which psychological principles are applied. The book aims at introducing psychological principles and research that are important in aviation; however, principles such as that of human perceptual and attentional capabilities and limitations can certainly be applied to contexts other than aviation. Second, we hope the research findings and applications in aviation psychology presented will be interesting and useful to academic researchers as well as professionals from the industry and government.

Chapters are arranged by important topics in aviation psychology, but they are by no means the only important topics. Given that selection is necessary, our focus is on the pilot in the cockpit (as opposed to the entire aviation system). The impact of technology on piloting and the strategies and approaches aviation psychology as a field can take to meet this ever-changing challenge in the future is another underlying theme.

Our goal is to present state-of-the-art knowledge in the various topics chosen for the book. To this end, we invited experts to author chapters in the area of their expertise. One concern was how the chapters might come together as a coherent treatment of the selected scope of the field; we think that the readers will find minimum redundancy across chapters and the many cross-references between chapters will illustrate how the various topics are interrelated. While the order of the chapters makes sense to us, we do not think it necessary to follow the order presented in the book as each chapter is a complete document on its own.

An important step in the creation of this book was the review process to which each chapter was subjected. We express our deepest appreciation to our reviewers for their generous, constructive contribution of their expertise. Their high-caliber reviews have greatly enhanced the quality and clarity of the final product. We thank: Herbert Bell, Charles Billings, Eugene Burke, Bob Cheung, Francis Durso, Robert Ginnett, Ralph Haber, David Hardy, Keith Hendy, David Hunter, Paul Jacques, Richard Jensen, Raymond King, Steve Landry, William Levison, Monica Martinussen, Grant McMillan, Todd Nelson, Tomas Nygren, Richard Pew, Fred Previc, Amy Pritchett, Roy Ruddle, Robert Shaw, Dominic Simon, Michael Skinner, Robert Taylor, and Christopher Wickens.

The project took much longer than we could have imagined, but we also learned far more than we could have hoped. We thank our authors for sharing their expert knowledge and professional experience. We are in debt to their guidance, cooperative spirit, and patience throughout the production process. We have learned greatly from them.

We are most happy that Captain Neil Johnston consented to write the foreword to this book. His considerable experience as a pilot and a researcher makes him a uniquely appropriate person to provide perspective on the field of aviation psychology.

We would also like to thank Barry Kantowitz, the series editor, Anne Duffy, and other editors and staff at Lawrence Erlbaum Associates for their cheery prompting, their patience, and their guidance and assistance.

Pamela gratefully acknowledges the grant support from the National Institute of Aging, the Ohio Board of Regents, and the National Science Foundation during the course of preparing the book for publication.

Foreword

As someone with a long-standing interest in aviation psychology, I was delighted to be asked to write the foreword to *Principles and Practice of Aviation Psychology*. My personal interest in aviation human factors dates back to my days as an airline copilot in the late 1960s, when I quickly became aware that there was more to aviation than basic stick and rudder skills. I was fortunate to become actively involved in various human factors endeavors in the early 1970s, before the roles of human factors and aviation psychology were accepted as valid by the aviation industry or even by most legislators and safety specialists. Reflecting back on that time, it is hard to believe the extent to which the aviation industry, with its predominant technical focus, was insulated from the outside world. As a result of this isolation, the internal battle to break down prejudices against aviation psychology and human factors took some time to complete. On the other hand, success, when it did arrive, was decisively achieved within a relatively short time.

The key initiative of that era was launched in 1975 at the 20th Technical Conference of the International Air Transport Association (IATA). The official report of the conference subsequently claimed that it had established "...a major and very important change in attitude in the industry towards certain classes of accidents." The world's aviation industry announced two important strategic policy changes at the conference: (a) the official end of the mindless attribution of "pilot error" following accidents and (b) a parallel need to involve those from outside the aviation industry who have expertise in the human sciences.

At that time there were no specialist conferences dedicated to aviation psychology or any specialist books on the subject. The first such book, by Stan Roscoe, appeared in 1980 and was intended for university students. The first dedicated aviation human factors textbook, by Frank Hawkins, followed some seven years later. Hawkins' book was intended to meet the needs of both university students and practitioners, and its appeal continues to this day. While it was an important and groundbreaking text in many ways, most practitioners felt Hawkins' book was too academic to enable them to draw practical training benefits from its contents. However, it was an important initiative, and it performed the valuable service of correctly anticipating the direction in which we needed to go.

Throughout the 1980s there was a steady growth of interest in human factors and a most welcome interaction between various research communities and those involved on a day-to-day basis within the industry. This interaction was greatly facilitated by events such as the active involvement of NASA in air carrier research projects and the inauguration of the first international aviation psychology conference (in 1981). This interaction has, I believe, continued to be of considerable benefit to both communities. Since the start of the 1990s, we have seen a rapid growth of interest in human factors and aviation psychology. The range of publications, conferences, and breadth of topics addressed has become almost overwhelming, and we have moved to the point where it would be a full-time job just to keep track of them all.

I make these introductory remarks by way of welcoming the publication of *Principles and Practice of Aviation Psychology* and with the specific objective of acknowledging just how far we have come in such a short period. These initial observations may also serve to remind us how easy it would be to take for granted the range and depth of the contents of this book. A brief perusal of the contents will quickly establish that it contains a comprehensive and wide-ranging treatment of a domain that has seen a progressive deepening of understanding and sophistication.

We may, for example, take it for granted that all the chapters incorporate credible and relevant aviation examples and applications, appropriately embedded within broad and solid theoretical foundations. That we have steadily advanced to a position where we can achieve this state of affairs should not blind us to the editors' success in actually achieving such a worthy goal. Each chapter is rooted in the "real world" of aviation and performs the valuable service of presenting us with solid theoretical discussions that combine applied relevance and practical insights. I also found helpful and instructive the discussion of the many issues or areas of contention that have not been tackled in the different domains of aviation psychology. By way of example, one might consider the comprehensive treatment of the "platform motion debate" in the chapter on flight simulation by Kaiser and

Schroeder or the discussion of issues relating to pilot selection in the chapter by Carretta and Ree.

As outlined in the first chapter, the development of the discipline of aviation psychology is intimately tied to the historical development of aviation and, in particular, the problems that this development has fostered. The capabilities of aircraft changed radically over the course of the last century, and the role of the pilot—along with others in the aviation system—has necessarily changed in parallel. New challenges for aviation psychology arise with each change in the role and function of pilots and others within the aviation domain. It has been well identified that this trend is seen at its most "cognitively opaque" in the context of the gradual distancing of pilots from direct control of their aircraft. Interaction with automation increasingly demands more sophisticated—or at least "different"—cognition and cockpit management techniques. The contents of this book pay due homage to this trend, especially in respect of developments and issues in automation design and cockpit teamwork.

As evidenced by the content of the different chapters, a healthy interaction between the world of practice and the world of theory works to the advantage of both practitioners and theorists. In this context I particularly welcome the overall orientation and tone of this book, especially because this orientation appeals directly to an enduring bias of my own. Aviation psychology that narrowly treats the applied world as simply providing empirical data to validate theoretical constructs is as impoverished as a world of practice that denies any value to the world of theory. The necessity and value of a dialogue between the two domains is thoroughly endorsed by what the authors of the various chapters have said. I remember only too well the total irrelevance to aviation of most of the sterile theoretical material on decision making available in the 1970s and 1980s. Those involved in university-based research were, for the most part, remarkably uninterested in the challenges presented by real-world problems (or, indeed, by any real-world data that might cast a jaundiced eye on their theoretical framework). The chapter by David O'Hare on aeronautical decision making provides ample testimony as to how far one can advance by virtue of a healthy interaction between the university, research, and applied communities. This example serves to underline the fact that each chapter provides ample evidence of the benefits of such a healthy interaction between these three communities. Furthermore, the contents provide ample testimony as to how this has worked to the advantage of all interested in aviation psychology and human factors.

Moreover, this book has manifestly benefited from the talents of a group of notably experienced and distinguished authors, each providing the reader with a comprehensive and valuable introduction to the various subject areas. In addition to the strengths discussed above, I might also add that much of the discussion within each topic area is set within an appropriate historical context. This assists the reader in understanding how the domain in question has developed in parallel with the demands of aviation itself.

There is a lot to be learned from this book. I congratulate all involved for their manifest success in creating such a comprehensive, relevant, and interesting text. I am grateful to the editors for giving me the opportunity to write this foreword, and I wish both the editors and this excellent book every success in the years to come.

Captain Neil Johnston
Aerospace Psychology Research Group
Trinity College, Dublin, Ireland
April 2002

Series Foreword

The domain of transportation is important for both practical and theoretical reasons. All of us use transportation systems as operators, passengers, and consumers. From a scientific viewpoint, the transportation domain offers an opportunity to create and test sophisticated models of human behavior and cognitions. This series covers both practical and theoretical aspects of human factors in transportation with an emphasis on their interaction.

The series is intended as a forum for researchers and engineers interested in how people function within transportation systems. All modes of transportation are relevant, and all human factors and ergonomic efforts that have explicit implications for transportation systems fall within the series perview. Analytic efforts are important to link theory and data. The level of analysis can be as small as one person or international in scope. Empirical data can be from a broad range of methodologies, including laboratory research, simulator studies, test tracks, operational tests, field work, design reviews, or surveys. This broad scope is intended to maximize the utility of the series for readers with diverse backgrounds.

I expect the series to be useful for professionals in the disciplines of human factors and ergonomics, transportation engineering, experimental psychology, cognitive science, sociology, and safety engineering. It is intended to appeal to the transportation specialist in industry, government, or academia as well as the researcher in need of a test bed for new ideas about the interface between people and complex systems.

This text on aviation psychology covers a domain that is central to transportation human factors. Many of the principles and practices of human factors originated in aviation psychology work as far back as World War II with the pioneering efforts of Paul Fitts and other historic figures in human engineering. Although the name of the discipline has since morphed into human factors and ergonomics, the system approach to aviation remains and is well illustrated in the present text. I particularly commend the editors for emphasizing the crucial relationship between theory and practice in aviation psychology. Theoretical models of pilot control, perception, decision making, selection, execution and control, attention, mental workload, situational awareness, and cognitive aging are prominently featured throughout the book. Theory and application are linked in each chapter, thus fulfilling a major goal of the series. Forthcoming books in this series will continue this blend of practical and theoretical perspectives.

Barry H. Kantowitz
University of Michigan Transportation Research Institute
Ann Arbor, Michigan

Abbreviations

AA	adaptive aiding
ACFS	Advanced Concept Flight Simulator
ADI	attitude directional indicator
ADM	aeronautical decision making
ADS-B	automatic dependent surveillance-broadcast
AFOQT	Air Force Officer Qualifying Test
AGATE	advanced general aviation transport experiments
AGL	above ground level
AI	attitude indicator
ALPA	Air Line Pilot Association
altitude AGL	altitude above the ground
ANCS	aviate–navigate–communicate–systems management
ANOVA	analysis of variance
APA	American Psychological Association
APAMS	Automated Pilot Aptitude Measurement System
APU	Applied Psychology Unit
AQP	advanced qualification program
ARCS	attention, relevance, confidence, and satisfaction
ASRS	Aviation Safety Reporting System
ASVAB	Armed Services Vocational Aptitude Battery
ATA	adaptive task allocation
ATA-H	ATA from the machine to the human
ATA-M	ATA from the human to the machine
ATC	air traffic control
ATP	airline transport pilot
Auto-GCAS	Automatic Ground Collision Avoidance System
AVOR	angular vestibulo-ocular reflex
BAT	Basic Attributes Test
BFT	basic flying training
CAPSS	Canadian Automated Pilot Selection System
CDI	course deviation indicator
CDTI	cockpit display of traffic information
CDU	control and display unit
CFIT	controlled flight into terrain

CGI	computer-generated imagery
CMAQ	Cockpit Management Attitudes Questionnaire
CNP	central nervous processor
CPL	commercial pilot license
CRM	crew resource management
CRT	cathode ray tube
CTAS	Center Tracon Automation System
CTM	cockpit task management
DA	descent advisor
DECIDE	detect, estimate, choose, identify, do, evaluate
DME	distance measuring equipment
DMS	decision making styles
DMT	defense mechanism test
ECAC	European Civil Aviation Conference
EEG	electroencephalograph
EGPWS	Enhanced Ground Proximity Warning System
EICAS	Engine Indicator and Crew Alerting System
ERP	event-related potentials
ERP	eye reference point
EVA	extravehicular activity
FAA	Federal Aviation Administration
FAR	Federal Aviation Regulation
FFOV	forward field of view
FL	flight level
FMAQ	Flight Management Attitudes Questionnaire
FMC	flight management computer
FMS	flight management system
FO	first officer
FOQA	flight operations quality assurance
FOR	field of regard
FOV	field of view
g	gravity
g	general mental ability, intelligence
G-force	gravity force
G-loading	gravity loading
GA	general aviation
GAT	general aviation trainer
GFOV	geometric field of view
GIF	gravito-inertial force
GIM	global implicit measure
GPS	Global Positioning System
GPWS	Ground Proximity Warning System
GWS	graphical weather service

HMD	helmet-mounted display
HQSF	handling qualities sensitivity function
HTA	hierarchical task analysis
HUD	head-up display
IA^2	Introduction to Aircraft Automation
IFR	instrument flight rules
IJAP	*International Journal of Aviation Psychology*
ILS	instrument landing system
IMC	instrument meteorological conditions
IPISD	interservice procedures for instructional systems development
IPO	input-process-outcome
ISD	instructional systems development
JAA	Joint Aviation Authority
KSAO	knowledge, skill, abilities, and other
LCD	liquid crystal display
LLC	line and line-oriented simulation checklist
LOD	level of detail
LOFT	line-oriented flight training
LQG	linear, quadratic, gaussian
LTM	long-term memory
LTWM	long-term working memory
LVOR	linear vestibulo-ocular reflex
MAT	Multi-Attribute Task
MCP	mode control panel
MDA	minimum descent altitude
MFD	multifunction display
MIDAS	Man-Machine Integrated Design and Analysis System
MIDIS	Microcomputer-Based Flight Decision Training System
MVSRF	Man-Vehicle Systems Research Facility
NAS	National Aerospace System
NASA	National Aeronautics and Space Administration
NATO	North Atlantic Treaty Organization
NDM	naturalistic decision making
NOE	nap-of-the-earth
NTSB	National Transportation Safety Board
OCM	optimal control model
OTTR	otolith-tilt-translation-reinterpretation
OVAR	off-vertical axis rotation
PAN I	Positional Alcohol Nystagmus I
PAN II	reversed PAN I

PC	personal computer
PCATD	personal computer-based aviation training devices
PCSM	pilot candidate selection method
PFT	preliminary flying training
PIO	pilot induced oscillation
PIRIP	risk profile
POC	performance-operating characteristic
PPL	private pilot license
PSD	power spectral density
RA	resolution advisory
R/M	recognition/metacognition model
RMS	root-mean-square
RPD	recognition-primed decision model
RSI	response-stimulus interval
SA	situation (or situational) awareness
SAGAT	Situation Awareness Global Assessment Technique
SD	spatial disorientation
SDT	signal detection theory
SME	subject-matter expert
SMS	space motion sickness
SOP	standard operating procedure
SPARTANS	Simple Portable Aviation Relevant Task-Battery and Answer-Scoring System
SPV	subjective postural vertical
STM	short-term memory
SVV	subjective visual vertical
TCAS	Traffic Alert and Collision Avoidance System
TRACES	Technology in Retrospect and Critical Events in Science
UAV	unmanned air vehicle
URET	user request evaluation tool
VAMP	variable anamorphic motion picture
VFR	visual flight rules
VMC	visual meteorlogical conditions
VOR	vestibulo-ocular reflex
VORs	very-high-frequency omnidirectional range navigational beacons
VR	virtual reality
VSD	vertical situation display
WAD	Workload Assessment Device

1

Introduction to Aviation Psychology

Pamela S. Tsang
Wright State University

Michael A. Vidulich
Air Force Research Laboratory

THE CHANGING ROLE OF THE PILOT

Early Beliefs of the Pilot's Role in Aviation

Heavier-than-air aviation started about 100 years ago with the first remarkable heavier-than-air powered aircraft flight by the Wright brothers in 1903. Many aviation researchers besides the Wright brothers also contributed to the birth of aviation. Otto Lilienthal, Samuel Langley, Octave Chanute, and Glenn Curtiss are just a few of the aviation pioneers that were either predecessors or contemporaries of the Wright brothers. Yet, the Wright brothers' contributions to the birth of modern aviation stand out, not only because of their ultimate success in being the first to demonstrate powered heavier-than-air flight, but also because of the spectacular balance they achieved in combining scientific principles and effective practice in achieving their goal.

For example, the Wright brothers made seminal contributions to the field of aerodynamics in their creation and use of the first wind tunnel to develop the wing shape for their aircraft, the Flyer. Later, they generalized from that knowledge to design propellers of unprecedented effectiveness for the Flyer's propulsion and incorporated the first effective three-axis control system into their 1902 glider (Hallion, 1978; Jakab, 1997).

With all of those signal accomplishments and the compelling image of their dramatic first flight, it is perhaps not surprising that another important contribution of the Wright brothers is often overlooked. The Wright brothers had a much clearer view of the role of the pilot than did many of their contemporaries. Many of the early aviation researchers of the time period believed that the challenging uncertainty of atmospheric conditions required the creation of an inherently stable aircraft that would respond to a human operator's navigational commands, but fly automatically otherwise (Crouch, 1978). In contrast, perhaps due to their background as bicycle mechanics, the Wright brothers saw the pilot of an aircraft as a skilled active controller of an unstable vehicle (Culick & Jex, 1987; Tsang & Vidulich, 1989). This insight guided their use of experimental gliders as a means of not only testing the aerodynamic properties of their designs and control systems, but also as a means for building their own skills as pilots. The central role of skill in their conception of piloting also led the Wright brothers to become pioneers in the field of educating others to fly. It might be overreaching to claim the Wright brothers as the first aviation psychologists, but their keen appreciation for piloting as a skill was certainly a harbinger of things to come.

The Birth of Scientific Aviation Psychology

Roscoe (1980) points out the difficulty in identifying a single founder of scientific aviation psychology, but identifies World War II as a seminal event in the emergence of the field. This is not surprising, given that during the war aircraft developed into weapons platforms that could travel at unprecedented speeds and attempt to deliver bombs with precision from unprecedented altitudes. Controlling these sophisticated machines, navigating them accurately, and using them effectively all required that equipment be developed that was well designed for human use. Also, the fact that many thousands of crew members were required for these aircraft throughout the war encouraged innovative training research. In England, Sir Frederick Bartlett of Cambridge University inspired studies of aviation human factors in support of the war effort at the Applied Psychology Unit. Early research on human vigilance by Norman Mackworth and compatibility between controls and displays by Kenneth Craik were representative of the valuable work conducted by this group (Roscoe, 1997).

Ross McFarland. McFarland's interest in the effects of altitude on mountain dwellers dated from 1928 and led him into scientific investigations in aviation in the 1930s. Probably his best-known work is two of the earliest volumes on human factors in aviation. In the first volume, *Human Factors in Air Transport Design* (1946), McFarland was mostly concerned with the physical variables that should be controlled to meet human requirements and that could be specified as design criteria for aeronautical engineers. Recognizing that not all the problems of flying could be solved by aeronautical engineering, in his second volume,

Human Factors in Air Transportation (1953), McFarland turned his attention to the integration of the human operator with the equipment. Issues regarding selection and training of flight personnel took center stage. At a time when history first began to witness aging airline pilots, McFarland was concerned with prolonging the useful lives of highly skilled operators. McFarland's approach to understanding ranged from laboratory studies to extensive field observations on aircrews in all parts of the world. McFarland believed that "air transports can be operated safely and efficiently only in so far as the human variables are understood and controlled. The contributions of many specialized scientists are needed to solve these ever-changing problems. In addition, basic research must be continued and the results successfully interpreted and integrated with the practical aspects of airline operations" (1953, p. ix).

Paul Fitts. In the United States, the wartime research on human factors encouraged the May 1945 formation of the Psychology Branch of Aero Medical Laboratory at Wright Field, under the leadership of Paul Fitts (Fitts, 1947; Grether, 1995). Among Paul Fitts' many contributions to aviation psychology were his detailed studies of pilot errors (Fitts & Jones, 1961a, 1961b), pilot eye-scanning behaviors (Fitts, Jones, & Milton, 1950), and the human engineering aspects of air traffic control (Parsons, 1972). However, his contributions extend well beyond the domain of aviation psychology to include producing one of the early, and very influential, taxonomies for function allocation between humans and machines (Fitts, 1962) and evaluation of human manual control characteristics (Fitts, 1954). It is fair to consider Paul Fitts as one of the founders of the entire field of modern human factors, not just aviation psychology.

Alexander Williams. Although Roscoe (1980) was quick to acknowledge the contributions of pioneers such as Bartlett, McFarland, and Fitts, the individual that he nominated as the "Father of Aviation Psychology" was Alexander Williams. After service as a naval aviator during World War II, Williams founded the Aviation Psychology Laboratory at the University of Illinois in 1946. As part of his own research legacy, Williams performed valuable investigations into task decomposition, pilot training, and display and control design (Roscoe, 1980; Williams, 1980). However, even more important than his own individual research contributions, the laboratory founded by Williams has flourished, albeit while undergoing changes in name and directorship. Williams led the laboratory for the first decade following its 1946 founding. Jack Adams and Stanley Roscoe were two of his successors. Roscoe led the lab during a highly productive period in the 1970s (Roscoe, 1980). The current head of the Aviation Research Laboratory at the University of Illinois is Christopher Wickens (author of chap. 5 and 7), and the laboratory's activities remain at the forefront of aviation psychology research.

The Modern Role of the Pilot

The evolution of aviation from the Wright brothers' time until the present day has seen remarkable changes in the capabilities of aircraft. The initial flight by the Wright brothers did not even last a single minute and accomplished no purpose other than to show that powered heavier-than-air flight was possible. Now commercial aircraft flights routinely last many hours and connect cities separated by thousands of miles over entire oceans. Commercial aviation not only provides transportation for people traveling to far-distant lands, but also moves mail and cargo at speeds unattainable any other way. Military aviation has seen even larger increases in capabilities (see, e.g., chap. 2; Shaw, 1985).

As exciting and as important as the increases in aviation capabilities have been, they have not been acquired without affecting the person most responsible for bringing these capabilities to fruition, that is, the pilot. An early pilot received almost all of the vital flight information from his or her own senses. Vision and vestibular functions helped keep the aircraft oriented safely, and navigation was accomplished by sighting landmarks, or even signs, on the ground (e.g., see chap. 3). The pilot had little more than a throttle and a joystick to control the aircraft and often aspired to no more than getting into the air and returning to earth safely. The pilot's world has become much more complex since those early days.

In Fig. 1.1, Wilbur Wright sits in a completely open cockpit, illustrating the simplicity enjoyed by early pilots. All of the information used to control his flight was gained through direct observation of the world. The simple and few controls available to him were connected directly to the aircraft control surfaces with no intermediary.

Buck (1994) details many of the technological changes that have confronted pilots over the course of his own flying career that spanned most of the 20th century. Buck noted the appealing simplicity of early flight. Here are his memories of a 1931 flight in a biplane Pitcairn Mailwing aircraft (23 years after the photo of Wilbur Wright in Fig. 1.1 was taken):

> Imagine flying across New York City through the northeast corridor with no traffic, no ATC, no two-way radio, not a thing to think about except those key basics: fly the airplane, navigate, and avoid the terrain. It was a beautiful, simple life. (p. 8)

Continuing his flying career through World War II, through the formative years of modern commercial air travel, and up to current times, Buck outlined the increasing list of concerns and issues that have been added to the pilot's burden. In many cases, items added to increase safety or effectiveness became potential problems that required additional effort to monitor and control. The same proliferation of additional systems to be controlled confronted the military pilot, along with changes in weapons technology from first controlling the settings for the

FIG. 1.1 The simplicity of early flight. Wilbur Wright on a triumphal tour of Europe in August 1908. Sitting at the controls of the Wright 1907 Flyer at Les Hunaudieres race course near Le Mans. (Courtesy of Special Collections and Archives, Wright State University)

increasingly sophisticated sighting systems for guns and progressing to controlling the launching of air-to-air missiles (Coombs, 1999).

Until the 1960s, the additional information presented to the pilot usually took the form of another single-purpose instrument (Coombs, 1990). Fig. 1.2 illustrates a cockpit from this time period. The numerous single-purpose indicators could exert a serious visual scanning and cognitive integration load on the pilot (see chap. 5). Additionally, as aircraft became increasingly complex and the amount of information that needed to be presented to the pilot increased, the finite "real estate" of the cockpit became a limiting factor. As Coombs put it (1990), "A point in evolution was reached in which the number of pointers, numeral counters and flags could no longer be increased or their display characteristics improved" (p. 12).

The breakthrough in the appearance of the cockpit occurred during the 1960s when cathode ray tube (CRT) displays became practical for use in the cockpit (Coombs, 1990). Because the appearance of the display was no longer limited by

FIG. 1.2 Typical cockpit design of the early 1960s, as illustrated by the X-15. The X-15 research rocket aircraft provided much-needed information on high-speed flight for NASA's space program. The X-15 cockpit shows the use of individual instruments to display data to the pilot. (Courtesy of NASA)

the physical constraints of moving the electromechanical indicator, CRTs allowed innovative display formats to be used. And, by allowing different displays to appear on the same CRT at different times, CRTs were a great boon for alleviating the cockpit real estate problem. CRT technology was followed by liquid crystal displays (LCD) that also allowed flexible multifunction formats.

The prominent role of multifunction displays in the modern cockpit is well illustrated by the Boeing 777 cockpit shown in Fig. 1.3. The change from single instruments and gauges to flexible multifunction displays makes the modern cockpit look much different than earlier cockpits (such as that illustrated in Fig. 1.2). However, much more has changed than just the appearance of the cockpit. In addition to the changes in display technology, the 1960s ushered in the use of computerized automation to assist pilots. For example, the Boeing 777 computer systems incorporate more than 2.6 million lines of software code to support the autopilot, flight management, navigation, and maintenance functions (Norris & Wagner, 1996).

FIG. 1.3 The Boeing 777–300 flight deck: The newest member of the 777 family is a large, modern commercial air carrier. The 777 flight deck illustrates the use of large multipurpose displays to efficiently convey a great deal of information to the crew. (Courtesy of the Boeing Company)

Thus, the current commercial and military pilot has become increasingly removed from direct contact with the aircraft control surfaces. More and more of the tasks require that the pilot work cooperatively, not only with other crew members, but also with advanced computerized technologies (Buck, 1994; Coombs, 1990, 1999). Unfortunately, these advanced computerized technologies often behave in surprising or inexplicable ways to the human pilot (Sarter & Woods, 1992, 1994). On the other hand, researchers (e.g., Weiner, 1993) have pointed out that automation often has positive effects, even on the pilot's cognitive processing, by enabling effective responses in emergencies that would not be possible without it.

Nevertheless, Buck (1994) closed his book arguing that more consideration needs to be paid to the pilot's role:

> Our plea, then, is to develop and then implement real ways to reduce rather than add to the pilot's burden, to simplify the pilot's job rather than complicate it. To respect, finally, that pilot judgment will always be needed—and to make room for it. (p. 233)

Visualizing the Future Pilot

Undoubtedly, the role of the pilot in commercial and military aviation will continue to be affected by the trend toward more highly automated systems. It is hoped that as these systems become more advanced, they will meet Buck's plea to reduce rather than to expand the pilot's burden. Some researchers believe that this happy result will occur as the automated systems become intelligent enough to become "electronic crew members" (see, e.g., Reising, Taylor, & Onken, 1999) and become more "aware" of the actual state of the pilot they are attempting to aid (e.g., see Taylor, Howells, & Watson, 2000; see also chaps. 4 and 9).

In a broader context, Fallows (2001a, 2001b) presents another intriguing vision of the future of aviation. Examining the increasing bottlenecks and delays inherent in the existing airline industry, Fallows proposed that a simple scaling-up of the existing system with more planes and more runways at existing airports would not be a practical or economically viable approach to keep pace with the projected increases in airline travel. Fallows made a compelling argument that the increased reliability of aircraft mechanical systems combined with innovative research on cockpit interfaces will not only revitalize general aviation, but will also lead to the emergence of a much more extensive small aircraft "taxi" service.

In the current context, it is important to emphasize the role that human factors is expected to assume in the development of the expanded air system. Fallows (2001a, 2001b) points out that current research, such as that performed by the National Aeronautics and Space Administration's (NASA) Advanced General Aviation Transport Experiments (AGATE) alliance, will decrease the difficulty of piloting and increase flight safety for a new generation of more user-friendly aircraft. NASA is aggressively pursuing the goal of making advanced cockpit technologies effective and affordable to even general aviation pilots. These technologies include highway-in-the-sky displays, head-mounted displays (Fiorino, 2001), and synthetic vision systems that use advanced sensors to present a view of the world to the pilot during degraded visual conditions (Wynbrandt, 2001).

Fallows (2001a, 2001b) presented his vision of aviation's future before the tragic September 11, 2001, terrorist attacks on the World Trade Center and the Pentagon. The ensuing turmoil, security increases, and economic downturn that the airline industry experienced following the attacks will, at least, slow movement toward such an open expansion of the U.S. and international aviation systems. However, it seems inevitable that in the long run civilization will continue its development of increased mobility for people and that the aviation system will change in ways to accommodate that increase. To this end, understanding the human pilot and building systems that best accommodate the human's cognitive strengths while supporting human frailties will remain a vital component of making those systems effective and safe.

THE ROLES OF BASIC AND APPLIED RESEARCH IN AVIATION PSYCHOLOGY

Basic and Applied Research in Meeting Future Needs

A major organizing principle of this book is the belief that a viable practice of aviation psychology must be based on a solid theoretical foundation. Maintaining the balance of basic and applied work within the field has always been important in both aviation psychology and in the broader domain of engineering psychology. This challenge was forthrightly faced 30 years ago by Jack Adams (Adams, 1972) in his 1971 Presidential Address to the Society of Engineering Psychologists, Division 21 of the American Psychological Society (APA). In the opening of his address, Adams stated, "Our research efforts have been and are insufficient. The future of engineering psychology is in jeopardy unless we examine realistically the state of our knowledge and ask what we must do to strengthen it" (p. 615). Adams contrasted Division 21's basic research activities with that of the other applied divisions of the APA (such as clinical psychology and educational psychology) and found Division 21 to be wanting. Adams found in his review several examples of engineering psychology research activities that had promising starts in World War II, but had failed to continue to a satisfactory conclusion. Interestingly, many of these examples involved early aviation psychology research on the design of aviation displays (e.g., attitude displays, circular versus linear displays, pictorial navigation displays, and contact analog displays for vehicular control).

Adams (1972) also considered the outcome of two major studies of the precursors to successful system designs as a means of contrasting the relative merit of basic and applied research. Project Hindsight, conducted by the Department of Defense, examined 20 successful and important military systems to determine what research and development innovations preceded them. The precursors were tracked back about 20 years, and 710 related research and development innovations were identified. The results strongly supported the contribution of applied research. The majority of the precursor innovations identified within Project Hindsight were developed in an applied setting in which the researcher was attempting to address a known deficiency rather than simply to extend domain knowledge. However, Adams pointed out that the research was most often done in support of another system development process, and the median timing was 9 years prior to application of the innovation. In other words, good research, including applied research, can have implications well beyond its current setting.

An even more interesting picture emerged when the 20-year window preceding innovations was expanded. A second study reviewed by Adams (1972) was conducted by the Illinois Institute of Technology and called Technology in Retrospect and Critical Events in Science (TRACES). In examining the essential precursors to five important technologies, 341 precursors were identified as far back as 50 years prior to the current technology. Seventy percent of the precursors were identified as originating in university basic research, and the most common time window was 20 to 30 years prior to the development of the ultimate system.

Adams (1972) concluded that engineering psychology must be more aggressive in examining its own knowledge store and seek to develop a stronger basis for science and its applications. Although applied research answering immediate questions would certainly be vital for the field, Adams (1972) suggested that:

> Furthermore, unfettered basic scientists, going where their imaginations take them, can give applied investigators a vision they otherwise would not have. I cannot believe that applied scientists of the 1930s and 1940s who were concerned with the efficiency of explosives would have ever invented the atomic bomb. (p. 621)

The Role of Basic Research and Theory in Aviation Psychology Practice

In a review, Gopher and Kimchi (1989) considered the implications of the increasing pace of technological changes on the field of engineering psychology in general. The rapidly increasing power of microcomputers and their increasing incorporation within almost every type of human-machine system (especially aircraft) was seen to be a fundamental problem for engineering psychology. It was not seen to be cost effective to engage in long-term applied research for systems that would only last for short periods. As an analogy, Gopher and Kimchi likened the situation confronting engineering psychology researchers to a controller tracking an input function exceeding its point-to-point tracking capabilities. The best response of the controller is to give up attempting to react to every momentary deviation and to focus instead on: (a) tracking any higher order, slow-moving changes within the input function and (b) attempting to predict future inputs. Gopher and Kimchi translated this analogy to the domain of engineering psychology:

> . . . the analogy emphasizes the role of theoretical models in practical work. Only with such models can we generate principles and predict the future. If there existed only a limited set of slow-moving technologies, strict empirical approaches could suffice. (p. 432)

Along the same vein, Kantowitz (1992) identified five major advantages that theory provided the real-world practitioner: (a) Theory can provide accurate and sensible extrapolation to new situations, (b) theory can be the basis for precise predictions of system performance before a system is built, (c) theory can encourage efficient generalization across a range of practical problems (i.e., theory can keep researchers from constantly reinventing the wheel), (d) theory can provide a normative baseline for judging human and system performance to determine whether additional effort might produce significant improvements, and (e) theory is the best practical tool.

The last point best summarizes Kantowitz's (1992) and Gopher and Kimchi's (1989) viewpoints and the organizing principle behind the present volume. As in the larger domain of engineering psychology, the pace of technological changes in aviation has long been too rapid to allow a totally reactive applied research strategy to succeed. The only rational option, then, is to make thorough use of basic research and theories as the foundation of the field of aviation psychology. There will be ample opportunity and compelling need for focused applied research to optimize the application of theory to emerging technologies, but this research will have its best chance to succeed if it makes good use of the theory for guidance.

A visionary, Donald Broadbent was one of the most important figures in experimental and applied psychology during the 20th century (Moray, 1995). As a student, he was mentored by Sir Frederick Bartlett and eventually became head of Cambridge's prestigious Applied Psychology Unit. Being a renowned theoretician himself, it is not surprising that Broadbent held the importance of theory to applications with the same regard as Gopher and Kimchi (1989) and Kantowitz (1992). In addition, Broadbent forcefully advocated the role of applied work in inspiring good theory. Results of applied research, in addition to being practically useful, should also provide tests and expansion of the underlying theory. In summarizing his views after a careful consideration of the role of applied problems in advancing psychological theory, he wrote, "applied psychology is the best basis for a genuine theory of human nature" (Broadbent, 1971, p. 29). To illustrate the point, Broadbent pointed to two attentional mechanisms that had been discovered in applied work and then became prominent in his attentional theories:

> Such mechanisms might be overlooked by purely theoretical psychologists, who can ignore the complex nature of most real-life situations. Once found, these principles are of use in understanding many other situations as well as those that gave rise to them; that is, they are "theoretical" in the best sense. (1971, p. 30)

This volume was inspired by the same beliefs expressed by Donald Broadbent, namely, that theories and applications are not two distinct goals of aviation psychology, but two essential facets of progress in any healthy science. Just as good psychological theories will help guide applications to aid the pilot, the challenging applications inherent in aviation will provide a fertile ground to grow robust theories.

PREVIEW OF THE CHAPTERS

Although the chapters are arranged by important topics in aviation psychology, readers will find pertinent psychological principles, their associated methodologies, and related empirical findings discussed throughout the book. The book adopts a psychological perspective toward understanding the demands on the pilot. Following the arguments presented earlier, we hold that systematic investigations with the rigor of scientific methods would be the most fruitful approach. This approach should yield principles and methodologies that would contribute to the understanding of human behavior in general and to enhancing performance in a variety of systems and settings beyond that of aviation. For example, a sound display principle would apply in the cockpit of an aircraft, a spacecraft, an air traffic control's console, or an automobile dashboard, as well as an office computer. The book also adopts the view that the understanding of human behavior can be greatly informed by applied problems. Issues related to the enhancment of pilot performance and to the improvment of system efficiency and safety should drive the selection of topics to study. For example, one important consideration that spans across the chapters is the impact of the ever-changing technology on the pilot's task. We believe the juxtaposition of theories and applications in each chapter will help us achieve the goal of attaining a better understanding of the pilot–aircraft system and being in a better position to contribute to the extant knowledge.

The chapter by Ralph and Lyn Haber, "Perception and Attention in Low-Altitude High-Speed Flight" (chap. 2), introduces some flying basics. The detailed analysis of low-altitude flight in military aviation shows the intrinsically unforgiving nature of aviation and the rich interplay of perceptual and attentional abilities and skills required of the pilot. The detailed account of the highly specialized operational tasks from a psychological viewpoint elucidates the role of psychologists in aviation. Haber and Haber present some of the most pertinent principles of psychology to flight but also illustrate the limitations of current theories. Appreciating the limitations of current theories is, of course, a necessary step towards advancing theories. This chapter exemplifies the fruitfulness of psychologists working hand in hand with operational personnel, applying basic knowledge to the understanding of operational demands, and validating theoretical understandings through detailed analysis of the operational performance. Haber and Haber's discussion foreshadows several major topics covered in the ensuing chapters.

Three chapters cooperatively cover a topic of the earliest concerns of flight—pilot control. Continuous vehicular control is often the pilot's main activity whenever the aircraft cannot be left to an autopilot. The demands of pilot control are illustrated by the examples of combat maneuvers deftly sketched by Haber and Haber. In chapter 8, Hess presents formal quantitative pilot control models. For those with less of a quantitative background, Wickens provides a more intuitive description of pilot control models in chapter 7. The models presented by Hess are mathematical constructs that can be used in descriptive or predictive fashion. The models can be used to describe or explain results observed from flight tests or simulation runs, contributing to the understanding of human pilot behavior. The pilot models also can be used to predict pilot/aircraft performance and handling qualities, thus contributing to improved aircraft designs. The advantages of adopting psychological constructs developed by psychologists in conjunction with the powerful analytical techniques used by engineers are amply evident in this chapter. Hess believes this to be an efficient approach to identifying aids that could be provided to the operator and determining the appropriate roles of humans and machines in pilot control.

Several chapters expand on the perceptual and attentional issues facing the pilots that Haber and Haber related to piloting. As discussed in Young's (chap. 3), Hess's (chap. 8), and Kaiser and Schroeder's (chap. 12) chapters, perception is accomplished through not just the visual system, but also the auditory, proprioceptive, and vestibular systems. Of note is that these authors emphasize the importance of the integration of the signals from the various sensory systems in contributing to spatial orientation, pilot control, and motion perception.

Similarly, with regard to the attentional issues, instead of focusing on isolated, single-task performance, Vidulich (chap. 4), Wickens (chap. 7), and Tsang (chap. 14) consider attentional management as a higher order cognitive skill with subcomponents that involve the executive control of limited resources. Vidulich presents the concepts of mental workload and situation awareness that are now frequently used to characterize the impact of complex aviation tasks on the pilot's information processing system. Wickens discusses how pilots cope with the temporal demands of multiple tasks (Cockpit Task Management, CTM)—a critical issue in the cockpit in which checklist activities might easily be time-shared with, or interrupted by, other tasks. Tsang focuses on two components of the executive control—attention switching and attention sharing—and how age and expertise might affect this control.

The impressive maneuver capability of modern aircraft demonstrated in Haber and Haber's chapter reflects the increasing sophistication of aviation technology that has far-reaching implications on an array of issues covered in the book. For one, it points to the opportunity and need for automation to support pilot control. But as researchers have long recognized (notably, Wiener & Curry, 1980) and as Parasuraman and Byrne have discussed in the present volume (chap. 9), although automation is largely helpful, enabling operations that would otherwise be impossible, it also has placed an additional burden on the pilot. With the advent of increasing automation, the role of the pilot shifts from primarily a manual controller of focused, isolated tasks to a supervisory controller overseeing and managing several subsystems concurrently. This has important implications on the level of psychological processes that require understanding, and new psychological concerns emerge. For example, concepts like mental workload and situation awareness (chap. 4) are relatively new to the psychological literature. These concepts actually originated from the operational personnel who felt a need for new vocabulary to describe the demands and challenges that they face. In addition to providing a thorough overview of the issues associated with the use of automation within the cockpit, Parasuraman and Byrne and Vidulich propose that future automation might take advantage of monitoring the human's mental state in order to act more adaptively and intelligently.

Although automation moves the pilot further and further from direct control, the pilot still requires information to keep abreast of the state of the aircraft in order to aviate, navigate, communicate, and manage all the subsystems. In chapter 5, Wickens advocates the use of psychological principles to shape the presentation of information. A well-designed display should seem "intuitive" to the pilot. The flexibility of information formats afforded by modern display technology provides great opportunity to achieve compatibility between display format and the pilot mental information processing.

The additional information afforded by technology (such as real-time traffic information) can be enormously helpful for many decision-making tasks. But additional sources of information, along with all their inherent uncertainty, could also increase the complexity of decision making. Further, technology can now effect a decision with a push of a button or a simple voice command that could produce quick and far-reaching consequences. In chapter 6, O'Hare argues that the understanding of human decision making is one of the primary issues confronting aviation psychology. He presents both normative and descriptive models of human decision making but found the current attempts to provide decision-making training to be not particularly effective. O'Hare advocates for continual development of theoretical understanding and the use of a better theory-based tool for more effective interventions.

The challenges inherent in the increasing technological capabilities and the anticipated increasing volume of air travel around the globe call for new consideration on pilot selection and training. These issues are covered in chapters 10 and 11, respectively. The early disastrous results of training unselected individuals for flying during the first years of World War I prompted intensive study of pilot selection across the globe (Armstrong, 1939). It also soon became clear that physical fitness alone was not sufficient to guarantee success in aviation. Tests of emotional stability, personality, and cognitive tests of memory, attention, and spatial ability were therefore implemented. In the present volume, Carretta and Ree (chap. 10) discuss the complexities of trying to identify individuals who are likely to succeed as a pilot. State-of-the art selection batteries for commercial and military selection from around the world are presented.

Though not covered extensively, Carretta and Ree rightly raise the issue of considering the advantages of selection methods to be able to predict training and job performance for members of both genders and members of different ethnic/racial groups. In 1934, Helen Richey was the first woman pilot to be hired by a regularly scheduled airline. However, the Federal Bureau of Air Commerce forbade her to fly in bad weather, and the all-male pilot's union denied her membership (Thornberg, 2000). In was not until 1973 that a U.S. airline would hire a second woman pilot (see Fig. 1.4). As of 1994, female airline pilots constituted 2.3% and African American airline pilots constituted 1.2% of the airline pilots' population (Hansen & Oster, 1997). Hansen and Oster (1997) urge a concerted effort to increase diversity in the aviation workforce to maximize human capital.

Along the same vein, to capitalize on the skill and experience of older pilots, there are ongoing fervent debates surrounding the FAA Age 60 Rule. This regulation prohibits pilots over the age of 60 to be the pilot-in-command of air carriers and commuter and air taxi operations that include 10- to 30-seat aircraft. Tsang (chap. 14) presents the debates on the potential benefits and compromises in safety with the possible revision of the Age 60 Rule. A scientific, objective approach that could contribute to the current debates is proposed.

Effective selection methods minimize but do not obviate the need for training. In chapter 11, Patrick provides an overview of the design, implementation, and evaluation of a training program for individuals and teams. In this endeavor, Patrick emphasizes equally the reliance on general principles found in the literature and on more specific lessons learned from training studies in aviation. Patrick considers the strengths and weaknesses of the traditional Instructional System Development model within the context of training of an individual pilot, as well as that of increasing automation and the need for crew coordination.

Another implication of the increasing capability made possible by technology is the need for additional personnel to control and manage the aircraft systems. Effective information flow among crew members becomes an important factor toward maximizing performance and safety. In chapter 13, Sherman reviews the impact of the demands of processing additional information as well as those of increased social interaction among crew members. Sherman shows how Crew Resource Management (CRM) emerged as an approach for improving team communication and coordination. Whereas technology had gradually transformed a single-pilot cockpit to a three-person cockpit for air carriers (with a pilot-in-command, a copilot, and a

FIG. 1.4 Bonnie Tiburzi, the second woman pilot hired by a U.S. air carrier and the first to be hired by a major airline. (Courtesy of the Ninety-Nines, Museum of Women Pilots)

flight engineer), the newer, more automated cockpits experience a reduction in crew size, with the flight engineer being eliminated from most cockpits. As discussed by Sherman, this turn of events does not reduce the need to understand teamwork. Rather, team communication and coordination research needs to be conducted to incorporate the electronic crew members.

Because flight simulation is frequently used for training and for the evaluation of training, the two topics are intricately related (e.g., What should be simulated in order to maximize transfer of training?). In addition, flight simulators can be used for developing and testing aircraft models (chap. 12) and mathematical constructs of pilot control such as those discussed by Hess (chap. 8). The many advantages of the use of simulation were recognized early. The first simulators were naturally very crude by today's standards, but the regimes of flight that could be reasonably emulated expanded with the present computer processing and video presentation capabilities. Although the degree of physical realism that can be achieved today is arresting, the most powerful computer in the world still cannot duplicate the world. The mere use of a realistic setting does not mean that a study is valid. Any number of unintended factors could render it impossible to draw any inferences from the results with confidence. Instead of aiming for surface realism, it is utterly important to first identify the critical variables that support piloting. This is one aspect in which guidance by a solid theoretical foundation rather than by technical feasibility is especially needed (see also chap. 11).

Several chapters also offer tutorials on some of the most common analytical methods used in aviation psychology. In chapter 4, Vidulich provides a tutorial on important techniques used for assessing mental workload and situation awareness. In chapter 7, Wickens presents a tutorial on control theory that would be most helpful for those who would need some introductory background to appreciate the more quantitative treatment presented by Hess in chapter 8. In chapter 10, Carretta and Ree provide an overview of the validation process, specific analytic strategies, and cautionary notes on the many potential pitfalls in drawing conclusions from improperly conducted analysis. The invaluable lessons found in this chapter should be applicable in a variety of settings. In chapter 11, Patrick presents the general steps taken in conducting a training program. In chapter 14, Tsang presents the common approaches to, as well as methodological challenges in, assessing age effects on pilot performance.

Finally, the treatment of aviation errors and accidents are more explicit in some chapters (e.g., chaps. 2, 3, 6, 9, 11, 13, and 14) than in others. But all chapters address issues that, through training (e.g., chaps. 2, 3, 11, and 13), by design (e.g., chaps. 5, 7, 9, and 12), by our ability to predict pilot performance (e.g., chaps. 6 and 8), or by capitalizing on pilots' skill and experience (e.g., chaps. 4, 10, and 14), aim at reducing errors and minimizing accidents, while enhancing safety and efficiency.

To conclude, the rapid metamorphosis of flight in a mere 100 years is nothing short of breathtaking. The early paths that pioneer aviation and engineering psychologists have set out for us appear to be good ones. The common approach among them is the need to continue basic laboratory research and bring it to bear on applied issues. But as Broadbent (1990) put it, "the world has features that current theory cannot handle" (p. 237). Consequently, basic research also should be driven by some important practical problems, and applied studies should be conducted as test of the theory. Further, the findings of such studies should be used to modify the underlying theory. The future of aviation psychology does not lie in comparing the merits of various subsystems. Rather, it lies in understanding the underlying principles in design alternatives for intelligent pilot–aircraft systems. Toward this end, as the pilot's task becomes more the managing of complex systems, the understanding of the performance of highly simplified tasks and processing isolated signals would not suffice. The information processing that typifies piloting requires a high level of comprehension, dynamic decision making, executive control, organization, and integration. Advancing the theoretical understanding of the pilot behavior in the complex, demanding aviation environment will require research to be conducted in an environment sufficiently rich to allow the elicitation of behavior that would be representative of today's piloting. This is the case for laboratory, simulation, or field experiments. But for this process to move forward, institutional and government recognition of, support for, and commitment to equipment and resources needs are necessary.

REFERENCES

Adams, J. A. (1972). Research and the future of engineering psychology. *American Psychologist, 27,* 615–622.

Armstrong, H. G. (1939). *Principles and practice of aviation medicine.* Baltimore: Williams & Wilkins.

Broadbent, D. (1971). Relation between theory and application in psychology. In P. B. Warr (Ed.), *Psychology at work* (pp. 15–30). Harmondsworth, England: Penguin.

Broadbent, D. (1990). A problem looking for solution. *Psychological Science, 1,* 235–239.

Buck, R. N. (1994). *The pilot's burden: Flight safety and the roots of pilot error.* Ames: Iowa State University Press.

Coombs, L. F. E. (1990). *The aircraft cockpit: From stick-and-string to fly-by-wire.* Derby, England: Patrick Stephens Limited.

Coombs, L. F. E. (1999). *Fighting cockpits 1914–2000: Design and development of military aircraft cockpits.* Shrewsbury, England: MBI Publishing.

Crouch, T. D. (1978). Engineers and the airplane. In R. P. Hallion (Ed.), *The Wright brothers: Heirs of Prometheus* (pp. 3–19). Washington, DC: Smithsonian Institute.

Culick, F. E. C., & Jex, H. R. (1987). Aerodynamics, stability, and control of the 1903 Wright Flyer. In H. S. Wolko (Ed.), *The Wright Flyer: An engineering perspective* (pp. 19–43). Washington, DC: Smithsonian Institute.

Fallows, J. (2001a). *Free flight: From airline hell to a new age of travel.* New York: Public Affairs.

Fallows, J. (2001b, June). Freedom of the skies. *The Atlantic, 287*(6), 37–49.

Fiorino, F. (2001, August 13). Tech trickle-down enhances GA safety. *Aviation Week & Space Technology, 155*(7), 52.

Fitts, P. M. (1947). Psychological research on equipment design in the AAF. *American Psychologist, 2,* 93–98.

Fitts, P. M. (1954). The information capacity of the human motor system in controlling the amplitude of movement. *Journal of Experimental Psychology, 47,* 381–391.

Fitts, P. M. (1962). Functions of men in complex systems. *Aerospace Engineering, 21,* 34–39.

Fitts, P. M., & Jones R. E. (1961a). Analysis of factors contributing to 460 "pilot-error" experiences in operating aircraft controls. In H. W. Sinaiko (Ed.), *Selected papers on human factors in the design and use of control systems* (pp. 332–358). New York: Dover.

Fitts, P. M., & Jones R. E. (1961b). Psychological aspects of instrument display: I—Analysis of 270 "pilot error" experiences in reading and interpreting aircraft instruments. In H. W. Sinaiko (Ed.), *Selected papers on human factors in the design and use of control systems* (pp. 359–396). New York: Dover.

Fitts, P. M., Jones, R. E., & Milton, J. L. (1950, February). Eye movements of aircraft pilots during instrument landing approaches. *Aeronautical Engineering Review, 9*(2), 24–29.

Gopher, D., & Kimchi, R. (1989). Engineering psychology. *Annual Review of Psychology, 40,* 431–455.

Grether, W. F. (1995). Human engineering: The first 40 years, 1945–1984. In R. J. Green, H. C. Self, & T. S. Ellifritt (Eds.), *50 years of human engineering* (pp. 1-9–1-17). Wright-Patterson Air Force Base, OH: Crew Systems Directorate.

Hallion, R. P. (1978). The Wright brothers: A pictorial essay. In R. P. Hallion (Ed.), *The Wright brothers: Heirs of Prometheus* (pp. 57–100). Washington, DC: Smithsonian Institute.

Hansen, J. S., & Oster, C. V., Jr. (Eds.). (1997). *Taking flight.* Washington, DC: National Academy Press.

Jakab, P. L. (1997). *Visions of a flying machine: The Wright brothers and the process of invention.* Washington, DC: Smithsonian Institute.

Kantowitz, B. H. (1992). Selecting measures for human factors research. *Human Factors, 34,* 387–398.

McFarland, R. A. (1946). *Human factors in air transport design.* New York: McGraw-Hill.

McFarland, R. A. (1953). *Human factors in air transportation: Occupational health and safety.* New York: McGraw-Hill.

Moray, N. (1995). Donald E. Broadbent: 1926–1993. *American Journal of Psychology, 108,* 117–121.

Norris, G., & Wagner, M. (1996). *Boeing 777.* Osceola, WI: Motorbooks International.

Parsons, H. I. (1972). *Man-machine system experiments.* Baltimore: Johns Hopkins University Press.

Reising, J., Taylor, R., & Onken, R. (1999). *The human-computer crew: The right stuff? Proceedings of the 4th Joint GAF/RAF/USAF Workshop on Human-Computer Teamwork* (Tech. Rep. No. AFRL-HE-WP-TR-1999–0235). Wright-Patterson Air Force Base, OH: Air Force Research Laboratory.

Roscoe, S. N. (1980). Preface: Roots of aviation psychology. In S. N. Roscoe (Ed.), *Aviation psychology* (pp. xi–xiv). Ames: Iowa State University Press.

Roscoe, S. N. (1997). *The adolescence of engineering psychology: Vol. 1. Human Factors History Monograph Series.* Santa Monica, CA: Human Factors and Ergonomics Society.

Sarter, N. B., & Woods, D. D. (1992). Pilot interaction with cockpit automation: Operational experiences with the flight management system. *International Journal of Aviation Psychology, 2,* 303–321.

Sarter, N. B., & Woods, D. D. (1994). Pilot interaction with automation II: An experimental study of pilots' model and awareness of the flight management system. *International Journal of Aviation Psychology, 4,* 1–28.

Shaw, R. L. (1985). *Fighter combat: Tactics and maneuvering.* Annapolis, MD: Naval Institute Press.

Taylor, R. M., Howells, H., & Watson, D. (2000). The cognitive cockpit: Operational requirement and technical challenge. In P. T. McCabe, M. A. Hanson, & S. A. Robertson (Eds.), *Contemporary ergonomics 2000* (pp. 55–59). London: Taylor & Francis.

Thornberg, H. B. (2000). Women in aviation. In T. Brady (Ed.), *The American aviation experience: A history* (pp. 367–412). Carbondale: Southern Illinois University.

Tsang, P. S., & Vidulich, M. A. (1989). Cognitive demands of automation in aviation. In R. S. Jensen (Ed.), *Aviation psychology* (pp. 66–95). Aldershot, England: Gower.

Wiener, E. L. (1993). Life in the second decade of the glass cockpit. In *Proceedings of the Seventh International Symposium on Aviation Psychology* (Vol. 1, pp. 1–7). Columbus: The Ohio State University.

Wiener, E. L., & Curry, R. E. (1980). Flight-deck automation: Promises and problems. *Ergonomics, 23,* 995–1011.

Williams, A. C. (1980). Discrimination and manipulation in flight. In S. N. Roscoe (Ed.), *Aviation psychology* (pp. 11–32). Ames: Iowa State University Press.

Wynbrandt, J. (2001, Spring). Seeing the flightpath to the future: Synthetic vision and heads-up displays are on the way to a cockpit near you. *Pilot Journal, 2*(2), 30–34, 94.

2

Perception and Attention During Low-Altitude High-Speed Flight*

Ralph Norman Haber
Lyn Haber
University of California at Santa Cruz

In this chapter, we undertake a perceptual and attentional analysis of exceedingly complex performance: military low-altitude flight. This performance comprises a constellation of tasks that represent the very edge of human perceptual and attentional capabilities. Our analysis extends the concept of automatic and controlled attention as it has evolved from primarily verbal and semantic tasks to automatic and controlled perceptual processes. This extension is critical because performing on the edge, where death is the presumptive consequence of every mistake, creates a massive conundrum: many of the automatic perceptual processes available to pilots often produce information that is sometimes misleading or simply wrong. As a result, at precisely those moments when perceptual and attentional demands on the pilot are most extreme, he is forced to switch from automatic to focused processes. Further, both perceptual and performance tasks must be prioritized by their importance (defined by likelihood of dying), the duration of the task itself, and the duration of the time interval before the task must be performed again. A major component in every prioritization is whether automatic processes have to be abandoned in favor of focused ones. Just considering this possibility requires focused attention.

---

*A brief note about language. We use the pronouns he and his exclusively when referring to a pilot of combat jet aircraft, because at that time there were no women flying jet fighters in low-altitude arenas.

Military pilots, maneuvering their jet fighters at high speed during low-altitude flight, must also complete their mission tasking and avoid threats. Low-altitude flying provides a laboratory for the analysis of perceptual and attentional processes under conditions where efficiency must be maximized, perception controls and is controlled by performance, and prioritization of tasking determines survival. Typical velocities are just up to the speed of sound, between 600 and 1,000 feet per second. A fighter plane can roll 360 degrees around its axis of travel in little more than a second, enter a turn at a 90-degree angle of bank from level flight in even less time, and complete a U-turn in less than half the time it can be done in a sports car. When these maneuvers are performed near the ground, they can produce angular velocities at the pilot's eyes exceeding 500 degrees per second. They also require the ability to perceive and comprehend objects and terrain at distances of 5,000 to 10,000 feet.

Performing these maneuvers at low altitude places a unique constraint on the pilot. Whatever his reason for flying low, his most important task at all times is to avoid hitting the ground. Successful ground avoidance requires that the pilot's direct and continuous visual attention be focused on observation of the ground. He must perceive the ground, perceive his location relative to the ground, and predict where he will be in relation to the ground during and at the completion of every maneuver.

By definition, a pilot enters the low-altitude environment whenever his primary task at the moment is to avoid hitting the ground (Miller, 1983). Thus, the definition of the low-altitude environment requires knowing the pilot's tasks rather than his absolute elevation over the terrain. Given the speed and maneuverability of modern jet fighters, for some maneuvers low altitude may extend to 5,000 or even 10,000 feet above ground level. For example, a "Split S" maneuver, in which the pilot reverses direction by diving under in order to come up and behind an approaching plane, requires the pilot to first roll inverted, then dive upside down toward the ground in a 180-degree half loop, so that he can pull out into level flight traveling in the opposite direction, though much closer to the ground. At normal maneuvering speeds, most fighters require a vertical diameter of about one mile for their diving loop. Therefore, when planning to begin this maneuver, even at a "safe" altitude of 5,000 feet above ground level, the pilot must be concerned with his ground clearance—he is flying at that moment in the low-altitude environment.

Flying near the ground (except when preparing to land) rarely occurs in general or commercial aviation, and heavy maneuvering near the ground is usually explicitly prohibited by regulation. However, in military aviation, flying near the ground is common for reasons of weather, protection, and mission.

Weather drives pilots low. Because the priorities of military missions take precedence over the momentary weather, a fighter pilot looking for a target he must detect visually may be forced to fly below the bottoms of the clouds, even if that means flying very close to the ground. The need for protection drives pilots low. A jet fighter is more difficult to detect by pilots of other planes, ground observers, or electronic surveillance technology when flying very close to the ground. Even if detected, it is much more difficult to successfully attack a jet flying very close to the ground, especially with heat-sensitive missiles. The mission also drives pilots low. Jet fighter pilots have to go where their targets are located, either to pursue an enemy plane that goes low or to attack a ground target.

In the four sections that follow, we analyze the perceptual and attentional components of avoiding collision with the ground while flying fast and low. In the first section, we describe the aerodynamics of low-altitude flight maneuvers. In the second, we examine the objective information in the physical world that can inform the pilot about his ground clearance. In the third, we explore how those sources of information, singly and in combination, are used by the pilot to perceive ground clearance. We conclude with an analysis of low-altitude military flight in which the perceptual and attentional demands of the components of flight are combined, and tasks are prioritized by their inherent demands as well as by mission and ground-avoidance requirements.

This chapter departs slightly from the others in this volume in that it is largely based on our experiences observing and participating in the training of U.S. fighter pilots to perform low-level flight tasks at the 162nd Fighter Weapons School of the Arizona National Guard at Tucson. The instructor pilots knew more about perception, attention, and performance in a practical sense than all the textbooks teach. In the time that we worked with these instructors, we discovered the problems of applying our land-bound theories and models to the remarkable airborne performance that we observed every day. We are particularly indebted to a young instructor, Capt. Milt Miller, who translated skills and understanding of the inherent problems of flying low and fast into a superb training program (Miller, 1983). In later chapters in this volume, models of pilot control are introduced by Wickens in chapter 7; a more formal analytic treatment is presented by Hess in chapter 8.

AERODYNAMICS OF MOTION: THE EFFECTS OF CONTROL INPUTS ON FLIGHT

Survival at low altitude requires that the pilot understand and continually consider the physical dynamics of the movement of his aircraft and the changes in those dynamics as a result of the control inputs that he makes. Specifically, the pilot needs to know how long it takes his aircraft to respond to his inputs.

Suppose a pilot is deliberately flying low, well below an upcoming ridge. How long can he delay changing his flight path until it is past that instant in time beyond which collision with the ridge becomes irreversible and inevitable? The answer, counting backward from the moment of impact, includes the time required for the pilot to perceive, decide, and respond to the upcoming ridge, as well as the time required for the plane to respond to the inputs from the controls. This instant in time, beyond which collision is unavoidable, is called *time-to-die.* Every maneuver performed near the ground has an associated time-to-die limit. Conversely, *free time* refers, in seconds, to all the time available to the pilot before he passes the time-to-die instant, when collision is irreversible. Free time describes how many seconds the pilot can use to attend to tasks other than ground clearance. Time-to-die and free time are critical concepts in our presentation of low-altitude flying.

In this section, we discuss how the physical dynamics of an aircraft are defined and specified; and then, through the consideration of some sample maneuvers, describe the relation between the aerodynamics of flight and time-to-die and free time when flying near the ground.

Descriptions of the Position and Movements of Aircraft in Relation to Control Inputs

Three parameters—G-force, altitude above the ground, and velocity vector—define the location of the aircraft in relation to the ground and its ongoing flight path through space.

G-force refers to movement of the aircraft with respect to gravity, that is, parallel to the lift axis, which is usually perpendicular to the axis of the wings. With wings level, initiating a climb exerts a downward force on the aircraft and its occupants proportional to the acceleration. The accelerating force of the climb is specified in Gs, or the multiple of the weight of the plane needed to produce the acceleration. A 3 G accelerating climb triples the weight of the plane (and the pilot within it as well) during the course of the acceleration. In a level turn, the lift axis remains perpendicular to the wings, as seen in Fig. 2.1. But the turning force generates centrifugal force, which requires more lift to maintain level flight; the steeper the bank angle, the more lift is needed. The forces resulting from lift in a turn, as in a climb, are specified in Gs.

Altitude above the ground (altitude AGL) specifies the perpendicular separation between the plane and the ground (or water) directly under it. Altitude AGL, the distance to the ground under the plane, is the most important referent for altitude in the low-altitude environment—reference to elevation above sea level is rarely used.

The *velocity vector* of the aircraft refers to the direction of travel through the air, not to the location of the aircraft with respect to the ground. The velocity vector is specified in three dimensions: the speed through the air, the angle between the path of flight and the horizontal (specifying the plane's elevation above or below the true horizontal), and the angle made between the path of flight and a

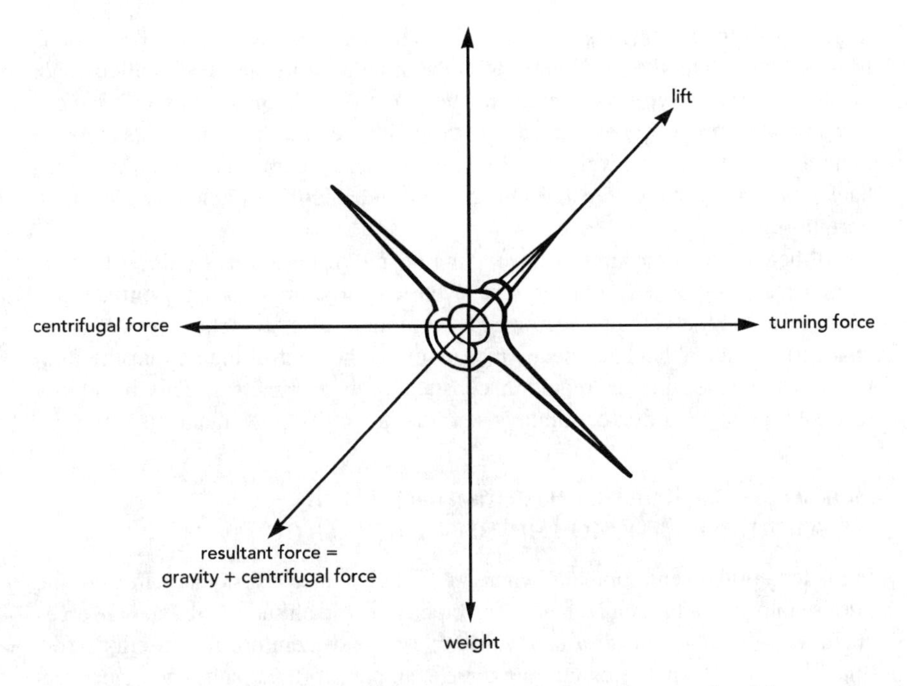

FIG. 2.1. A diagrammatic representation of the forces acting on an aircraft in a banked turn.

reference compass direction (azimuth), such as north. The second and third dimensions taken together define the flight path.

When the pilot applies some input to the aircraft's controls, the changes in G, altitude, and velocity vector define the aircraft's future position. Pushing the stick forward or backward controls the pitch axis of the aircraft; pulling the stick back lifts the nose, causing an ascent. Pushing the stick to the side controls the wing axis of the aircraft; moving the stick to the left lowers the left wing, initiating a descending turn to the left. To make a level turn to the left, the stick has to be moved to the left and simultaneously pulled back. The pull back adds lift to counteract the descent that would otherwise occur.

Aircraft also have a throttle, which changes the thrust exerted by the aircraft's power plant. Although adjustments of the throttle affect the aircraft's *velocity* through the air, military jet fighters are usually operated at a relatively constant (and high) throttle setting. Therefore, changes in aerodynamics of flight do not stem from engine power changes settings as much as from changes in stick position (e.g., pulling back on the stick, which initiates a climb, will, without additional thrust, also result in a decrease in velocity).

Except by momentary coincidence, the flight path of the plane is rarely the direction in which the nose is pointing. This means that the flight path is not aligned with the pitch (longitudinal) axis of the plane in level flight. The nose is usually higher than the flight path, with the amount of its elevation (called angle of attack) varying with the weight of the plane (which changes as fuel is consumed and weapons are expended), its G-loading, and its velocity. Therefore, to maintain level flight, the weight of the aircraft must be compensated for by a pull-back on the stick, which creates a nose-up attitude along its pitch axis (assuming upright flight).

All heavier-than-air aircraft have some degree of instability, with jet fighters being excessively unstable. Therefore, even when the pilot does not initiate new control inputs, the velocity vector and the altitude of a jet fighter continuously change unpredictably. This means that many of the control inputs that the pilot makes are needed just to maintain constant flight parameters. This instability forces frequent and often continuous attention to collision-avoidance tasks.

Some Sample Maneuvers and Their Associated Time-to-Die and Free Time

In the low-altitude environment, where ground-clearance tasks dictate most of the pilot's control inputs, small changes in velocity vector, altitude AGL, and G-forces create large changes in the time-to-die interval and therefore in the pilot's free time. We illustrate this by examining three examples of flight paths and maneuvers regularly performed at low altitude: straight and level flight, level turns, and ridge crossings as part of terrain masking. In each case, the concern is whether the aircraft is on a ground-collision course, and if so, how much time the pilot has while converging with the ground during which he can still initiate a recovery.

Straight and Level. Suppose while flying at 1,000 feet per second at 100 feet altitude AGL on a straight and level course over level terrain, the pilot stops attending to his velocity vector (because he is looking for his wingman, looking for a possible enemy aircraft, or changing a radio frequency, which requires a heads-down position in the cockpit). During this time his jet begins an unintended and minimal 1-degree descent toward the ground. This change in velocity vector starts slowly enough that unless the pilot pays attention to ground clearance or the flight path marker in his head-up display, the descent can continue without detection. This rate of descent is too small to be picked up quickly enough by most of his other flight instruments (e.g., attitude indicator, barometric altitude gauge, radar altimeter, vertical velocity indicator) or his vestibular system (see chap. 3).

Figure 2.2 shows the loss of altitude AGL as a function of time after the descent begins. This graph is pure physics and concerns the time it will take the aircraft to impact the ground. At this speed, a negative flight path angle of 1 degree with the horizontal produces a negative vertical velocity of 16 feet per second, so at 100 feet altitude AGL, impact with the ground occurs in just under 7

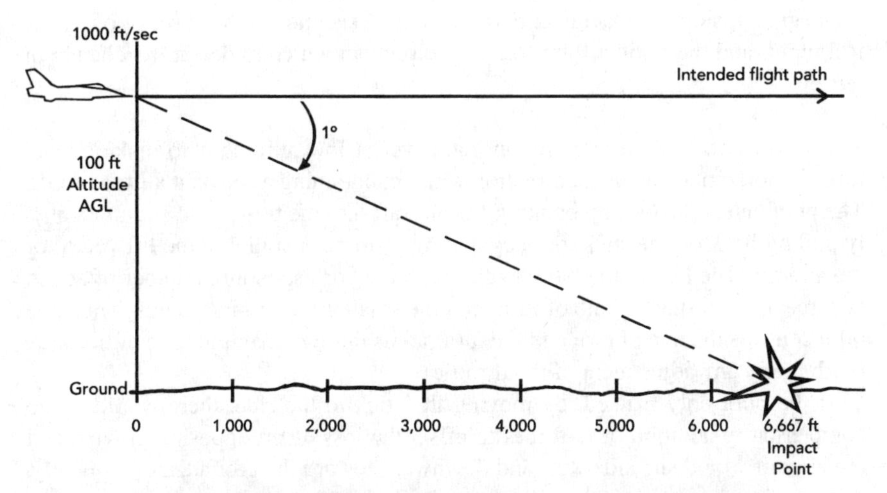

FIG. 2.2. An aircraft, traveling at 1,000 feet per second at 100 feet altitude AGL that begins a 1-degree unintended wings level descent will lose altitude at a constant 16 feet per second and will impact the ground in 6.67 seconds, at a location 6,667 feet from where the descent began.

seconds. Doubling the altitude AGL doubles the time until impact. Unfortunately, this is about the only maneuver in which the time until impact is linearly related to altitude AGL.

The free time for a 1-degree wings level descent from 100 feet altitude AGL is much shorter than the 6.7 seconds shown in Fig. 2.2. At what point in time does collision with the ground become irreversible, regardless of what the pilot does? Detecting the descent by the pilot takes some time, and after detection, a decision to act is needed. Then the pilot has to respond by pulling back on the stick, and then the aircraft itself needs to recover. Although the stick movement will produce a nearly instantaneous change in the orientation of the nose (angle of attack), the velocity vector changes from negative to positive more slowly; the plane continues mushing toward the ground, until the changed angle of the control surfaces of the aircraft begins to return the aircraft to a positive flight path angle. Miller's (1983) training manual shows that the detection, decision, and response times of the pilot and the response time of the aircraft consume 2.5 seconds, so the time-to-die is 2.5 seconds. Once the plane gets within 2.5 seconds of impact at this rate of descent, the pilot is dead meat, though he will live several seconds more. His free time at 100 feet altitude AGL in this example is just over 4 seconds—the time until he can no longer prevent impact.

The same physics and free time apply if, instead of a 1-degree change in the flight path angle of the velocity vector, the terrain changes from a level grade to a 1-degree upslope. Impact occurs in less than 7 seconds in the absence of a control input, and the pilot still has only 4 seconds in which to detect the change in terrain.

Level Turns. A very common maneuver at low altitude is to make a level turn—a horizontal change in direction while maintaining a constant altitude AGL. The pilot enters the turn by banking the aircraft into the turn while simultaneously pulling back on the stick, thereby adding G-forces parallel to the lift vector of the aircraft. The higher the bank angle (with the corresponding number of added G-forces), the faster the rate of turn, and the smaller the turning radius. Thus, the pilot controls the rate of turn and the turn radius through manipulation of the stick (with rarely any adjustment of the throttle).

If the pilot only banked, by moving the stick to the side, there would be no compensatory addition of G-forces to offset the loss of lift opposing gravity, and the aircraft would slip sideways and downward toward the ground. Consequently, to maintain a level turn with no loss or gain in altitude AGL, the pilot must exactly synchronize the degree of bank angle with the proper amount of G-forces added by also pulling back on the stick as he banks. For example, to go from wings level flight (1 G) to 30 degrees bank angle requires the addition of 0.15 G. An additional 30 degrees of bank needs another 0.85 G. Especially beyond 60 degrees of bank, Gs must be added in an ever-accelerating fashion, so that massive G-forces are needed to make very tight turns. (Because adding G-forces to the aircraft also adds them to the pilot, their physiological and psychological effects must also be considered. These are substantial and potentially overwhelming.)

The pilot can coordinate his bank angle and G-forces to maintain a level turn by watching the ground and the horizon out the canopy and adding or subtracting bank angle (or G) as needed to keep the turn level. To do this, he does not have to know the numbers, but he must monitor the ground or watch his flight path marker.

However, if he attends to some other flight maneuver or task, physics says the plane will descend if he has too much bank angle for the appropriate G or too little G for the appropriate bank angle. That descent is not linear with the amount of mismatch; altitude AGL is lost at an ever-increasing rate as the bank angle and G mismatch increases. For example, if a pilot enters an intended level 5-G turn while flying at 100 feet altitude AGL (which will remain level indefinitely if the bank angle [75 degrees] and G [5] match), but overbanks by 5 degrees, he will impact the ground in 3.7 seconds; with 10 degrees overbank, he will impact the ground in 2.6 seconds; and with 15 degrees overbank, impact occurs in only 1.9 seconds. These are time-to-die values; free time is much less. Miller uses 2.5 seconds for the time-to-die value for an intended level turn—the time to perceive the mismatch, decide what to do, react, and have the plane react. Therefore, for a 5-G intended level turn at 100 feet altitude AGL with a 5-degree overbank, if impact occurs in 3.7 seconds and time-to-die is 2.5, the free time is barely more than a second. It is zero for a 10-degree overbank (impact is in 2.6 seconds and time-to-die is 2.5 seconds), and no recovery is possible from a 15-degree overbank at 100 feet altitude AGL. Given these numbers and that overbanking is difficult to detect when performing other tasks, Miller teaches that there is no free time during a level turn at 100 feet altitude AGL or less and barely a second at 200 feet altitude AGL.

Overbanking can occasionally be intentional, as when a pilot wishes to make a rapid descending turn. What a pilot must know about this method of losing altitude, compared to merely entering a wings-level unaccelerating descent, is that the altitude AGL loss in a descending turn is an accelerating function of time, not the linear one shown in Fig. 2.2. The aircraft approaches the ground at an ever-increasing rate. Free time disappears at an ever increasing rate.

Ridge Crossing During Terrain Masking. The third example is a frequently performed maneuver used during terrain masking, in which the pilot maneuvers his aircraft to follow the contours of the ground, using combinations of straights, turns, and dives, as well as rolls, inverted flight, and other high-intensity maneuvers. The goal is to keep altitude AGL below 200 feet or less, regardless of the changes in the terrain, to minimize visual and electronic exposure.

To cross a ridge, however, the pilot has to expose himself by flying higher than the surrounding terrain. To minimize his risk, the pilot must keep his altitude AGL at the ridge line as low as possible without actually hitting the ridge. For aerodynamic reasons, this increases the chances of hitting the ground on the other side. The complexity of ridge crossing is illustrated in Fig. 2.3, which shows two different ways to perform the maneuver.

Figure 2.3 (top panel) shows in profile the flight path of a fighter approaching and crossing a ridge line head-on. The approach is with wings level, followed by a rapid wings level ascent up the ridge wall until the ridge top is just cleared (a few feet is sufficient!). At the ridgeline, the pilot pushes the stick forward to make a wings level descent down the other side. This direct approach creates three problems for the pilot: the plane balloons over the ridge, with too much exposure on the far side (it takes more time to push a plane down than to pull it up); the nose of the plane blocks his view of the other side of the ridge as he comes over, so he cannot see if there is a second ridge waiting to eat him; and the pull down on the far side exerts negative G-forces on his body, pulling his body out of his seat and his last meal into his throat (or higher). For each of these reasons, this method of ridge crossing is never preferred.

The lower panel of Fig. 2.3 shows one typical solution, one that takes advantage of the great maneuverability of jet aircraft. As he begins his approach to the ridge, the pilot both begins to pull up and to roll his aircraft about its velocity vector, so that he crosses the highest part of the ridge nearly upside down. To get back down to the valley bottom on the other side, he either rolls back or the rest of the way around (depending on the angle at which he approached the ridge), all the while continuing to pull back on the stick. Now gravitational forces continue to be positive throughout the entire maneuver, keeping his last meal where it belongs; being inverted, he can look out the top of the canopy so that the nose of his aircraft is not in his way. He can see the entire area over the ridge as soon as

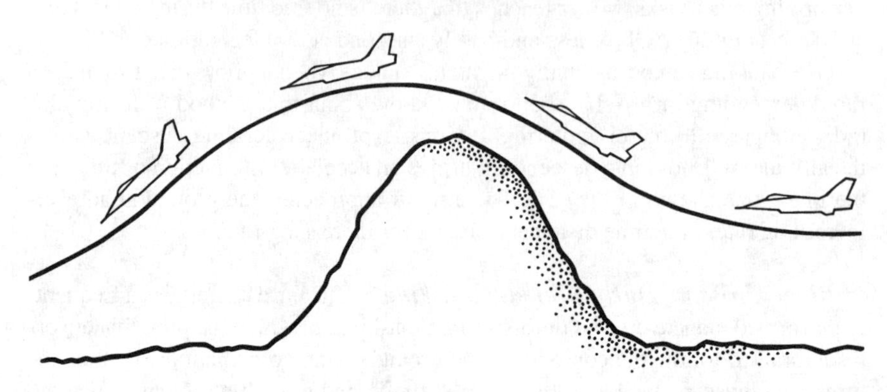

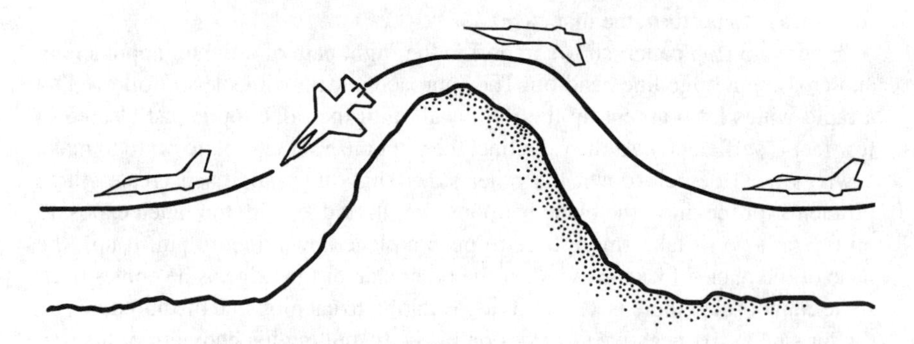

FIG. 2.3. Two examples of crossing a ridge perpendicular to the direction of flight. (a) The pilot makes a wings level climb up the ridge and then, after passing over the ridgeline, makes a wings level descent. (b) The pilot rolls inverted and pulls going up the front side and over the top of the ridge, and then rolls the remainder of the way around, pulling down the backside until he resumes level flight.

he clears the top of the ridge; most importantly, his greater maneuverability under positive Gs gets him back down faster without the great exposure. Partway down the backside of the ridge, he rolls out wings level to continue into the new valley. As can be seen from the lower panel of Fig. 2.3, the pilot has his velocity vector pointing toward the ground for nearly the entire time spent in this ridge crossing maneuver—a series of intentional dives toward the ground.

We have briefly sketched three maneuvers and for each shown the relation between time and altitude loss as determined by physical principles of aerodynamics. Each of these relationships defines a time-to-die for any given altitude AGL, as a function of planned or unintentional deviations from level flight. These three maneuvers have very different time-to-die values and are very different in their sensitivities to altitude AGL. For example, consider the two maneuvers designed to maintain level flight. Straight and level is a low-intensity maneuver, in the sense that the pilot has lots of free time to perform tasks other than monitoring his ground clearance and can easily buy more free time by adding altitude AGL. In contrast, a level turn is a high-intensity maneuver, without much free time and little gain in free time when performed at a higher altitude AGL. Any maneuver in which the plane actually converges with the ground at low altitude, such as an intentional dive, is of even greater intensity and allows the pilot no free time at all, as the remaining altitude approaches the minimum needed to recover from the dive.

These examples of maneuvers show how free time changes as a function of maneuver intensity and altitude AGL. When pilots know and understand the aerodynamics of the maneuvers of their aircraft, then they can always know how much free time they have for tasks other than ground clearance, and they should never fly into the dirt.

Although we discussed three maneuvers for illustrative purposes, they are not a random sample. The low-altitude environment over hostile territory involves primarily high-intensity maneuvers. One measure of intensity is the amount, variability, and onset rate of G-forces exerted on the pilot (and plane)—the higher the G-forces, the greater the variability; and the more rapid the onset time, the higher the intensity of the maneuver. As an example, Fig. 2.4 shows a 1-minute sample

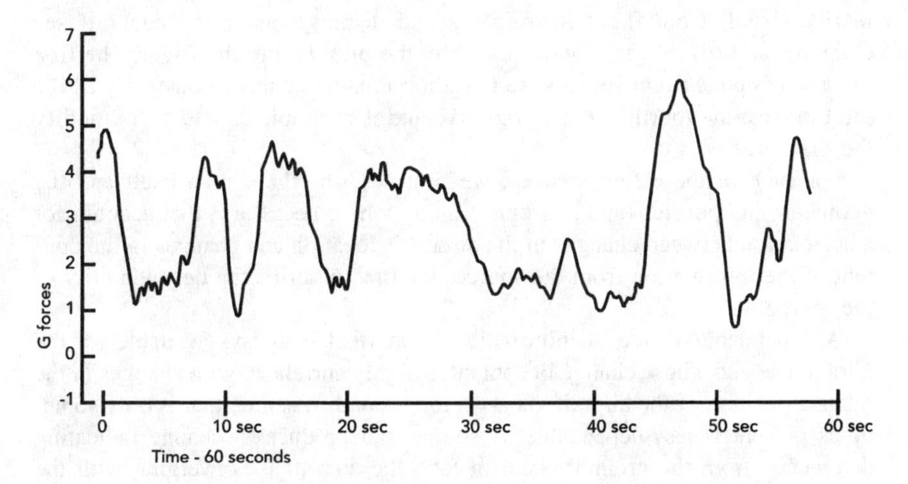

FIG. 2.4. A 1-minute record of changing G-forces exerted on an A-10 pilot during combat maneuvering at low altitude.

of the G-forces exerted on an A-10 pilot by his maneuvers over a 30-minute period of low-altitude attack and masking (Gillingham, Makulous, & Tays, 1982). The entire record is taken from less than 500 feet altitude AGL, with most of it from less than 150 feet altitude AGL. Notice that this A-10 pilot did not pull negative Gs but is always between +1 and +6 Gs, very rarely maintaining a peaceful 1 G for even a second. In any 10-second period, he dives, pulls, turns, and jinks each several times. The onsets are very rapid. This is typical of high-intensity maneuvering when flying a mission near the ground. This figure defines many of the problems for pilots flying in this environment.

SOURCES OF INFORMATION THAT SPECIFY GROUND CLEARANCE

If a pilot knew his altitude AGL and velocity vector, he could look at upcoming terrain and never impact it. In this section, we describe the sources of information available to the pilot that specify his altitude AGL, velocity vector, ground track, and distance to objects as he flies over natural terrain. As we show, sources of information are neither equally available nor equivalent in terms of accuracy, timing, or ease of being processed. To appreciate these differences, we focus here on the information content itself; in the following section, we describe how that information is processed and used.

The four sources of the relevant information that a pilot can use in order to perceive his ground clearance include (a) visual information from the surface of the terrain; (b) vestibular information from the changes in gravitational forces acting on the pilot's body (see chap. 3); (c) symbolic information (primarily visual) from flight instruments and displays inside the cockpit (see chap. 5); and (d) information acquired by the pilot before the flight. The first three stress concurrent sources, so that their content changes constantly as the pilot moves; the fourth is more cognitive and is available to select and modify the first three.

For each of these four sources, we first describe the source itself and the geometry and physics (and neurophysiology where necessary) that account for a correlation between changes in the aircraft's location and changes in the content of the information from the source. We then describe the dependability of the source.

A dependable source of information is one that is always available for the pilot to use and whose change in content is highly correlated with changes in the relative position of the aircraft vis-à-vis the ground. A source that is only sometimes present is less dependable. A source that produces a change indicating divergence from the ground when, in fact, the aircraft is converging with the ground is a widowmaker.

Visual Information From the Surface of the Terrain

The most important source of information for low-altitude flying comes from the way the terrain reflects unique patterns of light to the pilot's eyes. Pilots routinely fly over all naturally existing terrains, some with properties that reflect patterns of light that are highly informative of their distance, orientation, and regularity with respect to the pilot and some that provide little information about their position relative to him. Therefore, analysis of this source of information requires description of the relevant terrain properties. Four independent properties of terrains are critical for their informativeness about ground clearance: the degree of irregularity of projections above their surface, the amount of surface texture they possess, linear perspective arising from the angle of viewing the terrain, and the amount of fine detail present on objects on the terrain.

Terrain Contour Irregularity. Terrain contour varies from rugged to flat. Typical examples of rugged terrains include hills, mountains, and extents containing many vertically projecting objects, such as trees, buildings, or towers. Almost invariably, rugged terrain also has substantial surface texture (as explained later). At the other extreme, flat terrains are characterized by an absence of surface irregularities in height. Typical examples of flat terrain contours include an extended body of water, a large prairie or desert without trees or other vertical projections, and a dry lake bed. Flat terrains may or may not have much surface texture.

Moving over rugged terrain produces changes in the dynamic patterning of light reflected from the terrain to the pilot's eyes that are highly correlated with changes in the pilot's altitude AGL, velocity vector, and distance to particular objects on the ground. Therefore, rugged terrain is highly informative about ground clearance. The primary visual source of this information is the dynamic optical occlusion of far surfaces by the interposition of near ones in the pilot's line of sight as he flies over the irregularities (see Owen, 1981, for examples and Cutting, 1986, for a review).

For dynamic optical occlusion to occur, there must be differences in the elevation of parts of the terrain. A flat surface has nothing in the near ground that projects upward enough to occlude farther objects or terrain. Therefore, flat terrains are relatively uninformative to a pilot concerning his ground clearance.

To illustrate the geometric basis for the informativeness of dynamic occlusion (i.e., the high correlation between changes in occlusion and changes in ground clearance), Fig. 2.5 shows a side-profile view of a terrain in which a near ridge is interposed between the pilot and the higher far ridge. The dynamic (changing) optical occlusion of the far ridge precisely specifies the velocity vector of the aircraft and therefore whether the plane will clear the near ridge. As the pilot approaches the near ridge, if more and more of the far ridge becomes visible (as

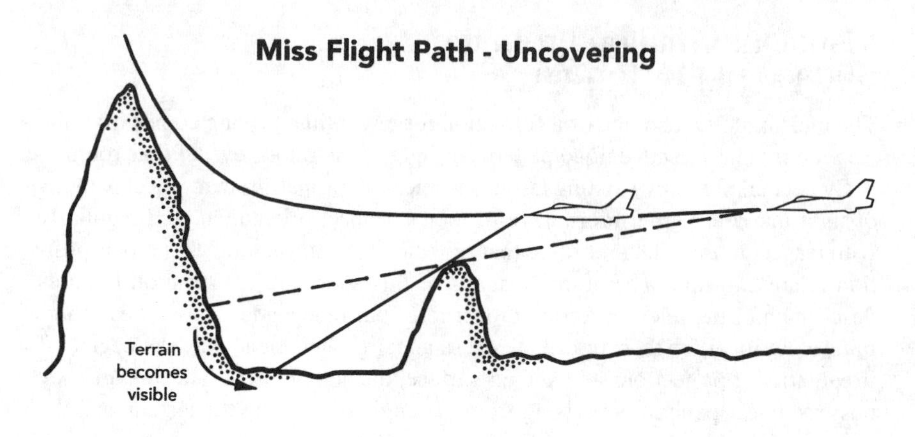

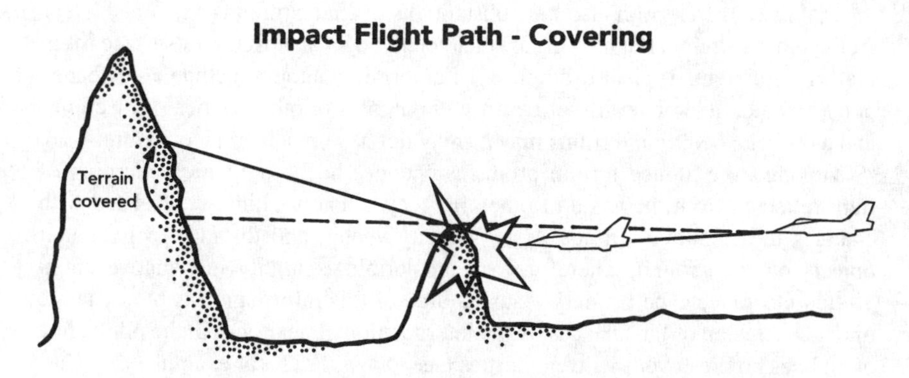

FIG. 2.5. A pilot can determine whether his velocity vector will carry him over a nearby hill by noting the nature of the changing visibility of any farther hill. (a) The flight path is above the near hill, and more and more of the far hill becomes *visible* as the pilot approaches. (b) In a flight path that will impact the low hill, the pilot should have noticed that more and more of the far hill was being occluded from view as it was approached.

in the top of Fig. 2.5), his velocity vector must be above the crest of the near ridge to clear it. In contrast, if more of the far ridge disappears from view as the pilot approaches it (as in the bottom of Fig. 2.5), his velocity vector must be below the crest of the near ridge, and the plane will impact it. A near surface covering up a far one as it is approached signals impending collision. This is probably the single most important source of information that allows a pilot to fly safely up an irregularly rising valley without colliding with the outcrops.

This example shows how dynamic interposition can provide information about an aircraft's velocity vector with respect to any aspect of rugged terrain. Dynamic interposition, producing a change in occlusion with the passage of the plane across the terrain, also provides information about the plane's relative distance from different features of the terrain; the rate of uncovering or covering of a partially occluded far surface by a near one is proportional to the distance between the two surfaces. Hence, in Fig. 2.5, increasing the separation of the higher and lower hills produces a slower rate at which the occluding edge moves. Flying over terrain with many near hills or other obstructions interposed in front of farther ones provides a myriad of changes in occlusion, combinations of which change in proportion to their distances to the aircraft.

The information available from terrain contour irregularity also depends on the elevation of the pilot as he crosses the ground. Crossing a low hill at 50 feet altitude AGL uncovers more distant terrain than when flying at 500 feet altitude AGL. In contrast, flying over flat terrain, such as extended bodies of water, never provides a pilot with any useful changes in occlusion, regardless of how close he flies to the wave tops.

Terrain Texture. Surface texture, the amount of visible fine-grain structure on the surface of a terrain, also provides the pilot with information about his ground clearance, especially his altitude AGL and changes in altitude AGL. Highly textured terrains include those with lots of boulders, irregular mountains, forests, and planted fields. Untextured surfaces include smooth water, snow fields, or desert sands.

Surface texture generates changes in patterned light that are highly correlated with ground clearance. A textured ground surface produces differential flow in the patterned light—an optic flow movement of the light-dark variation—across the pilot's eyes as he flies over the textured terrain. Holding both the pilot's speed of movement and the density of the texture constant (see the following for discussion of these limitations), the rate of the optic flow reflected from any part of the surface is correlated with the distance of that part of the surface to his eyes. Hence, the rate of the optic flow produced by the nearby ground directly below the plane is very rapid and falls off proportionately with distance. Mathematically, the pattern of optic flow past the observer exactly specifies the relative distances from the observer to all the reflecting surfaces on the terrain (Harrington, Harrington, Wilkins, & Koh 1980; Lee, 1974; Nakayama & Loomis, 1974; Warren, 1976).

Figure 2.6 illustrates this proportionality in the flow of textural details across the pilot's retinas during a segment of level flight over level terrain. Each reflected point on the terrain moves over the pilot's retinas at a rate proportional to its distance. The greatest angular rate arises from surface textures directly beneath the plane, with the rate falling off uniformly in all directions, reaching a zero rate of flow when looking through the plane's velocity vector.

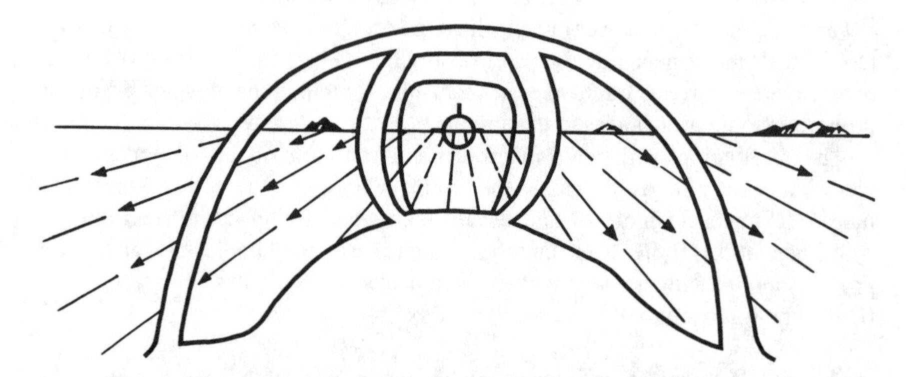

FIG. 2.6. The optic flow pattern reflected from the terrain during level flight. Each reflected point on the terrain moves over the pilot's retina at a rate proportional to its distance, except for the point at the velocity vector, from which there is no flow.

Changes in the rate of optic flow arising from textured surfaces informs the pilot about his altitude AGL, irregularities in the terrain, and his velocity vector. First, consider altitude AGL and its changes. Assuming a constant velocity and a constant density of surface texture, any change in rate of optic flow when looking through the same area of the windscreen means a change in the distance from terrain to plane. When looking straight down beneath the plane, if the optic flow increases, so it is faster now than it was a moment ago, the ground is closer, by an amount proportional to the amount of change in the rate of optic flow. This difference is directly correlated with a change in a plane's altitude AGL.

Changes in optic flow also inform the pilot about irregularities in the terrain, because the texture of a flat terrain stretching away in front of the pilot produces a uniformly decreasing rate of optic flow (mathematically, a second-order invariant). Any discontinuity in the change in rate of flow means a discontinuity in the surface itself—a change in overall slope or a local change in slope such as a ridge or depression. Thus, the terrain irregularities are mirrored in the contour envelop of the angular rate of movement. All this information becomes available to the pilot whenever the terrain has a visible surface texture.

The patterning in the optic flow from a textured ground is also informative of a pilot's velocity vector (Cutting, 1986). When a pilot looks anywhere other than through his velocity vector, the pattern of optic flow of the terrain texture moves along the windscreen from front to back, picking up speed (rate of flow) as the plane nears that piece of terrain being observed. However, when the pilot looks through his velocity vector at the piece of sky or terrain toward which he is flying, it produces no flow at all. The texture expands in detail as the pilot gets closer, but it does not move or flow across his eyes (or across the windscreen). Thus, not only is the pattern of the rates of flow proportional to distance, but that pattern also converges to a zero flow point, which is the momentary velocity vector (a point at infinity).

Linear Perspective. Information about the pilot's altitude AGL is also provided by the patterns of variation in light produced by surface features on the ground. These sources are present whether the observer moves or not. This static patterning arises from the geometric properties of linear perspective. For example, far textural elements produce smaller images at the eye than do near ones, and linear perspective effects translate parallel lines on the ground into converging lines at the retina. Such static cues to distance are the familiar content of most classical discussions of depth perception. They depend on a general perspective effect—equal-sized elements on the ground project unequal-sized images to the eye, the inequalities being proportional to their distances to the eye. This proportionality provides the correlation between changes within the pattern at the eye and changes from the eye to the reflecting surface.

Resolution of Fine Details of Objects on the Ground. Fine details observable by the pilot inform him of both changes in altitude AGL and absolute altitude AGL. Because a pilot rarely knows the absolute sizes of the bushes, boulders, cacti, trees, or other natural elements making up the ground texture, he cannot tell if a given optic flow arises from large bushes seen from a large altitude AGL or from small ones he is barely clearing. In the low-altitude environment, however, in the range of 25 to 250 feet, the pilot learns from extensive experience that if he can resolve fine details in an object, he is flying very close to the ground.

More importantly, this resolution can also be used as a check on absolute altitude AGL; in most cases, resolution of very fine detail is possible only within 100 feet of the ground. For example, although a pilot may have no idea how tall a tree stands, and therefore cannot use the tree's height to calibrate either his distance above it or the distances to other objects, if he can see the pattern of veins on the leaves of the tree, he knows that he must be very close to it. Similarly, cacti vary in sizes from inches to tens of feet, so the size of a particular cactus can provide no reliable information about its distance from the pilot; when he is able to see the individual thorns, he is within about 50 feet of it. The wave pattern on the ocean is unreliable as a source of information about a pilot's elevation over it, but resolving the windblown spray off the whitecaps can only be accomplished at a very low altitude.

The presence of fine detail also gives the pilot flying at very low altitudes a quick check on whether he is maintaining a constant ground clearance. Any change in the clarity of the fine details of objects signals a change in his altitude AGL. Many terrains have sufficient small details to be highly informative to the low-flying pilot. However, a glassy sea fails, as does snow cover, to provide enough details to resolve.

Limitations on the Informativeness of Patterned Light Sources. Two critical limitations restrict the usefulness of information arising from light reflected from the terrain: (a) Some terrains fail to provide information about ground clearance, and (b) some terrains "misinform" the pilot about his velocity vector.

Most pilots and many visual scientists describe particular kinds of terrains as sources of visual illusions, in the sense that a pilot misjudges his altitude AGL because of the optical information he receives. However, the term *visual illusion* has two quite different meanings by visual scientists and pilots. The first meaning is found widely in laboratory research literature (see Gillam, 1998, for a review) and refers to misperceptions caused by *misinformation* from particular features of the "scene." In the laboratory, these scenes consist of either pictorial displays (as distinct from 3-D scenes) or very impoverished displays. We know of no reasonable examples that are relevant to low-altitude flying.

The other meaning of visual illusion refers to misperceptions caused by *underinformativeness* of the terrain. This occurs frequently in low-altitude flying (see McNaughton, 1981–1984, for many important examples). Because of the absence of texture, objects, and details, a pilot flying over a featureless dry lake bed cannot visually determine how close it is. In such circumstances, the ground surface can sometimes appear closer or farther than it really is or simply be of indeterminant distance. All three may be dangerous.

Underinformativeness occurs frequently during low-altitude flight. For there to be patterned light available to the pilot's eyes, there must be irregularities and texture on the ground surfaces. However, pilots have to fly over snow, dry lake beds, deserts, and extended bodies of smooth water, none of which have sufficient texture and irregularities to produce patterning in the light that is correlated with distance. Because information obtained by the eyes while looking out the windscreen is the single most important means by which pilots avoid collisions with the ground, serious problems arise when this source is absent. As will be discussed at the end of this section and in the next, when dynamic optic flow is not available, pilots must rely on alternative sources to determine their altitude AGL and velocity vector. Acquiring ground-clearance information from these alternative sources is substantially more difficult.

Misinformation occurs during low-altitude flight under at least two common conditions. A normally informative terrain will misinform a pilot if either his velocity over the ground or the terrain texture itself changes without his detection of the change. Undetected velocity changes are unusual, but when they occur, collisions with the ground are likely when flying at low altitude (see Haber, 1987, for an example). More commonly, and more insidiously, the density of the texture on the terrain can change. These changes occur rapidly, given the vast expanses of terrain that a pilot crosses in short periods of time. In desert country, the density of sage bushes (their size and their distance from one other) is a function of water sources and soil nutriments, both of which can vary substantially over the course of a few miles. For example, the land near a river bed might have dense growth, which when viewed from the air, produces a high rate of optic flow, suggestive of a low-altitude AGL. After the pilot crosses the river and continues to fly away from it, the density of surface texture from the vegetation growth drops as the water table recedes, producing a lower rate of optic flow, falsely suggesting a gain in altitude AGL. Not only do these changes in surface texture mislead a pilot to perceive a gain in altitude AGL when none has occurred, in fact, there may be a *loss* in altitude AGL as the land slopes upward from the river bottom. If the pilot pushes the stick forward after crossing the river, he may collide with the rising land.

Normally, the pilot assumes a constant velocity and a uniform texture density. If the assumptions are wrong, the correlation between optic flow rate and terrain distance is neither perfect nor interpretable. Any perceptual processing based on optic flow under these conditions is fraught with danger for a pilot.

In sum, the presence of surface irregularities and a perceptible texture provides potentially useful information about altitude AGL, change in altitude AGL, and the velocity vector. It is not surprising that terrains without textures are very difficult to fly over and easy to fly into. It is also not surprising that many ground collision accidents occur when the assumptions of constant texture density and constant velocity are violated.

Vestibular Information Arising From Changes in G-Forces

Changes in centrifugal force can potentially inform the pilot about changes in his velocity vector. Any change in the vertical, horizontal, or lateral velocity of the aircraft produces a change in the centrifugal force acting on the aircraft and the pilot within it. The amount of change in centrifugal force is directly proportional to the change in velocity and opposite to the direction of the velocity change. Hence, changes in centrifugal force are proportional to changes in the direction of movement of the aircraft and very informative about such changes.

Acceleration or deceleration or entering a climb, dive, or turn each produces a change of G-forces acting on a pilot's body. Because changes in aircraft speed are less typical over the short term, the most important G-force is the one parallel to the lift axis of the plane. In level flight, 1 G is exerted (the weight of the plane and its contents being exactly matched by the lift provided). Beginning a wings level climb requires added G-forces, which the pilot experiences by feeling pushed down in the seat as the plane rises. A loss of G occurs when the plane begins a wings level descent from level flight, and the pilot floats out of his seat as the plane drops out beneath him.

The vestibular system senses changes in G-forces, and pilots typically report being sensitive to changes in G. In addition, this force is displayed in a G-meter, present in most jet fighters. However, current thinking is that G-force information may frequently misinform the pilot of the way in which the movement of his aircraft is changing during low-altitude maneuvering (though it may be useful in landing, formation flying, and some other tasks).

Although the correlation between objective changes in G-forces (as measured by a G-meter) and changes in velocity vector is high, the correlation as measured by the output of the human vestibular system is not high at all (Gillingham & McNaughton, 1977; McNaughton, 1985), mainly due to rapid adaptation. Further, even when measured precisely by the G-meter, there are circumstances in which the change in G-forces is easily misinterpreted by the pilot, leading him to make an incorrect response. A discussion of these limitations of G-force information is postponed until the next section, where processing is covered in detail.

Symbolic Information From Flight Instruments and Displays

Flight instruments provide information about several critical aspects of ground clearance, when changes in the instrument are correlated with changes in the position or movement of the aircraft. Some instruments are displayed as round gauges (usually analog) located in the instrument panel below the line of sight of the pilot. To acquire the information from a round gauge, the pilot must fixate it in foveal vision by shifting his gaze away from the ground through the canopy, and he must shift his accommodation from far to near focus. In a more recently developed version of some flight instruments, information is presented in a head-up display (HUD) located directly in the pilot's line of sight above the instrument panel. The symbols are projected at optical infinity superimposed on the ground or sky he is looking at through the canopy. Thus, the pilot shifts neither his direction of gaze nor his accommodation. We will consider those instruments most important for low-altitude maneuvering.

Altimeters. Round gauge altimeters either sense altitude by changes in barometric pressure or changes in radar signals. A barometric altimeter specifies elevation above sea level by measuring changes in air pressure. Except when flying over the sea, however, it provides no direct information about ground clearance (altitude AGL), because the elevation of the terrain under the plane is not specified. Further, a barometic altimeter is relatively slow to respond to changes in altitude, usually lagging several seconds behind the velocity vector. Consequently, barometric altimeters are of restricted use in low-altitude maneuvering—they do not provide information about altitude AGL. However, they become critical in those conditions that compromise the radar altimeter (as explained later).

A radar altimeter measures ground clearance under the aircraft, providing a direct, accurate, and instantaneous indication of altitude AGL. Its drawback is that it samples only the ground under the plane, not in front along its intended flight path. The sampling cone is about 60 degrees in solid angle, extending out from directly beneath the aircraft so that at 100 feet altitude AGL, the radar altimeter indicates the distance to the closest object or elevation on the ground within a circle with a radius of about 50 feet. If the plane is flying over the flat and level ocean, then the radar altimeter is predictive of altitude AGL ahead and provides a useful source of information to the pilot about ground clearance. However, if the terrain is irregular, the radar altimeter cannot predict clearance over individual ridges until the aircraft actually gets over each protrusion. By then, the plane has either collided with or cleared the ridge.

A further limitation of the radar altimeter becomes obvious during steeply banked turns. Because the focus of the cone of the altimeter does not reorient toward the ground while the aircraft is banked, it provides incorrect readings during steep turns. This limits its usefulness in monitoring unintended descents during turns!

Even with its limitations, the radar altimeter does provide direct information about absolute altitude AGL under the aircraft, making it an ideal device to aid the pilot in learning what particular altitude AGLs look like. Instructor pilots refer to this process as "calibrating the eyeballs," which is discussed in some detail in the next section.

The radar altimeter often has an auditory and/or light warning, which can be preset by the pilot to ring whenever altitude AGL drops below the selected value. This warning is very useful when flying over level terrain, especially water, but most pilots turn it off when flying over rugged terrain and especially when engaged in terrain masking, because the pilot's intentional approaches toward the ground set it off continuously.

Flight Path Marker. The most important instrument for low-altitude flight is the flight path marker, which exists only as a HUD component. The flight path marker projects a lighted spot on the HUD screen, which corresponds to the velocity vector of the aircraft, the actual path of the plane through the air, independent of the direction pointed by the nose. (The third dimension of the velocity vector—aircraft velocity through the air—is displayed separately in a round gauge, the airspeed indicator.) Having a flight path marker in the HUD is extremely helpful, as the velocity vector rarely corresponds to the direction in which the nose is pointing or the degree to which the wings are banked (or what is displayed in the attitude indicator). There is no way for the pilot to "see" his velocity vector by looking at the structure of the aircraft or any of the round gauges. When the pilot's head is in its normal position looking straight ahead, his velocity vector corresponds to what he sees when looking through the flight path marker. This provides a precise indication of the levelness of flight, clearance over upcoming ridges, or a fixed aiming point in dives.

However, there are three important limitations to the flight path marker. First, because the range of excursion of the indicator is limited by the size of the HUD, the marker can disappear from the pilot's sight at the beginning of rapid climbs and dives or tight turns. Second, it is useless if the pilot's head is off center, or he is looking elsewhere than straight ahead, as is frequently the case in tight turns. Finally, the flight path marker cannot provide a picture for the pilot of his future ground track for any maneuver other than flying straight. When the plane enters a turn, the velocity vector sweeps across the horizon as the turn progresses but does not show the specific part of the terrain the pilot has to clear during the turn.

In summary, although there are many flight instruments whose indications are correlated with changes in the position and movement of the aircraft, all have severe limitations that restrict the range of their usefulness for low-altitude flying. Serious debate exists in the research literature as to whether some of these instruments are useful for pilots flying at low altitude (Roscoe, 1980). As we will discuss in the next section, these instruments are, nevertheless, critically needed to supplement or replace direct visual information if accurate perception is to be achieved.

Information Acquired Before the Flight

Information provided to the pilot before the flight has quite different properties from the others just discussed; it is prior rather than concurrent information (used to anticipate), cognitive rather than perceptual (to be used to update or refine concurrent information), and does not change as the flight unfolds. Training has focused on three of these prior knowledge sources.

Specific Information About the Route. A properly briefed pilot has examined maps of the terrain to be traversed noting landmarks and potential hazards, such as the presence of a radio transmission tower, that others have reported to be very difficult to see. He will also have been warned of any difficulties he may have in perceiving his altitude AGL when flying over a desert approach to the target and advised to attend to his instruments. This kind of knowledge is terrain specific. Prior knowledge functions primarily allow the pilot to anticipate an upcoming event. The greater the amount of prior experience, the more detailed the map, or the better organized the briefing, the more likely the pilot will use this knowledge at the proper time.

Knowledge of the Redundancies in Natural and Artificial (Man-Made) Terrains. Another kind of prior knowledge arises from understanding the redundancies or predictabilities that are present in all terrains of a given kind. For example, natural terrain normally slopes downward toward a river, and rivers run downward. If pilots are aware of these predictabilities, they have information about terrain slope from the presence of rivers. Pilots are also taught to be sensitive to any change on the terrain that appears to be the result of man-made construction. For example, a break line through the trees of a forest may be due to a pipeline, a road, a firebreak, a power line, or natural erosion. Although most of these possibilities are of no concern for a low-flying pilot, power lines are widowmakers. Therefore, a break in the trees alerts the pilot that there *might* be wires that he will not be able to see; the break should alert him to pull up before he crosses it.

Knowledge About Objects on the Terrain. Pilots describe themselves as very sensitive to the presence of familiar-sized objects on the ground, and they treat that familiarity as an important source of information that helps them perceive the distance to such objects. They also claim that the presence of a familiar object helps them perceive their altitude AGL. Pilots also report being misled by familiar objects in the "wrong" size, such as stunted fir trees.

Familiar size of an object has long been considered a source of information allowing observers to determine the distance to an object (see Hochberg, 1971, for a review). Suppose the pilot recognizes a familiar object on the ground and knows its size or extent (e.g., a telephone pole, house, or a car). Given a known true size of an object in the distance, and given the size of the visual image of the object projected on each of the pilot's retinas (which in theory is also known to the pilot's perceptual system), these two sizes determine a unique true distance. Further, because the known true size remains constant regardless of its distance, changes in the size of the visual image of the pilot are perfectly correlated with changes in the true distance to the object as the pilot approaches it. Thus, knowledge of familiar size provides information that could allow a pilot to perceive both his altitude AGL and his change in altitude AGL.

However, current research evidence fails to support these anecdotal claims and possibilities. For example, Haber and Levin (2001) found that familiarity with an object has relatively little effect on the accuracy of perception of its distance. Further, in their experiments, they distinguished between familiar objects that only come in one size (e.g., Volkswagen beetles) and those that do not have a single fixed size (e.g., the width of a road). They found that familiarity with the objects in this second group had no effect whatsoever on the accuracy of perception of their distance. These results suggest that pilots may be incorrect in their assessment of the usefulness of familiarity with the sizes of objects as a source of information about their distance.

In summary, in this section, we have considered the major *sources* of information that covary with the position and movement of an aircraft and that may potentially aid the pilot in determining his ground clearance. We have also described the limitations on each source of information. These limitations are of two kinds: some sources are unreliably present, and some may misinform the pilot about his ground clearance. Both of these categories of limitations place further burdens on the *perceptual processing* of the information, the topic discussed in the next section.

PERCEPTUAL PROCESSES: HOW A PILOT PERCEIVES GROUND CLEARANCE

The previous section described the different sources of concurrent visual and vestibular information that can permit a pilot to perceive his altitude and direction of flight, as well as prior knowledge to permit him to anticipate. Given this information, by what processes, and how well, does a pilot actually achieve a perception of (a) a change in altitude AGL, (b) absolute altitude AGL, (c) velocity vector, (d) ground track, and (e) distance to other aircraft and objects on the ground? These five topics comprise the major subdivisions of this section.

Underlying most of the presentation that follows are issues of why and when perceptual processing fails during low-altitude flying. We have already identified in the previous section a number of limitations in the information itself, some of which concern its underinformativeness (not enough to lead to an adequate perception) and its misinformation (leading to the wrong perception). Underinformation and misinformation require the pilot to override normal perceptual processes and add focused attention to perceptual processing. In addition, many of the ground-clearance tasks cannot be handled only by automatic perceptual processing, given the intense demands of mission tasks, maneuvers, and the proximity of the ground.

Whether driven by information quality or competition for attention, the result is that automatic perceptual processes frequently fail or are inadequate during low-altitude flight. Pilots must learn the conditions under which these failures are to be expected and, in addition, learn to substitute alternative information sources. Overriding automatic processes requires controlled attention. The decision as to which alternative information sources to use requires controlled attention. And the alternative sources themselves require controlled attention. This constellation of demands to substitute a set of controlled processing actions for automatic perceptual processes places a huge attentional burden on the pilot. As we will show, the demand arises constantly and forms the crux of the difficulty of learning to fly safely and effectively at low altitude.

Automatic and Controlled Perceptual Processes

An automatic perceptual process as defined here includes three properties: (a) It does not require any conscious effort, (b) it does not require any decision or thought process to be completed, and (c) it normally functions without prior practice or learning (at least in adults). A controlled perceptual process requires some cognitive effort or conscious attention.

This distinction between automatic and controlled perceptual processes draws on an analogy from theoretical developments in research on attention. Recent theory in attentional processes has made a careful and useful distinction between automatic versus controlled processes (see Schneider & Shiffrin, 1977; for a detailed review, see Allport, 1989). For an attentional process to be or to become automatic, two conditions need to apply to the behavior in question: (a) There must be a stable and consistent relationship between that specific stimulus and its required response, and (b) either that relationship has been prewired in the nervous system or there must be massively extensive practice in performing that stimulus-response mapping. When these conditions are satisfied, then regardless of which stimulus might appear, the proper response is selected and performed without the need for conscious attention, decision making, or monitoring.

Automatic perceptual processes include sensitivity to contrast, particularly contrast resulting from edge discontinuity, sensitivity to relative movement among different parts of the visual field, sensitivity to edge discontinuities to segregate figure from ground as a way to perceive the presence of objects, sensitivity to changes in global optic flow to perceive a change in distance from the surface reflecting the patterned light, and sensitivity to differential optic flow generated by self-motion to specify the shapes and contours of surfaces and objects. All of these seem to be built-in perceptual processes that are either initially hardwired in the visual nervous system or develop with early perceptual experience into effectively hardwired systems (see Haber & Hershenson, 1980, for a review of the importance of this kind of early visual experience).

The great advantage of these kinds of automatic perceptual processes is that the information is acquired without any controlled, focused, or conscious thought. The perceptual processing does not interfere with other activities being carried out at the same time, and those other activities do not interfere with automatic perceptual processes.

In contrast, examples of controlled perceptual processes include reading digits from dials and gauges to learn about velocity, altitude, and the like; using the shadow of the plane as it moves across the ground under the plane to estimate elevation above the ground; and counting the number of seconds it takes for an object to move a fixed distance across the canopy as an estimate of distance from the plane to that object on the ground. Skill in performing these actions is achieved by practice. However, they never become automatic.

Many of the perceptual processes that function automatically at slower angular velocities and closer distances require controlled processes at the much higher speeds and flow rates typical of flying in the low-altitude environment. Low-flying pilots have to perceive and respond when moving at 1,000 feet per second, speeds that produce optical flow rates continually above 200 degrees per second; they require accurate distance perception out to several miles and accurate direction perception to a fraction of a degree. To perform these perceptual feats, the pilot cannot use the perceptual information he acquires automatically at the slower rates typical of terrestrial travel; neither can he acquire this information automatically by extensive practice of where to look and how to look.

Switching to controlled processing is critical when automatically processed information alone is either insufficient or potentially misleading, so effortful processing is required to compute or fill in information or compensate for misleading information. As we show in the subsections to follow, pilots cannot depend on their automatic perceptual processes during low-altitude flight because misperceptions are likely to occur. As a result, the pilot must engage in three processing steps, all of which require focused attention. First, he has to consciously override automatic processes when they potentially provide him with incorrect information; second, he has to consciously remember to refer to his instruments or other sources for that information; and third, he has to process the alternative sources of information, using focused attention.

Many of Miller's (1983) procedures are designed to provide exactly this kind of training, which provides practice to override the automatic processing of available information when it is inadequate or misleading and switch via controlled process to a specific alternative source of information. For example, through preflight briefings and terrain assessment, the pilot is taught when to expect inadequate or misleading information from smooth, featureless terrains; and at those instances is taught to switch to focused perceptual sources. When gravitational effects obtain, the pilot is trained to ignore his vestibular system for perception of his attitude and velocity vector and depend on what he can see outside the canopy; if vision is compromised, he is to use his instruments.

We now turn to the perceptual processing of five aspects of ground clearance: change in altitude AGL, absolute altitude AGL, velocity vector, ground track, and distance to other planes and objects on the ground. For each, we first examine automatic perceptual processes, then the limitations on information acquired through automatic processes, and finally, the controlled perceptual processes the pilot must employ to acquire the information he needs to maintain proper ground clearance.

Perceiving a Change in Altitude AGL

If a pilot could always detect a loss in altitude AGL, then he should never be surprised or dead. There are three automatic processes that enable a pilot to detect relative changes in altitude AGL: (a) Processing global optic flow patterning, (b) processing the level of the horizon, and (c) processing G-forces. We begin with optic flow.

Processing Global Optic Flow. The single most important source of information from which a pilot can perceive a change in his altitude AGL is the changing pattern of global optic flow, the pattern of stimulation across his retinas that arises from moving across a textured terrain. The third section of this chapter showed that the patterns of optic flow at the retinas geometrically reflect properties of the terrain, both its variation in contour and its distance from the observer.

Research evidence shows that global optic flow is picked up over the entire retinal surfaces of each eye and, hence, is not restricted to foveal processes in either eye—in fact, it probably does not use foveal processes at all (Haber, 1981; Liebowitz & Post, 1983). The processing of global optic flow does not require the presence of any visible aircraft structure for a comparison or contrast (thus, it is much more general than the seemingly related process of motion parallax that is usually studied in the laboratory (Cutting, 1986; Koenderink & van Doorn, 1976).

How the geometric relationship between optic flow and distance is processed in order to perceive the terrain contour and distance is not fully understood. There are three viewpoints on the processing of global optic flow: (a) a traditional Helmholtzian view closely linked to geometry (Hochberg, 1971), as described in the previous section; (b) Gibson's (1979) view that perceivers extract invariances in flow patterns (see also Lee, 1978); and (c) Warren and Owen's (1982) more active view that pilots make responses to adjust their position over the ground to maintain an invariant flow pattern. These three viewpoints agree that the processing of optic flow is hardwired and therefore automatic; current research is attempting to identify the processes by which the flow patterns are translated into perceptions. For our purposes, the details of the answer are unimportant. Optic flow reaching the eyes of a moving perceiver is both automatically processed and highly informative of distance and the contours of the terrain. Unfortunately, we will show that pilots must override this automatically acquired information to survive.

Five characteristics of optic flow rates limit their usefulness as a source of information about changes in altitude AGL.

The first limitation arises because the true rate of global optic flow (corresponding to the rules of geometry) can also change independently of a change in altitude AGL. For example, a global flow change will occur geometrically whenever there is a change in the plane's speed over the ground, a change in the flight path angle, or a change in the density of the objects scattered on the ground. In each case, the change in flow suggests a change in altitude AGL when none has occurred. If the pilot interprets the flow as a gain in altitude when none has occurred, he may initiate maneuvers from which recovery is impossible.

A second limitation is the converse of the preceding example. Little or no global optic flow is generated at the pilot's retinas by light reflected from terrains devoid of texture, such as snow or deserts. A pilot can lose altitude without any change in optic flow being generated from these terrains. Therefore, an invariant optic flow rate does not always mean an invariant altitude AGL.

A third limitation arises from changes in the pilot's physiology that may mask the correlation between flow rate and altitude AGL. The perception of the rate of global optic flow adapts rapidly, beginning within a matter of seconds. Even in the absence of any objective change in altitude AGL, speed, or ground texture (flow rate is optically constant), the subjective rate of optic flow seems to slow, so the pilot perceives the flow rate as decreasing. This results in a perception of altitude

AGL gain (or velocity decrease) and often causes a pilot to initiate a slow descent after flying near the ground to make up for the apparent but false altitude gain.

A fourth limitation of optic flow processing arises because the same physical flight and terrain conditions can produce substantially different perceptions, depending on what the pilot had just been doing. In an automobile, the same 50 miles per hour seems much slower when dropping from 70 mph than when speeding up from 30 mph. Similarly, in a plane, the ground rush at 100 feet altitude AGL seems much slower when coming up from 50 feet than when descending from 200 feet. Failure to take the prior context into account can lead a pilot when pulling up to feel he has gained more ground clearance than he really has. The problem occurs most seriously in pop-up bomb deliveries. Typically, the pilot has flown to the target area at very low altitude, using terrain masking to avoid detection. As he nears the target, he pulls up sharply and then enters into a dive at the target. His adaptation to the very high rate of flow during his very low approach, followed by the rapid altitude AGL gain in the pop-up, leads him to overestimate his actual altitude AGL gain. This may cause the pilot to be overly complacent in delaying his dive recovery unless he closely watches his altimeter (a controlled perceptual process).

A fifth limitation on automatic processing of global optic flow stems from the retinal mechanisms involved. The pickup of global flow can occur over the entire retinal surface and is dependent on the full extent of the retina being stimulated (Liebowitz & Post, 1983). However, nearly all maneuvers that add G-forces to the body of the pilot produce impairment of peripheral vision, with the amount of narrowing of vision proportional to the G-forces exerted (Gillingham & McNaughton, 1977). A typical 5-G level turn narrows effective vision from 150 degrees to about 60 degrees; 7 Gs reduces the tunnel to less than 20 degrees. This means that during heavy maneuvering creating 5 Gs or more, such as tight turns, climbs, inverted pulls, or dive recoveries, the pilot loses some of his ability to perceive global optic flow.

These five limitations on the accuracy of the perception of altitude AGL based on automatic processing of global optic flow mean that a pilot cannot always automatically react to a change in global optic flow: it can misinform him and either allow him or cause him to fly into the ground.

Miller (1983) provides specific training procedures to aid pilots in getting the most accurate information from global optic flow and prevent misperception. Each of these impose controlled processes, both attentional and perceptual, onto what would otherwise be automatic ones. We present four examples to illustrate Miller's training strategies to overcome global optic flow limitations.

First, Miller teaches the pilot to thoroughly assess the optic flow potential of the terrains over which he is going to fly, so he can anticipate when the flow rates may be imperfectly correlated with changes in altitude AGL, and when there might be no flow rate at all.

Second, Miller teaches the general strategy of switching information sources every few seconds, for example, from processing optic flow rates to processing horizon or radar altimeter information. A pilot can safely use global flow rate in the very short term—a few seconds at a time, a span of a mile or two of terrain at low altitude; in this short time, speed does not change rapidly (or without awareness), the density of objects on the terrain is more likely to be constant over small extents, and the pilot has not yet adapted. Hence, any change in flow within a few seconds is likely to be due to a change in altitude AGL or velocity vector, not one of the potentially confounding factors. Remembering to switch information sources is a controlled process.

Third, Miller suggests that during sustained low-altitude flying the pilot periodically make a maneuver to "recalibrate" global optic flow rate. A rapid vertical jink, climbing several hundred feet quickly and then just as quickly descending again, dramatically changes the flow rates, so that when the pilot returns to level flight, he is no longer adapted. Remembering to occasionally perform this kind of maneuver requires a controlled decision process.

A fourth procedure to perceive changes in altitude AGL is to attend to the shadow of the aircraft (assuming a sunny day and flying away from the sun). As altitude AGL decreases, the apparent size of the shadow, its clarity and sharpness, and the apparent speed with which the shadow moves over the ground all increase. Pilots are taught to look for their shadow and to attend to changes in its appearance to inform them about their ground clearance. Attending to these changes requires controlled processing. (We know of no evidence on the accuracy with which pilots can do this and whether it can be improved with training.)

It seems unlikely that practice would ever transform these kinds of controlled perceptual processes into automatic ones. First, given the sight of an upcoming dry lake bed, can the pilot automatically know that he must switch from the automatic processing of optic flow (which will be useless)? Second, can the pilot automatically know that he should switch to looking at his altimeter? And third, having switched, can he learn to automatically perceive his altitude AGL from the numbers? All of this seems unlikely, in theory, and certainly from what we have observed in practice. The sight of the dry lake bed shuts down automatic perceptual processing, even in the most experienced pilot. This should be the case with each failure to be able to maintain automatic perceptual processing.

Processing the Elevation of the Horizon. A change in altitude AGL can also be perceived from changes in the appearance of the distant horizon. Pilots comment that they use the changing location of the horizon as a guide to their changing altitude AGL without conscious effort, which suggests an automatic processing of information about changes in altitude AGL, especially when flying over regular ground.

However, if there is poor visibility, the true horizon may be obscured, and the bottom of the cloud deck may be misperceived as the true horizon. This false horizon is invariably perceived to be closer and lower in elevation than it really is. Low ceilings also render the horizon information misleading, often causing a pilot to reduce altitude AGL too much as a way to put the horizon back where he "expects" it to be. Further, if the cloud bottoms are not parallel to the true horizon, their slope can suggest a false bank angle, a changing altitude AGL, or an incorrect velocity vector. This is particularly dangerous during intended level turns, in which maintaining a constant elevation of the horizon fails to guarantee a level turn if the cloud deck is sloping. Whenever the visibility is low or the cloud deck is sloping, the pilot is trained to ignore horizon information and switch to instruments or other visual sources. This switch requires controlled processing.

Processing Vestibular Information. The pilot's vestibular system is finely tuned to register changes in altitude AGL (as well as in turning rate and acceleration) by sensing the G-forces imposed. As we described in a previous section, under some circumstances, the changes in G-forces are closely correlated with these physical changes in velocity vector, and the automatic processing of G-force changes provides a direct perception of changes in altitude AGL.

However, the vestibular system adapts rapidly to constant stimulation, so the physical G-forces imposed on the body are underrepresented by the output of the vestibular system (see chap. 3). For example, a prolonged constant turn in the absence of vision, such as at night or in a cloud, is incorrectly perceived as a gradual decrease in the rate of turning. This false perception can cause the pilot to increase his bank angle so as to maintain a sense of a constant gravitational pull on his body. Because the perception is wrong, the overbanking produces an unintended descent into the ground (see McNaughton, 1985; McNaughton 1981–1984 for examples of these crashes). As a second example, an accelerating wings level dive produces on the pilot both negative G-forces from the descent and positive G-forces from the acceleration. In the absence of good visual information, this kind of dive is often misperceived to be a climb.

Consequently, pilots must be trained that when sustained G-forces above 1 G are encountered, they cannot rely on their automatic vestibular perceptual processing but must override it with either visual processing (if available) or instruments. Miller argues that pilots should ignore vestibular information altogether to be safe in a low-altitude environment. Knowing to do this requires controlled processing.

Finally, the research evaluation of flight simulators (see chap. 12; Haber, 1987) suggest that G-force information might not be needed. Simulators cannot hope to match the changes in G-forces that the pilot experiences during real flight, yet pilots report that simulators can provide realistic flight-training experiences. Further, there are no demonstrated differences in the quality of training performance produced by the presence or absence of motion information in the simulator (see chap. 12 for a review).

Perceiving Absolute Altitude AGL

In general, being able to perceive or determine absolute altitude AGL allows the pilot to estimate his free time away from ground-clearance tasks more precisely, determine the acceptability of entering into certain maneuvers, and determine when to initiate dive recoveries.

It appears that even with extensive practice, pilots cannot automatically perceive absolute distance through the air to the ground beneath them (or targets in front of them). Whenever a pilot needs an absolute value of altitude AGL, he has to rely on his radar altimeter (or barometric altimeter if flying over terrain of a known height), both of which require controlled processing. For example, the only way to determine when to begin a dive recovery is by watching (through controlled processing) an altimeter unwind to a preselected value and then pulling.

There are several techniques to help a pilot estimate absolute ground clearance—all of them controlled processes. Some of these are validated by research, but most are based on anecdotal reports by pilots and instructors. As one example, pilots can compare their radar altimeter readings with the appearance of particular patterns of texture, optic flow, or object details from the ground. After such repeated practice, in theory, a pilot should be able to associate the learned altitude AGL with the view outside simply by looking. However, the experimental evidence is nonexistent that such absolute distance skills over long distances are actually learned or learnable.

As another example, pilots can be taught to use the known size of ground objects to perceive ground clearance over them. In theory, if a pilot can see an object that has a familiar and therefore known size, he can use that known size to compute or derive its absolute distance. Pilots report that they can use familiar-sized objects to perceive their absolute clearance over them. However, research results contradict these reports. Haber and Levin (2001) have shown that observers do not estimate distances to objects with familiar sizes more accurately than to unfamiliar-sized objects.

Although these techniques demand controlled attention, it is highly beneficial for a pilot to have access to them. A pilot spends most of his time during low-altitude maneuvering heads-up looking outside. If he can be taught to derive absolute altitude AGL from information from the terrain (rather than having to attend to his radar altimeter), then a change in visual attention from outside to inside is avoided. Even more important, a pilot would have access to his absolute altitude AGL during high G-turns, when the radar altimeter may be compromised by the high bank angle.

Perceiving the Velocity Vector in Straight and Level and in Turning Flight

There is no part of the structure of a plane, such as its nose or a mark on the canopy, that reliably corresponds to the velocity vector—the direction of the plane's travel through the air. Consequently, the pilot must use either visual information from the light reflected from the terrain or some instrumental display to determine his direction of travel.

The best and easiest source of information to process the velocity vector automatically comes from the flight path marker in the HUD. In addition, pilots get substantial information about their velocity vector automatically from the visual information reaching their eyes through the canopy. This is especially true when flying straight and level, but it also holds at least for level turns. We have already described the automatic processing of clearance along the flight path based on changes in occlusion that result from the continually changing interpositions of near terrain blocking the line of sight to further terrain (review Fig. 2.5 for illustrations). This dynamic interposition cue provides explicit information about relative position and therefore relative distance—which ridge is closer and whether it will be cleared (Cutting, 1986; Gogel, 1977; Hochberg, 1971). Consequently, dynamic interposition can inform a pilot about his clearance over the elevations nearest to him. Miller also teaches pilots how to perceive dynamic occlusion even from *low*-contrast hills. Figure 2.7 shows an example in which a low-contrast near hill would be invisible against a far hill to a stationary observer but produces an optical shearing as the pilot approaches it. This shearing, or movement along the ridgeline, defines a covering or uncovering of the higher background terrain and exposes the presence of the nearer hill. Without the plane's motion toward the hills, the nearer ridge would remain invisible under these low-contrast conditions. Not only is it exposed through motion, but the direction of the shear also defines whether clearance will occur or not. If the shear is an uncovering of the far terrain, the plane will clear the near hill; if the shear is a covering of the far terrain, the velocity vector is into the near hill. The flight path marker in the HUD is of no help in these low-contrast terrains, because the pilot assumes that the dirt he sees is from a far hill, not an unseen near hill. Miller notes that unless this information is pointed out to pilots, and they are trained, most pilots are unaware of this dynamic occlusion cue and do not know how to use it.

There is another important automatic perception of the flight path components of the velocity vector. Mathematically, optic flow is zero when looking though the canopy in the exact direction corresponding to the plane's velocity vector during straight and level flight. All that changes in the flow is optical expansion. Thus, if the pilot fixates (looks through) his velocity vector, it is the only point in the scene that he can track and keep fixated without having to move his eyes. If a pilot attends to the pattern of global optic flow, the velocity vector should stand out. Most pilots feel they can and do see their velocity vector without effort (by automatic processing).

Regan and Beverley (1979) describe a powerful automatic processing mechanism for perceiving velocity vector based on *binocular* processing of optic flow. Consider first the optics. Although monocular viewing of the velocity vector produces no flow in the single retina, there is a flow in binocular viewing (which, of course, is what a pilot always does). When a pilot flies directly toward an object while looking at it, the image of that object enlarges in size on each of his retinas as the plane gets closer to it. Each of the two enlargements is concentric, with the center of the image centered on the fovea. If the pilot continues toward the object, its image symmetrically expands to fill each retina, and when he gets quite close,

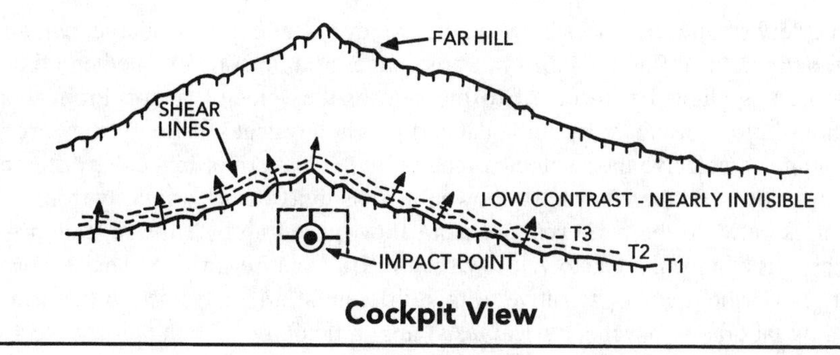

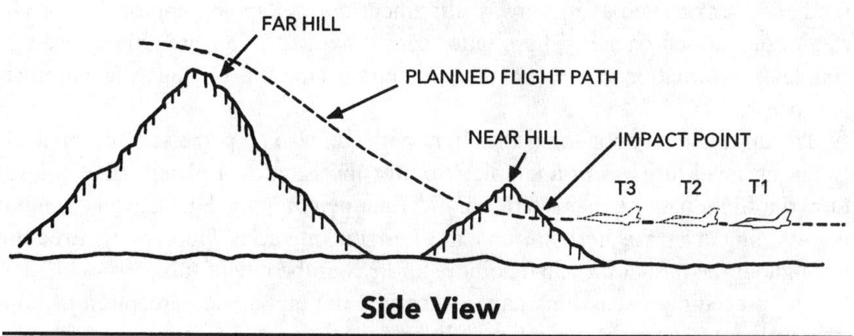

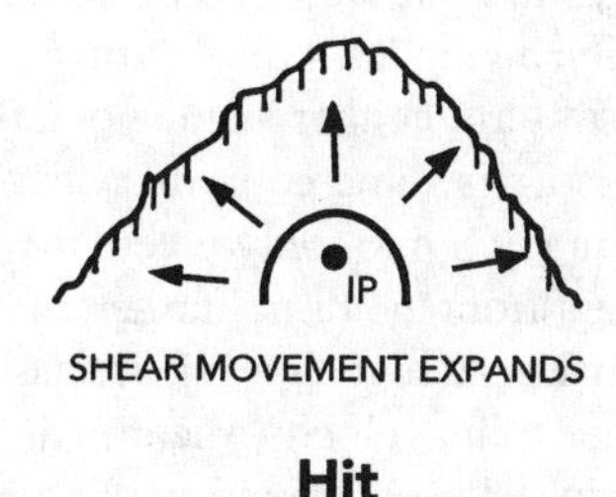

FIG. 2.7. Three views of the direction of shear movement during occlusion of far hills by near ones, even when the contrast is too low to perceive the ridgeline of the near hill clearly.

there is a flow in opposite directions on the two retinas (until the object hits the pilot between his eyes). In contrast, when a pilot fixates an object just off the velocity vector (an object that he will miss), the images on each retina enlarge in both eyes nonconcentrically and asymmetrically but larger in the eye closest to the object, so the center of enlargement drifts across each retina in the same direction, away from the object, with the greater drift in the far as compared to the near eye.

All of this is geometrical optics—the physics of available information. Regan and Beverley (1979) have shown that the human visual system is highly sensitive to the very small differences between the two eyes that occur in enlargement and especially in the differential drift rates of images as a function of whether a fixated object is aligned with or off to the side of the velocity vector. From their demonstrations of neurophysiological and psychophysical sensitivity, they argue that humans perceive their velocity vectors by this pattern of binocular differential optic flow. Further, they have shown this sensitivity comes primarily from differences between the two eyes, not from flow picked up by either eye alone—hence, it is a binocular cue to velocity vector. Kruk and Regan (1983) have shown that pilots who are very sensitive to this differential drift information (as measured by laboratory psychophysical tests) are particularly good at velocity vector prediction tasks, such as in runway alignment during landing approaches or air-to-air combat approaches. These latter data show that pilots are able to use this source of information to perceive their velocity vector. It is obviously an automatic process.

The up-down component of the flight path can also be perceived automatically during level turns from a simple stimulus invariant. If a plane makes a level turn, the flight path marker retains an invariant height in the HUD, even though it is sweeping over the horizon, and the horizon intersects the canopy structure throughout the turn at the same point regardless of the rate of turn.

The preceding discussions have concerned the automatic perception of constant velocity vector, as during straight and level flight. The mathematical relationships underlying changes in optic flow and expansion become quite complex when the velocity vector is changing, as it does during a turn. There is no point of zero optic flow in any kind of turn; neither vision nor a flight path marker can tell a pilot where he is going vis-a-vis some point in space. The momentary position of the flight path marker can tell a pilot only where the plane would go if at that moment he rolled out straight from the turn. Further, in heavy maneuvering, such as while in the process of making a high G level, ascending, or descending turn, a pilot may not be able perceive his velocity vector. All the pilot can see out the canopy is a rapidly sweeping sky or ground passing before his eyes. To look across his ground track in a steeply banked turn, the pilot has to look directly out the top of his canopy (rather than at his flight path marker, which may be temporarily off the HUD anyway). Hence, there is neither instrument in the cockpit, nor place outside the cockpit toward which the pilot can look and automatically perceive if he is on a collision course with the ground.

In low-altitude heavy maneuvering, a pilot is taught that to avoid a collision with the ground, he must know how long he can remain in that particular maneuver and still have enough time to recover. This requires both conscious decision making and controlled processing of his velocity vector. Because these controlled processing tasks consume both time and focused attention, Miller teaches his pilots that under several hundred feet altitude AGL, they have no free time at all during any kind of heavy maneuver: all their attention must be focused on ground clearance.

Perceiving Ground Track

Unless the pilot can perceive his track over the ground, he cannot determine what obstacles he has to clear. This perception is an automatic task when flying straight—the ground track is through the velocity vector (either directly perceived or with the aid of the flight path marker), so everything along the terrain under the velocity vector will be crossed by the plane. The task for the pilot with respect to ground clearance is only to perceive his altitude AGL clearance over each of the objects and terrain features along that track. But this applies only for straight and level flight. Whenever the aircraft turns, the velocity vector during the turn is independent of ground track and cannot be used to perceive or predict that track. More seriously, there are no terrain sources of information that tell a pilot where to look while turning in order to perceive his ground track.

Therefore, controlled processing to perceive ground track is essential during turns. This is akin to a prediction task: What route over the ground will the plane follow during the course of the present turn? If the terrain is level, with no vertical obstructions, the task is easy, mainly because as long as the pilot maintains adequate altitude AGL, he does not have to predict his ground track exactly. But if the terrain is irregular or has obstructions protruding into the potential airspace under the turn, then it is critical that the pilot know over which part of the terrain his plane will fly during the turn.

The pilot has to be trained where and how far ahead to look so he can be sure the turn is not carrying him into a collision course with an obstacle under his track. No instrument can help the pilot predict his path: he must be looking at the correct part of the terrain ahead during each step of the maneuver so he knows he will clear it.

In low-altitude flight, where to look during a turn is an important controlled processing component of ground track prediction. This problem is not usually important at higher altitude. There, the pilot is only interested in his final rollout point, the amount of turn room he has available versus the amount he requires, and the direction he will be flying when he completes the turn. He has no concern with the path he actually follows around to that rollout point—all he has to do is look at the point itself. This means that most pilots bring to the low-altitude environment a long history of looking through turns to their rollout point, even if it is 60, 90, or 180 degrees away. Although they may have lots of practice with trying to predict their flight path, especially in relation to what a bandit aircraft might be doing, none of that experience transfers very directly to knowing how to predict their ground track around the turn itself. Even with many hours of low-altitude flying experience, looking too far into turns remains a difficult problem for many pilots and requires conscious thought to correct.

In addition to failing to observe potentially fatal terrain features over which he might fly, looking too far into a turn may deprive the pilot of vital information about an unintended descent. With high bank angles, the pilot's viewing of the terrain during a turn is through the top of the canopy and is further deprived of global optic flow by the high G-forces in the turn. It is critical that a pilot detect an overbank during the first second of a 100 foot altitude AGL intended level turn in order to have enough time to correct the resulting unintended descent into the ground. It is much harder to detect an overbank if the pilot is looking straight out the top or rear of the canopy than if he is looking no more than 5 or 10 degrees into the turn. Low-altitude training teaches the pilot to make a quick visual sweep out to 90 or even 120 degrees of the potential flight path across the turn and then initiate turn watching in front of the nose within 30 degrees or less of the momentary velocity vector.

Perceiving Distance to Objects in Front of the Plane

Frequently, a pilot needs to know the absolute distance to another aircraft or to a target object being approached in a dive. We treat distance separately from altitude AGL, because looking downward is a very different task from looking outward. Furthermore, altitude AGL is usually only a few hundred feet in the low-altitude environment; in contrast, except for aerial refueling, distance outward to objects is frequently 1,000 feet and may extend to 10,000 or 20,000 feet.

For very near distance perception, such as needed for formation flying or aerial refueling, stereopsis provides distance perception automatically. Further, although pilots claim that familiar size provides accurate distance perception (automatically), Haber and Levin's (2001) results suggest that even in this task, familiar size would fail to be the basis of accurate distance perception. Hence, there may be some automatic perception of distance that is limited to only very near distances.

However, as with perceiving absolute altitude AGL, there is no evidence that pilots (or anyone else) can automatically perceive the absolute distances to targets or other objects in front of them for large distances. Even stereopsis fails to function beyond several hundred feet. The authors have informally tested pilots who claimed good absolute distance perception and failed to find any with absolute accuracy levels sufficient for low-altitude flying. Research findings confirm this result. Weist and Bell (1985) reviewed a large body of research on distance perception using stationary perceivers standing on the ground and estimating distances to targets attached to the same ground. Few experiments used distances in magnitudes of thousands of feet, presumably because they are of little relevance to terrestrial perceiving and traveling. The few studies that included ranges exceeding 1,000 feet all indicated systematic underestimation of true distance, with the amount of underestimation increasing with the true magnitude of the distance. Further, Haber and Levin (1989) argued that the sources of visual information and the visual mechanisms needed for the processing of this information are severely limited when range distance exceeds even a hundred feet. Estimating absolute distances through the air (where textural cues provided by continuous ground are absent) is an even more difficult task. Therefore, there would seem to be no automatic perceptual processes capable of providing a pilot with the perception of distance automatically, especially of the large extents that he needs.

The only instrument available to the pilot for controlled perceptual processing of distance information over large ranges is airborne forward-looking radar. It is particularly helpful in determining the distance to another plane, because other planes provide sharp and accurate return radar signals. However, forward-looking radar frequently fails to provide distance information about terrain features and objects on the ground, especially if the ground object is too small to show up on the radar screen.

In summary, in this section, we have analyzed how each of the sources of information available to a pilot about ground clearance is processed. As we have shown, automatic processing in the low-altitude environment presents a hazard as well as a help, because in many instances, the automatically processed information is misleading in fatal ways. To counteract this problem, pilots are taught to engage in three kinds of highly demanding focused tasks: first, to *remember* when to override their automatic responses to automatically perceived information; second, to *decide* which alternative source(s) of information to use; and third, to *employ* the focused attention required by those sources or techniques. For all these reasons, low-altitude flight challenges human perceptual and attentional capacities to perform close to their limit. It is obvious why such extensive, intensive training is necessary and why even experienced pilots have flown into the ground.

Improbable as it seems, we have only begun to describe the perceptual and attentional demands these pilots must learn to fulfill. The military purpose of flight is not merely to cross the terrain without colliding with it. The pilot is there because he has a mission, and he is flying that mission low because that mission requires him to be low, or threats or weather force him low. What happens when mission, threats, and weather demand more free time than a pilot has? This is the topic of our final section.

INTEGRATION OF TASKS AND PRIORITIES IN LOW-ALTITUDE FLIGHT

The pilot's number one task in the low-altitude environment, regardless of mission or threat, is the management of his altitude AGL and velocity vector so as to avoid collision with the ground. A pilot who flies into the ground is dead and has failed his present mission and all future missions more profoundly than if he failed to destroy his target. Keeping the priority for ground clearance first is theoretically correct but difficult in practice. The pilot who enthusiastically watches his missiles as they zoom toward the target may himself collide with the ground or with the blast from the explosion. The pilot who exhaustively continues to search the sky through the top of his canopy for a bandit, even though he himself has just initiated a sharp turn, may descend unheedingly into the ground. The problem is not just one of overcoming human nature. Part of what these pilots must learn and practice is specifically when the attentional demands of ground clearance are so high that they must attend only to avoiding the ground. They must also learn specifically which additional tasks (if any) can be combined with or performed in addition to ground clearance. In this section, we describe the factors that affect the difficulty and prioritization of flight tasks—the factors that the pilot must evaluate to determine whether he has enough time to perform a specific task and decide the sequence of ground clearance, threat avoidance, navigation, and mission tasks he must perform to always get his priorities right. We conclude with a brief description of a training program designed to teach pilots to integrate their tasks and priorities in low-altitude flight.

Factors That Affect the Difficulty of Flight Tasks

Low-altitude missions are fundamentally more difficult than those flown high. At high altitude, the tasks in most mission segments are performed in a fixed order, whereas at low altitude, the sequences of tasks often change. As a result of this difference, the priority of a given task at high altitude is highly correlated with its fixed place in the sequence, whereas the priority of each task at low altitude needs to be steadily reassessed. Further, at low altitude, but rarely at high altitude, the demands of ground clearance are themselves continually changing, usually in ways that have nothing to do with the unfolding of the mission, being dictated by the plane's proximity to the ground, changes in the terrain itself, and the particular maneuver being performed at the moment.

To illustrate the difference between predictable and unpredictable sequences of tasks, consider two brief examples of tasks during a mission over hostile territory: departure (takeoff) and threat evasion.

The flight tasks associated with departure have as their goal becoming airborne; the constellation of tasks to accomplish this goal occur during this segment only and rarely recur. Further, in the departure segment, the tasks occur in a fixed order, with little likelihood of varying the sequence. The pilot taxies to the proper location on the runway; requests and acknowledges clearance from the tower and from wingmen; initiates throttle, stick, and control procedures during the takeoff roll; checks and rechecks engine gauges, airspeed, and other indicators for rotation and liftoff; checks the position of wingmen after becoming airborne; and completes the first altitude and course corrections. When this cluster of tasking is completed, the departure segment terminates.

If a threat is detected, such as a hostile missile, then one of two alternative new sets of tasks is instantly triggered—either evade the threat or defeat it. For example, threat-evasion tasks include evasive maneuvers such as tight turns and jinks, avoiding collision with the ground during the heavy maneuvering, communications with wingmen as to the position and potential target of the missile threat, dispense countermeasures, and so forth.

Threat-evasion tasks differ from departure tasks in two critical ways: the tasks are repeated a number of times and may occur in any order. A large body of research literature has shown that tasks that occur in variable order are harder to learn, harder to perform, and virtually immune to time-sharing. Further, the variable sequence of tasks makes it much harder to keep priorities straight (e.g., balancing evading the missile with not flying into the ground).

Tasks also vary in difficulty depending on how close to the ground they are performed, the properties of the terrain over which the pilot is flying, and whether his wingmen are in tight formation. How frequently and at which points during a maneuver does the pilot need to check his ground clearance and ground track? How frequently and at which points during a maneuver does he need to check the location of his wingmen or the presence of a threat? The specific context in which the tasks are performed determines their sequence, which recur when, and their priority. Unlike the highly variable sequencing of tasks during threat avoidance, however, the contextual variables of altitude AGL, terrain difficulty, and proximity of wingmen can be defined, learned by pilots, and practiced in increasingly complex configurations until performance is smooth under the most difficult scenarios.

Finally, heads-down tasks pose special difficulties. Dial and button changes are often heads-down tasks, in which the pilot has to look down into the cockpit. Checking engine gauges, flight instruments, many navigation tasks, and arming missiles are typically heads-down tasks for most aircraft. Heads-down tasks, although usually simple in their own right, present three kinds of difficulty for the pilot. First, they inevitably compete with all other heads-up visual tasks that require the pilot to observe the ground or surrounding airspace. Second, every heads-down flight task requires focused attention and therefore cannot be time-shared with any other task. Third, as a consequence, heads-down tasks present serious prioritization problems for their performance.

Factors That Affect the Duration of Tasks

The variables that determine how long it takes to perform a task are relatively fixed. Once a skilled pilot has achieved a high level of training in a particular aircraft, has flown every day, and is attentive that day, then the time needed to complete each task remains fairly constant. Further, a pilot can learn these task durations, so for every task he knows how long it should take to complete each and every task. But with fatigue, stress, worry, reduced attention, or many days off from flying, the duration to perform many of a pilot's tasks will fluctuate substantially (mainly upward).

Task duration is an explicit component of Miller's low-altitude pilot training program. Because the available time for performance of currently demanding tasks is always limited, then the time required to perform a task determines whether it can be done at this instant. For example, pilots learn that when well practiced, they can change a radio frequency in an A-7 fighter in 3 seconds (this speed is achieved by practice in a part-task trainer). Because it is a heads-down focused-attention, manually performed task, they know that for those 3 seconds, they cannot carry out any other task. Similarly, they can "check 6" (verify that there is no hostile plane or missile directly behind them) in slightly less than 1 second with a snap glance, but during that second, they cannot do any other tasks.

When a pilot has a 5-second task to do in a 2-second time slot, he is in trouble. Miller teaches pilots that if they have only 2 seconds of free time available from ground-clearance tasking, they can do any set of tasks whose sum time is no more than 2 seconds. But they cannot make a radio frequency change in a 2 second free time slot, because it takes 3 seconds to complete; they could be irreversibly pointing at the ground by then.

Time Savings Through Time-Sharing. One way to gain time is to time-share or to perform more than one task in the same time slot (see chap. 7). Perfect time-sharing occurs when the duration and accuracy of performance of any of the component tasks is not impaired by the concurrent performance of the other. In theory, two tasks that require automatic processing can be performed perfectly at the same time (e.g., Schneider & Shiffrin, 1997; Shiffrin & Schneider, 1977). But we know of no examples in low-altitude flight in which two processing tasks can be performed simultaneously *and* perfectly.

Time-sharing of a controlled processing task with an automatic processing task is considered possible by many attention theorists. For example, as long as the pilot continues to look out the canopy at the ground sweeping past, he can pick up and process optic flow information and perceive whether he is converging with the ground at the same time he is performing some other task, such as talking to a wingman or looking for a ground target on the same part of the terrain as his ground track. But combining two controlled processing tasks or two tasks that both require focused attention generally reduces performance (e.g., Gopher, 1996; Kahneman, 1973; Wickens & Hollands, 2000). The amount of performance degradation of either task is a function of the attentional resources available, the availability of information sources to guide the performance, the amount of practice on the controlled task, the compatibility of the stimuli and the responses, and how carefully task prioritization is managed (see chap. 7).

Finally, time-sharing cannot occur when the two tasks each require incompatible sensory inputs (e.g., information to be acquired by looking both heads-down and heads-up), incompatible motor responses (using the same fingers to change a radio frequency and arm a missile), or focused attention for their performance. In these cases, the tasks must be performed serially, and there is always a significant performance cost to any attempt to do both in the same time slot.

Thus, there are serious limitations in time-sharing at low altitude. Because ground-clearance tasks in the low-altitude environment nearly always involve intensive visual observation of the ground outside the cockpit, any other task requiring vision creates interference with ground clearance. A pilot can only look in one direction at a time. Many of the mission and threat-evasion tasks also require the pilot's vision; these inevitably create interference with ground clearance. Another limitation on time-sharing occurs when the information available to the pilot that specifies his ground clearance is degraded in quality, especially at low altitude. A snap glance at the ground allows the pilot to perceive his altitude AGL easily (and automatically) if the ground is highly textured and irregular. He can make this snap glance to verify that he will clear an upcoming ridge while simultaneously talking to a wingman. However, if the terrain changes to smooth water or a dry lake bed, that snap glance does not pick up enough information to allow the pilot to perceive his upcoming clearance. To obtain the equivalent confidence about maintaining ground clearance, the pilot must check his flight instruments or the horizon. The pilot can no longer use a snap glance automatically to determine his ground clearance, and he is forced to use focused attention to switch to a different information source. Now he can no longer simultaneously talk with a wingman. A task that permitted time-sharing under one condition (informative terrain) does not permit time-sharing under a more difficult condition (uninformative terrain).

The Pilot's Current Level of Practice and Attentiveness. For most complex tasking, even highly trained personnel get rusty with time away from the job. Instructor pilots feel this is particularly true in low-altitude flying. They report that they can always tell if a pilot has not been in the air within the last 3 or 4 days by just watching his performance. He is slower to perform all tasks, loses smoothness, is poorer at integrating tasks, makes mistakes, or has to repeat tasks he normally gets right the first time. All of these increase the duration of task performance and the chance of serious errors.

Performance also varies in a well-trained, highly qualified pilot who flies every day. Miller estimates that even for well-trained pilots, the difference in available free time and in the time to perform tasks can vary by as much as two-to-one, as a function of current factors that affect the pilot's ability to focus all his attentional resources on his flying right now, such as how much beer and sleep he got the previous night or whether he is getting along with his wife (for a review, see Vaughan & Ritter, 1973).

Miller has explored ways for a pilot to assess his momentary variation in performance skill, so he can adjust his tasking accordingly. He asks a pilot to complete a brief checklist of easily observed errors or inefficiencies that may occur during the initial phase of a flight before hostile territory is reached. These include missed first attempts at switches, wasted movements, erratic aircraft control, momentary indecision, missed tasks, late responses to radio calls, and the like. If a pilot notices even a few instances, he knows that he will generate less free time in his ground-clearance tasking, and he will need more time for all other tasks—in other words, he needs more tolerances than normally.

Free Time

In our first section, we defined free time as time available to the pilot to perform a variety of tasks, including heads-down tasks, without entering an irreversible collision course with the ground. The free-time concept can also be applied to the tasks themselves: free time refers to how long after performing a task the pilot can delay until having to do it again. Ground-clearance tasks in low-altitude flying have to be repeated over and over or the pilot dies. It takes a pilot only a second's duration to make a snap glance to verify that he is not on a collision course with the ground, but during heavy maneuvering, he must repeat that task every few seconds to survive.

When flying in the low-altitude environment, ground-clearance tasks invariably have the shortest free time, and there is a very high probability of flying into the ground if the pilot fails to perform them as needed. Therefore, when low-altitude flight permits only a few seconds of free time, all the other mission-related tasks, such as evading threats and navigation, must be interspersed one by one, as their own task duration permits, among the ground-clearance tasks. Having checked the ground during a low-altitude maneuver to evade a threat, the pilot has to check it again, often before he can look for his wingman or verify that the threat was evaded.

The magnitude of free time delimits the tasks that can be performed at that moment. Proper performance requires the pilot to do as many of those as he can within that block of free time, then return to his ground-clearance tasking. In the next block of free time, he continues down the ordered list of remaining tasks, assuming that there have been no priority changes in any of the nonground-clearance tasks.

Unlike the *duration* to perform a task, which remains relatively constant over the course of a particular flight, the free-time limit of a task until it has to be performed again varies continually, primarily as a function of proximity to the ground and the intensity of the maneuver and secondarily with the informativeness of the ground and the other tasking in the mission. This variation defines the *priority* of a task, following what Miller calls the free-time rule.

The Free-Time Rule and Task Prioritization. During each segment of his mission, the pilot has a number of tasks that he needs to perform in addition to ground clearance. These tasks can be pictured as a stack. Assuming the pilot has sufficient free time to perform only a limited number of these tasks at a given moment, he must know which require immediate performance and which can be postponed safely. These priorities define the order of the tasks in the stack. As time passes and the highest priority task is completed, then the next highest priority task takes over, and so forth. The stack of tasks becomes reordered; the task just performed moves down in the task stack to a lower level.

The free-time rule says that each ground-clearance task has an associated free time, which specifies how much time can elapse until the task must be performed. The priority of each task in the segment is determined by its momentary free-time limit. The highest priority goes to the task with the shortest free time, the next highest to the next shortest free time. If all the free times are greater than zero, then the pilot can do something else besides ground clearance. However, as time ticks away, the free times of every ground-clearance task become less and less, until the shortest one reaches zero, when that task must be performed at that instant, or the pilot is dead. As soon as that task is performed, its free time resets to its default for the current altitude AGL, terrain, and maneuver. A new order of priorities among the stack of tasks is specified by the new free-time values. When the one with the shortest free time approaches zero, it is performed, and so forth.

For example, if a ground-clearance task has a free time of 3 seconds, then as the 3 seconds tick off, the priority of that ground-clearance task increases until it has the highest priority and must be performed before the end of the time interval. After the pilot performs the task, its free time resets to 3 seconds (assuming the altitude AGL and maneuver continue unchanged), and remaining tasks with shorter free times can be performed (as long as they can be accomplished in the remaining free time), until this ground-clearance priority again reaches the top of the stack. If the pilot should descend during this time, then what had been a 3 second free time decreases to 2 or 1 second, and the priorities in the stack cause a reordering.

Earlier, we described a number of maneuvers, with a value for free time for each one. Straight and level flight at 100 feet altitude AGL and 0.7 Mach has 4 seconds of free time; at 200 feet, it is 8 seconds. A level turn at 100 feet altitude AGL at 5 Gs has 1 second of free time; at 8 Gs, there is no free time. The aerodynamic parameters that determine free time for any maneuver are primarily altitude AGL, with velocity, G, and bank angle each contributing additional variance. The values differ slightly for an A-7 or an F-22, but pilots receive extensive retraining and practice if they switch to a different kind of jet fighter.

Miller teaches and discusses every maneuver. Under all possible circumstances, for each maneuver, he shows how its free time is computed. He shows why the pilot can consume up to 4 seconds of heads-down time during 100 feet altitude AGL, 0.7 Mach, and straight and level flight and never collide with level ground. But take 5 seconds, and there is an unacceptable probability that the plane is on an irreversible collision course with the ground. Miller demands from day one that his trainees understand the aerodynamics of their plane for every maneuver, understand the concept of free time as it applies to each maneuver, and memorize the free-time values of each maneuver for combinations of altitude AGL, velocity, G, and bank angle. They do not fly at low altitude until they master everything.

Free-time values quantitatively specify the ordering of task priorities in the low-altitude environment. If the pilot performs a maneuver that has 2 seconds of free time, then while continuing that maneuver, he can do the next task in the ordered stack, as long as that added task can be completed in 2 seconds and return to checking ground clearance as part of the present maneuver.

The pilot should never attend to a clock nor try mentally to estimate elapsed time. Rather, all he has to know is that he cannot change a radio frequency (no matter how life threatening this delay might be) if he is also making a high-intensity turn at low altitude. If the priority for the radio is extreme, then he must roll out or pull up enough to buy the additional seconds of free time.

Severity of consequences also provides a criterion for prioritization. Delaying ground-clearance tasking (when flying low) beyond its available free time results in highly probable death, so ground clearance is the highest priority task. Delaying a "check 6" for an attacker or missile when flying over hostile territory results in probable death, which makes "check 6" a high priority task. In contrast, delaying a navigation check results in probably getting lost, which, while serious, need not be fatal or even irretrievably compromise the mission. Hence, navigation usually has a lower priority. However, although consequences are important, the free-time rule provides a precise quantification for priorities.

Miller has developed strategies to help a pilot adhere to the priority discipline. We illustrate one that Miller calls "first pass success." Suppose a pilot has 3 seconds of free time, decides to change his radio frequency, and fumbles at the dials. Now this 3-second task needs more time—time he does not have. Miller says the pilot has to act as if the task was completed successfully and return to his ground clearance. Then, in his next free-time slot, he again tries to change the frequency. This rule assures the pilot that he will never spend more than the allotted time on any task. Human nature is to persevere until successful (or dead). Adherence to the first pass success rule counters this natural impulse.

How does a pilot handle occasions when there is more to do than the available free time allows, particularly when disaster threatens from both above and below? Free time becomes inadequate or even negative for two reasons, which usually occur together: (a) ground clearance may be too demanding, and (b) mission and threat demands are exceedingly high. Miller trains pilots to consider one of four possible responses and to choose the correct one instantly. The first is to *postpone* a required high-intensity maneuver; delaying a turn or a dive always produces free time, and the more intense the maneuver that is postponed, the more free time is generated. The second is to *reduce* the intensity of the maneuver itself; a shallow banked turn has more free time than a steeply banked turn; flying straight has more than turning. The third is to *gain* altitude. The increase in free time may not be linear with gain in altitude AGL, but it is always positive. And finally, the last resort is to *abort* the mission and turn tail. At least the pilot will live to fight another day.

A Training Program to Integrate Low-Altitude Tasks With All Other Mission Tasks

All fighter pilots (in the U.S. Air Force, at least) who enroll in low-altitude training have already undergone substantial jet fighter training and have usually completed at least one assignment with an operational squadron, usually one whose role is air-to-air superiority or ground support. Hence, the enrollees already have a vast amount of flying experience. However, except for landing and takeoff, they have probably not flown near the ground voluntarily. So each needs to be trained to perform the maneuvers that are unique to the low-altitude environment, such as terrain masking and ridge crossings. More important, each enrollee has to be trained to handle the priorities of ground clearance, because these were virtually irrelevant in their prior training and flight experience. And finally, the enrollees have to be trained how to integrate the mission tasks (which may be quite familiar already) with the ground-clearance tasks.

Lt. Col. Robert Cassaro developed a "comfort level" training philosophy that explicitly starts with low-altitude tasks under the control of the free-time rule. As low-altitude flying proficiency develops, mission tasks are added. Capt. Milton Miller, an instructor pilot in Cassaro's squadron, enlarged Cassaro's comfort level philosophy into a full-scale training program. The program differs fundamentally from more traditional "step-down" methods. In the latter, trainees perform all mission tasks simultaneously at very high altitude until they are smooth, and then step down successively to lower altitudes, practicing at each until they are smooth. In contrast, the comfort level approach focuses on integrating mission and navigation tasks one by one into ground-clearance tasking.

The comfort level term comes from Cassaro's awareness that pilots become uncomfortable when they fly too close to the ground, but as their skill with ground-clearance tasks develops, their comfort level improves. If they then are asked to fly lower, they again feel uncomfortable, until they have mastered their tasking at the lower altitude. Most critically, Cassaro noticed that when uncomfortable, most pilots pull up enough to reestablish comfort. Hence, the pull-up is a signal to the instructor that the pilot has not mastered the tasking required at that moment.

The distinctive feature of the comfort level approach is that the pilot always has to deal with ground-clearance tasking and can never sacrifice its priority in order to perform some other task. Every task must be learned, performed, and integrated with ground clearance, one at a time, until everything can be done within the time constraints of task duration and free time.

CONCLUSION

We started our work on low-altitude flying from a textbook interest in perceptual and attentional theory. Even when such theories have been applied to performance tasks, as distinct from verbal and symbolic tasks, they have primarily focused on stationary terrestrial human beings performing ground-based tasks or moving slowly over near-range distances. Further, they have concerned the performance of tasks in which the penalty for errors is usually inconsequential. Finally, they have been supported by research carried out in greatly simplified laboratory settings.

Theories of visual perception and attention have not considered the kinds of tasks that pilots routinely perform. Consequently, not only are such theories incomplete, more seriously, they also have not been of much help in trying to solve the perceptual and attentional problems faced by pilots or in attempting to design better displays, instrumentation, and training.

Though we are trained as cognitive scientists, steeped in research and theory, this chapter makes little reference to theory. It is practical in two senses: (a) it describes the details of a particular set of tasks that are performed daily by well-trained pilots, and (b) it is more sensitive to what these pilots can and cannot do than to what our theories say they can do. When we describe limits on information, this is based on what we and others can see out the cockpit at 100 feet altitude AGL at 1,000 feet per second while avoiding an upcoming ridge, not what we demonstrate on a 20-inch monitor in a darkened laboratory while sitting in an easy chair. We have noted many places where perceptual and attentional performance in flying low and fast seems contrary to what our theories predict and places where our theories are still silent on this performance. We hope these discrepancies prompt better theories and better understanding of these tasks.

We analyzed low-altitude flying in terms of the sources of information available to the pilot, their dependability and accuracy, how that information is processed, limitations on information acquired by automatic processing, the specific alternative information used and how it is processed, how that information is integrated in specific tasks, and how specific tasks are prioritized. Extending the concept of automatic and controlled attention to perceptual processes allowed greater insight into the complexity of the tasks confronting a pilot who is flying low and fast. Quantifying how priorities are determined is likewise a novel analysis and also illuminates the complexity of the pilot's cognitive tasking.

Our analysis explores low-altitude military flight but also applies to any complex human performance, especially those tasks for which the penalty to the performer of error is injury or death.

It is our hope that the theories of attention and perception presented here can be applied to improve the performance, and especially the safety, of low-altitude flying. Low-altitude flying accounts for more plane losses and pilot and crew casualties than any other flying activity. This is true in combat, where collisions with the ground during low-altitude missions account for more losses than does enemy action. It is true in peacetime, when collisions with the ground during low-altitude maneuvering account for more military losses than all other causes combined.

REFERENCES

Allport, D. A. (1989). Visual attention. In M. I. Posner (Ed.), *Foundation of cognitive science* (pp. 631–682). Cambridge, MA: MIT Press.

Cutting, J. E. (1986). *Perception with an eye for motion.* Cambridge, MA: MIT Press.

Gibson, J. J. (1979). *The ecological approach to visual perception.* Boston: Houghton-Mifflin.

Gillam, B. J. (1998). Illusions at century's end. In J. Hochberg (Ed.), *Perception and cognition at century's end* (pp. 98–137). San Diego, CA: Academic Press.

Gillingham, K. K., Makalous, D. L., & Tays, M. A. (1982). G stress on A-10 pilots during JAWS II exercises. *Aviation, Space, and Environmental Medicine, 53,* 336–341.

Gillingham, K. K., & McNaughton, G.B. (1977). Visual field contraction during G stress at 13, 45, and 65 degree seatback angles. *Aviation, Space, and Environmental Medicine, 48,* 91–96.

Gogel, W. C. (1977). The metric of visual space. In W. Epstein (Ed.), *Stability and constancy in visual perception: Mechanisms and processes* (pp. 129–182). New York: Wiley.

Gopher, D. (1996). Attention control: Explorations of the work of an executive controller. *Cognitive Brain Research, 5,* 23–38.

Haber, R. N. (1981). Locomotion through visual space: Optic flow and peripheral vision. In W. Richards & K. Dismukes (Eds.), *Vision research for flight simulation* (pp. 72–81). Washington, DC: National Academy Press.

Haber, R. N. (1987). Why jet fighters crash: Some perceptual factors in collisions with the ground. *Human Factors, 29,* 519–532.

Haber, R. N., & Hershenson, M. (1980) *The psychology of visual perception.* New York: Holt.

Haber, R. N., & Levin, C. A. (1989). The lunacy of moon watching: Some preconditions on explanations of the moon illusion. In M. Hershenson (Ed.), *The moon illusion* (pp. 299–320). Hillsdale, NJ: Lawrence Erlbaum Associates.

Haber, R. N., & Levin, C. A. (2001). The independence of size perception and distance perception. *Perception and Psychophysics, 63,* 1140–1152.

Harrington, T. L., Harrington, M. K., Wilkins, C. A., & Koh, Y. O. (1980). Visual orientation by motion-produced blur patterns: Detection of divergence. *Perception and Psychophysics, 28,* 293–305.

Hochberg, J. (1971). Perception II: Space and motion. In J. W. Kling & L. A. Riggs (Eds.), *Woodworth and Schlosberg's experimental psychology* (3rd ed., pp. 475–550). New York: Holt.

Kahneman, D. (1973). *Attention and effort.* Englewood Cliffs, NJ: Prentice Hall.

Koenderink, J. J., & van Doorn, A. J. (1976). Local structure of movement parallax of a plane. *Journal of the Optical Society of America, 66,* 717–723.

Kruk, R., & Regan, D. (1983). Visual test results compared with flying performance in telemetry-tracked aircraft. *Aviation, Space, and Environmental Medicine, 54,* 906–911.

Lee, D. N. (1974). Visual information during locomotion. In R. B. MacLoud & H. L. Pick, Jr. (Eds.), *Perception: Essays in honor of J. J. Gibson* (pp. 250–267). Ithaca, NY: Cornell University Press.

Lee, D. N. (1978). The functions of vision. In H. L. Pick, Jr. & E. Saltzman (Eds.), *Modes of perceiving and processing information.* New York: Wiley.

Liebowitz, H. W., & Post, R. B. (1983). The two modes of processing concept and some implications. In J. Beck (Ed.), *Organization and representation in perception* (pp. 343–364). Hillsdale, NJ: Lawrence Erlbaum Associates.

McNaughton, G. B. (Ed.). (1981–1984). Life sciences: Notes to the flight surgeon. *USAF Safety Journal.* Published quarterly by the Life Sciences Division, Directorate of Aerospace Safety, Norton Air Force Base, CA.

McNaughton, G. B. (1985). The role of vision in spatial disorientation. *Flying Safety.*

Miller, M. (1983). *Low Altitude Training Manual.* Tucson, AZ: 162nd Fighter Weapons School, Arizona Air National Guard.

Nakayama, K., & Loomis, J. M. (1974). Optical velocity patterns, velocity sensitive neurons and space perception: A hypothesis. *Perception, 3,* 63–80.

Owen, D. H. (1981). Transformational realism: An interaction evaluation of optical information necessary for the visual simulation of flight. In E. G. Monroe (Ed.), *Proceedings of the 1981 image generation/display conference* (pp. 385–400). Williams Air Force Base, AZ: U.S. Air Force Human Resources Laboratory.

Regan, D., & Beverley, K. I. (1979). Visually guided locomotion: Psychophysical evidence for a neural mechanism sensitive to flow patterns. *Science, 205,* 311–313.

Roscoe, S. N. (1980). *Aviation psychology.* Ames: Iowa State University Press.

Schneider, W., & Shiffrin, R. M. (1977). Controlled and automatic human information processing: I. Detection, search, and attention. *Psychological Review, 84,* 1–66.

Shiffrin, R. M. & Schneider, W. (1977). Controlled and automatic human information processing: II. Perceptual learning, automatic attending and general theory. *Psychological Review, 84,* 127–190.

Vaughan, H. G., Jr., & Ritter, W. (1973). Physiological approaches to an analysis of attention and performance: Tutorial review. In S. Kornblum (Ed.), *Attention and performance, IV* (pp. 129–154). New York: Academic Press.

Warren, R. (1976). The perception of egomotion. *Journal of Experimental Psychology: Human Perception and Performance, 3,* 448–456.

Warren, R., & Owen, D. (1982). Functional optical invariants: A new methodology for aviation research. *Aviation, Space, and Environmental Medicine, 53,* 977–983.

Weist, W. M., & Bell, B. (1985). Steven's exponent for psychophysical scaling of perceived, remembered, and inferred distance. *Psychological Bulletin, 98,* 457–470.

Wickens, C. D., & Hollands, J. G. (2000). *Engineering psychology and human performance* (3rd ed.). Upper Saddle River, NJ: Prentice Hall.

ACKNOWLEDGMENTS

Nearly 20 years ago, both of us assisted instructor pilots at the 162nd Fighter Weapons School of the Arizona National Guard at Tucson to implement a new training procedure in low-altitude flying. We have many people to thank for their encouragement and insights into flying, perception, attention, and training. Three have been especially important. Col. Grant McNaughton, M.D., Chief of the Life Sciences Directorate of the U.S. Air Force Inspection and Safety Center, showed us the intricacies of safety and mishap and gave unstintingly of his time, knowledge, and enthusiasm about flying. At the 162nd Fighter Weapons School, Lt. Col. Robert (Cass) Cassaro, the director of training, was responsible for the development of the most innovative training program we have ever seen. At that same time, Capt. Milt Miller was beginning the process of translating Cass' ideas onto paper as a Tactical Training Manual for Low Altitude Flight (Miller, 1983, and revisions). Milt was no mere scribe; he understood and enlarged on what he himself had been taught and translated it into a superb program. Milt allowed us to consult freely with him during the time he designed the then current version of the program.

In addition to the preceding individuals, we have benefited from comments on earlier drafts of this paper made by Capt. Ed Houle, U.S. Air Force, Prof. Edward Strelow (then at the Universeide of California at Riverside), Prof. James Cutting at Cornell, Dr. Rick Toye (then at the University of Illinois), and Maj. Robert Shaw (U.S. Air Force). Omissions and commissions are entirely our own.

We are both also indebted to the University of Illinois at Chicago, which granted us leave for the time to work in Arizona, and to the U.S. Air Force Office of Scientific Research, which supported all of Ralph's work there.

3

Spatial Orientation

Laurence R. Young
Massachusetts Institute of Technology

The pilot gently banks his airplane into a right turn. Twenty seconds later he unexpectedly enters the top of a cloud. As trained, he glances down at his artificial horizon, which shows him banked right, but he is *sure* that he is flying straight and level. Remembering the dictum to always believe the instruments, he turns the yoke briefly to the left to take out the right bank on the instrument, but now he's sure he has entered a left turn. What to do? Are the instruments failing? No, he has a classic case of spatial disorientation (SD). In this case, the explanation for the illusion of turning is simple and is based on the dynamic response of the semicircular canals. If he recognizes the problem and manages to fly the airplane on instruments despite his disorientation, the day may turn out well after all. But if he succumbs to the temptation to "fly by the seat of his pants," he may reenter a progressively tighter, sinking right turn and join those unfortunate 15 to 30% of all aircraft fatalities attributable to SD in flight (Braithwaite et al., 1998; Cheung, Money, Wright, & Bateman, 1995; Knapp & Johnson, 1996). Despite vigorous attention to the problem and considerable public attention following the fatal crash of the light airplane piloted by John F. Kennedy, Jr., in 1999, the percentage of class A accidents in the U.S. military remains alarmingly high, for both fixed-wing (19 to 25%) and rotary-wing (30%) aviation (Symposium on Recent Trends in Spatial Disorientation, 2000).

As Gillingham and Previc (1996) point out in their comprehensive review of the subject, this insidious killer occurs in all classes of aviation and is no respecter of age or experience. Among military fliers, about two thirds of all fixed-wing aircraft pilots and over 90% of all helicopter pilots have experienced "the leans" after leveling out from a turn (Benson, 1999).

Spatial orientation refers to one's perception of body position in relation to a reference frame. Although the usual frame is the surface of the earth, for pilots in formation flight it may be the lead aircraft, and for space travelers it may be the stars, another planet, or some part of the spacecraft. The usual reference frame for discussion of pilot reactions to forces and spatial orientation is shown in Fig. 3.1.

Spatial orientation normally entails both the subconscious integration of sensory cues and the conscious interpretation of external information. The visual and vestibular signals are the prime sensory orientation cues, but proprioceptive and auditory inputs also come into play. The external information may include the flight instruments, cultural features in the environment, or verbal instructions. Spatial orientation is limited to the *sense* of one's angular and linear position and velocity relative to some local reference coordinates, such as the earth's surface, the glide slope, or the runway threshold. The perception of *body position* in spatial orientation thus distinguishes it from geographic orientation, or one's *location* within the reference frame. SD is a loss of the correct sense of local orientation (such as feeling inverted, banked, low, or spinning), whereas geographic disorientation refers to being lost. Of course, SD can eventually lead to geographic disorientation and navigation errors through a mistaken sense of motion or direction.

SD always involves a false conscious or unconscious sense of orientation, according to Clark and Graybiel (1955) and Benson (1965, 1999), but its consequences vary widely, leading to the classification of its three types (Gillingham & Previc, 1996). The most common and most dangerous is Type I (unrecognized) SD, in which the pilot is unaware that the perception of orientation is incorrect and may unwittingly control the vehicle into a crash. Type II SD entails a conscious recognition of disorientation and the knowledge that there is some conflict between the veridical orientation (e.g., as indicated by the flight instruments) and the sense of motion. The increase in workload associated with continuing to fly the airplane while being knowingly disoriented can easily lead to a focusing of attention on just the orientation, thus ignoring other important aspects of flight, such as altitude, traffic, navigation, and communication. That was the problem faced by our pilot who entered the descending spiral. The problem falls under the term *situational awareness,* or simply SA, which can be defined as "a pilot's (or aircrew's) continuous perception of self and aircraft in relation to the dynamic environment of flight, threats, and mission, and the ability to forecast, then execute tasks based on that perception" (Carroll, 1992, p. 6).

While SD can contribute to a decrease in SA, the two should not be confused. The occurrence of Type II SD is high among aviators, so SA and response tech-

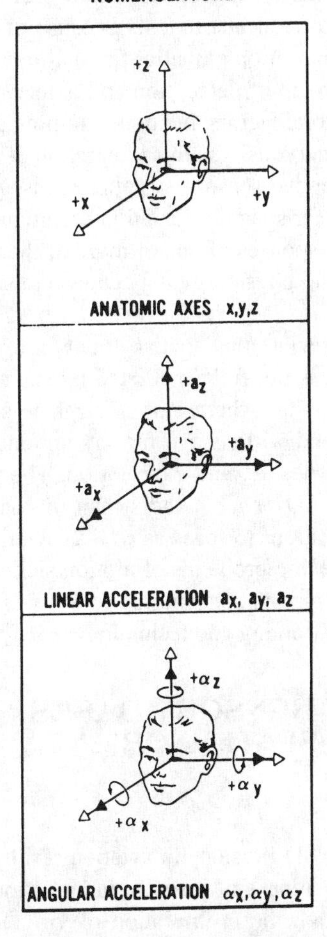

FIG. 3.1. Standard axes for description of human acceleration, following the accepted nomenclature for aircraft. Forward acceleration ($+x$) produces a backward inertial reaction force. Rotations about the x,y, and z axes produce roll, pitch, and yaw, respectively. *Note.* From *Fundamentals of Aerospace Medicine,* 2nd ed. (p. 316), by K. K. Gillingham and F. H. Previc, 1996, Baltimore, MD: Williams & Wilkins. Copyright 1996 by Williams and Wilkins. Reprinted with permission.

niques are stressed in SD training. The pilot with Type II SD might find it difficult to fight the conflicting sensations, might flip back and forth between different orientation states, and might ultimately hand over control to the copilot. Although "aviators' vertigo" has been used to describe Type II SD, we will restrict the use of *vertigo* to refer to illusions of rotation. Finally, Type III SD invokes a sense of helplessness and the inability to maintain control of the aircraft. It might entail an overwhelming sense of spinning or confusion about orientation, severe motion sickness, postural reactions to vestibular stimuli that override normal control actions, or the inability to read instruments because of the vigorous eye movements induced by vestibular or optokinetic stimulation.

Although SD is a problem of perception and is therefore influenced by many psychological and behavioral factors, including training and expectation, it is fundamentally rooted in the nervous system's integration of signals from the vestibular, visual, and proprioceptive sensory systems. Consequently, this chapter will discuss the sensory basis for spatial orientation before turning to the major classes of SD attributable to responses of one or more of these senses. The reader not interested in the underlying physiology may choose to skip ahead to the sections discussing SD illusions.

It is important to appreciate that spatial orientation, though based to a large extent on sensory inputs, is not fully predicted by such stimuli. The orientation sensation for a given stimulus, whether in an airplane or in a carnival ride, will vary among individuals and will change for any one individual as a function of experience, recency of exposure, and expectation. The pilot in command is far less likely to experience SD or airsickness than the passenger or the backseat electronics officer. Applications to space as well as aviation are considered in this chapter, along with the vexing problems of motion sickness. Finally, this chapter also considers the major factors contributing to SD and some means of dealing with the problem through training and technology.

SENSORY BASES FOR SPATIAL ORIENTATION

Visual System

Many of the most vexing SD illusions in aviation result solely from visual phenomena. The public's awareness of visually induced motion illusions has been heightened recently through the proliferation of virtual reality technology and increased exposure to wide-screen interactive visual displays, previously available only in advanced and very expensive flight simulators.

A more detailed discussion of the visual systems and the physiological basis for visual perception is presented in chapter 12, "Flights of Fancy: The Art and Science of Flight Simulation." The reader interested in a full treatment of the physiological basis of visual illusions is referred to one of the standard visual psychology books, such as Cornsweet (1970). For the purpose of this chapter, it is sufficient to recall that we possess two visual systems—one for determining *where* things are and the other for recognizing *what* is present. The first system, the *ambient system,* makes use of the peripheral as well as the central visual field and is principally concerned with detecting the motion of large objects in the field or of self-motion with respect to the visual environment. Orientation judgments from the visual field using the ambient system employ such natural cues as parallax and perspective. Ambient vision is employed during good outside visibility, or Visual Meteorological Conditions (VMC). The second system, the *focal system,* relies primarily on discrimination of fine detail in the central visual field and is related to SD through size and shape constancy, perspective, and other cues that lead to appreciation of the object's size and character. The focal system is used during Instrument Flight Rules (IFR) flight, in which the pilot "reads" the instruments in order to perceive aircraft orientation accurately. Spatial disorientation is far more likely to occur during instrument flight.

Vestibular System

The real "sixth sense" is more properly the sense of body motion than some mysterious extrasensory capability. The key organ for sensing body motion and for postural control is the vestibular system—a small, primitive fluid-filled inertial system the size of a pea and located in the inner ear. In place of gyroscopes to measure angular motion of the head, we have a set of three roughly orthogonal semicircular canals in each inner ear. And in place of accelerometers to measure gravity and linear acceleration, we have a pair of otolith organs in each ear. Instead of a guidance computer, we have primitive neural machinery, largely in the brain stem and cerebellum, which combines information from these sensors and other sources to produce our sense of spatial orientation. Finally, to extend the analogy to stabilizing the airplane against wind gust disturbances, we produce muscle commands of vestibular origin to stabilize the eyes and head in space and to maintain our erect posture in a gravity field. Because of the importance of the vestibular organs in normal spatial orientation in flight and their involvement in so many SD illusions, they are discussed in more detail than the other sensory systems. (The reader looking for an in-depth discussion of vestibular sensation should consult more extensive reviews of the topic [Goldberg & Fernandez, 1984; Guedry, 1974; Howard & Templeton, 1966; Peters, 1969; Wilson & Melvill-Jones, 1979; Young, 1984].) The major anatomical elements of the vestibular system (Fig. 3.2) relate to its function in SD.

The entire apparatus is located close to the midline of the head and enclosed in a hard, protective shell called the "bony canals," embedded snugly in the dense temporal bone. The sensors rest inside a delicate membranous labyrinth that is filled with *endolymph,* a "shear thinning" fluid with density and viscosity much like water. Head movements displace the fluid in the semicircular canals and shift a portion of the otolith organs, which in turn bends the cilia of sensitive hair cells to produce frequency coded neural signals.

Semicircular Canals

The labyrinth contains three narrow semicircular canals filled with the sodium-rich endolymph. The canals lie in planes that are mutually orthogonal to within 5 to 10° and respond to angular accelerations containing a component perpendicular

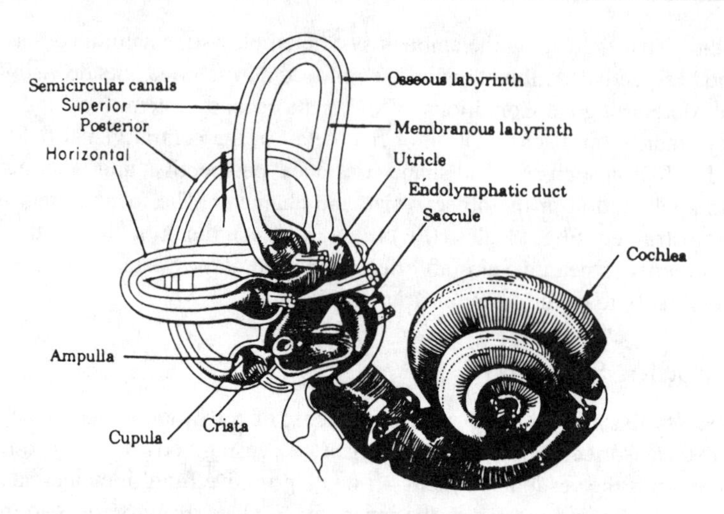

FIG. 3.2. The vestibular system in relation to the auditory structure in the inner ear. *Note.* From *Dorland's Illustrated Medical Dictionary,* 29th ed. (p. 562), by W. A. Newland Dorland (ed.), 2000, Philadelphia, PA: W. B. Saunders Company. Copyright 2000 by W. B. Saunders Company. Reprinted with permission.

to their plane. One end of each of the canals, near the entrance to the utricular sac, is widened to form the ampulla, which is normally completely blocked by a gelatinous wedge, the cupula. The cupula bends slightly, but prevents endolymph from flowing through the ampullary space. The base of the cupula widens into the crista, which contains cilia projecting from hair cells in the underlying neuroepithelium. Bending the hairs in the direction of the kinocilium increases the cell depolarization and increases the frequency of hair cell firing. In this manner, pressure changes in the endolymph, causing the cupula to deflect slightly, are transformed into changes in the firing rate of afferent nerve fibers traveling from the semicircular canals toward the brain.

First consider *intuitively* what happens to the ring of endolymph in the horizontal canal when the head is suddenly accelerated to the left, or *counterclockwise* as seen from above. The horizontal membranous canal turns with the skull, of course. However, the endolymph within that canal initially remains stationary because of its inertia, because no external force acts on it. The ring of endolymph is therefore initially displaced *clockwise* relative to the canal. Meanwhile, the forces of viscous shear between the fluid and the membrane oppose this small motion. The shear force and associated torque are proportional to the flow of endolymph and act to oppose the motion of endolymph relative to the canal. Newton's law for rotation states that the angular acceleration (of the endolymph, in this case) times its moment of inertia equals the external torque applied to it. Were it not for the presence of the cupula, therefore, the endolymph flow rate would eventually be proportional to the skull angular acceleration, and the relative displacement of endolymph in the canal would consequently end up proportional to the skull angular velocity. That is how the semicircular canals, stimulated by *angular acceleration,* can signal *angular velocity* to the brain. However, the actual endolymph flow is blocked by the cupula, so that the early endolymph displacement merely bends the cupula slightly, like a diaphragm, in the direction of flow. For brief head accelerations, such as those that occur commonly in natural activity, endolymph and cupula displacement are indeed proportional to skull velocity in the plane of that canal. The cupula bending is magnified in the subcupula space at the crista and produces hair cell displacement with exquisite sensitivity, signaling angular velocity by increasing or decreasing the hair cell firing rate from its resting discharge frequency. During prolonged rotation, however, the weak *elastic restoring force of the cupula* presses back against the endolymph. It tends to bring the cupula and endolymph back to the rest condition after some tens of seconds, even when the head is spinning at constant angular velocity.

This simple mechanical explanation shows how the semicircular canals correctly indicate head velocity for brief head movements, but decay to zero for sustained constant velocity turns. For extremely short disturbances, the cupula displacement is actually proportional to head displacement, but the intervals are too brief to play any role in spatial orientation. The dominant time constant for the human horizontal semicircular canals, which separate short from long stimuli, is about 5 to 10 sec. In addition to this mechanical cupular time constant, there are important neural processes that act to extend the effective time constants. One of these, known as "velocity storage" (Raphan & Cohen, 1985), maintains the ongoing subjective velocity in the absence of information to the contrary and effectively lengthens the time constant for angular velocity sensation to about 16 sec, rather than the 5 to 10 sec of the semicircular canals. Because of their different dimensions, the vertical canals have shorter time constants and lower gains than the horizontal canals. The subjective response to yaw (z-axis) rotation is about 16 sec, compared to 7 sec for pitch (y-axis) rotation (Guedry, Stockwell, & Gilson, 1971). These results, using retrospective judgment of displacement, are consistent with earlier research showing shorter time constants for pitch and roll than for yaw, based on the "subjective cupulogram" measurements of the duration of postrotation sensation of turning (Melvill-Jones, Barry, & Kowalsky 1964). For exposures to constant stimuli exceeding 20 to 30 sec, a neural adaptation becomes apparent, decreasing the response to a sustained stimulus even below the level signaled by the cupula deflection (Young & Oman, 1969). The subjective velocity during prolonged constant angular acceleration therefore not only plateaus, but also begins to decay toward zero after about 30 sec. Similarly, subjective velocity during a constant velocity turn about a vertical axis decays through zero after about 25 sec and may produce a reversal in the perceived direction of turning about 30 sec after beginning the turn.

The semicircular canals exhibit an effective *threshold* in the human perception of rotation, even though no such threshold appears at the level of the hair cell firing rate. The lowest threshold levels are found in laboratory tests when subjects are fully attentive to the stimulus and are given a long time to indicate its presence. If a lighted point is visible, it will appear to move when the subject is accelerated by as little as 0.1°/sec/sec, thus illustrating the Oculogyral Illusion, which is discussed following. Rotation in the dark produces an *effective angular acceleration threshold* near 0.2°/sec^2 for yaw, with higher values (about 0.5°/sec^2) for pitch or roll. Higher levels of angular acceleration are detected more rapidly. A reasonable approximation to a *velocity threshold* is given by the "Mulder's law" relationship that a constant angular acceleration α (greater than 0.5°/sec^2) will be detected after a time τ, when the angular velocity reaches a minimum value of ω_{min} according to the formula $\tau = \omega_{min}/\alpha$. Typical values for rotation about the vertical axis are 2.5°/sec for ω_{min} and 10 sec for τ.

In addition to the important function of the semicircular canals in the perception of angular rate, they play a critical role in several reflexes. Through the vestibulo-ocular reflex they move the eyes to compensate for brief passive or active head movements, thereby keeping the eyes stable in space. Compensatory eye movements and nystagmus, discussed later, compensate for 0.7 to 1.0 of the horizontal and vertical head velocity and about 0.5 of roll movements of the head. Other important semicircular canal reflexes are responsible for stabilization of the head (vestibulo-collic reflex) and the trunk, as part of the important "righting reflexes."

Although the semicircular canals primarily detect angular head motions, under some circumstances they also respond to linear acceleration and to gravity. To the extent that the cupulae are not precisely the same density as the surrounding endolymph, they will tend to sink or float slightly under the pull of gravity or when subjected to linear acceleration. In this case, the semicircular canals are also slightly sensitive to head position. The cupula may also be substantially unbalanced by caloric stimulation used in clinical vestibular testing or by alcohol ingestion, as discussed later. Both of these conditions will cause a false sensation of spinning that depends on head position relative to gravity.

A mathematical modeling approach to semicircular canal function requires knowledge of the dimensions of the canal, the density and viscosity of endolymph, and the elasticity of the cupula. All but the last have been measured directly. The torsion pendulum model, which has been accepted as standard, was introduced by Van Egmond, Groen, and Jongkees (1949), based on the suggestions of Steinhausen (1931), and elaborated and summarized by Young (1974).

Otolith Organs

The otolith organs indicate the orientation of the head relative to gravity. Like the carpenter's plumb bob, they usually indicate the direction of the vertical and are crucial in maintaining human balance and equilibrium. However, just as with any other physical accelerometer, they follow Einstein's equivalence principle and cannot distinguish between gravity and the inertial reaction force to any linear acceleration, so they actually indicate the orientation of the head relative to the gravito-inertial force (GIF). This is the *apparent vertical,* or the direction in which a plumb bob would hang.

Each labyrinth contains two otolith organs, the saccule and the utricle. The otolithic membranes in each are analogous to the seismic mass of an accelerometer and contain numerous dense calcium carbonate crystals. When the head is erect, the utricle lies mostly in a horizontal plane (except for its anterior end, which tips up about 25°), and the saccule lies primarily in a vertical plane. Cilia from hair cells in the underlying maculae extend into the otolithic membranes and respond to each displacement of these structures. Unlike the semicircular canals, however, the hair cells in the maculae are not all oriented in the same direction. There are some otolithic hair cells that are maximally discharged—and optimally sensitive—for any orientation of the head. However, generally speaking, the utricle responds optimally to head tilts in pitch or roll, starting from a head-level orientation. The precise function of the saccule remains open to question at this time. Some evidence suggests its primary role is that of sound detector, both for low frequencies and clicks (McCue & Guinan, 1997), whereas other evidence indicates it might also function as a gauge of vertical acceleration (Malcolm & Melvill-Jones, 1974).

The intuitive explanation for otolith function is simple and is illustrated in Fig. 3.3. Because the calcium carbonate otoconia embedded in each otolithic membrane make the membranes denser than the surrounding endolymph, they are only partially supported by its buoyant forces. When the head is accelerated or tilted, the utricular macula slides slightly "downhill" under the force of gravity, resisted only by the elastic bending forces from the supporting cells and opposed slightly by damping from the surrounding endolymph. It comes to rest at a displacement depending on the angle of tilt. This displacement is a measure of both the magnitude and direction of head tilt and is signaled to the brain by the complex of hair cells. Obviously this organ works best when the head is initially upright, or tipped slightly forward in a normal gravity environment, and it becomes close to useless when the head is tipped close to 90° to the vertical. The adequate stimulus to the utricular otolith appears to be the "shear component" of GIF, or the portion of the gravity and acceleration vector that lies in the dominant plane of the macula (Schöene, 1964). The function of the saccule is less clear. Although it is positioned so that it could also signal head angular changes from the supine or horizontal position, its spatial orientation function seems more likely to be related to detection of vertical acceleration or falling.

The otolith organs indicate the direction of the postural vertical to within a threshold of about 2°, under ideal conditions. However, they are also quite sensitive to the inertial forces produced by linear acceleration. With the head upright, the threshold for detection of sustained horizontal lateral (y) or fore-aft (x) linear acceleration is approximately 5 to 10 cm/sec^2 (0.005–0.01 g). It rises to 10 to

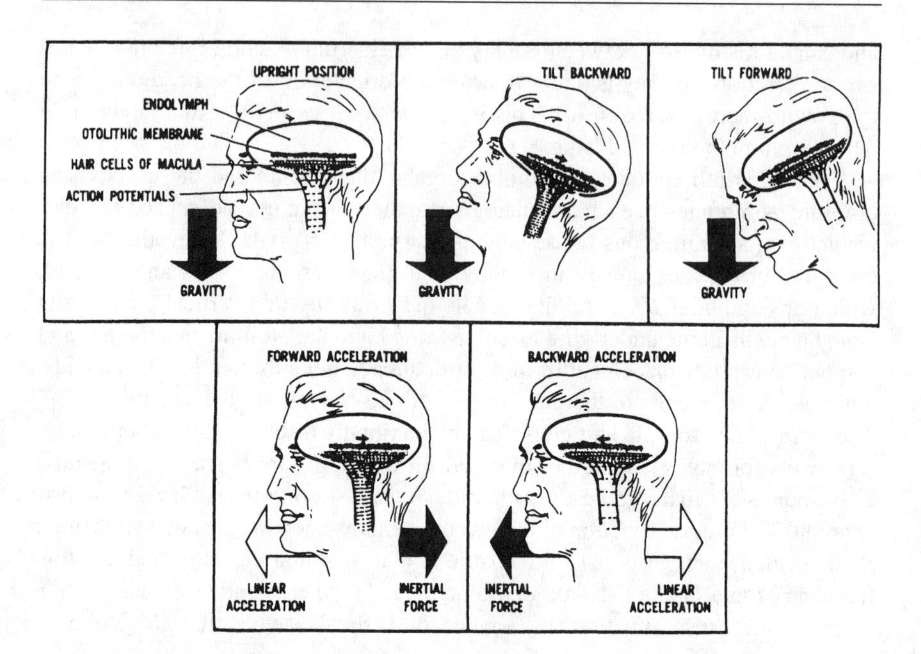

FIG. 3.3. Mechanism of the otolith organs, showing their sensitivity to head tilt and to linear acceleration. *Note.* From *Fundamentals of Aerospace Medicine,* 2nd ed. (p. 332), by K. K. Gillingham and F. H. Previc, 1996, Baltimore, MD: Williams & Wilkins. Copyright 1996 by Williams and Wilkins. Reprinted with permission.

20 cm/sec^2 (0.01–0.01 g) for horizontal acceleration along the head longitudinal (z) axis. Thresholds for vertical acceleration, parallel to gravity, are somewhat higher. Larger linear accelerations are detected in shorter times, leading to an effective linear velocity threshold of approximately 20 cm/sec for horizontal movements (Melvill-Jones & Young, 1978). This ambiguity between linear acceleration and tilt is used to advantage in moving base flight simulators, which exploit "G-tilt" to produce the illusion of acceleration.

Vestibulo-spinal reflexes as well as otolith-ocular reflexes attempt to compensate partially for head shifts relative to gravity. Their role in spatial orientation is principally to indicate the direction of the vertical. However, it is obvious that they cannot fulfill this role in other than a 1-G field. In weightlessness, of course, they will not slide "downhill," and they no longer serve as an indicator of "down" in an orbiting spacecraft. In an aircraft or a centrifuge "pulling 2 Gs," for example, they will falsely signal an excessively large head tilt. Finally, during any linear acceleration they will indicate orientation of the head with respect to the *apparent vertical* (the vector difference between gravity and linear acceleration), rather than the true vertical.

The means by which the brain resolves the inherent ambiguity of otolith signals to discriminate between tilt and translation has been a subject of discussion for some years. One theory is that the frequency content of the signals is used to interpret low-frequency portions as head tilt and high-frequency cues as head translation (Mayne, 1974; Paige & Tomko, 1991). An alternate theory is that multisensory integration of information from different senses, along with expectation, is used to generate an estimation of the direction of gravity that in turn allows separate estimation of tilt and translation (Hess & Angelaki, 1999).

The time it takes for a subject to notice that he or she is moving depends on the direction and magnitude of the imposed acceleration step. When the head is erect the time, *t,* to detect a step of linear acceleration, *a,* is given by the relationship $t = B/\alpha + t_r$, where t is the time required to correctly detect the movement, α is the linear acceleration in m/s^2, and t_r is the minimum reaction time, of about 0.4 sec. B is the "velocity constant" that determines the speed at which the acceleration is noticed and is approximately 22 cm/sec for either horizontal or vertical motion with the head erect. However, errors in the direction of motion are far more common for the vertical direction, in which the saccule rather than the utricle is primarily stimulated.

For sinusoidal acceleration the transfer function relating the actual velocity (Young, 1984) to the perceived velocity is given by

$$\frac{\text{perceived velocity}}{\text{actual velocity}} = \frac{1.5\,(s + 0.076)}{(s + 0.19)\,(5 + 1.5)}.$$

The perceptual and postural reactions to sustained tilt all show a measure of adaptation, with a tendency toward adopting the steady state GIF as the new vertical reference. Ocular counterrolling, which is the reflexive torsion of the eyes in the direction toward maintaining their original orientation during head tilt, also shows a tendency to return to the rest position during sustained tilt. The extent to which the otolith signals decay back toward the resting discharge rate with prolonged stimulation is unclear, although some first-order afferents clearly do demonstrate such adaptation (Goldberg, Desmadryl, Baird, & Fernandez, 1990). Regional differences in the dynamic response characteristics of hair cells innervating different parts of the sensory epithelium suggest that some frequency separation of the otolith afferent response may occur at the end organ itself (Holt, Vollrath, & Eatock, 1999).

Nystagmus

Any interference with eye fixation can lead to SD. When the head turns, the eyes are driven in the opposite direction by the angular vestibulo-ocular reflex (AVOR, or simply VOR). A rhythmic series of slow compensatory eye movements, which tend to stabilize the eye in space, rather than in the cockpit, alternates with fast resetting eye movements in the opposite direction in a pattern called *nystagmus.* Normally the VOR is measured in terms of its "gain," defined as the ratio of eye velocity to head velocity. Gain depends on the imagined task, even for rotation in the dark, and is typically about 0.7 when the alert subject imagines an object at eye level, or as high as one if the subject imagines an object fixed in space. The time constant for decay of vestibular nystagmus is about 12 to 16 sec for horizontal nystagmus, induced by yaw rotation about the *z*-axis vertical, but only about 7 sec for vertical nystagmus, induced by pitch rotation about the *y*-axis vertical (Benson, 1999; Guedry et al., 1971; Melvill-Jones et al., 1964). Torsional nystagmus, induced by head rotation about the roll axis, is generally smaller and causes much less difficulty in attempting to fixate an object.

The eyes also move to follow moving objects in the visual field. Individual small objects are tracked by pursuit eye movements, up to a maximum velocity of 30 to 60°/sec. When the entire field is in uniform motion, the eyes follow up to a speed of 60 to 100°/sec, interrupted by fast return movements, in a pattern known as *optokinetic nystagmus.*

When both vestibular and optokinetic stimuli are present, the eye movement patterns are complex, but generally demonstrate a dominance of vestibular nystagmus at higher frequencies and optokinetic nystagmus at lower frequencies. Because nystagmus tends to stabilize the retinal image of the external world, instruments inside the cockpit become difficult to read. In that case, the pilot is faced with the challenge of visually suppressing nystagmus, which can sometimes be accomplished, though with considerable effort and occasional motion sickness and associated discomfort.

A less-well-known vestibulo-ocular reflex is the linear one (LVOR). Compensatory eye movements in the direction opposite to lateral head motion are stimulated by signals from the utricular otolith organs. The gain of the LVOR depends on the distance of the real or imagined target and appears to be controlled by the vergence angle between the two eyes (Crane & Demer, 1998; Musallam & Tomlinson, 1999; Paige & Tomko, 1991).

Other Proprioceptive Sensors

A number of other sensory systems play a role in the perception of orientation and in the sense of the *relative position* of the body parts. Although these contributions are rarely important in the issues of SD, they are relevant to flight simulation. Furthermore, when loading conditions change, as in weightlessness or during a high-G turn, these sensors provide misleading information affecting limb position and control. It is important to recognize that although nonvestibular proprioception plays a very strong role in judgment of perceived postural verticality, it is virtually uninvolved in the separate judgment of the subjective visual vertical (Bronstein, 1999; Mittelstaedt, 1999).

Sensors in the joint capsules and elsewhere measure limb angles. Head orientation relative to the trunk is of particular importance in relating body position to the orientation signals from the vestibular and visual sensors (Burgess, 1976; Skoglund, 1973).

Muscle receptors, in the form of the muscle spindles and the Golgi tendon organs, also contribute to the perception of limb position. Muscle and joint receptors play an indirect role in spatial orientation through their involvement in manual control. By providing the pilot with feedback about the aircraft control commands, issued through force or displacement of the control stick or rudder pedals, the proprioceptive system contributes to the expectation of vehicle movement, and thereby influences the interpretation of other sensory signals.

Pressure applied to the skin can serve as a significant orientation cue. Both the direction and magnitude of forces applied to the skin generate the perception of orientation and of acceleration (Zotterman, 1976). Reliance on these cues alone is sometimes referred to as "flying by the seat of your pants." Although the thresholds for localized pressure changes are low, especially in nonhairy skin areas, the rapid adaptation characteristics normally make the cutaneous receptors insignificant contributors to overall body orientation. In the absence of any vestibular function, however, labyrinthine defective subjects can judge the postural vertical while in the dark nearly as well as those with full vestibular function (Guedry, 1974). Scrubbing motion of a surface over the skin can contribute to a sense of self-motion, and localized small areas of higher pressure can produce the perception of increased total force, as employed in the simulator "G-seat." The deliberate use of haptic stimulation to provide the pilot with a substitute orientation cue is the basis of a "tactor vest." This device, developed by the U.S. Navy for avoidance of SD, employs a matrix of small vibrators that help to indicate motion and attitude in a natural manner (Rupert, McTrusty, & Peak, 2000).

Long before the vestibular system was identified as the primary organ for equilibrium in the 19th century, many scientists speculated about other possible sources of sensory information concerning the vertical. Mach (1875) disposed of most of these through a series of decisive demonstrations. However, Mittelstaedt has more recently indicated an important supplementary source of information used by humans in postural perception, which he calls the truncal receptors. These inertial receptors, which he speculates are located in the kidneys and possibly in the mass of the blood, influence the sense of the postural vertical, especially in subjects lacking any vestibular function, but do not affect the subjective visual vertical (Mittelstaedt, 1996).

Auditory Spatial Orientation

The ability to use auditory localization for isolating sound sources is important in large areas: it contributes both to human navigation and to the "cocktail party effect"—the discrimination of one conversation from among many. The difference in intensity of lower frequency sound waves arriving at the two ears and the difference in arrival time for higher frequency and broadband sounds both contribute to auditory localization in the azimuth plane of the head. The shape of the pinna also contributes to sound localization in both elevation and azimuth. In a small closed environment such as a cockpit, however, this capability is severely limited because of the high ambient noise level, substantial internal sound reflections, and the relative isolation from external noise sources. (Sound cues are, however, important to the pilot in other ways as subtle indications of aircraft speed, angle of attack, and engine performance.)

Auditory localization is further affected by the orientation of the body relative to gravity. When a subject is tilted, sound sources are sensed as coming from the direction of the "down" ear. Even self-motion illusions, generated by an optokinetic stimulus, can displace the apparent source of a sound in the direction of the self-motion (Lackner, 1978).

Multisensory Interaction

Semicircular Canal–Otolith Interaction

During body rotation about a horizontal axis, multisensory interactions occur among signals from the otoliths, vertical semicircular canals, and visual systems. Visual cues indicate the apparent horizontal and vertical as well as whole field motion. At constant rotation in the dark, known also as "barbecue spit rotation," nystagmus and subjective rotation do not decay to zero but instead continue as long as the rotation lasts, with some cyclical modulation of the eye velocity. Nystagmus and subjective velocity extinguish rapidly following cessation of a prolonged rotation about a horizontal axis, in contrast to the extended post-rotatory decay following vertical axis rotation. During post-rotatory nystagmus and turning sensation following sustained vertical axis rotation, if a subject should suddenly pitch or roll the head to horizontal position, the nystagmus and rotation sensation would rapidly disappear. This phenomenon is known as nystagmus and sensation "dumping." It is attributable to the unchanging orientation signals from the otolith organs, which fail to confirm the rotation cues signaled by the vertical semicircular canals. (The dumping phenomenon fails to occur in weightlessness, where otolith cues concerning the direction of gravity are nonexistent [Oman & Balkwill, 1993].) Otolith signals can even provide horizontal axis rotation information to the brain in the absence of canal signals (Hess & Angelaki, 1999). In general, when tilted about the horizontal axis, the short-term, transient motion sense is dominated by the semicircular canals, and the longer-term resolution of body position is resolved to be consistent with otolith signals. For example, during sustained pitching motion about a horizontal axis, most subjects change their spatial orientation sensation of an initial pitching motion to a sustained sense of vertical plane circular motion, as on a Ferris wheel, with a constant body orientation relative to the vertical (Meiry, 1965). During off-vertical axis rotation (OVAR), most subjects sustained motion around the surface of a cone, with vertex down, and no rotation. Furthermore, the axis of perceived body rotation and nystagmus tends to align with the direction of the apparent vertical during sustained centrifugation (Dai, Raphan, & Cohen, 1991; Merfeld & Young, 1995; Merfeld, Young, Paige, & Tomko, 1993).

Visual–Vestibular Interaction

Visual information normally dominates spatial orientation for a pilot, but it frequently conflicts with motion or orientation signals emanating from the vestibular system. Several characteristics of the interaction are important for a consideration of the physiological basis of some of the SD phenomena (see chap. 12). For constant or very low frequency rotation about a vertical axis, below about 0.1 Hz, the visual cues largely determine the perception of velocity. When there is a sudden change in vestibular stimulation, such as at the initiation of a turn, the semicircular canal signal dominates at first and seems to inhibit the visual influence on the interaction. In terms of motion perception, the "onset cues" seem to come chiefly from the semicircular canals. This is a practical consideration in aviation when determining the need for motion drives in a wide-field-of-view flight simulator. Visual–vestibular interaction at higher frequencies (above 1 Hz) is driven almost entirely by the vestibular input. In the intermediate range of frequencies, from 0.1 Hz to 1.0 Hz, both visual and vestibular motion signals play a role, with an apparent nonlinear interaction relating the two (Henn, Cohen, & Young, 1980; Zacharias & Young, 1981).

Some of the models for this interaction involve the notion of an "internal model," driven in part by past sensory signals and in part by expectation based on copies of motor commands ("efferent copy"). One implementation of this concept is the "velocity storage" model, which assumes that an ongoing estimate of head velocity is developed on the basis of past visual and vestibular sensory signals and then maintained for several seconds in the absence of any new sensory information to the contrary. It effectively lengthens the time constant for eye velocity and subjective sensation (Raphan & Cohen, 1985). The subjective self-motion thresholds are reduced when the visual signal is consistent with the vestibular cue and increased when they are in conflict (Young, Dichgans, Murphy, & Brandt, 1973).

For linear acceleration, stimulating primarily the otoliths but also acting on any truncal graviceptors, the story is similar. Linear visual–vestibular interaction shows visual self-motion dominance at low frequencies and vestibular and proprioceptive dominance at higher frequencies. As for rotation, the eventual conscious perception of self-motion depends on the expectation of orientation as well as the peripheral interactions.

Although each of the sensory modalities has the capability of detecting motion in any direction, or of signaling any orientation, there is a special role played by the vertical, or at least the *apparent vertical.* Recognition of objects is made much more difficult if they are tilted more than 30° from the retinal vertical, and object recognition is also disturbed somewhat when the head is tilted more than 45° from the vertical. Not only do visual signals influence vestibular information, as discussed earlier, but vestibular signals also modify the representation of visual signals in the brain (Horn & Hill, 1969). The ability to sense the vertical is strongly degraded when the head is away from the erect position.

Proprioceptive–Vestibular Interaction

The skeletal muscle spindle receptors normally act in conjunction with the otolith organs to confirm voluntary movements, such as performing a deep knee bend. Under an altered GIF, however, the usual relationship is disturbed, both because of differences in otolith reactions to linear motion and because of differences in the weight of body, head, or limb.

SPATIAL DISORIENTATION ILLUSIONS

Illusions Attributable Primarily to the Semicircular Canals

Nearly all the SD somatogyral illusions associated primarily with the semicircular canals are the result of low-frequency rotation, or of rotation for a period that exceeds the canals' dominant time constant of 7 to 10 sec. For short-duration head movements, lasting only a few seconds, the semicircular canals perform the task of transducing head angular velocity reasonably well. Consequently, head position changes are also correctly perceived. For sustained rotation longer than several seconds, however, the restoring force of the cupula comes into play, as discussed earlier. As a result, the semicircular canal signals no longer transduce only velocity, but instead send a signal more closely related to head acceleration. The nervous system, in the absence of any other information to the contrary, still interprets the semicircular canal afferent signals as velocity, thereby producing motion illusions. Obviously this is a problem only when humans place themselves in sustained rotation, as in an airplane, automobile, or a carnival ride.

Sustained Constant Turn

When a pilot banks the airplane into a right wing down constant rate level turn in the clouds, the information is registered by the horizontal semicircular canals, which register the initial yaw rate, and by the vertical canals, which measure the brief roll rate on entry into the turn, as shown in Fig. 3.4. Assuming a coordinated turn with no sideslip, the otolith organs and the haptic receptors continue to register the net gravito-inertial force direction along the z-axis of the aircraft, although at a magnitude greater than 1 G. (The direction, rather than the magnitude of the GIF, is principally involved in orientation. The magnitude of the apparent vertical acceleration, which is the vector sum of gravity and the centripetal acceleration,

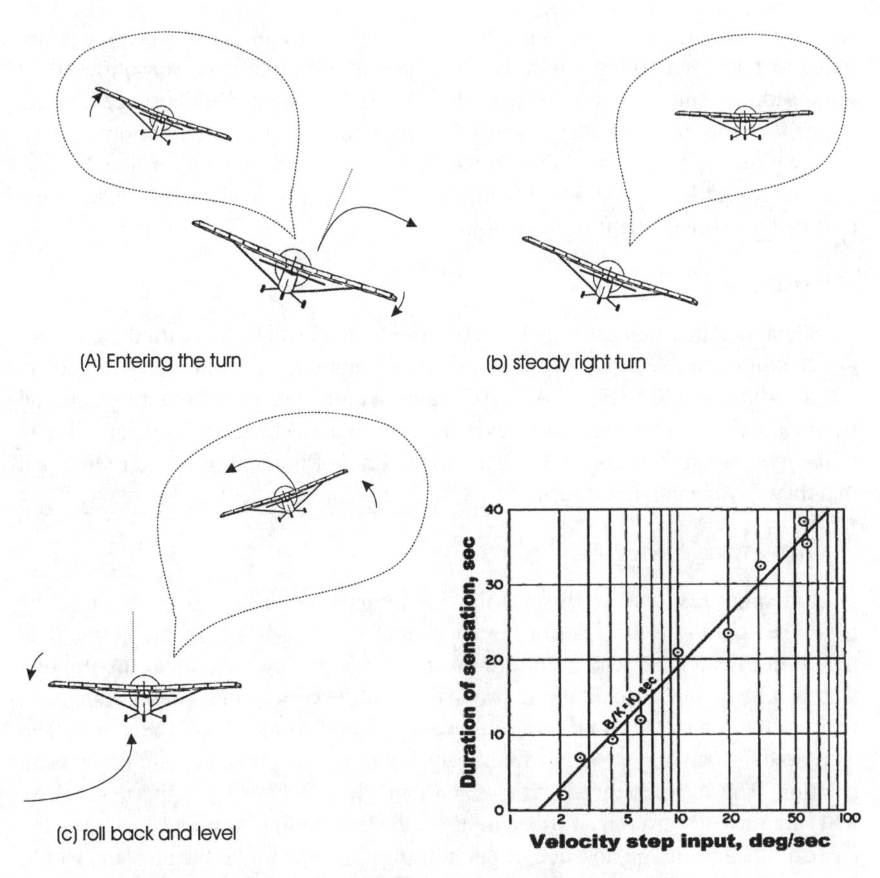

FIG. 3.4. Development of illusions during and following a sustained turn. In (a) the pilot's semicircular canals reflect the actual bank and yaw of the airplane, but after some time the sensation is replaced by level flight, shown in (b). The post-rotatory sensation of turning in the opposite direction is shown in (c). The duration of the post-rotatory sensation depends on the rate of turn, as shown in (d). [Figure 3.4 reproduced from Van Egmond, A. A., Groen, A. J., & Jongkees, L. B. W. (1949).]

comes into play in the somatogravic illusions considered later.) Several seconds into the turn, the horizontal semicircular canals signal a steadily reduced yaw rate, which finally drops below subjective threshold at a time determined by the initial turn rate. The yaw rate sensation theoretically begins at the original yaw rate and decays exponentially with a time constant of 10 to 20 sec. It finally disappears when it drops below the threshold level of 2°/sec. The duration of the turning sensation depends on the yaw rate, as shown in Fig. 3.4(a), for a nominal time constant of 10 sec. (The situation is complicated somewhat by the flying experience of pilots. Fighter pilots with recent practice have reduced vestibular responses and might lose the sensation of a 30°/sec turn in as little as 5 sec, compared to pilots not flying regularly, for whom an initial 30°/sec sensation might not disappear until 20 sec have passed [Aschan, Bergstedt, Goldberg, & Laurell, 1956].) At this point, in the absence of any confirming out-the-window visual cues, a passenger would feel the airplane to be neither banked nor turning, but flying straight and level. And so would the pilot if no reference were made to the artificial horizon or the turn indicator.

Postturn Illusion

When the pilot then rolls back left to straight and level flight, with the GIF still directed into the seat, the passenger's vertical semicircular canals correctly detect a brief roll to the left [Fig. 3.4, (c)]. The horizontal canals, however, which had been indicating no yaw rate, now experience a sudden change in angular velocity to the *left,* which they now duly signal to the brain. Pilot and passenger alike feel that they have begun a left turn.

Graveyard Spiral

If the pilot responds to this SD illusion by pushing the control column to the right, the conditions are ripe for the appropriately named "graveyard spiral." The post-turn illusion can lead a pilot to resume a descending turn under the impression that he or she is maintaining level flight, while continuing a descending turn with all the aircraft's gravito-inertial acceleration directed along the z-axis. The increased G-load may produce a false illusion of roll or pitch, depending on head position. The inexperienced pilot who attempts to recover from the descent by first pitching upward only further tightens the descending turn and increases the G-load. Because in the absence of pitch or throttle adjustment the airplane would lose altitude in the turn, this spiral to the earth can be fatal. The situation is especially aggravated when the rates are high, as in the attempt to recover from a spin. On reduction of the turn rate, the pilot may attempt to reestablish his attitude reference by looking at the artificial horizon or the turn and bank indicator, but may be unable to read the instruments because of the vigorous postrotatory nystagmus.

In the absence of strong visual cues, pilots attempt to maintain a steady bank angle following a substantial roll rate; they have a tendency to command a continuing roll in the same direction. This can lead to a diving impact with the ground at a step bank angle, known as the Gillingham Illusion (Ercoline, Devilbiss, Yauch, & Brown, 2000).

The Leans

One of the consequences of the somatogyral illusion can be a common and disturbing illusion of a sustained bank angle. Following one or more turns in the same direction, such as in a holding pattern, the pilot may be left with only the recent memory of the roll back to level flight, which leaves the impression of flying in a constant bank, yet not turning. In the absence of any clear view of the ground or horizon, the leans may continue for some time, and the pilot is likely to physically lean in the opposite direction in an attempt to maintain his head and trunk upright. The asymmetry can also be produced by a series of turns that are subthreshold in one direction and superthreshold in the other. Any number of other factors, both visual and vestibular, can contribute to "the leans," which would be described more accurately as a disturbing symptom than as a single, well-defined instance of SD. The pilot who continues to fly through a sustained case of the leans is likely to be fatigued and operating at near maximum workload.

Oculogyral Illusion

The oculogyral illusion is closely related to the somatogyral illusion. It is the illusion that objects in the visual field, including the aircraft and fixed body parts, are in motion *relative to the subject.* The threshold for oculogyral illusion is slightly lower than that for the subjective illusion of rotation in the dark. The apparent motion of a fixed target light may serve as the first indication of self-rotation. The oculogyral illusion does not require eye movements or retinal slip of the object's image, as once thought, because it can be developed with a stationary retinal image or in the absence of eye movements during visual suppression of vestibular nystagmus. It is possible, of course, that the effort to overcome the vestibular drives to the oculomotor system in turn creates the perception that the object being fixated is actually in motion, even though it never changes its relative position. A slight amount of oculogyral illusion, just enough to start the perception of object motion, might be associated with the onset of the purely visual autokinetic illusion (Clark, Graybiel, & MacCorquodale, 1948), which is discussed in the following section.

Coriolis Cross-Coupling Illusion

There are two confusing aspects of the Coriolis Cross-Coupling Illusion—its name and the physics involved. The phenomenon occurs when the head is tilted while continuing to rotate around an axis that is not parallel to the axis of the tilt. It is commonly experienced in flight when a pilot in a turn looks down to adjust some navigation setting or looks up to check a switch. It is familiar to those who have undergone flight physiological training involving a Barany chair or other rotating device. After a period of some tens of seconds of rotating at constant speed about a vertical axis with the head erect (z-axis rotation), the subject makes a rapid head-down pitching movement (y-axis rotation). The surprising (and possibly sickening) sensation is a complex one, including a transient roll acceleration (about the head's x-axis) followed by a continued sense of rotation about this axis, which is now vertical, and an acceleration opposite to the original rotation direction about the head's z-axis, which is now horizontal. This training demonstration is effective in alerting flight candidates to the reality of disorientation and to the importance of believing the flight instruments. It is also exceedingly useful in showing how the physics and physiology of the movement explain SD. The consequences of Type I Coriolis disorientation can be fatal, as in the case of a student pilot who broke out of his traffic pattern in a climbing left turn toward a dark mountain range. When the runway supervisor asked the student for a fuel check, requiring a 110° head turn to the right, the pilot apparently experienced the illusion of pitch up and right roll. He pushed his T-38A into a steep left dive into the mountain (Simson, 1971).

Coriolis forces refer specifically to the forces and linear accelerations that occur when a body attempts to move in a straight line within a reference frame that is in rotation with respect to a fixed (inertial) frame of reference. Early on this was referred to as the Coriolis phenomenon, even though no such explanation is required if one considers the semicircular canals as point sensors of angular acceleration. The name is retained in conjunction with the more applicable term *cross-coupling.*

It is helpful to consider the torques acting on each semicircular canal during a cross-coupling maneuver in the Barany chair. Prior to the head movement, the horizontal canal had been in the plane of rotation long enough for its cupula to return to center. It signals no rotation to the brain, yet it retains all the angular momentum of its ring of endolymph rotating about a horizontal axis. When the head is pitched forward, the horizontal canals are suddenly removed from the plane of chair rotation. Their rings of endolymph are thereby suddenly decelerated, leading to cupula deflection in a direction opposite to the original sense, with the consequent decaying sense of rotation about the head's z-axis, which now lies horizontal. (The fact that this post-rotatory sensation decays more quickly than post-rotatory sensation about a vertical axis is related to "dumping," which was discussed in the section on "Semicircular Canal–Otolith Interaction.")

Next, consider the vertical canals, formerly at rest, as they are pitched by the forward head movement into the plane of chair rotation. Although neither the anterior nor posterior canals are aligned with the pitch or roll axes, they can be considered to operate cooperatively, so that together they register head pitch or roll acceleration. The roll axis equivalent vertical canal (with its sensitive axis along the naso-occipital axis) is suddenly exposed to a step change of angular velocity as it is pitched into the plane of chair rotation. Its cupula is deflected and slowly returns, reinitiating the veridical subjective sensation of chair velocity, which will again decay after several seconds.

The most difficult aspect to understand is the transient cross-coupled angular acceleration that is sensed by the roll axis component of the semicircular canals during the forward head movement. Before the head movement, the angular velocity vector was along the head z-axis; during the movement, it shifts to the head x-axis without changing magnitude. In the head reference frame, the angular velocity has changed its direction by 90° (from z to x) at a rate equal to the speed of the head pitching movement. Therefore, the head has undergone an (cross-coupled) angular acceleration, and the axis about which it has accelerated is the head's x- (roll) axis. The magnitude of the acceleration is the vector cross-product of the two velocities (the product of the two angular velocities times the sine of the angle between their axes of rotation). The acceleration is directed along an axis orthogonal to both the head velocity and the reference frame angular velocity, according to the right-hand rule (Hixson, Niven, & Correia, 1966).

The Coriolis cross-coupling phenomenon is often disorienting, surprising, and the cause of air sickness, but it is not necessarily an illusion. The orthogonal axis rotation is real and correctly detected by the semicircular canals. If the subject is not aware of the ongoing rotation before the head movement, the previously discussed sensations of roll during and following pitching come about unexpectedly. They are not consistent with the cockpit visual surround or with the otolith cues, both of which support the notion that the head is stable relative to the vertical and directed down toward the seat. If the pilot senses the yaw motion before making the pitch head movement, however, the surprise is minor, and motion sickness is unlikely to occur. But if the pilot has been turning in yaw for 10 sec or more without outside reference cues, so that the yaw rate sensation has substantially decayed, then the cross-coupling is unexpected, leads to a sensory conflict, and often provokes motion sickness if repeated.

Ancillary Effects on SD Illusions

Habituation and Adaptation

Subjective vestibular responses to repeated angular acceleration, as well as nystagmus strength and duration, decline with repeated frequent stimulation. This habituation may be quite specific, for example, applying only to a specific direction of turning or to a specific frequency of oscillation. On the other hand, the habituation might occur more generally, thereby reducing vestibular responses to all motion stimuli. Pilots who are active and frequently exposed to intense stimuli demonstrate reduced subjective and nystagmus response to clinical rotatory tests, but return to normal values when they have been away from flying for as little as a week or two (Aschan, 1954; Mann & Cannella, 1956). Adaptation, in contrast to habituation, refers to the development of an appropriate new response to a novel environment. A certain amount of adaptive response occurs whenever a subject changes eyeglass prescription, since higher magnification requires a larger VOR to maintain image stability. The fact that people adopt the appropriate VOR gain with glasses on or off, or even when looking up or down through the two different sections of bifocals, shows our ability to develop and maintain "context specific dual adaptation." Pilots will normally adapt to a wide variety of distortions in the visual field or to unusual motion environments, provided they have been sufficiently exposed to such circumstances. Inappropriate or misleading visual or motion cues in training can similarly produce a maladaptive orientation response and lead to disorientation in flight. This has been the case for certain kinds of wide-field-of-view, limited-motion flight simulators, leading to both simulator sickness and dangerous post-training "flashbacks," as discussed later.

Alcohol Effects

Alcohol is a leading cause of general aviation accidents. Even a modest amount of alcohol, which might not be very intoxicating at sea level, has much greater effects on judgment and coordination at higher altitudes and lower cabin pressure. Beyond the cognitive and motor aspects, however, there are some important direct effects on the semicircular canals and on the central compensation for vestibular asymmetry.

Alcohol alters the density first of the cupula and subsequently of the endolymph. Normally the cupula floats almost neutrally in the surrounding endolymph, so that it is nearly unaffected by the orientation of the canals relative to gravity. Alcohol ingestion upsets this balance, however. Because its density is only 0.8, and the density of the endolymph and cupula is just above 1.0 (Steer, Li, Young, & Meiry, 1967), the presence of alcohol in the cupula quickly alters the balance of the cupula by making it lighter (Money & Myles, 1974). When the head is tilted to the side, therefore, the cupulae in the horizontal semicircular canals tend to rise slightly, just as though the subject were rotating in the direction of the ear that is down. Nystagmus (Positional Alcohol Nystagmus I, or PAN I) occurs, with its fast phase directed down, and is accompanied by a consistent, disturbing sensation of rotation, the degree of which depends on both the head orientation relative to gravity and the alcohol level. Following a period of 4 to 6 hours after alcohol ingestion, when the blood alcohol has returned to a low level, a reversed direction of nystagmus (known as PAN II) appears, again attended by the spinning sensation. This second stage occurs when the alcohol in the cupula has metabolized but the endolymph has become less dense, causing the cupula to sink.

Alcohol also interferes with visual suppression of vestibular nystagmus, which normally allows pilots to retain visual fixation on a flight instrument while the airplane is turning, producing a decrement in the ability to read instruments (Collins, Schroeder, & Hill, 1973). Furthermore, alcohol disturbs normal habituation (Aschan, 1967) and the compensatory mechanisms that allow us to make up for inherent asymmetry in labyrinthine function (Berthoz, Young, & Oliveras, 1977), thus leading to a variety of SD illusions.

Pressure Vertigo

Pressure changes in the inner ear brought about by sudden changes in middle ear pressure may stimulate the semicircular canals. On occasion, blockage of the eustachian tubes prevents pressure equalization during ascent. The sudden change in middle ear pressure when the pilot finally succeeds in clearing the ears, or during a Valsalva maneuver to counteract positive acceleration, can produce a rapid deflection of the semicircular cupulae. The resulting severe nystagmus and vertigo normally last only the 10 to 15 sec required for the cupulae to return to their rest position but may, on occasion, remain for several minutes (Benson, 1999).

Illusions Attributable Primarily to the Otolith Organs

The principal SD problem associated with the otolith organs lies in the inherent inability of any physical accelerometer to distinguish between gravity and linear acceleration. Whenever the brain interprets otolith outputs in terms of gravity alone, neglecting to take linear acceleration into account, an erroneous illusion of orientation to the vertical occurs. Furthermore, the otolith organs are only accurate for determining the direction of the vertical with the head upright; they do so optimally when the head is pitched about 25° forward and the plane of the utricular macula is thereby aligned almost normally to the gravity vector. In this position any slight pitch or roll of the head produces a shearing component of specific force on the utricle, which is easily detected and leads to the low thresholds of about 2°, as described earlier. The interpretation of otolith organ signals seems to rely almost entirely on the *direction* and not the *magnitude* of the applied gravito-inertial force.

Aubert-Müller Effect

When a subject is tilted to the side by an angle greater than about 70°, a truly vertical luminous line in a dark room will appear to be tilted in the *opposite* direction from the body. When the line is adjusted so that it is apparently vertical, it is actually tilted toward the body axis. This effect, named for its discoverer, Aubert (A-effect), is consistent with a perceptual underestimation of body tilt angle at large angles. This consistent judgment error has further been shown to require the contribution of the intact otolith system and to be relatively uninfluenced by somatosensory cues. Moreover, the magnitude of the percieved tilt of the visual vertical (the subjective visual vertical, or SVV) depends almost entirely on the shear component of gravito-inertial acceleration acting in the plane of the utricle, even under increased g's on a centrifuge. Some of the A-effect could be explained by ocular counterrolling, which torts the eyes opposite to head tilt by up to 6 deg and therefore reduces the angular deviation of the retinal image. However, the A-effect magnitudes (in excess of 45° of error in setting the visual vertical) are far too large for this to be a major factor.

To further complicate matters, the subjective postural vertical (SPV) is dissociated from the SVV and has been shown to be greatly dependent on somatosensory cues. The SPV also tends to align with the direction of the major axis of the trunk, known as the *ideotropic vector.* The subjective judgment of body tilt angle is quite accurate and does not show the A-effect (Bronstein, 1999).

For smaller tilt angles, of less than 30 to 45°, most subjects perceive just the opposite visual effect. The luminous vertical line appears tilted *toward* the subject, and it must be tilted away from the body in order to appear vertical. This phenomenon, called the Müller effect, or E-phenomenon, does not occur for everyone, however.

A complete and satisfactory explanation for these effects must entail more than the shear component of force on the utricular macula, but as yet there is no agreement among researchers on this issue. One attempt, by Mittelstaedt (1999), is based on trigonometric analysis with different weightings applied to signals from the utricle, the saccule, and the ideotropic vector. It has been used to explain both the A- and E-effects of the judgment of the SVV.

In terms of SD for the pilot, the A-E phenomena are important because they explain a large misjudgment of the angle of visual indicators of the horizontal or vertical when the pilot's head is tilted away from the gravito-inertial vector.

Oculogravic Illusions

Mach first pointed out that objects outside a train that is going around a sharp curve appeared tilted toward a passenger, and that the estimate of the vertical is based on the net gravito-inertial acceleration (Henn & Young, 1975; Henn, Young, & Scherberger, 2000; Mach, 1875). Clark and Graybiel (1955) investigated the factors affecting the tilt illusion, which is also named the *oculogravic illusion.* It should be distinguished from the related perception of self-tilt, called the *somatogravic illusion.* In general, a subject adopts the resultant steady inertial force direction as the vertical, although it may take some time to do so if signals from other sensors fail to confirm the implied tilt.

Pitch-Up Illusion on Takeoff

In aviation the most common somatogravic illusion is the sense of pitching up excessively when taking off into poor visibility. As the pilot accelerates during the takeoff roll, or even more extremely during a catapult-assisted takeoff from an aircraft carrier, the forward acceleration produces a backward inertial reaction force (Cohen, Crosby, & Blackburn, 1973). When combined with the 1 G downward gravitational force, the net gravito-inertial vector is rotated backward. For a catapult launch of 3 to 5 Gs, the rotation might produce a pitch-up illusion of about 5° that can last for a minute or more (Benson, 1999). For a high-performance noncatapult launch of 1 G, the gravito-inertial vector angle reaches 45°, and even for a transport aircraft accelerating down the runway at 0.1 G the vector angle is displaced backward by about 6°. When combined with the actual pitch up associated with takeoff rotation and the continued forward acceleration as the gear is raised and the flaps are retracted, the illusion of an excessive nose-high attitude can become very strong. In the absence of a strong external visual horizon, the pilot may react by pushing the nose forward to achieve more level flight. This action can result in diving into the ground or into the water following a night carrier takeoff or a takeoff over a shoreline presenting a false visual horizon (Gillingham & Previc, 1996). Even a standard overshoot procedure for a missed approach, involving sudden forward acceleration associated with full throttle and reduced drag, can produce an illusion of pitch up in a transport airplane. In a fatal Vanguard crash in London in 1965, the pilot reacted to this illusion by pitching the nose down 35° while only 100 ft above the ground. Delayed reaction in the rate of climb indicator and pressure altimeter confounded the problem by showing a positve climb rate (Barley, 1970).

G-Excess Phenomena

Most of our daily activities are carried out in a background of earth's gravity. In aviation, however, the pilot can be exposed to prolonged periods of "G-excess," in which the steady gravito-inertial pull exceeds 1 G. As long as the orientation perception is based on an assumption of a 1 G background, these sustained accelerations are likely to produce one of the following G-excess phenomena.

Elevator Illusion. Even when a subject is accelerated parallel to the G-vector, so that there is no rotation of the gravito-inertial vector, somatogravic and oculogravic illusions occur. During upward acceleration in an elevator, for example, the utricular otolith (which is tilted up in front) slides backward on its macula and produces an illusion of pitching upward. An object fixed in front of the subject also seems to rise, relative to a fixed reference. For a pilot in a coordinated turn, this increase in $+G_z$ can produce an illusion of pitching up. A pilot also experiences upward acceleration when leveling out following a descent and may react to the nose-high illusion by resuming the dive.

Top-of-Climb Inversion Illusion. A similar illusion can occur to a pilot who levels out suddenly after a climb, producing reduced $+G_z$, or even "negative Gs" ($-G_z$) at the time. The airplane is also likely to accelerate forward ($+G_x$), since the thrust now serves to drive forward, rather than causing a climb. The combined effect of the forward acceleration and reduced $+G_z$, illustrated in Fig. 3.5, is to create a gravito-inertial vector that rotates to the pilot's back, or even up toward his or her head, if the centripetal acceleration associated with the pushover is large enough . The pilot is likely to feel inverted, as though he had pitched upward, and to react by pushing forward on the controls. This action, which sends the plane into a dive, only further accentuates the $-G_z$ acceleration and exacerbates the inversion illusion. (Another "inversion illusion," experienced when entering weightlessness and discussed under "Space Flight Illusions," is similar but requires only an elimination of the steady 1 G gravitational acceleration.)

G-Excess Tilt Illusions. When a pilot makes a real head tilt during a coordinated turn, the excess gravito-inertial force acting on the otolith organs produces a sensation of a greater head movement. It is primarily the shear component

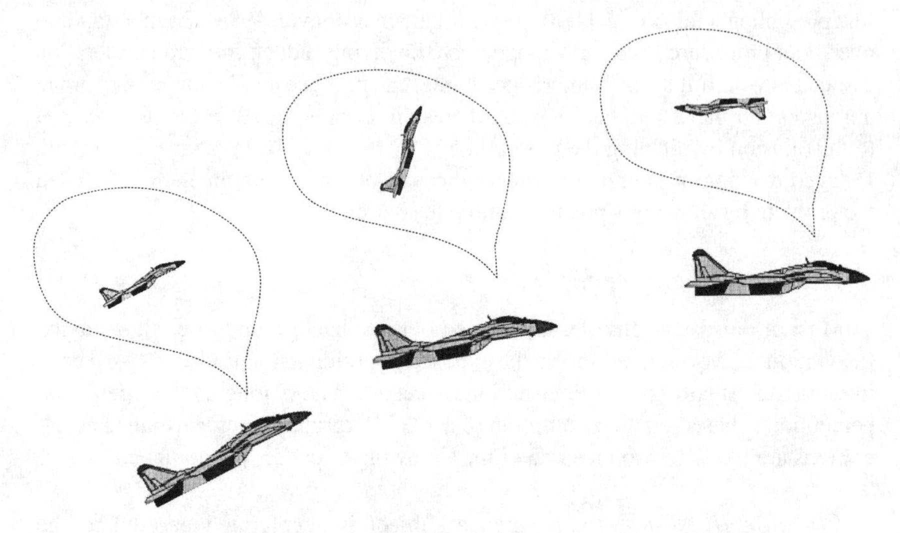

FIG. 3.5. Top-of-climb inversion illusion. As the airplane ends its climb and accelerates forward the pilot may feel a strong pitch upward, even to the point of inversion, as the downward pull of gravity is cancelled in part by the downward acceleration.

of gravito-inertial force that stimulates the otolith organs. For example, a pitch or roll of the head by 30° during a 2 G turn will produce a shear component equal to that produced by a head movement of 90° while flying level in a 1 G field: ($2g \times \sin 30° = 1g = 1g \times \sin 90°$). If the head tilt were performed passively and slowly in the dark, a subject might feel tilted by up to 90°. For a pilot in a turning airplane, however, the cues from the semicircular canals, neck proprioceptors, and local visual field all support the perception of the real head movement. Consequently, this excess in tilt signal is frequently attributed to additional tilt of the airplane in the direction of the head tilt. Errors in bank estimation of 10 to 20° were reported for this condition (Schöene, 1964).

Whether the G-excess effect in a turn causes the airplane bank illusion to be underbank or overbank depends on the direction of the pilot's head movement. For example, as Gillingham and Previc (1996) illustrate, a pilot whose head is turned toward the inside of a turn and is then elevated perceives a greater than actual tilt up of the head. That excess tilt is attributed to an underbank of the airplane and may cause the pilot to tighten the turn.

Visual Illusions

The enormous literature on visual illusions is beyond the scope of this chapter, though quite important in understanding the associated disorientation phenomena. Several of the most important aviation SD situations are discussed next.

Fascination and Fixation

When a pilot becomes so fixated on a given instrument or element in the scene that a regular scan is no longer maintained, SD may occur. The related phenomenon of "perceptual narrowing" is a literal loss of attention to anything in the peripheral visual field during high stress and high workload situations. In the case of a target, this loss of situational awareness is termed "target hypnosis" and can result in a controlled descent into the terrain. A novice pilot under stress may fail to note this descent while concentrating on the artificial horizon. The causes of such fixation include fatigue, anxiety, hypoxia, drugs, and personality (Gillingham & Previc, 1996). A well-known example concerns a pilot returning from a very long mission who landed perfectly, performed the postflight check, got into his car, and proceeded to back right into a clearly visible car parked directly behind his own.

Formation Flying

While flying wing in formation, the pilot is likely to lose sense of his own attitude and position relative to the ground and to concentrate solely on maintaining a fixed distance from the lead airplane. It is difficult to determine whether a given change in relative position or attitude results from movement of the lead plane or of one's own. When no ground cues are visible, the lead plane can fill a large part of the visual field and produce vection illusions, as discussed later. This can easily lead to a loss of orientation and a period of disoriented flying after breaking away from the formation.

Transition Between Instrument Flight Rules (IFR) and Visual Flight Rules (VFR)

A particularly dangerous period for the pilot occurs when making the transition from instrument flight to flying by external visual cues. There is not a specific illusion associated with the transition, but rather a period of uncertainty concerning orientation. A pilot who has been concentrating on the instruments in lining up for a landing may easily experience SD during the several seconds after looking up and trying to find the runway and the horizon through broken clouds. Just as disturbing is the loss of orientation when a pilot in a turn enters a cloud and must reorient on the instruments. The delay in distance accommodation, which becomes more severe with age, is another factor in this problem. A common practice to avoid this disorientation in two-pilot operations is to have one pilot look outside the cockpit while the flying pilot concentrates on the instruments.

Autokinesis

A fixed star or a single dim light on the ground, when viewed by a pilot against a dark and otherwise empty background, may appear to wander about erratically at speeds up to 20°/sec over angles as large as 15°. This phenomenon of autokinesis can be quite compelling, especially if the object is fixated for several seconds from a dark cockpit. It has stimulated pilots to attempt to rendezvous with a star. It is not dependent on eye movements, although the unconscious attempt to suppress ocular drift might be responsible. Interior lighting and frequent changes in eye fixation can reduce the likelihood of autokinesis.

Sloping Runway

The simple parallelogram of a runway as seen from an approaching aircraft does not imply any unique combination of position, altitude, or attitude. The runway may appear low in the windscreen, either because the airplane is pitched nose up or because it is higher than expected. Only the presence of a visual horizon for attitude or a cultural reference, such as buildings or trees of known height, can resolve the ambiguity. Shape constancy implies an invariant shape of the runway when seen from a certain approach angle, but that may be misleading. The opportunity for spatial disorientation is increased when the runway is not level, in which case the perspective is further thrown off from what the pilot anticipates. If the runway slopes up, the pilot is likely to feel too high and to duck under the desired glide slope; if the runway slopes down, the approach is likely to be too high. Because the visual angle of the runway is quite insensitive to altitude, simulation pilots have been shown to descend more rapidly and approach very low when coming into an airport on an upward-sloping terrain (Kraft, 1978).

Sloping Cloud Bank

The tops of clouds often are aligned along approximately the same altitude and can provide a horizon reference. Unfortunately, sloping cloud banks are also quite common. The pilot who assumes that the cloud bank is the horizon reference may end up in an unsuspected turn or with an excessive opposing rudder. Certain shapes on land can lead to a similar illusion. For example, an oblique line of lights along a shore can lead to an illusion of bank (Kraft, 1978).

Sun as Zenith

A related attitude disorientation stems from an illusion that the sun is directly overhead Earth. When attempting to fly level above clouds or with no clear horizon, there is sometimes a tendency to keep the sun overhead, even though that would be correct only at local apparent noon over the equator.

Black Hole

The ambient visual system, integrating information from a variety of peripheral sources, is critical to judgment of both the attitude and altitude of an airplane, and errors are likely if one is forced to rely on the central focal visual system alone. In the absence of sufficient ambient visual cues, the pilot lacks any spatial reference for localized lights and may experience the black hole illusion (Kraft & Elworth, 1969). The runway lights alone, or the lights of a city beyond the airfield, may easily be misinterpreted. The pilot may also perceive the runway lights as being in motion. When approaching a seacoast runway over dark water, the pilot is likely to perceive the rising ground beyond the airfield as level. Any attempt to maintain a constant glide slope angle relative to this rising slope would cause the actual approach to be low and shallow. Errors in estimation of altitude are also common when approaching a runway at night over a dark area and have been blamed for controlled flight into the ground short of the runway threshold. Similar problems can occur during landings on an aircraft carrier in limited visibility. The absence of ground cues when landing a helicopter over snow or sand leads to difficulty in height judgment, as was evident in the well-publcized helicopter landing incidents during the failed attempt to rescue the American hostages in Iran.

Distance and Height Judgment

Size Constancy. Judgment of the distance of an object involves numerous cues, including an estimate of the object's size. By assuming that an object is a known size, its visual angle corresponds to a certain distance from the observer. The common illusion that a jumbo aircraft is barely moving through the sky is explained by size constancy. (One assumes a normal airplane length, which places the plane closer than it really is, and in turn makes its angular motion through the field seem like a slower linear motion.) The very large Vehicle Assembly Building at NASA's Kennedy Space Center fools many observers in the air (and on the ground) by appearing closer than it is. A wider than normal runway will make one feel too low on an approach and may cause an overshoot. An approach over stunted trees, which are assumed to be of normal height, may convince the pilot that the plane is higher than it is and result in a low approach.

Other Distance Cues. Judging one's distance from an object is as difficult as estimating its size. Although the apparent size of an object is the most obvious distance cue, parallax, color, brightness, and clarity also influence depth perception. Artists and architects since the Renaissance have made good use of all these techniques to create depth perception. For the pilot, distance judgments are critical and can be impeded by any of these factors. For example, a mountain range or runway lights may appear farther away if they are seen dimly through rain or mist, and converging roads will produce a false distance cue through perspective judgment.

Attitude and Distance Confusion. It is all too easy for those not used to navigating and controlling in six dimensions to confuse a vehicle's rotation with a linear displacement. The position of an outside object relative to the windscreen of an airplane depends on the airplane position as well as its orientation. For example, the real land horizon, when viewed from high altitude, is depressed from the actual horizontal projection of the airplane and can cause the pilot to believe the plane is flying with a higher pitch angle. If the depressed horizon is viewed out a side window, a false sense of bank, with the opposite wing low, is likely to occur. During approach, the runway could appear toward the right side of the windscreen whether the pilot is displaced to the left of the extended runway line and flying parallel to it or on the correct approach path but pointed to the left. In the latter case, the airplane could be headed to the left of the runway either because it is crossing the approach line from right to left or because it is on the approach path and subject to a crosswind from left to right. By combining an estimate of the airplane's true velocity, derived from the divergence point in the moving scene, with observations of changes in the relative position of the runway, the pilot could determine whether he is displaced or headed erroneously. A particularly confusing situation arises when the runway appears directly ahead of the pilot, although the airplane is displaced left and headed to the right of the threshold. In time, as the picture evolves, the interpretation will force the pilot to deal with the influence of the crosswind.

Break-Off Phenomenon

Pilots have often reported a sensation of isolation from the ground when flying at very high altitudes. The break-off phenomenon includes this sensation as well as the psychological disturbance of dissociation from the surroundings and even from the airplane. Pilots flying level at altitudes above 30,000 ft are more likely to report it when the curved earth fails to provide a clear horizon or demarcation between clouds and sky. Some pilots describe pleasure in the sensation, wheras others have expressed feelings of anxiety. Attitude changes in the aircraft may be overestimated, and occasional false perceptions of attitude may occur. A return to normalcy usually follows some increase in pilot workload.

Vection

Visually induced self-motion, or vection, is a common phenomenon produced by the nearly uniform motion of a large part of the visual field (Dichgans & Brandt, 1978). When the entire field moves, subjects soon begin to feel that the relative motion is their own. The audience in a wide-field movie feels they are moving to the right as the camera pans slowly to the left (circular vection). Multiwindow flight simulators or virtual reality goggles can produce a very compelling sensation of self-motion in the absence of any inertial motion cues. In each case the vection illusion is enhanced and comes on more quickly when vestibular cues confirm the initiation of the self-motion.

In flight, linear vection and circular vection can produce compelling disorientation illusions. If the motion of clouds is seen out of only one side window, it may produce a strong sense that the airplane is turning toward the other side. Even the relative movement of the light from a rotating beacon on one's own aircraft can create the sense of turning in the opposite direction. Rain or snow, seen to stream rapidly past the windows, may produce a sudden sense of increased speed. The movement of the tanker and its boom, seen by the pilot of a small plane during aerial refueling, may produce the illusion of unanticipated self-motion. For low-flying helicopters attempting to hover, the self-motion generated from the effect of the rotor wash on waves in the water or motion of grass can be disorienting (Benson, 1999). The compelling sense of self-motion produced in flight simulators can lead to "simulator sickness" as discussed later, when it is in conflict with other sensory cues or with the motion anticipated by the pilot.

MOTION SICKNESS

For many fliers, passengers and pilots alike, the first awareness of SD or of the existence of a motion-sensing organ occurs with the onset of airsickness (Oman, 1990; Reason & Brand, 1969). A substantial part of the flying public will experience symptoms during a prolonged bumpy ride in a small airplane during turbulence. Individual variability in susceptibility is quite large. Although some people claim never to experience it, anyone with a functioning vestibular system can be made motion sick. The conditions that bring it on are all too familiar—turbulence, vigorous maneuvering, head down in the cockpit, concentration on some visual task such as reading, or head movements during turns. Vestibular overstimulation does not completely explain motion sickness. Symptoms may also be induced by a moving visual scene, as presented in a wide-field-of-view movie or with virtual reality goggles. The pattern of symptoms may vary among individuals, but usually includes headache and drowsiness, proceeds through cold sweating, pallor, and stomach awareness, and eventually results in vomiting, which generally provides temporary relief.

Although nearly any large motion can produce motion sickness, certain kinds of movement are particularly provocative. Vertical oscillations are worse than horizontal, especially for low frequencies near 0.2 Hz. Combinations of vertical motion and pitch are disturbing, which explains the preference of many passengers in planes and ships for places near the center of rotation.

At one time it was suspected that motion sickness is caused by inadequate vestibular responses or by some directional preponderance favoring one ear. Although normal balance and postural control are characteristics of successful pilots, there is no firm evidence of any correlation between vestibular sensitivity and motion sickness susceptibility. On the other hand, patients who are absolutely lacking in vestibular function (labyrinthine defectives) are known to be immune to motion sickness, even to motion sickness stimulated by moving visual scenes.

Causes of Motion Sickness

Although neither the physiological mechanisms nor the evolutionary development of motion sickness are well understood, an adequate functional description of the causes has emerged. Neither movement of the body's fluids nor overstimulation of the vestibular apparatus accounts for the problem. Like other forms of motion sickness, airsickness is produced by motions that entail some conflict between sensory signals or between what is felt and what is anticipated. It may be caused by a conflict within the vestibular system or between the vestibular and visual systems. For example, a roll head movement during a steady yaw rate, producing the cross-coupled angular acceleration described earlier, implies an intravestibular mismatch between the semicircular canals signaling pitch and the otolith organs signaling only roll. (This is the most common demonstration of airsickness and disorientation and is easily performed in a spinning Barany chair or in a flight simulator.) Even pure visual stimuli, especially when they are wide-field and invoke the ambient visual system, can provoke motion sickness—as will be discussed under "Simulator Sickness." The conflict between the vestibular sense of orientation and the visual reference is apparent in the motion sickness that is produced by attempting to read or operate navigational or other equipment requiring fixation inside the cockpit during turbulence. The vestibulo-ocular reflex developed in response to the airplane acceleration tends to stabilize the eyes in space, whereas the reading of navigation material requires the eyes to be stabilized relative to the cockpit, producing a visual-vestibular conflict. (If one looks out the window, at the horizon, or at a distant object on the ground, the conflict and the motion sickness are likely to abate.) Finally, the expectation of the aircraft motion plays a major role in the development of conflict with the sensory feedback. It is rare for the pilot of an airplane to get sick, even during acrobatic maneuvers, although that same person might very well experience discomfort if someone else were flying. The sensorimotor conflict theory includes the notion of an internal model in the brain that processes environmental variables and voluntary motor commands to produce an expectation of what the sensors will signal. Habituation consists of updating this internal model so that it no longer produces signals in conflict with measurements.

A large number of secondary factors have been associated with motion sickness susceptibility, including personality factors and anxiety, but the linkage is speculative. Age and gender both play a role. Women were considered more susceptible, but recent studies have cast doubt on this difference. Children generally experience increased susceptibility until puberty and then become gradually more resistant—but large individual differences exist for all of these factors. It is currently impractical to use any simple airsickness susceptibility test to select pilot candidates.

Seeing a boat or plane, smelling the sea or fuel—let alone the odor of vomitus—may trigger a conditioned response generating motion sickness symptoms. Early exposure of pilot candidates to extreme motion without drug protection is therefore not advisable. The aftereffects of alcohol ingestion increase the likelihood of motion sickness, both through the physical imbalance created in the semicircular canals and because of probable interference with the process of habituation.

Degraded Performance

Pilots suffering from airsickness are also likely to fly less well. They will be reluctant to make brisk head movements to help locate traffic, because doing so might also exacerbate the sickness. The loss of visual suppression of vestibular nystagmus, common to vigorous maneuvers, makes the task of reading instruments more difficult. Finally, even without overt symptoms of airsickness, a subtle but dangerous condition, the Sopite syndrome can produce lethargy and reluctance to double-check difficult tasks such as navigation or fuel calculations (Graybiel & Knepton, 1976).

Habituation

Normally the airsickness problem disappears with repeated exposure, provided that the aviator is willing to undergo the discomfort. The use of anti–motion sickness drugs during the adaptation stage probably does not inhibit the development of habituation. For certain cases, special rehabilitation programs have proven quite successful in returning a pilot to duty. In addition to counseling and encouragement, it is often helpful to determine the possible existence of a psychological component to the problem. Techniques using biofeedback can sometimes be useful in training pilots to detect motion sickness symptoms early and control the stimulus to prevent the cascade of problems (Cowings, Naifeh, & Toscano, 1990).

Pilots who fly regularly, especially those who practice aggressive maneuvers or parabolic flight, usually become immune to airsickness when they are active. However, following a layoff of 2 weeks or so, the typical pilot must undergo a reconditioning process to recover this immunity. The vestibulo-ocular reflexes, which had been markedly reduced while actively flying, gradually return to normal during the layoff.

Simulator Sickness

Simulator sickness occurs most commonly with the use of wide-field-of-view visual systems, with or without true motion. The visually induced motion (vection) created by an out-the-window scene or presented to the pilot through virtual reality goggles can produce a sensory conflict unless the vestibular cues from a simulator motion base are synchronized and consistent with the visual scene. In some cases the transfer of training may be negative. The pilot then becomes accustomed to reacting to certain visual cues in the absence of motion and, as a result, may react inappropriately in the air. On occasion flashbacks of simulation experiences may occur to the pilot hours after the session. To avoid such mishaps, it has been the practice in some installations to prohibit pilots from flying the real aircraft for at least a day after a simulator session.

The root of simulator sickness appears to be the absence of appropriate motion cues in the presence of strong visual cues. Rapid scene rotations need to be accompanied by some platform motion to stimulate the vestibular sensors in order to avoid the transient conflict that produces motion sickness. A timing mismatch of greater than 100 msec can easily create such a mismatch. The problem of simulator sickness appears to be most serious in helicopter operations, where motion cues play an important role in the inner loop attitude control task. It is generally the experienced pilots, rather than the novices or the simulator engineers, who suffer most from simulator sickness, presumably because of their enhanced anticipation of the motion normally associated with maneuvers and complex visual motions.

Drug Treatment

The use of drugs to treat motion sickness has a long and not particularly successful history. Because of their side effects they are normally prohibited for solo pilots. Many of the common drugs are antihistamines, which act as central nervous system depressants and possibly reduce direct linkages to the vomiting center. Drowsiness is invariably a side effect that makes their use inadvisable for pilots. A combination of scopolamine and dextroamphetamine is popular and avoids some of the drowsiness, but it entails risks of its own from the "speed." The other popular alternative to over-the-counter anti–motion sickness remedies is promethazine, with or without the addition of ephedrine. Alternative means of applying scopolamine, via a transdermal patch, has gained some popularity and typically involves fewer cognitive or visual side effects. Rapid treatment is available with a nasal spray or intravenous injection of scopolamine, with the use of promethazine suppositories, or with a direct intramuscular injection of promethazine, as commonly applied in space flight. Direct drug effects on the vestibular end organ have not been successful in treating motion sickness to date.

SPACE FLIGHT[1]

In anticipation of newer challenges to spatial orientation beyond that of the pilot in a cockpit, a section on space flight is included in this chapter. An entirely new class of spatial disorientation problems confronts the astronaut. The human factors implications of the SD incidents and of space motion sickness are profound and are reviewed in the order they might occur, from launch to landing.

[1]The Space Flight section of this chapter is based on work previously published by the author (Young, 1999).

The fundamental issue in all the vestibular reactions is the altered gravito-inertial stimulus to the otolith organs, which causes disorientation and motion sickness. No longer do the otolith organs function primarily as indicators of the direction of the vertical. Instead, they react only to linear acceleration, forcing a major reorganization of their interpretation by the central nervous system (Bloomberg et al., 1998). If artificial gravity is employed as a countermeasure against deconditioning for a long space mission, the unusual vestibular stimuli associated with a rotating environment cause further human factor problems.

Launch

With the astronauts launched in their seats, leaning backward, the thrust is nearly entirely in the g_x direction. For the Space Shuttle, it reaches a peak of 3 Gs and lasts about 8 minutes. The strong internal visual field is sufficient to inhibit the elevator illusion and oculogravic illusion, which might otherwise occur. Additional high amplitude vibration associated with the launch can produce vestibulo-ocular responses that may interfere with the ability to read instruments clearly under certain conditions and calls for special attention to legibility and brightness of critical displays.

Early-On Orbit

Space Motion Sickness

Approximately 75% of first-time fliers experience space motion sickness (SMS) symptoms much like seasickness, which normally appear in the first hours on orbit and may last up to 3 or more days if untreated (Glasauer & Mittelstaedt, 1998; Oman, 1998). Vomiting is not uncommon. Head movements, particularly nodding or tilting of the head, usually bring on episodes of SMS. Unusual visual patterns may also elicit SMS. These include viewing another crew member inverted, seeing the earth at the top of a window, or entering a new part of the spacecraft in an unfamiliar orientation. Some of the more common incidents occur when emerging from the relatively uniform tunnel between two parts of a spacecraft or when getting out of a bunk in the "inverted" orientation. The absence of a distinct vertical, as usually indicated by ceiling or floor, seems to contribute to the problem. All the space motion sickness episodes are encompassed within the "sensorimotor conflict" theory of motion sickness, discussed previously.

Repeat fliers are less likely to suffer from SMS, and test pilots are less susceptible than science astronauts, although the differences are not great, and attempts to predict individual SMS susceptibility have not generally been successful. Reduction of SMS occurrence by preflight adaptation has not yet been demonstrated. Current treatment of SMS generally depends on the intramuscular injection of promethazine when symptoms occur and gives rapid relief in about 90%

of cases. However, the side effects, including drowsiness, are of concern and remain under investigation. The human factors implications of SMS include the reduced ability to perform effective work during the first day or two on orbit. Space missions generally plan for a reduced workload during the first day or two on orbit and avoid planned extravehicular activity (EVA) during the first days.

Orientation Illusions

The experience of SD on transitioning to weightlessness is the rule, rather than the exception. The 0-G inversion illusion was first discovered in parabolic flight, where most subjects, when deprived of visual cues, felt a backward tumbling to an inverted position on entering weightlessness (Kornilova, Mueller, & Chernobyl'ski, 1995). The vestibular phenomenon presumably is initiated by the sudden unloading of the otolith organs. Removal of the 1-G compressive load on the utricle displaces the otoconial membranes as though subjects were tilted back to an inverted orientation. The footward load is also removed from the sacculus, consistent with standing inverted, although the role of the sacculus in spatial orientation has again been called into question. Unusual visual scenes may maintain the inversion illusion. Some subjects report a disturbing sense of being upside down, even without any vertical reference, for several days. For most subjects, however, the inversion illusion, with eyes open or closed, lasts only a few minutes or hours. Astronauts in darkness or with their eyes closed are also likely to lose their sense of orientation relative to the spacecraft. Their dependence on rotation cues from the semicircular canals fares much worse than for rotation about the Earth's vertical axis on the ground (Glasauer & Mittelstaedt, 1998), even though otolith or other graviceptors are equally useless in both cases.

Reliance on Visual, Tactile, and Internal Cues

After exposure to weightlessness of more than a day or so, the brain begins to seek and use alternative sources of information regarding a local vertical reference. The absence of meaningful body position signals from the otoliths and other inertial graviceptors leads to the substitution of other sensory information. Visual indications of the spacecraft walls, ceiling, and floor are adopted as a reference, and many astronauts report that after several days in space they adopt the local surface beneath their feet as representing the "down" direction, be it floor, ceiling, or wall. Circular vection and linear vection are similarly enhanced (Kornilova et al., 1995; Lackner, 1993; Young, Mendoza, Groleau, & Wojcik, 1996). Over the period of a week or so, these cues are diminished in importance. Similarly, early on the presence of localized tactile cues contributes to the sense that the place where the feet are in contact with a surface represents a stable vertical. Squeezing into a corner and pressing one's feet and trunk against a firm, stable supporting surface can sometimes ameliorate SMS. The use of stretched elastic cords, of the type sometimes used in a treadmill exerciser, can apparently stabilize the astronaut's local world and reduce SMS. After a week or more in weightlessness, however, even these tactile cues become less of a factor in the sense of spatial orientation. Finally, most astronauts become dependent primarily on their own internal "ideotropic vertical" and refer all orientation cues to the alignment of their trunk.

The appropriate placement of foot restraints for all internal workspaces can aid in the suppression of SMS as well as helping in the ergonomic design of astronaut tasks.

Extravehicular Activity

Disorientation

Unlike the familiar scene inside the spacecraft, the initial view of the sky and the distant earth when emerging for EVA may be surprising and produce momentary anxiety. Unless one is concentrating on a nearby large fixed object, the absence of a close, unambiguous visual reference may lead to disorientation and, according to some reports, a sense of falling. Standing on a convex surface, without the encompassing sight of the surrounding structure, can also be disorienting at first. Generally, EVA astronauts who have found such phenomena on their first excursions quickly learn to locate a local visual reference, such as the remote manipulator arm, and use it as a basis for orientation. As the spacecraft passes from light to dark through the earth's shadow during each orbit, the visual scene changes dramatically and thus compels reorientation. Needless to say, useful work is severely hampered during such disorientation episodes.

Artificial Gravity

One of the suggested means of dealing with the debilitating effects of long-duration weightlessness, as might be encountered en route to Mars, is to provide artificial gravity. Whether the rotating device is to be a large torus, a tethered dumbbell, or a small centrifuge, the crew must adapt to the Coriolis forces and gravity gradient. The interior walls will have to be distinctive and unambiguous in order for the crew to avoid SD. Unexpected Coriolis forces will be a constant issue during movement, especially if the spacecraft rotation rate exceeds 6 rpm. A gradual period of adaptation over many days will be advisable both when the spacecraft begins to turn and again when it decelerates (Young, 1999).

A much smaller centrifuge, several meters in radius and rotating at higher speeds, can be used for intermittent astronaut conditioning. This "spinning gym" would need to provide for some means of acquiring and retaining context-specific dual adaptation in order to avoid problems of space sickness and SD each time one transitions between it and the 0-G spacecraft.

Reentry

As the reentry thrusters fire and steady acceleration builds up after a week or more in orbit, the astronaut is likely to feel much more acceleration than really exists. Because even 0.5 Gs feel like 1 or 2 Gs after extended weightlessness, a slight pitch or roll of the head feels like a much larger angle. Presumably, the semicircular canals continue to function almost normally in indicating head angular velocity (neglecting any Coriolis forces in a turn). Thus, a conflict is assured between the otolith and canal signals concerning head angle. Many astronauts have reported the vague disorientation illusion that "my gyros have tumbled" during head movements while reentering, but scientific study of the phenomena is still lacking. Only rarely do returning astronauts report any oscillopsia, or sense that the world shifts around during head movements. Shuttle pilots are usually cautioned to avoid large and sudden head movements during reentry, and several have reportedly experienced disorientation of some concern prior to landing.

Postflight Disturbances

Postflight Postural Control

Following space flight the ability of astronauts to maintain the usual stable control of eye, head, and trunk is compromised (Reschke et al., 1998). Astronaut postflight locomotion is often disturbed, sometimes bordering on clinical ataxia. Following long-duration flights the integration of head, eye, trunk, and lower limb motion is also impaired during locomotion (Bloomberg et al., 1998).

Other astronauts appear to have no problems at all in moving around after return. Typically, however, astronauts feel somewhat lightheaded and dizzy when first getting out of the seat . In addition to fluid pooling in the legs there is probably a vestibular-autonomic aspect that inhibits the normal cardiovascular compensation for changes in body posture in Earth's gravity. There may be some effects of muscle loss and reprogramming as well . According to the Otolith-Tilt-Translation-Reinterpretation (OTTR) hypothesis (Parker, Reschke, Arrott, Homick, & Lichtenberg, 1985; Young, Oman, Watt, Money, & Lichtenberg, 1984), the nervous system adapts to weightlessness by reinterpretation of the meaning of low-frequency otolith signals and requires time to re-adapt to the conditions on earth. When a subject attempts to stand quietly or to react to a disturbance after return from weightlessness, the utility of the vestibular system is severely impaired (Black, Paloski, Doxey-Gasway, & Reschke, 1995). A typical subject takes about three days to return to normal postural responses following a flight of about two weeks.

Earth Sickness

A small number of astronauts feel acute motion sickness discomfort following return to Earth. The clearest indication that "Earth sickness" is in part vestibular comes from the observation of astronauts who show no symptoms until they are placed in a completely dark room and forced to rely entirely on vestibular orientation cues. Vomiting may follow quickly.

Any emergency egress must be made as simple and easy as possible, given the problems of disorientation and posture control that will affect at least some of the crew. Adequate lighting and assistance are needed to help the crew in such conditions.

MAJOR FACTORS CONTRIBUTING TO SD

Although SD can strike any pilot at any time, certain conditions are known to precipitate it. Awareness of these conditions may help to alleviate or prevent the problem.

Night Flying

Night flying, with its absence of clear references, is especially conducive to SD. Factors include the lack of a clear horizon and the absence of familiar objects of known size to help judge height and speed. Erroneous false horizons, such as a row of lights on a shoreline or a sloping cloud bank, are insidious at night. Furthermore, the possibility of autokinesis is increased when looking at a single star or a point of light on the ground. Formation takeoffs or rejoins at night are particularly likely to produce SD.

Shifting Reference Frames and Instrument Flying

Instrument flight is conducive to SD. In particular, instruments that are difficult to read, especially at night or during maneuvers that might produce nystagmus, are partly to blame. Flying conditions most likely to produce SD are those that require the pilot to shift the frame of reference, either from instruments to outside cues, or between outside references. Formation flying in poor weather is a prime example, where the pilot has little time to check his or her own instruments and must refer entirely to the lead aircraft. Any procedure that requires the pilot to look away from the flight instruments for several seconds is likely to induce disorientation, particularly if rapid head movements are involved. Of course, the pilot lacking instrument flying skills, or recent practice, is quite likely to become disoriented on entering clouds, where a rapid shift to instrument reference is required.

Flight Maneuvers and Head Movements

The angular motions most likely to cause difficulty are prolonged constant rate turns, especially if they are initiated at subthreshold accelerations, and the rapid recovery from them, which is falsely detected as a turn in the opposite direction.

Head movements during turns are particularly disorienting. Sustained linear accelerations are likely to produce false pitch or roll sensations, especially if combined with head movements. Of course the performance of aerobatic maneuvers and recovery from them is highly likely to contribute to SD, particularly for a passenger.

Workload and Capacity

Anything that hampers the pilot's ability to concentrate on monitoring flight parameters and flying the plane can contribute to SD. Additional communication or navigational tasks, concern over icing or fuel, or other such demands may impose high workload. Impaired capacity may come about through inadequate training or lack of recent practice, the effects of alcohol or other drugs, fatigue, circadian disrhythmia (jet lag), or emotional stress. Any of these may interfere with the pilot's training to concentrate foremost on monitoring and flying the airplane.

METHODS TO MINIMIZE SD

Training to Deal With Spatial Disorientation

Considerably more can be done to prepare pilots for dealing with SD than merely instructing them to disregard their sensations and attend to the instruments. The fallibility of vestibular cues and the preponderance of situations that can prompt SD are important reasons for familiarizing the pilot with the potential for SD, which can be demonstrated in simple simulators or in the air.

Training on the causes of disorientation, with emphasis on the conditions that are likely to induce it, can be very helpful. Many illusions can be shown in SD demonstrators, in which the subject is passively exposed to motion or visual conditions. The demonstrations should be combined with emphasis on avoiding those conditions and on constant monitoring of aircraft state, either through instrument scan or out-the-window reference.

Because many of the SD situations are predictable, the pilot can be exposed to them in the airplane or in closed-loop disorientation simulators, so that the illusions become familiar and the appropriate recovery becomes automatic. This training is helpful when it is extended to requiring the student pilot to recover level flight when disoriented. Rather than espousing the old adage "believe your instruments," Gillingham and Previc (1996) advise telling the pilot to "make the instruments read right, regardless of your sensation."

Training to deal with SD once it has occurred is valuable and can be incorporated into pilot exercises at various levels. Both Benson (1999) and Gillingham and Previc offer practical advice to crew members for dealing with SD. For single-airplane situations, the pilot is advised to concentrate on the instruments, avoid looking back and forth between the cockpit and the external world, and return to straight and level flight. Head movements are usually to be avoided, although a quick head shake may serve to overcome minor SD if the airplane is not in a turn. For formation flights, specific procedures for separation or for transferring the flight lead position are appropriate.

Cockpit and Instrument Design to Minimize SD

Some benefit can be derived from cockpit layouts to minimize the kinds of situations that are associated with SD. Location of nonflight displays and controls, such as communication equipment, should minimize the need for head movements and reduce the time required to look away from the outside view or the primary flight instruments. Integration of the instruments into a "basic T" arrangement clustered in front of the pilot is important (Brathwaite et al., 1998). Beyond that, the use of integrated flight displays, providing all the geometric variables at one glance, is highly desirable. The individual instruments should be easy to read, even at night and during nystagmus episodes. To move the judgment of attitude from the focal to the ambient visual system, a large and unambiguous horizon is desirable. To date, extended horizons and peripheral vision displays (Malcolm, Money, & Anderson, 1975) have not proven acceptable. The presentation of essential flight parameters, including the artificial horizon, touchdown point, altitude, and airspeed, has been included in the head-up-display (HUD), allowing the pilot to look out the window while still attending to the flight instruments. For both military and commercial aviation, the HUD has been a major advance in avoiding many SD situations associated with shifting frames of reference and unnecessary head movements. Further advances in the development of a virtual reality visor display hold considerable promise for helping pilots avoid many of the problems of head movement and shifting references associated with spatial disorientation. Additional efforts toward inclusion of 3-D audio displays (Nelson et al., 1998) and vibrotactile displays (Raj, Kass, & Perry, 2000) are devoted to combatting the serious consequnces of spatial disorientation.

REFERENCES

Aschan, G. (1954). Response to rotatory stimuli in fighter pilots. *Acta Otolaryngologica (Supplement), 116,* 24–31.

Aschan, G. (1967). Habituation to repeated rotatory stimuli (cupulometry) and the effect of antinausea drugs and alcohol on the results. *Acta Otolaryngologica, 64,* 95–106.

Aschan, G., Bergstedt, M., Goldberg, L., & Laurell, L. (1956). Positional nystagmus in man during and after alcohol intoxication. *Quarterly Journal of Studies in Alcohol, 17,* 381.

Barley, S. (1970) *The search for air safety.* New York: William Morrow.

Benson, A. J. (1965). Spatial disorientation in flight. In J. A. Gilles (Ed.), *A textbook of aviation physiology* (pp. 1086–1129). New York: Pergamon Press.

Benson, A. J. (1999). Spatial disorientation—general aspects, Spatial disorientation—common illusions, & Motion sickness. In J. Ernsting, A. Nicholson, & D. Rainford (Eds.), *Aviation Medicine* (3rd ed., pp. 419–471). Oxford, England: Butterworth Heinemann.

Berthoz, A., Young, L. R., & Oliveras, F. (1977). Action of alcohol on vestibular compensation and habituation in the cat. *Acta Otolaryngologica, 84,* 317–327.

Black, F. O., Paloski, W. H., Doxey-Gasway, D. D., & Reschke, M. F. (1995). Vestibular plasticity following orbital space flight: Recovery from post-flight postural instability. *Acta Otolaryngologica (Supplement), 520,* 450–454.

Bloomberg J. J., Mulavara, P. V., McDonald, C. S., Layne, C. S., Merkle, L. A., Sekula, B., Cohen, H. S., & Kozlovskaya, I. B. (1998). The effects of long-duration spaceflight on sensorimotor integration during locomotion. *Society for Neuroscience,* Abstract No. 8299.

Brathwaite, M. G., Durnford, S. J., Groh, S. L., Jones, H. D., Higdon, A. A., Estrada, A., & Alvarez, E. A. (1998). Flight simulator evaluation of a novel flight instrument display to minimize the risks of spatial disorientation. *Aviation, Space, and Environmental Medicine, 69,* 733–742.

Bronstein, A. M. (1999). The interaction of otolith and proprioceptive information in the perception of verticality. In B. J. M. Hess & B. Cohen (Eds.), *Otolith function in spatial orientation and movement* (Annals of the New York Academy of Sciences, Vol. 871, pp. 324–333). New York: New York Academy of Sciences.

Burgess, P. R. (1976). General properties of mechanoreceptors that signal the position of the integument, teeth, tactile hairs and joints. In A. Iggo & O. B. Ilyinsky (Eds.), *Somatosensory and visceral receptor mechanisms (Progress in Brain Research,* Vol. 43, pp. 205–214). Amsterdam: Elsevier.

Carroll, L. A. (1992). Desperately seeking SA. *TAC Attack (TAC SP 127–1), 32*(3), 5–6.

Cheung, B., Money K., Wright, H., & Bateman W. (1995). Spatial disorientation-implicated accidents in Canadian Forces 1982–92. *Aviation, Space, and Environmental Medicine, 66,* 579–585.

Clark, B., & Graybiel, A. (1955). Disorientation: A cause of pilot error. In *Bureau of medicine and surgery research report* (No. NM 001 110 100.39). Pensacola, FL: U.S. Naval School of Aviation.

Clark, B., Graybiel, A., & MacCorquodale, K. (1948). The illusory perception of movement caused by angular acceleration and by centrifugal forces during flight, II: Visually perceived motion and displacement of a fixed target during turns. *Journal of Experimental Psychology, 39,* 298–309.

Cohen, M. M., Crosby, R. H., & Blackburn, L. H. (1973). Disorienting effects of aircraft catapult launchings. *Aerospace Medicine, 44,* 37–39.

Collins, W. E., Schroeder, D. J., & Hill, R. J. (1973). Some effects of alcohol on vestibular responses. *Advances in Otorhinolaryngology, 19,* 295 +.

Cornsweet, T. N. (1970) *Visual Perception.* New York: Academic Press.

Cowings, P. S., Naifeh, K. H., & Toscano, W. B. (1990). The stability of individual patterns of autonomic responses to motion sickness stimulation. *Aviation, Space, and Environmental Medicine, 61*(5), 399–405.

Crane, B. T., & Demer, J. L. (1998). Human horizontal vestibulo-ocular reflex initiation: Effects of acceleration, target distance, and uniliateral deafferentation, *Journal of Neurophysiology, 80,* 1151–1166.

Dai, M., Raphan, T., & Cohen, B. (1991). Spatial orientation of the vestibular system: Dependence on optokinetic after-nystagmus on gravity. *Journal of Neurophysiology, 66,* 1422–1439.

Dichgans, J., & Brandt, T. (1978). Visual-vestibular interaction: Effects of self motion perception and postural control. In R. Held, H. W. Leibowitz, & H. L. Teuber (Eds.), *Handbook of sensory physiology, VIII* (pp. 755–804). New York: Springer-Verlag.

Ercoline, W. R., Devilbiss, C. A., Yauch, D. W., & Brown, D. L. (2000). Post-roll effects on attitude perception: The Gillingham Illusion. *Aviation, Space, and Environmental Medicine, 71,* 489–495.

Gillingham, K. K., & Previc, F. H. (1996). Spatial orientation in flight. In R. DeHart (Ed.), *Fundamentals of aerospace medicine* (2nd ed., pp. 309–397). Baltimore: Williams & Wilkins.

Glasauer, S., & Mittelstaedt, H. (1998). Perception of spatial orientation in microgravity. *Brain Research Reviews, 28,* 185–193.

Goldberg, J. M., Desmadryl, G., Baird, R. A., & Fernandez, C. (1990). The vestibular nerve of the chinchilla, IV: Discharge properties of utricular afferents. *Journal of Neurophysiology, 63,* 781–790.

Goldberg, J. M., & Fernandez, C. (1984). The vestibular system. In I. D. Smith (Ed.), *Handbook of physiology: The nervous system, III* (pp. 1023–1066). Bethesda, MD: American Physiological Society.

Graybiel, A., & Knepton, J. (1976). Sopite syndrome: A sometimes sole manifestation of motion sickness. *Aviation, Space, and Environmental Medicine, 47,* 873–882.

Guedry, F. E., Jr. (1974). Psychophysics of vestibular sensation. In H. H. Kornhuber (Ed.), *Handbook of sensory physiology, vestibular system, VI (2): Psychophysics, applied aspects and general interpretations* (pp. 3–154). New York: Springer-Verlag.

Guedry, F. E., Jr., Stockwell, C. W., & Gilson, R. D. (1971). Comparison of subjective responses to semicircular canal stiulation produced by rotation about different axes. *Acta Otolaryngologica, 72,* 101–106.

Henn, V., Cohen, B., & Young, L. R. (1980). *Neurosciences research program bulletin: Visual-vestibular interaction in motion perception and the generation of nystagumus, 18*(4). Boston: MIT Press.

Henn, V., & Young, L. R. (1975). Ernst Mach on the vestibular system 100 years ago. *Annals of the Oto-Rhino-Laryngol, 37,* 138–148.

Henn, V., Young, L. R., & Scherberger, H. (2001). *Fundamentals of the theory of movement perception.* New York: Kluwer Academic/Plenum Publishers. (Forthcoming translation of Mach, E. [1875]. *Grundlinien der lehre von den bewegungsempfindungen.* Leipzig, Germany: Wilhelm Engelmann.)

Hess, B. J. M., & Angelaki, D. E. (1999). Inertial processing of vestibulo-ocular signals. In B. J. M. Hess & B. Cohen (Eds.), *Otolith function in spatial orientation and movement* (Annals of the New York Academy of Sciences, Vol. 871, pp. 148–161). New York: New York Academy of Sciences.

Hixson, W., Niven, J., & Correia, M. (1966). *Kinematic nomenclature for physiological accelerations with special reference to vestibular applications* (Monograph 14). Pensacola, FL: U.S. Naval Aeromedical Institute.

Holt, J. R., Vollrath, M. A., & Eatock, R. A. (1999). Stimulus processing by type II hair cells in the mouse utricle. In B. J. M. Hess & B. Cohen (Eds.), *Otolith function in spatial orientation and movement* (Annals of the New York Academy of Sciences, Vol. 871, pp. 15–26). New York: New York Academy of Sciences.

Horn, G., & Hill, R. M. (1969). Modification of receptive fields of cells in the visual cortex occurring spontaneously and associated with body tilt. *Nature (London), 221,*186–188.

Howard, I. P., & Templeton, W. B. (1966). *Human spatial orientation.* New York: Wiley.

Knapp, C. J., & Johnson, R. (1996). F-16 Class A mishaps in the U.S. Air Force, 1975–93. *Aviation, Space, and Environmental Medicine, 67,* 777–783.

Kornilova, L. N., Mueller, C., & Chernobyl'ski, L. M. (1995). Phenomenology of spatial orientation reactions under conditions of weightlessness. *Fiziolgiya Cheloveka, 21,* 50–62. Translated in *Human Physiology, 21,* 344–351.

Kraft, C. L. (1978). A psychosocial contribution to air safety: Simulator studies of visual illusions in night visual approaches. In H. Pick, H. W. Kuebowitz, J. R. Singer, A. Steinschneider, & H. W. Stevenson (Eds.), *Psychology from research to practice* (pp. 363–385). New York: Plenum Press.

Kraft, C. L. & Elworth, C. L. (1969) *Flight deck work load and night visual approach performance* (AGARD CP No. 56). Neuilly-sur-Seine, France: Advisory Group for Aerospace Research and Development of the North Atlantic Treaty Organization.

Lackner, J. R. (1978). Some mechanisms underlying sensory and postural stability in man. In R. Held, H. W. Leibowitz, & H. L. Teuber (Eds.), *Handbook of sensory physiology, VIII* (pp. 805–845). New York: Springer-Verlag.

Lackner, J. R. (1993). Orientation and movement in unusual force environments. *Psychological Science, 4,*134–142.

Mach, E. (1875). *Grundlinien der lehre von den bewegungsempfindungen.* Leipzig, Germany: Wilhelm Engelmann.

Malcolm, R., & Melvill-Jones, G. (1974). Erroneous perception of vertical motion by humans seated in the upright position. *Acta Otolaryngologica, 77,* 274–283.

Malcolm, R., Money, K. E., & Anderson, P. J. (1975). Peripheral vision artificial horizon display. In *AGARD Conference Proceedings, 145,* B20–1–B20–3.

Mann, C. W., & Cannella, C. J. (1956). *An examination of the technique of cupulometry* (NSAM-501). Pensacola, FL: U.S. Naval School of Aviation Medicine.

Mayne, R. (1974). A systems concept of the vestibular organs. In H. H. Kornhuber (Ed.), *Handbook of sensory physiology, vestibular system, VI (2): Psychophysics, applied aspects and general interpretations* (pp. 493–580). New York: Springer-Verlag.

McCue, M. P., & Guinan, J. J. (1997). Sound evoked activity in primary afferent neurons of a mammalian vestibular system. *American Journal of Otololaryngology, 18,* 355–360.

Meiry, J. L. (1965). *The vestibular system and human dynamic space orientation.* Doctoral dissertation, Massachusetts Institute of Technology.

Melvill-Jones, G., Barry, W., & Kowalsky, N. (1964). Dynamics of the semicircular canals compared in yaw, pitch and roll. *Aerospace Medicine, 35,* 984–989.

Melvill-Jones, G., & Young, L. R. (1978). Subjective detection of vertical acceleration: A velocity dependent response? *Acta Otolaryngologica, 85,* 45–53.

Merfeld, D. M., & Young, L. R. (1995). The vestibulo-ocular reflex of the squirrel monkey during eccentric rotation and roll tilt. *Experimental Brain Research, 106,* 111–122.

Merfeld, D. M., Young, L. R., Paige, G. D., & Tomko, D. L. (1993). Three-dimensional eye movements of squirrel monkeys following post-rotatory tilt. *Journal of Vestibular Research, 3,* 123–139.

Mittelstaedt, H. (1996). Somatic graviception. *Biological Psychology, 42,* 53–57.

Mittelstaedt, H. (1999). The role of the otoliths in perception of the vertical and in path integration. In B. J. M. Hess & B. Cohen (Eds.), *Otolith function in spatial orientation and movement* (Annals of the New York Academy of Sciences, Vol. 871, pp. 334–344). New York: New York Academy of Sciences.

Money, K. E., & Myles, W. S. (1974). Heavy water nystagmus and effects of alcohol. *Nature, 247,* 404–405.

Musallam, W. S., & Tomlinson, R. D. (1999). Model for the translational vestibuloocular reflex (VOR). *Journal of Neurophysiology, 82,* 1010–2014.

Nelson, W. T., Bolia, R. S., McKinley, R. L., Chelette, T. L., Tripp, L. D., & Esken, R. (1998). Localization of virtual auditory cues in high + GZ environment. In *Proceedings of the Human Factors and Ergonomics Society 42nd Annual Meeting* (pp. 97–101). Santa Monica, CA: Human Factors and Ergonomics Society.

Oman, C. M. (1990). Motion sickness: A synthesis and evaluation of the conflict theory. *Canadian Journal of Physiology and Pharmacology, 68,* 294–303.

Oman, C. M. (1998). Sensory conflict theory and space sickness: Our changing perspective. *Journal of Vestibular Research: Special Issue on Vestibular Autonomic Regulation, 8,* 51–56.

Oman, C. M., & Balkwill, M. D. (1993). Horizontal angular VOR nystagmus dumping, and sensation duration in Spacelab SLS-1 crew members. *Journal of Vestibular Research, 3,* 315–330.

Paige, G. D., & Tomko, D. L. (1991). Eye movement responses to linear head motion in the squirrel monkey: Part I, basic characteristics. *Journal of Neurophysiology, 65,* 1170–1182.

Parker, D. E., Reschke, M. F., Arrott, A. P., Homick, J. L., & Lichtenberg, B. K. (1985). Otolith tilt translation reinterpretation following prolonged weightlessness: Implications for preflight training. *Aviation, Space, and Environmental Medicine, 56,* 601–607.

Peters, R. A. (1969). *Dynamics of the vestibular system and their relation to motion perception, spatial disorientation and illusions* (NASA CR-1309). Hawthorne, CA: NASA Ames Research Center.

Raj, A. K., Kass, S. J., & Perry, J. F. (2000). Vibrotactile displays for improving spatial awareness. In *Proceedings of the IEA 2000/HFES 2000 Congress* (Vol. 1, pp. 181–184). Santa Monica, CA: Human Factors and Ergonomics Society.

Raphan, T., & Cohen, B. (1985). Velocity storage and the ocular response to multidimensional vestibular stimuli. *Reviews of Oculomotor Research, 1,* 123–143.

Reason, J. T., & Brand, J. J. (1969). *Motion sickness.* London: Academic Press.

Reschke, M. F., Bloomberg, J. J., Harm, D. L., Paloski, W. H., Layne, C., & MacDonald, V. (1998). Posture, locomotion, spatial orientation, and motion sickness as a function of space flight. *Brain Research Reviews: Special Issue on Space and Neuroscience, 28 (1/2),* 102–117.

Rupert, A. H., McTrusty T. J., & Peak J. (2000). *Haptic interface enhancements for navy divers.* Houston, TX: NASA Johnson Space Center.

Schöene, H. (1964). On the role of gravity in human spatial orientation. *Aerospace Medicine, 35,* 764–772.

Simson, L. R., Jr. (1971). Investigation of fatal aircraft accidents: “Physiological incidents.” *Aerospace Medicine, 42,* 1002–1006.

Skoglund, S. (1973). Joint receptors and kinesthesis. In A. Iggo & O. B. Ilyinsky (Eds.), *Handbook of sensory physiology, II* (pp, 111–136). New York: Springer-Verlag.

Steer, R. W., Li, Y. T., Young, L. R., & Meiry, J. L. (1967). Physical properties of the labyrinthine fluids and quantification of the phenomenon of caloric stimulation. In *Proceedings of the Third Symposium on the Role of the Vestibular Organs in Space Exploration* (NASA SP-152, 409–420). Washington, DC: NASA.

Steinhausen, W. (1931). Über den Nachweis der Bewegung der Cupula in der intakten Bogengangsampulle des Labyrinthes bei der naturalichen rotatorischen und calorischen Reïzung. *Pfluergers Arch, 228,* 322.

Symposium on Recent Trends in Spatial Disorientation. (2000, November). Retrieved from http://www.spatiald.wpafb.af.mil.

Van Egmond, A. A., Groen A. J., & Jongkees, L. B. W. (1949). The mechanics of the semi-circular canal. *Journal of Physiology, 110,* 1–117.

Wilson, V., & Melvill-Jones, G. (1979). *Mammalian vestibular physiology.* New York: Plenum Press.

Young, L. R. (1974). Role of the vestibular system in posture and movement. *Medical Physiology, 1,* 704–721.

Young, L. R. (1984). Perception of the body in space: Mechanisms. In I. D. Smith (Ed.), *Handbook of physiology: The nervous system, III* (pp. 1023–1066). Bethesda, MD: American Physiological Society.

Young, L. R. (1999). Artificial gravity considerations for a Mars exploration mission. In B. J. M. Hess & B. Cohen (Eds.), *Otolith function in spatial orientation and movement* (Annals of the New York Academy of Sciences, Vol. 871, pp. 367–378). New York: New York Academy of Sciences.

Young, L. R., Dichgans, J., Murphy, R., & Brandt, T. (1973). Interaction of optokinetic and vestibular stimuli in motion perception. *Acta Otolaryngologica, 76,* 24–31.

Young, L. R., Mendoza, J. C., Groleau, N., & Wojcik, P. W. (1996). Tactile influences on astronaut visual spatial orientation: Human neurovestibular experiments on Spacelab Life Sciences 2. *Journal of Applied Physiology, 81,* 44–49.

Young, L. R., & Oman, C. M. (1969). Model for vestibular adaptation to horizontal rotation. *Aerospace Medicine, 40,* 1076–1080.

Young, L. R., Oman, C. M., Watt, D. G. D., Money, K. E., & Lichtenberg, B. K. (1984). Spatial orientation in weightlessness and readaptation to earth’s gravity. *Science, 225,* 205–208.

Zacharias, G. L., & Young, L. R. (1981). Influence of combined visual and vestibular cues on human perception and control of horizontal rotation. *Experimental Brain Research, 41,* 159–171.

Zotterman, Y. (Ed.). (1976). *Sensory functions of the skin in primates: Proceedings of the international symposium held at the Wenner-Gren Center, Stockholm.* New York: Pergamon Press.

4

Mental Workload and Situation Awareness: Essential Concepts for Aviation Psychology Practice

Michael A. Vidulich
Air Force Research Laboratory

Few issues in aviation psychology, or in the larger arena of engineering psychology, generate the controversy that is commonly associated with the topics of mental workload and situation awareness. Serious arguments have arisen not only about how either of these concepts can be defined or measured, but even about whether either concept is useful or just a waste of time. It is doubtful that this chapter could succeed in converting the severest skeptics; nevertheless, the organizing principle behind this chapter is that the concepts of mental workload and situation awareness are key concepts for aviation psychology. It will be argued that these concepts reflect the practical impact of the pilot's cognitive processes in the cockpit. Mental workload and situation awareness have already played a vital role in inspiring interface designs and system evaluation tools. It will further be argued that these concepts will continue to influence the application of new technologies and provide major improvements for pilots of the future.

DEFINING MENTAL WORKLOAD AND SITUATION AWARENESS

It may seem reasonable that scientific research must commence with a careful definition of the topic to be studied. Indeed, much effort has been expended in trying to generate definitions of mental workload and situation awareness. Such a multiplicity of definitions is consistent with Wiener's earlier observation that the definition of situation awareness is so fuzzy that panel sessions on situation awareness could be brought to a standstill by merely asking for a definition (Wiener, 1993). It is also possible that overzealous arguments about precise definitions may actually obscure the broad consensus among researchers and practitioners. For the purposes of this chapter, common "working definitions" of mental workload and situation awareness will be adopted as starting points. As discussed by Reason (1990), a working definition is a serviceable but not necessarily ideal definition. The goal of a working definition is merely to provide a basis for starting research and not to be an exclusive and exhaustive formal definition.

In this spirit, the definition for mental workload adopted here is the definition advanced by Hart and Wickens (1990, p. 258), "Workload is a general term used to describe the cost of accomplishing task requirements for the human element of man–machine systems." The cost may be manifested as subjective discomfort, behavioral response capabilities, or physiological reactions. This starts from the assumption that pilots, like all humans, possess limited attention to support information processing and that flying an aircraft requires information processing. Mental workload is assumed to be high as the amount of information processing approaches, or exceeds, the information-processing resources available. Defining mental workload in terms of information processing demanded versus what is available is a common approach (e.g., Gopher & Donchin, 1986; O'Donnell & Eggemeier, 1986; Tsang & Wilson, 1997). In this approach, mental workload is the mental analogy to the easily understood concept of physical workload. Just as it is physically more demanding to move a series of heavy book-filled boxes than boxes filled with down comforters, it is more mentally demanding to do mental multiplication and division of multidigit numbers than it is to balance a checkbook. In either case, the demand should also be assessed with regard to the individual's total capability. Thus, carrying a normal suitcase or balancing a checkbook might be low workload for an average adult, but high workload for an average child.

A common concern in the aviation environment is that excessive workload might be inflicted on the pilot, and performance is likely to suffer. This is of particular concern in cases of emergencies or when complex maneuvers are required under trying conditions (see chap. 2). Consequently, the reduction of mental workload is often cited as a justification for system redesigns and upgrades. For example, Honeywell Systems developed an automated system that identified the information needed by a pilot, in order to avoid presenting unneeded information and thereby avoid inflicting unnecessary mental workload (Nordwall, 1995).

Although, as alluded to earlier, situation awareness has had numerous definitions, many researchers favor Endsley's (1990a) definition that describes situation awareness as "The perception of the elements in the environment within a volume of time and space, the comprehension of their meaning, and the projection of the status in the near future" (p. 1-3). In other words, situation awareness is the pilot's current understanding of the positions and significance of all relevant actors in the mission environment. Elements of this conception of situation awareness reach well beyond the human factors research community. For example, U.S. Air Force General Chuck Horner, the commander of the allied air forces during the Persian Gulf War, expressed this view of situation awareness (Clancy & Horner, 1999):

> The trait I admire most in great pilots is "situational awareness." It is the ability to keep track of what is going on around you and to project that awareness into an accurate mental image of what is about to happen during the next few moments; and it is extremely rare. (p. 355)

Situation awareness as a concept is not directly concerned with the attentional load inflicted by the task's information processing requirements; instead, situation awareness is more concerned with the quality of the information apprehended by the pilot. At any given moment, the typical pilot is concerned with satisfying several goals (e.g., flight safety, navigation, communications, tactical plan). Of course, to do this it must be assumed that the pilot possesses sufficient skill to perform the required activities. But skill is the general ability of an individual to deal with a multitude of situations. At the time of a specific performance, it is assumed that skill must be informed by an accurate "picture" of the specific current situation. It is this current picture that is typically referred to as situation awareness. And as General Horner pointed out, situation awareness is not a raw sense impression; it also involves some interaction with the pilot's knowledge and includes projections of likely future states.

Just as some aircraft interface upgrades are justified by purported decreases in mental workload, other upgrades are defended by their purported increases in situation awareness. Both are expected to lead to performance and safety improvements. For example, proposed upgrades to the display systems of fighter pilot helmets (Kandebo, 2000), the head-down displays in the U.S. Air Force F-117 fighter (Wall, 2000), or the displays of altitude information in commercial aircraft (Proctor, 2000) have all been linked to expected increases in pilot situation awareness. A proposed upgrade to the Space Shuttle cockpit was even predicted to simultaneously increase situation awareness while decreasing mental workload (Mecham, 1999).

To recap, in examining mental workload and situation awareness, it is appropriate to adopt working definitions that capture the broad consensus regarding them. In a nutshell, mental workload is the mental cost placed on the pilot by performing the necessary mental processing to accomplish the mission, and situation awareness is the momentary understanding of the current situation and its implications.

Types of Theoretical Concepts

Perhaps more important than the exact definition of either mental workload or situation awareness is the category of theoretical concept to which either is assigned. Lachman, Lachman, and Butterfield (1979) distinguished between two main approaches for theorizing in cognitive psychology. One approach was referred to as "black box theories." A black box theoretical concept relates inputs to the cognitive system to its outputs. It provides a name for observed or hypothesized relationships between inputs and outputs, but does not attempt to explain any causal mechanisms that account for the relationship. In psychology, a classic example of black box theorizing was the attempts of behaviorists to relate stimuli to responses while avoiding any discussion of any causal mental processes between the two. For example, Hilgard and Marquis (1940), in their influential text *Conditioning and Learning,* defined learning as "Change in the strength of an act through training procedures (whether in the laboratory or the natural environment) as distinguished from changes in the strength of the act by factors not attributable to training" (p. 347).

The second category of theoretical concepts discussed by Lachman et al. (1979) was a "structural" theory or model. A structural model, "describes at least at a logical level the inner workings of the 'box' whose output it seeks to explain" (p. 107). Just as behaviorists most commonly theorized at the black box level, cognitive psychologists typically aspire to using a structural modeling approach to specify the stages of information processing involved in task performance. For example, in the procedure, now commonly referred to as the "Sternberg task," participants are given a number of items (e.g., digits, letter, or words) that constitute a memory set. The participants are then asked to judge as quickly as possible whether a presented probe item is in the memory set (a "yes" response) or not (a "no" response). Sternberg (1966) found that the reaction time to the probe item linearly increased as the number of items in the memory set increased. This was interpreted as suggesting that a separate, serial memory scanning stage of information processing occurred after the participant perceived the stimulus item and before the participant generated the response. Within Lachman et al.'s (1979) terminology, Sternberg's work can be considered evidence of one stage in a structural model of the information processing involved in performing the Sternberg task.

To apply the Lachman et al. (1979) terminology to the present topics, there appears to be a consensus regarding the classification of mental workload. Mental workload has generally been considered to fit in the black box category of theories, although the exact term used to describe this status varies. Gopher and Donchin (1986) invoke MacCorquodale and Meehl's (1948) concept of a hypothetical construct as the best classification for mental workload. The hypothetical construct concept seemed appropriate due to the term *mental workload* carrying excess meaning beyond simple specification of the operations that produce it. Or, as Gopher and Donchin (1986) put it:

> We imply that we are describing some entity or some property of entities that is not given entirely by the relationship between our empirical observations. At the same time, we assume that this excess meaning can be captured, studied, and measured in ways that would advance our understanding of the system and make it possible to use the concept for practical activities. (p. 41-4)

In a similar fashion, Kantowitz (1986, 2000) defined mental workload as an intervening variable that, like attention, modulates allocation of the human's information-processing resources to task demands. Kantowitz and Casper (1988) explored this issue in greater detail within the context of aviation psychology. They noted that there were numerous plausible starting points for developing relevant workload theories and focused on attentional resource theories as a starting point. Hollnagel (1998) identified mental workload as one example of a measure that derives from a "folk" model in psychology. Folk models were argued to produce measures of intervening variables that were assumed by consensus to be linked to performance. All of these researchers reflect the consensus that mental workload is an observable and measurable phenomenon, but not an actual stage of information processing that would fit within a structural model of a pilot's cognition.

In contrast to mental workload, the theoretical categorization of situation awareness is less certain. Fig. 4.1 illustrates possible theoretical approaches to situation awareness. In Fig. 4.1(a), situation awareness is one part of a structural model of human information processing. Here, situation awareness is an intermediate stage that integrates the current perception and the pilot's knowledge to create a representation of the current situation that is used in the pilot's decision-making processes. Endsley (1995b), for example, presented a much more detailed model of situation awareness that placed situation awareness within the information-processing path in this fashion.

However, other researchers have employed situation awareness in a manner that is more in keeping with the black box approach. This approach is illustrated in Fig. 4.1(b). Here, situation awareness is portrayed as circumscribing a set of information-processing stages and without any specific inputs or outputs to situation awareness per se. According to this approach, situation awareness is not really a part of the structural model itself, but is instead an emergent property of the structural model's operation. Wickens (1996) discussed situation awareness as an emergent property of human information processing and presented it in this fashion in his more elaborate model of human information processing.

Despite the availability of structural models such as Endsley (1995b), in practice, situation awareness is more often treated as a black box model. Most real-world evaluations of situation awareness do not attempt to directly measure situation awareness, but implicitly infer changes in situation awareness from changes in overall performance (Vidulich, 2000b). This may seem an unhappy outcome for cognitive psychologists more used to, and satisfied by, structural models similar to Sternberg's (1966) model of the memory scanning stage of information

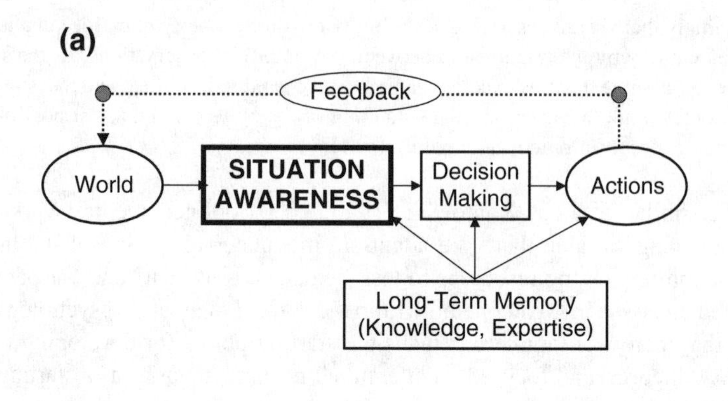

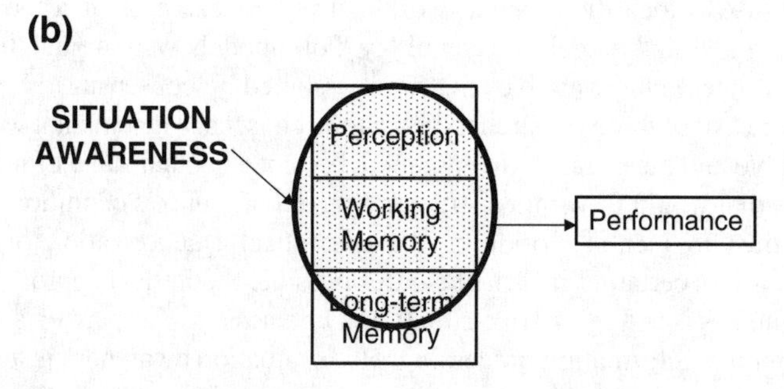

FIG. 4.1. Two theoretical approaches for understanding situation awareness as a theoretical concept: (a) a structural model concept and (b) a black box model concept.

processing or Endsley's (1995b) model of situation awareness, but it does not prohibit researchers from doing useful work with real-world applications.

Another way to conceptualize the scientific roles of mental workload and situation awareness would be to invoke the idea of a "framework" theory as advanced by James Reason in his work to understand human error (Reason, 1990). Reason contrasted framework theories with what he referred to as "local" theories. In general, a local theory was a relatively small, well-focused theory that could be tested and potentially disproved in typical laboratory research guided by the scientific method. Applying this concept to Lachman et al.'s (1979) distinction between black box and structural models, it is plausible that a well-specified structural model would readily fit into local theory. For example, Sternberg's (1966) study of memory scanning was already used as a good example of a structural model, and it is also a good example of local theory and research.

In contrast to the specificity and testability of local theories, Reason (1990) conceived of framework theories as reflecting the broad understanding of and consensus about a domain. Being rather broadly conceived, they will be difficult, perhaps impossible, to disprove by any single or small group of empirical results. Inconsistent results are more likely to be assimilated as representing limits or special cases of previous predictions. For example, Hull's (1943) or Guthrie's (1952) theories of learning could be classified as framework theories in that they provided an organized view for interpreting a great deal of prior research and made general predictions to guide future research. No single empirical result was likely to undermine either theory. Framework theories rise and fall based on whether they fit the general consensus of people working in the field as adding to their understanding of the relevant domain and making a useful basis for further research and applications.

Both mental workload and situation awareness seem to best fit the notions of framework theories and black box models. They both capture a consensus among researchers and subject-matter experts about something important in the human reaction to dealing with the complex, goal-directed, and potentially stressful situations that are all too often a part of being a pilot. Neither term fits nicely into the confines of a local theory, nor has research into either topic been dominated by structural models (although Endsley's efforts show that it can be attempted).

MODELS AND MEASURES OF MENTAL WORKLOAD AND SITUATION AWARENESS

Hollnagel (1998) pointed out that the models used to represent concepts and the measures used to assess those concepts are inextricably linked. Therefore, it is worthwhile to examine the methods used to measure mental workload and situation awareness to glean insights concerning their theoretical nature and potential application within aviation psychology.

The process of relating measures of mental workload and situation awareness to their theory and application is complicated by the multitude of measures that are sometimes used. Both mental workload and situation awareness have been measured by diverse subjective, performance-based, and physiological techniques. The strengths and weaknesses of all the various techniques have been discussed and debated many times (see, for example, O'Donnell & Eggemeier, 1986, or Tsang & Wilson, 1997, for reviews of mental workload measurement, and Endsley, 1995a, or Vidulich, 2000b, for reviews of situation awareness measurement, or Gawron, 2000, for listings of both types of measurement techniques). However, for both mental workload and situation awareness, certain measurement techniques exist that are so closely linked to the commonly accepted definition and understanding of the concepts that they might be considered "prototypical" measures. In the case of mental workload, the prototypical measurement technique that most directly exemplifies the concept of measuring the amount of load placed on the limited resources of an information-processing system is the secondary task methodology. In situation awareness, the prototypical measurement of a pilot's current understanding of a situation is usually conceived of in terms of responses to memory probes. In other words, mental workload can be seen as primarily an attentional phenomenon, whereas situation awareness is perceived to be more strongly associated with memory. Certainly, this one-to-one linking is simplistic because the relationships between both concepts and other cognitive processes have often been discussed (e.g., Gopher & Donchin, 1986, for mental workload; Durso & Gronlund, 1999, for situation awareness). Nevertheless, focused consideration of these primary relationships can be instructive.

Secondary Task Measurement of Mental Workload

The logic of the secondary task is straightforward. The human pilot is conceived of as an information processor with limited processing resources. For an aviation task to be performed effectively and safely, the information-processing resources demanded must not exceed the total resources that the pilot possesses. The aviation task that the aviation psychologist wishes to assess is referred to as the primary task. The difference between the demands of the primary task and the total resources available from the pilot is referred to as *reserve capacity.* The secondary task is an additional task that the pilot is asked to perform while also performing the aviation task. The notion is that the pilot will only allocate resources from the reserve capacity to perform the secondary task, so that secondary task performance will mirror primary task demands. That is, when primary task demands are low, the reserve capacity will be relatively great, and the secondary task will be performed relatively well. Alternatively, if the primary task demands are high, the reserve capacity is expected to be somewhat meager, and secondary task performance will be relatively poor.

Pertinent issues associated with the secondary task technique are discussed next. They include: What are the underlying resources demanded by both the primary and secondary tasks? How should a secondary task assessment be conducted? What determines whether a task would be a good or bad secondary task?

Theories of Attention and the Secondary Task Technique. The fundamental theoretical question behind the use of the secondary task technique is the nature of the resources demanded by both the primary and secondary tasks. In one of the earliest discussions of a secondary task technique, Knowles (1963) conceived of the human information processor as a single-channel processor. That is, the human was considered to be processing (and performing) only one task at any given time. Multiple task demands were thought to require switching all attention from one task to another. In this conception, the limited resource that was competed for was processing time. Presumably the single-channel processor would only switch to the secondary task when there were no demands from the primary task. The more free time (i.e., reserve capacity) allowed by the primary task, the better the performance would be on the secondary task.

Knowles' secondary task approach was consistent with the dominant trend in attention theory at the time. Only a few years earlier than Knowles' work, Donald Broadbent (1958) proposed his influential filter theory of attention. Filter theories were sometimes referred to as "bottleneck" theories, because the presumption was that at some point within the information-processing flow through the human, there was a bottleneck through which only the selected information could proceed while the rest was filtered out. Following Broadbent, other attention researchers adopted the basic notion of a filter, but debated the thoroughness of the suppression of the nonselected stimuli (e.g., Treisman, 1969) or the location of the filter within the processing system (e.g., Deutsch & Deutsch, 1963; Pashler & Johnston, 1998). However, the implications of these debates on the practical use of the secondary task technique of measuring mental workload were unexplored.

Explanation of secondary task results started to shift in 1967 when Moray provided the impetus to envision a different mechanism as responsible for observed attentional limitations. Moray suggested that rather than considering human attention to be limited by a filter or bottleneck somewhere in the information-processing flow, human attention could be likened to the active memory of a computer. Moray recounted the use of a computer with a main memory of 8 kilobytes (K). The 8 K of memory had to contain all of the program instructions along with all of the data to be processed. If the program was large and complicated, then the amount of data had to be limited. On the other hand, if there was a large data set, then the program had to be smaller. By analogy, the human was considered to have general-purpose information-processing resources that could be applied to any task's information processing. Resource theories of attention largely supplanted filter theories as a basis for understanding the results of experiments using the secondary task technique.

Kahneman (1973) developed the first detailed resource theory of attention. Amplifying on Moray's (1967) idea of a store of general-purpose attentional resource that could be allocated to multiple tasks, Kahneman considered the influences on attention allocation. Kahneman suggested that the overall amount of resources would vary depending on the human's arousal level. A groggy, tired person would have less attentional resources to allocate to task processing than an alert one. Some resources would be allocated based on enduring dispositions. For example, an unexpected loud noise will typically cause an orienting response in which the human allocates attention to determine the cause of the noise. In Kahneman's model, attention can be voluntarily allocated to tasks relevant to the human's momentary intentions (e.g., immediate goals or missions). Finally, Kahneman hypothesized that a closed-loop feedback mechanism would monitor the performance of tasks and changes in task demands and reallocate attentional resources as needed.

Figure 4.2 illustrates the application of a resource-based model of attention to the secondary task technique. In this case, there is a hypothetical upper limit on the information processing that the human can do at any moment supported by

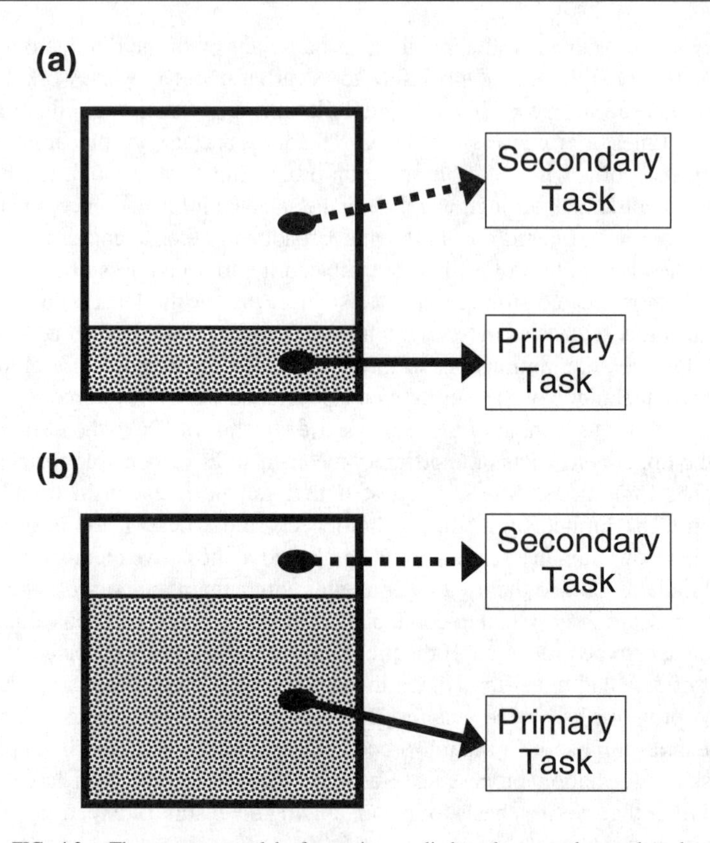

FIG. 4.2. The resource model of attention applied to the secondary-task technique. The complete rectangle represents the human's full information-processing resource capacity. The shaded area represents the resources demanded by a hypothetical primary task. (a) With a relatively easy primary task, there is a large reserve capacity (the white space) to support relatively good secondary task performance. (b) With a relatively difficult primary task, there is a smaller reserve capacity and presumably relatively degraded secondary-task performance.

attention. Assuming that the human will allocate the attentional resources required by the primary task to it, the available reserve capacity will reflect the demand of the primary task. So, the higher the primary task's demand, the less spare capacity of resources that can be allocated to the secondary task, and the worse the performance of the secondary task.

The simple general-purpose resource model possessed a very attractive theoretical elegance, but in practice there was soon a collection of anomalous results. In his seminal 1980 paper, Wickens reviewed relevant multiple-task research and noted at least three classes of anomalous results. First, there were cases in the literature of two tasks of appreciable demand being performed together with no detectible interference (i.e., cases in which performing the tasks together did not degrade either task's performance). Second, there were cases of increasing the difficulty of one task in a dual-task pair without increasing the amount of interference between the tasks. And, third, there were cases of changing the structure of a task (e.g., changing from visual to auditory presentation) without changing its difficulty, causing the interference with another task to change. All of these results were inconsistent with the notion that both tasks were competing for the same general-purpose attentional resources.

In examining the pattern of when these results occurred versus when the expected competition between tasks occurred, Wickens noted that the pattern was consistent with the idea that human attentional resources can be subdivided. Wickens (1980) proposed a multiple-resource model in which attentional resources were divided along three main dimensions: (a) stages of processing (perceptual/central and response), (b) codes of processing (spatial and verbal), (c) and input/output modalities (visual and auditory, manual and speech). (Note: chap. 7 presents the latest formulation of the multiple-resource theory.)

The fact that task interference is affected by task features and not just the sum of the competing tasks' total difficulty is a blessing for the system designer attempting to maximize joint performance. This implies that interfaces can now be designed to distribute demands across the hypothesized resources. However, the separate resources pose a challenge to system evaluators planning to use a secondary-task workload metric. Adopting the multiple-resource model, the logic of the secondary task dictates that the secondary tasks compete for the same attentional resources required by the primary task. Consequently, the evaluator must decide which demands of the primary task must be assessed and select the secondary task accordingly. A complete assessment of the primary task's demands may require multiple assessments, with several secondary tasks designed to tap differing attentional resources.

In addition to the challenges associated with separate resources, an evaluator using a secondary-task procedure must also consider the attentional allocation strategies of the participants. Traditional secondary-task instructions request the participant to allocate sufficient attention to the primary task to perform it at the same level of performance as would be possible if it were performed alone (i.e., without degradation from the single-task level). Any attentional resources not required to maintain the primary task at the single-task level are to be allocated to the performance of the secondary task. To use the secondary-task performance as a mirror of primary task difficulty, it must be assumed that participants can and will conform to the evaluator's instructions. But do the participants achieve this goal?

Unfortunately, all too often, the presence of the secondary task causes a degradation of primary-task performance. In other words, the secondary task "intrudes" on primary-task performance. For example, Tsang, Shaner, and Vidulich (1995) demonstrated that the control movements for a continuous tracking task were disrupted by the occurrence of periodic discrete tasks regardless of whether the tracking task was designated the primary or the secondary task on a given trial. The magnitude of the disruption was much smaller if the tracking was the primary task, but it was still present.

Gopher and Donchin (1986) carefully considered the impact of multiple resources, allocation nonoptimality, and other issues on the use of the secondary task procedure to assess mental workload. They suggested that the expectation that the secondary task would not intrude on primary-task performance requires a host of implausible assumptions concerning human attentional allocation. To remedy this, they advocated that rather than using the traditional set of primary-secondary task pairs, evaluators should have participants perform the two tasks several times under a variety of instructed levels of task priorities. These data could then be portrayed within a Performance-Operating Characteristic (POC) diagram that would describe the performance trade-off between the tasks much more completely. Gopher and Donchin (1986) acknowledged that the use of POC methodology would significantly increase the complexity of evaluation designs and make both data collection and data analysis more time consuming. Despite that, they suggested that "when one considers the richness and importance of the information, and the present state of our knowledge of workload, the additional effort seems worthwhile" (Gopher & Donchin, 1986, p. 41-9).

Although the original logic of the secondary task had an attractive and elegant simplicity, recent knowledge of the structure and operation of human attention somewhat undermined its validity or practicality. Attempting to deal with the impact of both multiple resources and attentional allocation optimality might require that the to-be-measured task be successively paired with a battery of tasks selected to represent a spectrum of resource demands and perform each task pair several times under different priority instructions. If measuring mental workload required all of this, it would not be surprising to find system evaluators in despair. This prompts the question, Is the secondary task procedure ever successfully used to assess mental workload for real-world aviation systems?

Application of the Secondary Task to Assess Mental Workload. Attempts to use the secondary-task procedure to measure the mental workload associated with cockpit tasks have been myriad and varied, but have they been successful?

In an ambitious attempt to make a general-purpose mental workload assessment tool for in-flight use, Schiflett, Linton, and Spicuzza (1981) made use of the Sternberg task as a secondary task. The tool, called the Workload Assessment Device (WAD), required pilots to memorize memory sets of one, two, or four letters. Subsequently, probe letters were presented one at a time on the pilot's head-up display (HUD) at a rate of one probe letter every 7 sec. The pilot responded "yes" (i.e., the probe letter was part of the memory set) or "no" (i.e., the probe letter was not part of the memory set) to the displayed probe letter by pressing the appropriate button on the aircraft control stick. Baseline secondary task data were collected with the aircraft sitting on the ground to allow the pilot to focus all attention on the Sternberg task. In-flight Sternberg task data were collected while the pilot performed touch-and-go landing exercises. Difficulty of the touch-and-go landings was varied by changing the display format of the HUD and by manipulating the handling qualities of the aircraft through alterations of the aircraft's pitch response. Baseline primary-task data were collected by performing touch-and-go landings without any secondary-task presentation. The results were encouraging. Primary-task performance, as measured by flight data from a digital recorder and pilot ratings of handling qualities, did not vary as a function of HUD format, aircraft pitch response, nor secondary-task presence or difficulty level. The authors inferred that this indicated that the pilots maintained an acceptable level of performance by compensating for changes in the difficulty of the primary touch-and-go landing task. The secondary Sternberg task data, on the other hand, showed a pattern of expected changes in response to the manipulations of both its own difficulty (increases in memory set size) and the difficulty of the primary touch-and-go landing task. The reaction time and error rate to the Sternberg task probe items increased as delay was introduced into the aircraft's pitch response and when the conventional HUD format was used. Landings conducted with the more pictorial HUD format that was under evaluation were associated with faster and more accurate Sternberg task performance, presumably because this HUD format required less mental workload to understand. As the authors pointed out, all of these results were only trends because the limited sample size (two pilots) prohibited the use of inferential statistical tests. Nevertheless, the work was an intriguing demonstration of secondary task methodology applied to a real-world flight task.

Kantowitz, Hart, and Bortolussi (1983) carefully evaluated the utility of the secondary-task technique for measuring pilot mental workload in a study conducted in a general aviation trainer simulator. The secondary task was a choice reaction time task. The secondary-task performance was sensitive to differing levels of workload associated with different segments of the simulated flight. Of particular note, given the theoretical concern of secondary-task intrusiveness to primary-task performance, Kantowitz et al. compared the primary flight control task performance obtained under two levels of secondary-task difficulty to that when no secondary task was presented. Flight control error did not increase as a result of the secondary-task presentation, suggesting that the pilots successfully defended the performance of the primary flight task from secondary-task intrusion.

Wickens, Hyman, Dellinger, Taylor, and Meador (1986) reviewed both previous uses of the Sternberg task as a secondary task and new studies using the Sternberg task as a secondary task in simulator-based aircraft evaluations. The results showed that not only was the Sternberg task sensitive to the overall demand of the primary flight task, but that changes in Sternberg task performance were diagnostic regarding the primary task's specific attentional resource demands. However, the authors cautioned that the use of a secondary task was counterindicated if primary task difficulty was extremely high. In cases of very high primary-task difficulty, there was presumably little reserve capacity to measure, and primary-task performance itself becomes a measure of mental workload.

More recently, Draper, Geiselman, Lu, Roe, and Haas (2000) and Ververs and Wickens (2000) demonstrated the evaluative utility of the secondary task technique in the aviation environment. Draper et al. evaluated various display formats to assist teams of operators controlling a simulated Predator Unmanned Air Vehicle (UAV) performing a reconnaissance mission. The results strongly supported the efficacy of one of the suggested improved displays; not only was the primary Predator UAV control task performed better, but the performance on a secondary task was better as well. Ververs and Wickens (2000) used a set of secondary tasks to assess simulated flight path following and taxiing performance with different sets of HUD symbology. One set of symbology presented a "tunnel-in-the-sky" for the participants to follow during landing approaches. The other display was a more traditional presentation of flight director information. The tunnel display reduced the participant's flight path error during landing. Participants also responded more quickly and accurately to secondary-task airspeed changes and were more accurate at detecting intruders on the runway. However, the other display was associated with faster detections of the runway intruder and quicker identification of the runway. The authors concluded that although the tunnel display produced lower workload during the landing task, it also caused cognitive tunneling that reduced sensitivity to unexpected outside events.

Status of the Secondary-Task Technique. The foregoing review of secondary-task theory and application cannot be considered exhaustive in any sense. Nevertheless, the examples presented provide a basis for considering the status of the secondary-task technique as a mental workload assessment tool. On the one hand, if the secondary task is judged by the rigorous standard of local theory, it will probably be found to be wanting. The demands of designing a secondary-task assessment that would withstand a close scientific scrutiny are daunting. The issue of primary-task intrusiveness provides a case in point. As Gopher and Donchin (1986) cautioned, an accurate assessment of the demand of a primary task requires that not only must the secondary-task performance be evaluated, but also that the impact of the secondary task on primary-task performance be considered as well. To accomplish this, they suggested that secondary-task methodology be expanded to include priority manipulations between tasks, so that the continuous trade-off between tasks (if any) can be thoroughly mapped out. This suggestion is rarely followed in applied settings. In some cases, the impact of the secondary task on primary-task performance is not discussed at all (e.g., Wickens et al., 1986). In other cases, there is a test that the presence of the secondary task did not lead to a detectable degradation of primary-task performance (e.g., Kantowitz et al., 1983). Sometimes the experimental manipulations improve performance on both the primary task and the secondary task simultaneously (e.g., Draper et al., 2000; Ververs & Wickens, 2000). This is a happy outcome that strongly supports the reduction of mental workload by the experimental manipulation, but it is not a common outcome.

On the other hand, if the uses of secondary tasks are judged as practical tools within the confines of a framework theory of mental workload, their use and usability is much easier to defend. Given that care is taken in the selection of the secondary task to represent the relevant resources demanded by the primary task and that instructions aimed at minimizing the potential of primary-task intrusion are given and enforced, the secondary-task technique is likely to provide valuable information (see Tsang & Wilson, 1997, for more specific guidance). Within the broad conception of the human having limited information-processing resources and that performance of a primary task uses some of those resources while leaving some in reserve, the secondary-task technique is a useful tool for determining the approximate reserve capacity of resources. In general, a system that allows the pilot to have a comfortable reserve capacity is better than one that fails to do so.

Memory Probe Measurement of Situation Awareness

The Logic of Memory Probe Measures. Just as the secondary task can be viewed as the prototypical measurement technique for mental workload, memory probes are the prototypical measure of situation awareness. This reflects both the natural connection between testing the contents of the human's current information processing and the view that situation awareness is a momentary picture of the situation. It also reflects the historical fact that the first popular and standardized procedure for measuring situation awareness was a memory probe procedure. That procedure was the Situation Awareness Global Assessment Technique (SAGAT), as formulated by Endsley (1988b, 1990a).

SAGAT (Endsley, 1988b) entails a series of steps. The first step is to identify the situation awareness requirements for the individual in the task being assessed. In a complex, real-world, dynamic task, there is likely to be a relatively large set of information requirements. For example, based on goal-directed task analysis of an air-to-air combat mission and subject-matter expert interviews, Endsley (1993) identified over 200 items at least "somewhat important" to situation awareness. The next step is to collect the situation awareness data during simulated performance of the task. At random times during the task performance, the simulation is frozen, and the human is asked to answer questions from memory about the status of specific information requirement parameters (e.g., bearing to the nearest enemy, current altitude, wingman's tactical status). At any single data-collection point, it would be infeasible to assess all of the previously identified information requirements, so a randomly selected subset of information requirements would be assessed. This sampling procedure would be repeated within and across participants until a sufficient data set for statistical analysis was collected. Finally, the SAGAT score would be calculated by comparing the collected responses to the true situation. Based on this procedure, Endsley (1988b) proposed that SAGAT offered the advantages of immediacy of measurement, globalness of information assessed, objective measurement, and face validity. The major limitation that she identified was the requirement to pause the simulation to acquire the data. Although the pausing of the simulation is clearly intrusive in the sense that it disrupts the flow of performance in a manner that would be impossible in actual flight, Endsley's (1995a) data suggested that the pauses did not influence subsequent performance in a simulated air-to-air engagement.

Vidulich (2000b) found that memory probe investigation using a SAGAT-style approach with unpredictable measurement times and random selection of queries from large sets of possible questions were generally sensitive to interface manipulations designed to affect situation awareness. In contrast, as memory probes were made more specific or predictable, the sensitivity to interface manipulations appeared to be diminished.

Memory probe measures of situation awareness, such as SAGAT, are not as strongly tied to a specific cognitive theory, as was the case of the tie between the secondary task technique and resource theories of attention. It is reasonable to expect that the use of memory probe measures of situation awareness should be tied to some aspect of memory theory, but which aspect?

At first glance, interrupting the task for an immediate probing of the momentary contents might lead to the expectation that memory probe techniques such as SAGAT are essentially recording a dumping of the contents of short-term memory. But this notion will not withstand even minor scrutiny. The best estimate of the capacity of short-term memory is seven items, plus or minus two items, based on the seminal work of Miller (1956). The number of questions routinely used in a SAGAT-type assessment will often exceed this presumed capacity of short-term memory. Also, Endsley (1995a) reports that simulation pauses of up to 5 to 6 minutes do not appear to degrade the human's ability to answer SAGAT queries. These observations support a rich interplay between specific memory of the current situation and the trained pilot's long-term memory.

The role of efficient and adaptive use of long-term memory to support expert performance has been demonstrated in well-controlled laboratory studies of skilled memory. For example, Ericsson and Staszewski (1989) demonstrated that human could be trained to recall long strings of random digits. The ability to reliably recall strings of 50 digits or more far exceeds the normal human performance in this task. To accomplish this, the participants in these studies developed strategies to efficiently encode the digits into long-term memory. At the end of a session, it was possible for a participant to relate back the earliest strings of digits, despite several intervening list presentations and recalls. Results of studies like these formed the empirical basis for the development of the theory of long-term working memory (Ericsson & Kintsch, 1995). In long-term working memory, information regarding the current situation are stored in long-term memory but kept accessible by means of retrieval cues in short-term memory. Ericsson and

Kintsch (1995) supported the theory of long-term working memory through careful examination of expert performance in the domains of mental calculation, medical diagnosis, and chess.

Long-term working memory is expected to emerge in expert performance in real-world tasks and, thus, serves as the basis for answering memory probe queries. This would explain the relative lack of effect of delay observed by Endsley (1995a) and the participants' general ability to respond to large numbers of unpredictable queries.

The role of long-term working memory in answering probe questions makes the use of memory probes for situation awareness assessment a joint test of the participant's skill and the information provided by the interface. The role of long-term working memory in responding to memory probes also highlights the inherent imprecision of the labels put on the phenomenon. In the case of situation awareness, the word *awareness* might be problematic for the very literal minded. Information accessed from long-term working memory would probably not fit the common view of something within awareness. Yet, this information might very well be used to successfully respond to memory probe questions. However, within cognitive psychology, the relationship between awareness and memory processes remains an imposing question (e.g., Adams, Tenny, & Pew, 1995; Cowan, 1995; Klatzky, 1984), and the final resolution does not appear to be close. Nevertheless, within the long-term working memory model, situation awareness appears to be a viable concept and the careful use of memory probes could assess the information available to the participant at the time of probing. Whether all of that information was in the participant's "online awareness" at the task pauses or constructed from skill-based knowledge at the presentation of the probe question may well be noncritical to the utility of the measurement. In either case, the information was effectively available to the person during the performance of the task.

Application of Memory Probes to Assess Situation Awareness. Some of the earliest research in situation awareness involved using memory probes to assess pilot situation awareness. Endsley (1988b) reviewed the results of two preliminary studies of situation awareness using the memory-probe-based SAGAT. Both of these studies were conducted in an air-to-air combat simulation. The first study was primarily a test of whether pausing the simulation to collect SAGAT data periodically during a simulated mission affected overall mission performance. No evidence of any detrimental effect was detected when the SAGAT trials were compared to non-SAGAT trials. The second simulator evaluation used SAGAT to compare two different sensor system concepts on situation awareness. Although the nature of the differences between cockpits was not explained (perhaps due to proprietary or classified information concerns), one cockpit was demonstrated to provide the pilots with a more accurate picture of enemy aircraft locations.

In another study, Endsley (1990b) examined the relationship between SAGAT performance and mission outcome. Performance was measured by air-to-air kills and losses during the simulated engagements between a "Blue" team of two pilots and a "Red" team of four pilots. The complexity of the simulated engagement probably mitigated the relationship between situation awareness and performance. The Red team's actions were stringently constrained by the rules of engagement that they were instructed to follow, and their mission performance was not related to their SAGAT scores. In contrast, the Blue team's performance was significantly better when their SAGAT scores were good.

Amar, Hansman, Vaneck, Chaudhry, and Hannon (1995) used a memory probe measure to assess the potential value of Global Positioning System (GPS) information presented in a chart display for improving situation awareness during a simulated aircraft taxiing task. The results showed that responses to the situation awareness memory probes were more accurate and faster when the taxiing display provided GPS-based location information with an accuracy of 50 meters or less, compared to information accurate only to a 100-meter resolution.

The effective use of SAGAT demands that a relatively large group of questions be randomly sampled and unpredictably presented during task performance. This does impose a pragmatic consideration of sampling enough to have a good data set, but not sampling so often or for such prolonged periods as to interfere with the rest of the evaluation. The challenge of meeting these needs, and the risks associated with failing to do so, might be illustrated by contrasting two studies of the effects of including a tactical situation display in a simulated air-to-air combat task. Hughes, Hassoun, Ward, and Rueb (1990) contrasted performing an air-to-air combat task using a traditional fire-control radar display to performing it with a tactical situation display. The fire-control radar display only showed the radar returns from the pilot's own aircraft radar system. This limited the information on the display to other aircraft located in front of the pilot's aircraft. The tactical situation display simulated a display that used information data-linked from other aircraft as well as the pilot's aircraft. In this display the pilot's aircraft was located in the center of the display. In addition to information about aircraft in front of the pilot's aircraft, information about any aircraft to either side or behind the pilot's aircraft was also displayed. Although the tactical situation display obviously provided more information about the current situation and performing the missions with the tactical situation display was accomplished with less reported mental workload, the SAGAT scores were not significantly different for the two display conditions. The authors speculated that the 12 participants used in the study might have provided insufficient power for the statistical analysis.

Vidulich, McCoy, and Crabtree (1995) also used a SAGAT-like memory probe procedure to evaluate the effect of a tactical situation display in a simulated air-to-air combat task. The study differed from the Hughes et al. (1990) study in a number of respects, but perhaps most notably in sample size. Vidulich et al. (1995) used 32 participants. The results of the study were also different, in that a significant benefit of the tactical situation display was detected in the memory probe test of situation awareness.

Status of the Memory Probe. As with the secondary task technique discussed previously, the memory probe procedure appears to have met the threshold to be considered a viable measure associated with a framework theory of situation awareness. Although unexpected failures to detect effects of some experimental manipulations have occurred (e.g., Hughes et al., 1990), the overall picture is that a well-designed memory probe evaluation can capture whether the pilots in an evaluation generally had an accurate picture of the task situation with which they were confronted. Memory-probe evaluations of situation awareness are also akin to secondary-task evaluations of mental workload in that they are very demanding to perform. Acquiring a sufficient data set can require a great deal of time and effort. Also, the standard approach to collecting memory probes involves freezing a simulator to provide time to collect the pilot's response to the memory probe queries. Such freezes are not possible in flight. An interesting attempt has been made to adapt the SAGAT approach to real-time situation awareness measurement in a simulated air operation center (Jones & Endsley, 2000). Participants were asked questions about the simulated exercise while it continued to run. The results were promising. The participants' performance on the real-time probes gave a very similar picture to obtained by a more traditional SAGAT evaluation. However, this approach to assessing situation awareness is still in its infancy, and much work remains to be done before it can be considered a viable alternative. Furthermore, even if such real-time memory probes prove to be a dependable measure of situation awareness, safety concerns may severely restrict or even prohibit their use in flight.

IS THERE A RELATIONSHIP BETWEEN MENTAL WORKLOAD AND SITUATION AWARENESS?

As situation awareness research began to flourish in the early 1990s, it was probably inevitable that comparisons to the concept of mental workload and mental workload research would be made. One of the earliest comparisons was made by Wickens (1992):

> In the same way that the seeds of workload research were planted and nourished in the late 70s and grew to full bloom in the 80s, so, a decade later, the seeds of applied interest in *situation awareness* were planted in the mid 80s, and, I forecast will grow and bloom in the 90s (I'll leave the "withering and dying on the vine" part of my analogy for others to speculate). There are other features of situation awareness (SA) that make it analogous to workload. It too deals with a hypothetical or inferred "mental construct," that cannot be directly observed in behavior, but which we know has many consequences for the potential of that behavior to succeed or fail. Just as important performance consequences follow when workload become excessive, so

> important performance consequences are assumed to follow when SA is lost. Furthermore, issues of both prediction and assessment have been very much at the forefront of SA research. Finally, while the engineer may sometimes feel uncomfortable with such a purely "mental" construct as SA, both the psychologist and the operational pilot realize the important validity of the concept. (p. 1)

In other words, although mental workload and situation awareness are distinct concepts, they both play similar roles in aviation psychology. Both of them serve as organizing concepts for system evaluation and as guidance for system design. Many proposed system changes have been justified on the basis of reducing the pilot's mental workload and/or increasing the pilot's situation awareness. Many system evaluations incorporate measures of mental workload and/or situation awareness. Yet, even if the individual values of both mental workload and situation awareness as concepts are accepted, there is another obvious question as to whether the two concepts have any necessary relationship to each other.

Endsley (1993) considered the relationship between mental workload and situation awareness. Her basic hypothesis was, "SA and workload, although inter-related, are hypothesized to be essentially independent constructs" (Endsley, 1993, p. 906). Endsley pointed out that plausible cases could be made for any combination of extremely high or low situation awareness and extremely high or low mental workload. So, in a simple sense, it is impossible to predict the level of one from the other. The lack of connection appeared to become even more convincing when Endsley considered the relationships of both mental workload and situation awareness with performance. In both cases, the concepts have been hypothesized and demonstrated to dissociate from the quality of performance under some conditions. Endsley's discussion of the relationship between mental workload and situation awareness was accompanied by an empirical study. She collected both subjective workload ratings and a memory-probe measure of situation awareness from six experienced pilots in a simulated air-to-air combat task. An overall analysis revealed no significant relationship between the mental workload and situation awareness data. Examining the data for individual pilots showed no relationship for four pilots, but a significant relationship was detected in the other two pilots' data. The pilots demonstrating a relationship showed lower situation awareness scores at higher levels of mental workload. Overall, the results seemed consistent with Endsley's (1993) emphasis on the independence of the two concepts as opposed to their interrelatedness.

The practical independence of mental workload and situation awareness was also supported by Vidulich (2000b). In a review of situation awareness measurement sensitivity to interface manipulations, 15 studies that measured both situation awareness and mental workload were identified. The "ideal" outcome of the typical interface modification was defined as a measurable increase in situation awareness and a measurable decrease in mental workload. Although the interface manipulations did generally improve situation awareness, there was no reliable effect on mental workload.

In another review, Vidulich (2000a) added 8 relevant studies to the original 15. As in the earlier investigation, when evaluating the data set as a whole, there was no consistent trend relating situation awareness and mental workload. However, a different picture emerged when the data set was subdivided by the type of interface manipulation performed. In 9 of the 23 studies, the original researchers examined the effect of adding new information to the operator's interface. Another 9 studies examined the effect of reformatting existing data to make it easier to use. The remaining 5 studies were automation studies that defied easy classification in terms of adding information or reformatting it and were not included in the comparison.

The comparison of the new information studies to the reformatting studies suggested a trend that might be relevant to the issue of a relationship between mental workload and situation awareness. If new information is added to one display condition to potentially add something to the operator's situation awareness, it is difficult to reasonably predict the mental workload effect. If the new information must be processed in addition to all of the other information, then the processing resources required should be increased, and a workload increase should be incurred. But it is also possible that the new information could allow a change of strategy that would eliminate the impact on mental workload or even allow a reduction to occur. On the other hand, if a designer attempts to improve situation awareness by reformatting information that is already available in the operator's interface, it is reasonable to expect that the interface manipulation should also reduce mental workload. After all, there is no additional information to be processed, and the reformatting is usually intended to reduce the cognitive processing that must be done with the previously available data. That is, the new format is presumably intended to provide an equal or better situation awareness picture, while placing less demand on the mental workload related attentional processes.

The results are summarized in Table 4.1. Experimenters were generally successful in significantly improving situation awareness by adding new information to the display (seven out of nine studies). However, the situation awareness improvement was associated with a significant decrease in mental workload in only four studies. In all nine reformatting studies, there were significant increases in situation awareness. More to the point, in six of the nine studies, mental workload decreased as situation awareness improved.

Overall, the pattern of results appears to support the contention that reasonable expectations about the relationship between situation awareness and mental workload can be generated based on the nature of the experimental manipulations under consideration. Of course, 18 studies divided into two categories cannot be considered conclusive, but the seemingly different patterns in the two categories appear suggestive enough to warrant consideration and further research.

TABLE 4.1
Interface Manipulation Effects in Vidulich (2000a)

Interface Manipulation	*Number of Studies*	*Situation Awareness Outcomes*	*Mental Workload Outcomes*
Added information	9	7 + 2 NC 0 −	3 + 2 NC 4 −
Reformatted information	9	9 + 0 NC 0 −	0 + 3 NC 6 −

Key: + = statistically significant increase, NC = no change, − = statistically significant decrease.

Alexander, Nygren, and Vidulich (2000) conducted a study to assess the relationship between mental workload and situation awareness in a simulated air-to-air combat task. Seven pilots flew simulated air intercepts against four bombers supported by two fighters. The experiment contrasted two cockpit designs: conventional and candidate. The conventional cockpit used traditional independent gauges for flight and tactical information. The virtually augmented candidate cockpit used advanced interface concepts designed by a subject-matter expert to enhance situation awareness and reduce mental workload. The task scenario consisted of four mission phases that were designed to influence pilot mental workload and situation awareness. A statistically significant negative correlation was observed between mental workload and situation awareness as influenced by cockpit design and phase of mission. This result would add to the "increase situation awareness—decrease mental workload" combination of the reformatted information portion of Table 4.1 and also shows that a negative relationship can result from changing task demands. In the Alexander et al. study, as the mission phase became more complex, the pilots were more likely to experience reduced situation awareness and likely to simultaneously experience greater mental workload as they attempted to cope with the increased mission demands.

The complex relationship between mental workload and situation awareness represented in Table 4.1 and in the Alexander et al. (2000) results demonstrates the importance of assessing both mental workload and situation awareness when assessing system effectiveness. Both concepts capture something important about the human's reaction to performing complex tasks. In some cases, they may both improve together in reaction to a task change (as in many of the reformatting studies), whereas in other cases the pattern of results may be less predictable (e.g., the added information studies). It would be dangerous to assume the status of either mental workload or situation awareness solely on the basis of measuring the other. However, seeing both types of measures reacting in an understandable fashion to an experimental manipulation or system design change provides good evidence that the impact of the changed environment on the human are really understood.

FUTURE DIRECTIONS FOR MENTAL WORKLOAD AND SITUATION AWARENESS: RESEARCH AND APPLICATIONS

Meta-Measures and Their Uses

The foregoing discussion of mental workload and situation awareness supports the notion that the two concepts are most often characterized as black box models of emergent properties of human information processing while performing demanding and complex tasks, such as piloting. Although many measurement techniques other than secondary tasks for mental workload and memory probes for situation awareness exist and are in common use, these techniques capture something fundamental about the underlying concepts. It was also suggested that the main theoretical drivers behind these concepts were consistent with the broad consensus-driven framework theory approach as opposed to the focused falsification-driven local theory approach (Reason, 1990). In examining the relationship between the two concepts in the previous section, it was argued that although the two concepts are indeed different in important ways, it is more appropriate to focus on their interrelationship rather than on their independence. Understanding how system parameters affect both mental workload and situation awareness should lead to a much better understanding of how to help the operator or pilot.

Other researchers have reached similar conclusions. For example, Hardiman, Dudfield, Selcon, and Smith (1996) and Selcon, Hardiman, Croft, and Endsley (1996) suggested that it might be worthwhile to consider situation awareness and mental workload as "meta-measures" in system evaluation. Selcon et al. pointed out that it is seldom possible to evaluate all of the mission-relevant uses of a display. Thus, a metric approach is needed that provides a generalized link to likely performance in future use. Hardiman et al. and Selcon et al. argued that rather than focusing on task-specific performance, it might be preferable to examine meta-measures that encapsulate the cognitive reaction to performance with a given interface. One proposed meta-measure was mental workload. Presumably, if a task can be performed with one interface with less mental workload, this should provide for more robust transfer to using that interface in more challenging conditions. Situation awareness was another suggested possible meta-measure. Presumably, the interface that provides a better understanding of the situation will provide the better chance of reacting appropriately to that situation. Consistent with this hypothesis, a review of situation awareness measurement in studies of interface manipulations found that the manipulations of interface design undertaken to improve situation awareness tended to improve task performance (Vidulich, 2000b).

Inasmuch as the acid test for a framework theory is its utility, if mental workload and situation awareness are meta-measures associated with the quality of the cognitive processing that a human experiences during task performance, then it follows that improvements in mental workload and situation awareness should lead to likely system performance improvements. In the past, this has been attempted, and largely demonstrated, within the contexts of design inspiration and system test and evaluation (see, for example, Endsley, 1988a; Hart & Wickens, 1990; Vidulich, Dominguez, Vogel, & McMillan, 1994).

Research in these areas will probably remain active and productive for the foreseeable future, but perhaps the most exciting prospect for the mental workload and situation awareness concepts is that they may be about to transform from tools for system designers and evaluators into tools present in the cockpit itself to help the pilot in real-time mission performance.

Real-Time Human Engineering

An early, but still very relevant, definition of human engineering was given by Alphonse Chapanis in his seminal 1965 book, *Man-Machine Engineering* (p. 8), "Human Factors Engineering, or human engineering is concerned with ways of designing machines, operations, and work environments so that they match human capacities and limitations." Much of the early work in human engineering, including human engineering of the aircraft cockpit, was concerned with the design of enduring features of the interface that could improve performance and reduce errors. This included standardization of control layouts in different aircraft, or shape coding of aircraft controls (Grether, 1995). Naturally, this often led to compromises. For example, an ideal layout of displays and controls for landing might be very different than the ideal layout for air combat. However, as computers and multifunction displays and controls become more and more prevalent in aviation systems, it seems desirable to explore the possibility of not just creating a good compromise design for a cockpit or interface, but to have the cockpit or interface intelligently change as circumstances change. In other words, the opportunity might exist to move from traditional static human engineering design to real-time human engineering. This is not an entirely new idea; many automated systems are examples of real-time human engineering (see chap. 9). However, in keeping with the theme of the present chapter, the question is whether the concepts of mental workload and situation awareness can contribute to the implementation of real-time human engineering.

An ideal metric suitable for guiding real-time human engineering should be real-time available, nonintrusive, sensitive, and diagnostic. The need for the metric to be real-time available is obvious, but is important to consider because it potentially eliminates metrics that could be useful for system test and evaluation. For example, in test and evaluation settings, it might be acceptable to use intrusive metrics, such as a secondary task to measure workload or a SAGAT-like memory probe procedure to measure situation awareness. However, the procedural demands of using these tools would be incompatible with safe and effective real-world operations. The need for a real-time human engineering metric to be sensitive refers to the requirement that the metric must react in a reliably detectable way to changes in the pilot's state. Finally, an adaptation metric would ideally be diagnostic. That is, the metric should provide some indication of what is causing a problem for the pilot and thereby point to an efficient and effective aid for the pilot.

This section considers whether any measures of mental workload or situation awareness can meet this criteria and potentially play a role in guiding real-time human engineering.

Mental Workload Adaptation Metrics. This condition has been referred to as a "redline" condition and has been linked to an increased propensity for performance errors (Reid & Colle, 1988; Rueb, Vidulich, & Hassoun, 1994). One class of mental workload measures that meets the real-time availability requirement for measures to guide real-time human engineering are physiological measures.

Physiological mental workload metrics are based on the fact that the strain of high workload conditions has detectable effects on various physiological parameters (e.g., heart activity, eye blinks and movements). Physiological metrics can be collected in real time and have been investigated not only for their appropriateness for guiding adaptation, but have even been demonstrated to effectively guide adaptation in some laboratory tasks (Hilburn, 1997; Kramer, Trejo, & Humphrey, 1996; Parasuraman, Mouloua, & Hilburn, 1999; Scerbo, Freeman, & Mikulka, 2000; Wilson, Lambert, & Russell, 2000). Parasuraman and Byrne provide a detailed discussion of this research domain in chapter 9.

Situation Awareness Adaptation Metrics. Inasmuch as the typical memory probe metric of situation awareness would be incompatible with the real-time availability requirement for measures to guide real-time human engineering, are there metrics that would represent situation awareness meaningfully in real time? Although physiological measures have a long history of use as mental workload measures, there has been relatively little research into their potential use as situation awareness measures (Wilson, 2001). However, there has been consideration of assessing real-time performance as an indication of situation awareness (Durso et al., 1995, 1998; Pritchett & Hansman, 2000; Vidulich & McMillan, 2000).

As an example of a real-time, performance-based measure of situation awareness, the Global Implicit Measure (GIM) will be considered (Vidulich & McMillan, 2000). The GIM is based on the assumption that the pilot is attempting to accomplish known goals at various known priority levels. Therefore, it is possible to consider the momentary progress toward accomplishing these goals as a performance-based measure of situation awareness. Development of the GIM (Brickman et al., 1995, 1999; Vidulich, 1995) was an attempt to develop an approach for creating real-time situation awareness measurement that could guide effective automated pilot aiding. In the GIM approach, a detailed task analysis is used to link measurable behaviors to the accomplishment of mission goals. The goals will vary depending on the current mission phase. For example, during a combat air patrol, a pilot might be instructed to maintain a specific altitude and to use a specific mode of their on-board radar, but during an intercept the optimal altitude might be defined in relation to the aircraft being intercepted and a different radar mode might be appropriate. For each phase, these measurable behaviors that logically affect goal accomplishment are identified and scored. The scoring was based on contribution to goal accomplishment. The proportion of goals being successfully accomplished in the GIM algorithms related to a specified mission phase identified how well the pilot was accomplishing the goals of that phase (leaving aside the issue of weighting behavioral components of different priorities, for simplicity sake). More importantly, the behavioral components scored as failing should identify the portions of the task that the pilot was either unaware of or unable to perform at the moment. Thus, GIM scores could potentially provide a real-time indication of the person's situation awareness as reflected by of the quality of task performance and a diagnosis of the problem if task performance deviates from the ideal, as specified by the GIM task analysis and scoring algorithms.

Vidulich and McMillan (2000) tested the GIM metric in a simulated air-to-air combat task using two cockpit designs that were known to produce different levels of mission performance, mental workload, and rated situation awareness. The participants were seven U.S. military pilots or weapons systems officers. The real-time GIM scores successfully distinguished between the two cockpits and the different phases of the mission. No attempt was made to guide adaptation on the basis of the GIM scores, but the results suggested that such an approach has promise.

CONCLUSION

As computerized and automated systems in military and commercial aircraft have proliferated over the years, the pilot's task has become increasingly complex and difficult. In many settings, automation has been used as a means of keeping the task demands within a reasonable level (see chap. 9). However, wide use of automation has introduced difficulties of its own. Pilots often find automation doing unexpected things for no obvious reasons (Sarter & Woods, 2000). The lack of understanding might be caused by the pilot being unaware of the current mode of the automation or possibly not having a good mental model of what the automation in a given mode is going to do under various circumstances. Such problems will often be attributed as pilot error when they occur and may be addressed with additional training or alarms to "assist" future pilots. On the other hand, it is equally reasonable to suspect that some of the problems in modern aircraft might be due to the tendency of the automation to effectively ignore the state of the pilot. A good human crew member, particularly a copilot, not only assesses the situation surrounding the aircraft and what the pilot appears to be trying to do, but also assesses the toll being placed on the pilot by the task. For example, a pilot that was being overwhelmed by simultaneous high-priority tasks might have a fine understanding of the relevant goals for the moment, but might be unable to accomplish them all. A responsive copilot might assist by identifying subtasks that could be off-loaded from the pilot. However, a copilot that detected a fundamental mismatch between a situation and a pilot's current activity might just try to bring the pilot's attention to the situation. In short, a good copilot's situation awareness includes relevant information about the state of the pilot.

What about the automated system's situation awareness? Presumably, the automation is provided with appropriate information about the situation to guide its activity. However, it also is presumably susceptible to "mode error" if it does not understand the operating "mode" of the pilot. An understanding of the pilot should logically be a piece of the information provided to the automated aiding system. This is a serious challenge that real-time human engineering research should be directed to address.

An approach for implementing real-time human engineering in the cockpit is illustrated in Fig. 4.3. To be successful, real-time human engineering will need to close all the loops between the task environment, the aircraft, and the pilot. Human engineering research has provided designers with a sizable toolbox for helping the pilot (e.g., customized display format, new display modalities, automation) cope with well-identified problems. If there is any glaring gap in the state of knowledge for supporting a fully adaptive real-time human engineering system as portrayed in Fig. 4.3, it is uncertainty about how to implement the "pilot state" estimator. The critical question is how the pilot's needs

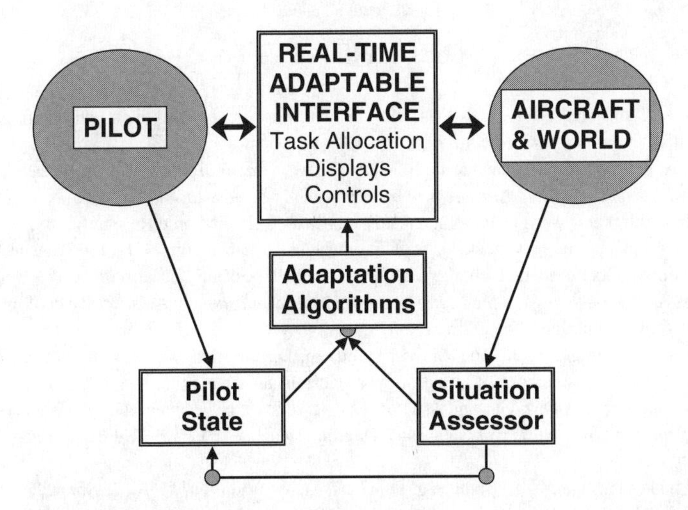

FIG. 4.3. Illustration of the potential future role of mental workload and situation awareness metrics in an adaptive cockpit. Real-time mental workload and situation awareness metrics would reside within the pilot state estimator.

can be identified accurately in real-time so that beneficial, rather than disrupting, changes in the interface can be made.

Based on the years of research and thinking about mental workload and situation awareness, the most defensible current suggestion seems to be that the pilot state estimator must contain viable real-time metrics of both. The prior work on psychophysiological measures of mental workload and performance-based measures of situation awareness are pointing the way, but neither is likely to be sufficient on its own. As described earlier, the concepts of mental workload and situation awareness capture important aspects of the pilot's reactions to the current task environment and his or her performance. A complete understanding of the pilot's (or, more generally, the operator's) state requires that both mental workload and situation awareness be measured. Research combining these approaches as part of an operator state estimator within the larger context of a full real-time human engineering system has barely begun. So far, although there has been discussion of the need for such an operator state estimator capability in future highly automated aircraft, there has not been a demonstration of the capability in an operational setting nor any realistic simulation of an operational setting (e.g., Haas & Repperger, 1998; Taylor, Howells, & Watson, 2000). Such research is important not only because it may be a harbinger of human–machine aviation systems of unprecedented complexity, effectiveness, and safety, but also because it will provide the most powerful test of the utility of mental workload and situation awareness as a meta-measure for characterizing the demands of aviation on human cognitive capabilities.

REFERENCES

Adams, M. J., Tenny, Y. J., & Pew, R. W. (1995). Situation awareness and the cognitive management of complex systems. *Human Factors, 37,* 85–104.

Alexander, A. L., Nygren, T. E., & Vidulich, M. A. (2000). *Examining the relationship between mental workload and situation awareness in a simulated air combat task* (Tech. Rep. No. AFRL-HE-WP-TR-2000–0094). Wright-Patterson Air Force Base, OH: Air Force Research Laboratory.

Amar, M. J., Hansman, R. J., Vaneck, T. W., Chaudhry, A. I., & Hannon, D. J. (1995). A preliminary evaluation of electronic taxi charts with GPS derived position for airport surface situational awareness. In *Proceedings of the Eighth International Symposium on Aviation Psychology* (Vol. 1, pp. 499–504). Columbus: The Ohio State University.

Brickman, B. J., Hettinger, L. J., Roe, M. M., Stautberg, D., Vidulich, M. A., Haas, M. W., & Shaw, R. L. (1995). An assessment of situation awareness in an air combat task: The Global Implicit Measure approach. In D. J. Garland & M. R. Endsley (Eds.), *Experimental analysis and measurement of situation awareness* (pp. 339–344). Daytona Beach, FL: Embry-Riddle Aeronautical University Press.

Brickman, B. J., Hettinger, L. J., Stautberg, D., Haas, M. W., Vidulich, M. A., & Shaw, R. L. (1999). The Global Implicit Measurement of situation awareness: Implications for design and adaptive interface technologies. In M. W. Scerbo & M. Mouloua (Eds.), *Automation technology and human performance: Current research and trends* (pp. 160–164). Mahwah, NJ: Lawrence Erlbaum Associates.

Broadbent, D. E. (1958). *Perception and communication.* London: Pergamon Press.

Chapanis, A. (1965). *Man-machine engineering.* Monterey, CA: Brooks/Cole.

Clancy, T., & Horner, C. (1999). *Every man a tiger: The Gulf War air campaign.* New York: Berkley Books.

Cowan, N. (1995). *Attention and memory: An integrated framework.* New York: Oxford University Press.

Deutsch, J. A., & Deutsch, D. (1963). Attention: Some theoretical considerations. *Psychological Review, 70,* 80–90.

Draper, M. H., Geiselman, E. E., Lu, L. G., Roe, M. M., & Haas, M. W. (2000). Display concepts supporting crew communication of target location in unmanned air vehicles. In *Proceedings of the IEA 2000/HFES 2000 Congress* (Vol. 3, pp. 3-85–3-88). Santa Monica, CA: Human Factors and Ergonomics Society.

Durso, F. T., & Gronlund, S. D. (1999). Situation awareness. In F. T. Durso, R. S. Nickerson, R. W. Schvaneveldt, S. T. Dumais, D. S. Lindsay, & M. T. H. Chi (Eds.), *Handbook of applied cognition* (pp. 283–314). New York: Wiley.

Durso, F. T., Hackworth, C. A., Truitt, T. R., Crutchfield, J., Nokolic, D., & Manning, C. A. (1998). Situation awareness as a predictor of performance for en route air traffic controllers. *Air Traffic Control Quarterly, 6,* 1–20.

Durso, F. T., Truitt, T. R., Hackworth, C. A., Ohrt, D., Hamic, J. M., Crutchfield, J. M. & Manning, C. A. (1995). In D. J. Garland & M. R. Endsley (Eds.), *Experimental analysis and measurement of situation awareness* (pp. 189–195). Daytona Beach, FL: Embry-Riddle Aeronautical University Press.

Endsley, M. R. (1988a). Design and evaluation for situation awareness enhancement. In *Proceedings of the Human Factors Society 32nd Annual Meeting* (Vol. 1, pp. 97–101). Santa Monica, CA: Human Factors Society.

Endsley, M. R. (1988b). Situation awareness global assessment technique (SAGAT). In *Proceedings of the IEEE 1988 National Aerospace and Electronics Conference—NAECON 1988* (Vol. 3, pp. 789–795). New York: Institute of Electrical and Electronics Engineers.

Endsley, M. R. (1990a). A methodology for the objective measurement of pilot situation awareness. In *Situation Awareness in Aerospace Operations* (AGARD-CP-478, pp. 1-1–1-9). Neuilly-sur-Seine, France: Advisory Group for Aerospace Research and Development.

Endsley, M. R. (1990b). Predictive utility of an objective measure of situation awareness. In *Proceedings of the Human Factors Society 34th Annual Meeting* (Vol. 1, pp. 41–45). Santa Monica, CA: Human Factors Society.

Endsley, M. R. (1993). Situation awareness and workload: Flip sides of the same coin. In *Proceedings of the Seventh International Symposium on Aviation Psychology* (Vol. 2, pp. 906–911). Columbus: The Ohio State University.

Endsley, M. R. (1995a). Measurement of situation awareness in dynamic systems. *Human Factors, 37,* 65–84.

Endsley, M. R. (1995b). Toward a theory of situation awareness in dynamic systems. *Human Factors, 37,* 32–64.

Ericsson, K. A., & Kintsch, W. (1995). Long-term working memory. *Psychological Review, 102,* 211–245.

Ericsson, K. A., & Staszewski, J. J. (1989). Skilled memory and expertise: Mechanisms of exceptional performance. In D. Klahr & K. Kotovsky (Eds.), *Complex information processing: The impact of Herbert A. Simon* (pp. 235–267). Hillsdale, NJ: Lawrence Erlbaum Associates.

Gawron, V. J. (2000). *Human performance measures handbook.* Mahwah, NJ: Lawrence Erlbaum Associates.

Gopher, D., & Donchin, E. (1986). Workload—an examination of the concept. In K. R. Boff, L. Kaufman, & J. P. Thomas (Eds.), *Handbook of perception and human performance: Volume II. Cognitive processes and performance* (chap. 41, pp. 41-1–41-49). New York: Wiley.

Grether, W. F. (1995). Human engineering: The first 40 years, 1945–1984. In R. J. Green, H. C. Self, & T. S. Ellifritt (Eds.), *50 years of human engineering* (pp. 1-9–1-17). Wright-Patterson Air Force Base, OH: Crew Systems Directorate.

Guthrie, E. R. (1952). *The psychology of learning* (rev. ed.). New York: Harper & Brothers.
Haas, M. W., & Repperger, D. W. (1998). Automated adaptive control of display characteristics within future air force crew stations. In *Proceedings of the 1998 American Control Conference* (Vol. 1, pp. 450–452). Piscataway, NJ: American Automatic Control Council.
Hardiman, T. D., Dudfield, H. J., Selcon, S. J., & Smith, F. J. (1996). Designing novel head-up displays to promote situational awareness. In *Situation awareness: Limitations and enhancement in the aviation environment* (AGARD-CP-575, pp. 15-1–15-7). Neuilly-sur-Seine, France: Advisory Group for Aerospace Research and Development.
Hart, S. G., & Wickens, C. D. (1990). Workload assessment and prediction. In H. R. Booher (Ed.), *Manprint: An integrated approach to systems integration* (pp. 257–296). New York: Van Nostrand.
Hilburn, B. (1997). Dynamic decision aiding: The impact of adaptive automation on mental workload. In D. Harris (Ed.), *Engineering psychology and cognitive ergonomics: Volume One* (pp. 193–200). Aldershot, England: Ashgate.
Hilgard, E. R., & Marquis, D. G. (1940). *Conditioning and learning.* New York: Appleton-Century.
Hollnagel, E. (1998). Measurements and models, models and measurements: You can't have one without the other. In *Collaborative crew performance in complex operational systems* (RTO-MP-4, pp. 14-1–14-8). Neuilly-sur-Seine, France: NATO-RTO.
Hughes, E. R., Hassoun, J. A., Ward, G. F., & Rueb, J. D. (1990). *An assessment of selected workload and situation awareness metrics in a part-mission simulation* (Tech. Rep. No. ASD-TR-90–5009). Wright-Patterson Air Force Base, OH: Air Force Systems Command.
Hull, C. L. (1943). *Principles of behavior.* New York: Appleton-Century.
Jones, D. G., & Endsley, M. R. (2000). Examining the validity of real-time probes as a metric of situation awareness. In *Proceedings of the IEA 2000/HFES 2000 Congress* (Vol. 1, p. 278). Santa Monica, CA: Human Factors and Ergonomics Society.
Kahneman, D. (1973). *Attention and effort.* Englewood Cliffs, NJ: Prentice Hall.
Kandebo, S. W. (2000, March 27). Advanced helmet in U.S. tests. *Aviation Week & Space Technology, 152*(13), 56–57.
Kantowitz, B. H. (1986). Mental workload. In P. A. Hancock (Ed.), *Human factors psychology* (pp. 81–121). Amsterdam: Elsevier.
Kantowitz, B. H. (2000). Attention and mental workload. In *Proceedings of the IEA 2000/HFES 2000 Congress* (pp. 3-456–3-459). Santa Monica, CA: Human Factors and Ergonomics Society.
Kantowitz, B. H., & Casper, P. A. (1988). Human workload in aviation. In E. L. Wiener & D. C. Nagel (Eds.), *Human factors in aviation* (pp. 157–187). San Diego, CA: Academic Press.
Kantowitz, B. H., Hart, S. G., & Bortolussi, M. R. (1983). Measuring pilot workload in a moving-base simulator: I. Asynchronous secondary choice-reaction task. In *Proceedings of the Human Factor Society 27th Annual Meeting* (Vol. 1, pp. 319–322). Santa Monica, CA: Human Factors Society.
Klatzky, R. L. (1984). *Memory and awareness: An information-processing perspecive.* New York: Freeman.
Knowles, W. B. (1963). Operator loading tasks. *Human Factors, 5,* 155–161.
Kramer, A. F., Trejo, L. J., & Humphrey, D. G. (1996). Psychophysiological measures of workload: Potential applications to adaptively automated systems. In R. Parasuraman & M. Mouloua (Eds.), *Automation and human performance* (pp. 137–162). Mahwah, NJ: Lawrence Erlbaum Associates.
Lachman, R., Lachman, J. L., & Butterfield, E. C. (1979). *Cognitive psychology and information processing: An introduction.* Hillsdale, NJ: Lawrence Erlbaum Associates.
MacCorquodale, K., & Meehl, P. E. (1948). On a distinction between hypothetical constructs and intervening variables. *Psychological Review, 55,* 95–107.
Mecham, M. (1999, August 16). Shuttle teams plan sustained upgrades. *Aviation Week & Space Technology, 151*(7), 63–64.
Miller, G. A. (1956). The magical number seven, plus or minus two: Some limits of our capacity for processing information. *Psychological Review, 63,* 81–97.
Moray, N. (1967). Where is attention limited? A survey and a model. *Acta Psychologica, 27,* 84–92.

Nordwall, B. D. (1995, February 6). Military cockpits keep autopilot interface simple. *Aviation Week & Space Technology, 142*(6), 54–55.

O'Donnell, R., & Eggemeier, F. T. (1986). Workload assessment methodology. In K. R. Boff, L. Kaufman, & J. P. Thomas (Eds.), *Handbook of perception and human performance: Volume II. Cognitive processes and performance* (chap. 42, pp. 42-1–42-49). New York: Wiley.

Parasuraman, R., Mouloua, M., & Hilburn, B. (1999). Adaptive aiding and adaptive task allocation enhance human-machine interaction. In M. W. Scerbo & M. Mouloua (Eds.), *Automation technology and human performance: Current research and trends* (pp. 119–123). Mahwah, NJ: Lawrence Erlbaum Associates.

Pashler, H. & Johnston, J. C. (1998). Attention limitations in dual-task performance. In H. Pashler (Ed.), *Attention* (pp. 155–189). East Sussex, England: Psychology Press.

Pritchett, A. R., & Hansman, R. J. (2000). Use of testable responses for performance-based measurement of situation awareness. In M. R. Endsley & D. J. Garland (Eds.), *Situation awareness analysis and measurement* (pp. 189–209). Mahwah, NJ: Lawrence Erlbaum Associates.

Proctor, P. (2000, April 24). VNav profile to boost situational awareness. *Aviation Week & Space Technology, 152*(17), 68–69.

Reason, J. (1990). *Human error.* Cambridge, England: Cambridge University Press.

Reid, G. B., & Colle, H. A. (1988). Critical SWAT values for predicting operator overload. In *Proceedings of the 32nd Annual Meeting of the Human Factor Society* (Vol. 2, pp. 1414–1418). Santa Monica, CA: Human Factors Society.

Rueb, J. D., Vidulich, M. A., & Hassoun, J. A. (1994). Use of workload redlines: A KC-135 crew-reduction evaluation. *International Journal of Aviation Psychology, 4,* 47–64.

Sarter, N. B., & Woods, D. D. (2000). Team play with a powerful and independent agent: A full-mission simulation study. *Human Factors, 42,* 390–402.

Scerbo, M. W., Freeman, F. G., & Mikulka, P. J. (2000). A biocybernetic system for adaptive automation. In R. W. Backs & W. Boucsein (Eds.), *Engineering psychophysiology: Issues and applications* (pp. 241–253). Mahwah, NJ: Lawrence Erlbaum Associates.

Schiflett, S. G., Linton, P. M., & Spicuzza, R. J. (1981). Evaluation of a pilot workload assessment device to test alternate display formats and control handling qualities. In *The impact of new guidance and control systems on military aircraft cockpit design* (AGARD-CP-312, pp. 24-1–24-12). Neuilly-sur-Seine, France: Advisory Group for Aerospace Research and Development.

Selcon, S. J., Hardiman, T. D., Croft, D. G., & Endsley, M. R. (1996). A test-battery approach to cognitive engineering: To meta-measure or not to meta-measure, that is the question! In *Proceedings of the Human Factors and Ergonomics Society 40th Annual Meeting* (Vol. 1, pp. 228–232). Santa Monica, CA: Human Factors and Ergonomics Society.

Sternberg, S. (1966). High-speed scanning in human memory. *Science, 153,* 652–654.

Taylor, R. M., Howells, H., & Watson, D. (2000). The cognitive cockpit: Operational requirement and technical challenge. In P. T. McCabe, M. A. Hanson, & S. A. Robertson (Eds.), *Contemporary ergonomics 2000* (pp. 55–59). London: Taylor & Francis.

Treisman, A. M. (1969). Strategies and models of selective attention. *Psychological Review, 76,* 282–292.

Tsang, P. S., Shaner, T. L., & Vidulich, M. A. (1995). Resource scarcity and outcome conflict in time-sharing performance. *Perception & Psychophysics, 57,* 365–378.

Tsang, P. S., & Wilson, G. F. (1997). Mental workload. In G. Salvendy (Ed.), *Handbook of human factors and ergonomics* (2nd ed., pp. 417–449). New York: Wiley.

Ververs, P. M., & Wickens, C. D. (2000). Designing head-up displays (HUDs) to support flight path guidance while minimizing effects of cognitive tunneling. In *Proceedings of the IEA 2000/HFES 2000 Congress* (Vol. 3, pp. 3-45–3-48). Santa Monica, CA: Human Factors and Ergonomics Society.

Vidulich, M. A. (1995). The role of scope as a feature of situation awareness metrics. In D. J. Garland & M. R. Endsley (Eds.), *Experimental analysis and measurement of situation awareness* (pp. 69–74). Daytona Beach, FL: Embry-Riddle Aeronautical University Press.

Vidulich, M. A. (2000a). The relationship between mental workload and situation awareness. In *Proceedings of the IEA 2000/HFES 2000 Congress* (Vol. 3, pp. 460–463). Santa Monica, CA: Human Factors and Ergonomics Society.

Vidulich, M. A. (2000b). Testing the sensitivity of situation awareness metrics in interface evaluations. In M. R. Endsley & D. J. Garland (Eds.), *Situation awareness analysis and measurement* (pp. 227–246). Mahwah, NJ: Lawrence Erlbaum Associates.

Vidulich, M., Dominguez, C., Vogel, E., & McMillan, G. (1994). *Situation awareness: Papers and annotated bibliography* (Tech. Rep. No. AL/CF-TR-1994–0085). Wright-Patterson Air Force Base, OH: Armstrong Laboratory.

Vidulich, M. A., McCoy, A. L., & Crabtree, M. S. (1995). The effect of a situation display on memory probe and subjective situational awareness metrics. In *Proceedings of the Eighth International Symposium on Aviation Psychology* (Vol. 2, pp. 765–768). Columbus: The Ohio State University.

Vidulich, M. A., & McMillan, G. (2000). The Global Implicit Measure: Evaluation of metrics for cockpit adaptation. In P. T. McCabe, M. A. Hanson, & S. A. Robertson (Eds.), *Contemporary ergonomics 2000* (pp. 75–80). London: Taylor & Francis.

Wall, R. (2000, November 20). Upgrades emerge to keep F-117 flying. *Aviation Week & Space Technology, 153*(21), 27–28.

Wickens, C. D. (1980). The structure of attentional resources. In R. Nickerson (Ed.), *Attention and performance, VIII* (pp. 239–257). Hillsdale, NJ: Lawrence Erlbaum Associates.

Wickens, C. D. (1992, December). Workload and situation awareness: An analogy of history and implications. *Insight: The Visual Performance Technical Group Newsletter, 14*(4), 1–3.

Wickens, C. D. (1996). Situation awareness: Impact of automation and display technology. In *Situation awareness: Limitations and enhancement in the aviation environment* (AGARD-CP-575, pp. K2-1–K2-13). Neuilly-sur-Seine, France: Advisory Group for Aerospace Research and Development.

Wickens, C. D., Hyman, F., Dellinger, J., Taylor, H., & Meador, M. (1986). The Sternberg memory search task as an index of pilot workload. *Ergonomics, 29,* 1371–1383.

Wiener, E. (1993). Life in the second decade of the glass cockpit. In *Proceedings of the Seventh International Symposium on Aviation Psychology* (Vol. 3, pp. 1–7). Columbus: The Ohio State University.

Wilson, G. F. (2001). Strategies for psychophysiological assessment of situation awareness. In D. J. Garland & M. R. Endsley (Eds.), *Experimental analysis and measurement of situation awareness* (pp. 175–188). Daytona Beach, FL: Embry-Riddle Aeronautical University Press.

Wilson, G. F., Lambert, J. D., & Russell, C. A. (2000). Performance enhancement with real-time physiologically controlled adaptive aiding. In *Proceedings of the IEA 2000/HFES 2000 Congress* (Vol. 3, pp. 3-61–3-64). Santa Monica, CA: Human Factors and Ergonomics Society.

5

Aviation Displays

Christopher D. Wickens
University of Illinois at Urbana–Champaign

Flying an aircraft can often be described in terms of performing four "meta tasks" (categories of tasks) whose priority, from highest to lowest, is in the following order: *aviate* (keeping the plane in the air); *navigate* (moving from point to point in 3-D space and avoiding hazardous regions of terrain, weather, traffic, or combat vulnerability); *communicate* (primarily with air traffic control and with others on board the aircraft); and *manage systems.* All four of these tasks require that information be presented (displayed) to the pilot or air traffic controller in order for them to be carried out successfully, and in this chapter, we will describe the characteristics of effective displays that will support their performance. In particular, the first two tasks, aviate and navigate, are closely linked, and hence their display issues are also closely entwined.

A major reason for the close linkage between aviation and navigation is that satisfying the goals for aviating often affect achievement of navigational goals as follows. In order to keep an aircraft from "stalling" (and falling out of the sky—i.e., in order to aviate), the pilot must keep an adequate flow of air over the wings, at a particular angle relative to the wing's surface, an angle known as the *angle of attack.* This angular flow is influenced collectively by the plane's airspeed, by the *pitch* of the aircraft upward or downward along the axis of motion, and by the roll or *bank* of the aircraft to the left or right. (Too much pitch, too much roll, or too low an airspeed will all increase the likelihood of stalling.) However, all three of these variables also influence navigation. Changing airspeed will change the time at which destinations are reached (and may influence altitude as well). Pitching upward or downward will increase or decrease altitude, respectively; banking to the left or right will lead to a left or right turn, respectively.

Because the two tasks of aviating and navigating therefore have overlapping information requirements and because they also stand at the top of the task hierarchy, they will receive the greatest focus of attention in our discussion of aviation displays. In the following chapter, we will first describe seven principles of effective aviation display design, in the process offering some aviation relevant examples, particularly for the *aviate* subtask. We then describe how navigational displays have evolved to incorporate hazard representation and alerting and then present some of the human factors problems associated with these "multipurpose" hazard displays related to reliability, attention, and clutter. A brief discussion of procedural displays is followed by a detailed examination of the human factor implications of three trends in advanced display technology: 3-D displays, head-up and helmet-mounted displays, and display automation.

DISPLAY PRINCIPLES

The cockpit designer confronts several constraints when trying to decide what information to present to the pilot, where it should be presented, and how it should be configured (i.e., formatting). Many of these constraints relate to such factors as cost and weight and, as such, have little to do directly with human factors, except as the latter can be entered into a cost–benefit analysis (Hendrick, 1998; Wickens, Gordon, & Liu, 1998). However, these constraints are very real and often can account for the reason some important design principles that are described in this section may not be incorporated. Bearing this important qualification in mind, we proceed to describe seven critical principles of display design, that relate to, and flow from, an understanding of the psychology of information processing (Wickens & Hollands, 2000). In doing so, we place these principles in an aviation context, illustrating them by specific examples. Then we refer to these more extensively in the later sections of the chapter, as different principles become relevant for the design of different kinds of displays.

Principle of Information Need

It would appear to be obvious that pilots should only be presented with the information necessary to carry out their required tasks. Yet, such an approach has not always been adhered to in practice, leading, for example, to the proliferation of displays characteristic of the L1011 cockpit, with several hundred different indicators. To avoid problems of visual overload, an effective display design should be proceeded by a careful *task analysis* (Kirwan & Ainsworth, 1992; Wickens, Gordon, & Liu, 1998), which, in turn, should spawn an *information needs analysis* (what information is required for each potential task; see Wickens, Vincow, Schopper, & Lincoln, 1997), and this latter analysis can then specify what information to display. If certain kinds of information are more frequently needed, they should be displayed in locations that are more accessible (e.g., closer to the pilots' forward line of sight, as he or she sits in the cockpit). If other kinds are not needed at all, they should not be displayed.

Principle of Legibility

It is also self-evident that displays must be legible to be useful (although legibility alone does not guarantee usefulness). Display size must be large enough so that details can be resolved. Contrast and brightness must be adequate for all levels of illumination and glare, and for auditory displays, adequate loudness and masking avoidance represent corresponding concerns. These are critical issues, but we do not consider them further in this chapter because, generally, they apply to all displays, whether or not these are in an aviation context. Hence, the reader is referred to their discussion in books such as Boff and Lincoln (1988), Salvendy (1997), Sanders and McCormick (1993), or Wickens, Gordon, and Liu (1998).

Principle of Display Integration/Proximity Compatibility Principle

When the eye scans separate sources of information, as, for example, it moves across the typical cockpit instrument panel shown in Fig. 5.1, this demands mental effort in order for the pilot to direct the visual scan (via eye movement and head movement) to the appropriate location at the right time. For the skilled pilot, this process is usually relatively automatic, but still takes time (Bellenkes, Wickens, & Kramer, 1997; Fitts, Jones, & Milton, 1950; Harris & Christhilf, 1980). The amount of effort that is required for scanning is amplified by two factors, which both have critical implications for display design: (a) when the scanning destinations (relevant instruments) are farther apart and (b) when pairs of destinations contain information that must be related, compared, or integrated in performing the task at hand. The first characteristic dictates the importance of keeping frequently used instruments directly in front of the pilot (Wickens et al., 1997) so that minimal long-distance scanning is required to access them repeatedly. Instruments contained within a visual angle of approximately 20 degrees can be scanned without head movement (Previc, 1998, 2000), a desirable feature if vestibular disorientation is to be avoided. The second characteristic is described by the *proximity compatibility principle* of display design (Wickens & Carswell,

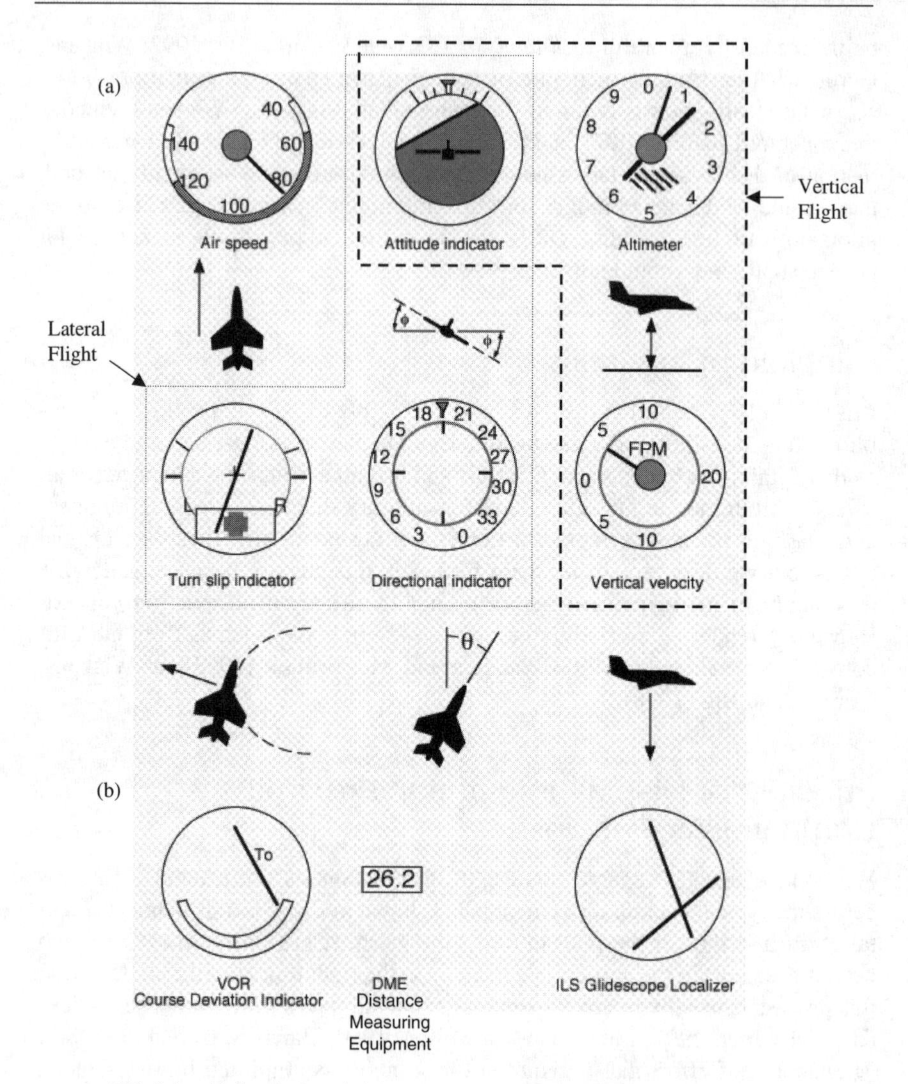

FIG. 5.1. (a) Standard cockpit instrument panel. Under each instrument is represented the flight variable whose information is conveyed. The enclosures represented by the solid and dashed lines represent the set of instruments used for lateral and vertical flight respectively. (b) Navigational instruments, present on a radio navigational aircraft, qualified for instrument flight rules (IFR). The configuration of the ILS needles on the right shows the aircraft (the center of the circle) above the glideslope and to the left of the localizer.

1995). That is, when information sources need to be integrated or compared by the pilot (close "mental proximity"), they should be positioned close together on the display (close "display proximity"), therefore requiring little scanning activity. Thus, there should be a "compatibility" or agreement between closeness or relatedness in the mind and closeness on the display, sources of displayed information that need to be integrated (mental proximity) should be closer together or more integrated on the display than sources that do not.

In Fig. 5.1(a), we use two aspects of the cockpit instrument panel to illustrate this principle. The first aspect relates to the layout of the pilots' most important cockpit instruments. A pilot's task can often be associated with control along the three primary axes of flight: lateral (turning), vertical (climbing or descending), and longitudinal (airspeed changes). The layout of the six primary flight instruments has been configured so that the sets of instruments most relevant for each axis are located in adjacent positions, as suggested by the lines surrounding each subgroup. Thus, for example, consider lateral control (solid line). Aircraft bank and turn information, indicated respectively by the bank angle of the artificial horizon (attitude indicator) and the turn coordinator (turn slip indicator), both lie directly next to the heading indicator (directional indicator or compass), whose value is determined by both bank and turn parameters. In a coordinated turn to reach a desired heading, the pilot must frequently consult all three of these instruments, relating their changing values to each other. Hence, it is effective design that they be placed in adjacent locations. Now, consider the vertical control necessary to change altitude (dashed line). Aircraft pitch information and vertical speed information, conveyed directly by the moving horizon and the vertical velocity display, are both positioned adjacent to the altimeter. Because pitch attitude directly affects airspeed, those two indicators are also adjacent.

The second aspect of the instrument panel demonstrates how display proximity can be achieved through the actual integration of related information rather than spatial proximity (Wickens & Carswell, 1995). This is reflected by the format of the moving horizon in the attitude indicator (AI), which is shown at the top center of the panel. This critical indicator incorporates two channels of information—pitch and bank (or roll)—integrated into the two dimensional movement of a single object—the horizon line. Such integration is valuable in reducing the mental effort of scanning, both because these two sources are of critical importance to the primary task of aviating (preventing stall) and navigating (choosing heading and altitude), and hence must be frequently scanned, and because, in a turn, these two dimensions of aircraft attitude must be integrated in the pilot's mind. This integration is required because a turning (banked) attitude causes a corresponding loss of lift and a downward pitch of the aircraft, which must be compensated by an appropriate pitch-up control. In later sections of this chapter, we will encounter other instances of both confirmation and violation of the proximity compatibility principle.

While the proximity compatibility principle dictates display closeness for information that needs to be integrated, it also dictates that for information channels that do *not* need to be integrated but should be the sole focus of attention, close proximity (to other information) should be avoided, since such proximity produces unwanted clutter (Wickens & Carswell 1995).

Principle of Pictorial Realism

The *principle of pictorial realism,* proposed by Roscoe (1968), is closely related to, but distinct from, the proximity compatibility principle. It dictates, quite intuitively, that a display should "look like" or be a pictorial representation of the information that it represents. In Fig. 5.1(a), one can see that the integrated characteristics of the moving horizon and aircraft symbol in the AI make it a good picture of the true horizon that it represents. However, pictorial realism can also be applied to single sources of information. For example, it can be argued that the round dial altimeter violates the principle of pictorial realism to a greater extent than does a vertical display, because one's mental "picture" of altitude runs up and down, not around in a circle (Wickens & Hollands, 2000). A vertically represented altimeter, like many of the "moving tape" displays in modern aircraft, would conform better to the principle of pictorial realism. As another example, the circular compass, or directional indicator, nicely conforms to the principle of pictorial realism, because heading is a circular concept, not a linear one.

Principle of the Moving Part

The *principle of the moving part,* also associated with Roscoe's (1968) work, proposes that the moving element on a display should correspond with the element that moves in the pilot's "mental model," or mental representation of the aircraft, and should move in the same direction as that mental representation. Because the pilot's mental model represents the aircraft moving in a world that is stable (Johnson & Roscoe, 1972; Kovalenko, 1991; Merryman & Cacioppo, 1997; Previc, 2000; Previc & Ercoline, 1999), this principle suggests that the aircraft should be depicted as the moving element on a display, and, for example, when the aircraft climbs, this element should move upward. The moving horizon of the attitude indicator in Fig. 5.1(a) violates this principle, because the moving element (horizon) moves upward to signal that the aircraft is pitching downward and rotates to the right (or left) to signal that the aircraft is rotating left (or right). To conform to the principle, the airplane should move around a stable horizon. However, the conventional design to configure the attitude indicator as a moving horizon, rather than as a moving aircraft, is one that can be argued to better satisfy the principle of pictorial realism, because, when the pilot looks outside, the moving horizon display better corresponds with what is seen looking forward from the cockpit. Like the conventional moving-horizon attitude indicator, the moving tape altimeter, which has replaced many conventional round dial altimeters in modern cockpits, also violates the principle of the moving part: The tape moves downward to signal an altitude increase. The consequences of such violations of the principle of the moving part are possible confusions leading to spatial disorientation when, for example, a downward display movement is incorrectly interpreted as a downward aircraft movement or when a clockwise rotation of the horizon on the attitude indicator is misinterpreted as a clockwise rotation (right bank) of the aircraft (Kovalenko, 1991; Roscoe, 1968).

The results of earlier studies by Roscoe and his colleagues (Beringer, Williges, & Roscoe, 1975; Roscoe & Williges, 1975), and Kovalenko (1991) comparing moving aircraft with moving horizon attitude indicators has led to considerable debate about the ideal configuration of this important instrument (Previc & Ercoline, 1999) to conform to the one principle (moving part) or the other (pictorial realism). A review of the literature suggests that the moving aircraft display is more effective than the moving horizon display for novice pilots with no prior experience on either and is no less effective, even for experienced pilots who have *had* prior flight experience with the moving horizon display (Cohen, Otakeno, Previc, & Ercoline, 2001; Previc & Ercoline, 1999). This overall benefit for a stable horizon may result from the pilot's reflexive tendency to orient the head to the outside horizon (i.e., to visually stabilize the horizon), a phenomenon called the *optokinetic cervical reflex* (Merryman & Cacioppo, 1997; Previc, 2000; Smith, Cacioppo, & Hinman, 1997; see also chap. 3). If the head rotates counter to the airplane when the latter banks, then the eyes will actually see a stable horizon, and those aspects of the airplane that are in the visual field will be seen to rotate.

Although the debate is not likely to lead to the redesign of the head-down attitude indicator with a moving aircraft, it is possible that that certain design solutions can capture elements of both principles. The concept of a frequency separated display is one such solution, which will not be discussed here (but see Beringer et al., 1975; Wickens & Hollands, 2000, for a discussion). Another solution is the 3-D *pathway display,* to be addressed later in the chapter. This display provides a prediction of the future location of the aircraft that will conform to the principle of the moving part, even as the attitude indicator itself conforms to the principle of pictorial realism (Roscoe, 1997). To examine the role of prediction in more detail, which underlies the pathway display, we turn now to our sixth principle, the principle of predictive aiding.

Principle of Predictive Aiding

Accurate prediction of the future behavior of dynamic systems is often a cognitively demanding task (van Breda, 1999; Wickens & Hollands, 2000) and becomes more demanding the farther into the future that such prediction is needed. To accomplish such prediction in aviation, pilots or air traffic controllers must consider the current state of the aircraft, all of the anticipated forces that will be acting on it, and in essence "run" this information through their own mental model of the aircraft dynamics in order to anticipate the future state of the aircraft. Although demanding of cognitive resources or mental effort (Wickens, Gempler, & Morphew, 2000), this anticipation is also quite necessary in aviation, given the importance of planning and the fact that the heavy and sluggish characteristics of many aircraft (e.g., the wide-bodied 747 or military transport) mean that control must be based on anticipated *future* changes in or deviations from a desired flight path, rather than current ones. A control that is intended to correct a current deviation will be too late to effectively correct it and can often lead to unwanted *pilot-induced oscillations* (see chap. 7). What this means from a design standpoint is that pilots and controllers may greatly benefit from any source of predictive information regarding future aircraft state, *as long as that prediction is reasonably accurate* and is understood (Fadden, Ververs, & Wickens, 2001; Jensen, 1981; Steenblik, 1989; van Breda, 1999; Wickens, Gempler, & Morphew, 2000; Wickens, Mavor, Parasuraman, & McGee, 1998). The cockpit instrument display of Fig. 5.1(a) contains some predictive information, in that the current state of both the vertical position of the moving horizon and the vertical velocity indicator *predict* the future state of altitude. Similarly, the current bank of the horizon predicts the future heading.

Principle of Discriminability: Status Versus Command

It goes without saying that similar displays are often confused (the displayed words *attitude* and *altitude* are clear examples). Therefore, designers should assure that a displayed element in a certain context never looks (or sounds) similar to another element that *could* occur in the same display context; that is, that they are *discriminable.* For example, an accident near Strassbourg, France, was inferred to occur because pilots confused the digital setting of flight path angle (2.5 degrees) with the nearly identical appearance of the setting of descent rate (2.5 thousand ft/min) in the same window (Dornheim, 1995).

The principle that displays should be discriminable is directly relevant to the important distinction between status and command displays. A status display tells the pilot where the plane is at the moment (or how it is oriented and moving relative to target values), whereas a command display informs the pilot what action he or she should take in order to reach a desired state. All of the cockpit instruments in Fig. 5.1 are status displays. However, the *flight director* in many aircraft is a command display because it tells the pilot how he or she should move the control stick in order to obtain or recover a desired flight path.

With regard to the principle of discriminability, it should be evident that a confusion between status and command information can sometimes have very serious consequences (imagine a pilot confusing status information that he *is* left of the flight path with command information that he should *steer* left to regain the flight path). Hence, depiction of command or status information must be clear and unambiguous in informing the pilot as to which is represented, so that no confusion is possible. It should also be noted that command information is more likely to be based on some level of automation, because the command indicator typically requires the display software to make some inference as to what should be done by the pilot given the current status (i.e., inference as to the pilot's goals). Hence, it is quite important to ensure that the automation algorithms that make the inference are reliable, an issue we discuss later in the chapter.

Summary and Conclusion

In summary, research has identified a number of important principles that support effective display processing. Unfortunately, for various reasons, not all principles can be readily incorporated into aviation display design. Sometimes departure from a principle will produce only minor costs to performance, and these can be fairly easily compensated for by pilot training. At other times, as we have seen in the case of the attitude indicator, pairs of principles will trade off against each other in competing designs. In such cases, research may reveal the conditions in which one principle or the other is more important or may reveal design solutions, such as the frequency separated display that can avoid the trade-offs and satisfy both competing principles.

CLASSIC NAVIGATION: THE HISTORIC PERSPECTIVE

As we noted at the outset, the pilot's instrument panel supports two somewhat different goals. Most critically, it supports the task of aviating, or preserving adequate airspeed and attitude to maintain lift and prevent stalling. But also high on the task hierarchy, the instrument panel provides partial information necessary to navigate. The attitude indicator helps the pilot to fly "straight and level" (which generally results when the aircraft symbol and the moving horizon are aligned), and the altimeter, vertical velocity indicator and the directional gyro or compass can guide the aircraft along 3-D flight paths. Yet, the six instruments shown in Fig. 5.1(a), sometimes referred to as the "sacred 6" are insufficient to support full navigation. For example, only the altimeter directly depicts the aircraft's location (as opposed to direction) in 3-D space, and this depiction is only in one of the three axes of space (the vertical). There is no indicator of where the aircraft is over the ground.

Historically, navigation has been supported by three primary sources—the map, the navigational flight instruments, and air traffic control. We discuss these sources first, before considering the way in which present and advanced display technology can and has addressed the human factors deficiencies associated with the three historic sources of navigational support.

The classic aviation map or chart provides a 2-D representation of important ground features as well as aviation navigation aids and air routes. When pilots are flying in visual meteorological conditions (VMC), and particularly when they are flying with visual flight rules (VFR), the map or navigational chart becomes a critical navigational aid, as pilots cross-check features that they can see on the terrain in front and beside the aircraft (in the "forward field of view" or FFOV) with those depicted on the map. Such comparisons, although necessary for careful navigation and for avoiding ground hazards, can be a source of error and high mental workload, because the FFOV is a 3-D perspective view, whereas the standard paper map display is usually a 2-D "God's eye" look down, hence violating the *principle of pictorial realism* (Aretz, 1991; Hickox & Wickens, 1999; Olmos, Liang, & Wickens, 1997). This violation becomes greater as the pilot flies lower because of the greater difference between the "God's eye" 2-D perspective of the map and the forward-looking 3-D view that the pilot sees at lower altitudes (Schreiber, Wickens, Renner, Alton, & Hickox, 1998).

During the 1950s and 60s, more specialized navigational instruments were incorporated into most aircraft to supplement the six primary instruments that were shown in Fig. 5.1(a). These navigational instruments, coupled with a series of radio procedures, helped the pilots navigate to fixed locations or electronic navigational beacons over the ground, rather than to visual landmarks, thereby enabling navigation to be done at night or in the clouds when the ground was not visible. These conditions are defined as Instrument Meteorological Conditions (IMC) and can be distinguished from the Visual Meteorological Conditions (VMC) in which visual navigation by ground reference is possible. In IMC, pilots must then navigate by reference to navigational instruments and Instrument Flight Rules, or IFR, requiring more specialized pilot training and licensing.

Examples of these IFR navigational instruments are shown in Fig. 5.1(b). For cross-country flight, these navigation instruments include (a) distance measuring equipment (DME), a digital readout of distance to or from the navigational beacons (VORs) on the ground and (b) a course deviation indicator or CDI, that depicts the deviation of the aircraft off straight line segments between these VORs. Furthermore, for landing at most commercial airports, the Instrument Landing System (ILS) glide slope and localizer needles present the vertical and lateral deviations, respectively, of the aircraft off a straight line, extending outward and upward from the runway. When the pilot is centered on the desired flight path, these needles depict a perfect cross. Unfortunately, both the VOR and ILS systems violate the *principle of the moving part,* because both show the aircraft as the stable reference point on the display and the desired flight path (the static part of the world) as the moving element. Furthermore, the DME and CDI together violate two other principles. The DME is digital, hence violating the *principle of pictorial realism,* because distance is an analog spatial concept, and although both instruments are displayed separately, the pilot needs to integrate the two quantities (position along and beside the desired course) in order to achieve a 2-D understanding of the current flight location relative to the desired ground track. This separated representation of quantities that need to be mentally integrated violates the *proximity compatibility principle.* We see next how developments in electronic display design addressed some of these principle violations.

Finally, in addition to the map and the standard flight instruments, the communication services provided by air traffic control (ATC) can be conceived as a "display" in classic navigation. This auditory display, the controller's voice, provides navigational information to the pilot regarding directions to go (heading vectors and altitudes), as well as those hazards to avoid (other aircraft and, on occasion, terrain or other ground hazards). ATC services may also provide information on weather to avoid, although this information is more often obtained during preflight planning, and updated from other sources in flight.

In closing this section regarding the historical presentation of navigational information and in setting the stage for current and future developments in display design, it is important to note the general lack of integration of much of the relevant information. In particular, information regarding the three most critical hazards for civilian pilots to avoid—other air traffic, bad weather, and terrain—often comes from very disparate and often incomplete sources (ATC, preflight planning, and paper maps), despite the critical role that such information plays in flight safety and the frequent need for pilots to mentally integrate this information in planning (and flying) safe routes.

PRESENT NAVIGATIONAL DISPLAY ENVIRONMENTS

The evolution of navigational display support is being aided by two important developments. One is the availability of digital display technology on the flight deck, which freed display designers from many of the electromechanical constraints imposed in the earlier generation of displays (e.g., round dial "steam gauge" displays or linear moving tape/moving needle displays). The second is the evolution of increased surveillance and communication technology that can identify, with great accuracy, the precise 3-D location and trajectory of an aircraft relative to the ground and to other airborne elements (weather, traffic). Much of this technology depends on digital radar and satellite navigational systems, such as the Global Positioning System. We consider in the following section three important navigational display developments that have occurred over the past few decades. Some of these have produced equipment that is now standard equipage for commercial aviation, but is less so in general aviation and in some military aircraft.

Display Integration: The Navigational Display

One of the greatest success stories in display automation has been the development of the 2-D navigational display "electronic map," or "horizontal situation display," shown in Fig. 5.2. Such a display simultaneously addresses at least six human factors concerns inherent in the previous generation of paper and electronic navigational systems, which were described previously.

1. *Integrated dynamic representation of position.* Unlike the paper map, the navigational display provides a current updated representation of the aircraft's location relative to the flight path, (the triangle at the bottom of the figure), a feature that makes a substantial contribution to improved situation awareness (see chap. 4). Furthermore, unlike the combined VOR and DME displays shown in Fig. 5.1(b), the navigational display provides a spatial *integration* of lateral error and along-track position, hence satisfying the *proximity compatibility principle.* (It should be noted, however, that the current navigational display has a fairly impoverished representation of altitude. The issue of vertical situation displays [Fradden, Braune, & Wiedemann, 1993; Oman, Kendra,

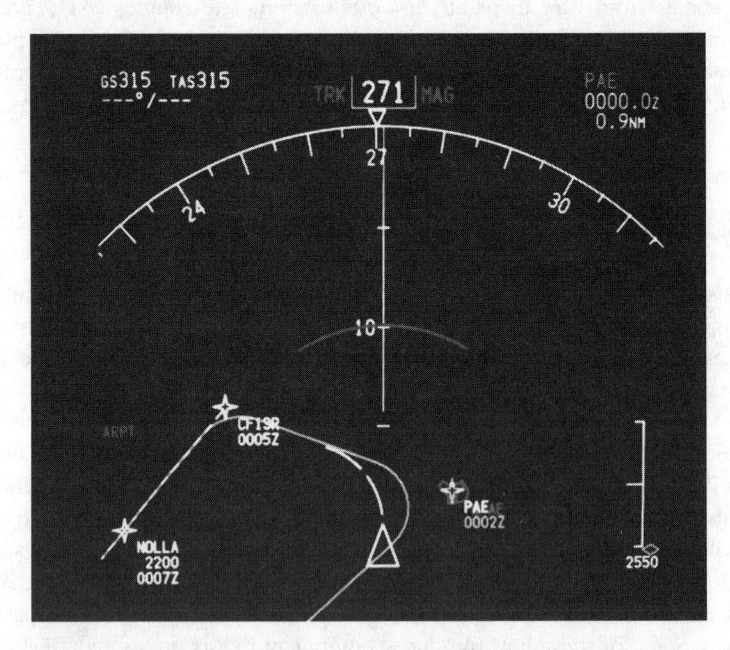

FIG. 5.2. Electronic map display, or navigation display, or horizontal situation indicator (HSI). Note the "noodle," or curved predictor line, on the end of the pilot's aircraft, located in the bottom center of the display. Courtesy of the Boeing Company.

Hayashi, Stearns, & Burki-Cohen, 2001] will be addressed at the end of this chapter.)

2. *Principle of predictive aiding.* The small dashed "noodle" at the front end of ownship's symbol in Fig. 5.2 represents an estimate of the future position of the aircraft, currently in a turn. As discussed earlier, such prediction, if accurate, can substantially reduce pilot workload in accurate maneuvering (Morphew & Wickens, 1998; Wickens, 1986).

3. *Map rotation.* The paper map must be either held in a north-up orientation (to make text reading easier) or manually rotated (upside down when heading south), if it is to be congruent, so that left on the map corresponds to left in the viewed world (and controlled aircraft). A north-up map orientation creates incongruence when the pilot is heading south, and hence requires cognitively demanding *mental rotation* when the pilot must compare features on the map with those in the world (Aretz, 1991; Clay, 1993; Wickens, Liang, Prevett, & Olmos, 1996; Williams, Hutchinson, & Wickens, 1996). If the pilot rotates the paper map orientation when the aircraft is on a southerly heading, this makes it difficult to read the upside-down text. Features of electronic vector map design allow map rotation without text inversion, hence enabling the display of spatially congruent track-up maps with readable symbology.

4. *Flexibility of representation.* The pilot has the option of choosing a fixed (north-up) or rotating (track-up) map mode, as well as selecting from a range of map scales (range of coverage). This flexibility allows a greater opportunity to match the map features to the needs of the navigational task. For example, if the pilot is communicating with other people in a mission in which directions are referenced in terms of compass directions (west, east), then the north-up representation might be best (Delzell & Battiste, 1993; Hofer & Wickens, 1997). Flying at a faster speed might require a greater range of coverage (smaller map scale) than flying at a slower speed, because in faster flight, a greater distance is covered in the same time.

5. *Updating the database.* As new "permanent" features of a terrain database change (e.g., a new radio tower is built) or as other aspects of the airspace are changed (e.g., a new standard arrival route), it is often easier to update this in the software of an electronic map than it is to modify paper maps.

6. *Integration of databases.* The single electronic "platform" of geographical space characterized by the electronic map allows for the possibility of integrating the three hazard data bases described—traffic, weather, and terrain—into a single spatial frame of reference in a manner that was previously impossible. Such integration supports the proximity compatibility principle whenever two or more of these hazards must be considered jointly (i.e., integrated) in planning or executing a flight path change (Kroft & Wickens, 2001; O'Brien & Wickens, 1997).

Considering these factors collectively, then, it is no surprise that the electronic navigational display has been a proven success story in the introduction of cockpit automation and continues to provide a valuable framework for augmenting the pilot's geographical information needs, in particular those related to ground, air, and weather hazards. In the following pages, we address three related topics that bear on the issue of geographical hazard representation. First, we describe the systems underlying the display of traffic and terrain hazards, the two systems most advanced in terms of their evolution. Then, we address certain features that are characteristic of all hazard alerting and warning systems. Finally, we address the human factors issues of integrating the displays of such information with each other and with the electronic map.

Ground Proximity Warning System (GPWS)

The GPWS is a system for alerting the pilot when the aircraft is dangerously close to the ground and is flying toward (rather than parallel) to it in a nonlanding configuration. Although it had been proposed for several years, the active mandate to require the GPWS in all commercial aviation cockpits only followed a tragic crash that a GPWS would very likely have prevented. The disaster was the 1975 crash of a commercial aircraft into the mountains west of Dulles Airport in Virginia, as the pilot heeded an ATC instruction for a direct vector to the airport. The pilot was apparently unaware that the mountain ridge intervened between the aircraft and the airport at a higher altitude than the vectored approach path (Wiener, 1977).

The GPWS is based on automation algorithms that sense and integrate sink rate, altitude above ground, and other parameters to provide a distinctive auditory alert and a command auditory display that says "pull up." Following its installation, the GPWS greatly reduced the number of controlled flight into terrain (CFIT) incidents, but was also plagued by many *alarm false alarms* or "nuisance alarms" (Pritchett, 2001). The algorithms were designed to detect a predicted ground collision even when the pilot was very much aware of the current state of the aircraft and had readily planned a maneuver to avoid it. As a result, the warnings and commands would sound, much to the annoyance of the pilot. Although such false alarms still do occur in the GPWS (Bliss, Freeland, & Millard, 1999) at least two solutions were subsequently pursued to reduce their frequency. First, algorithms have become increasingly sophisticated in order to better discriminate truly dangerous conditions from those for which the pilot is (accurately) aware that there is no danger of terrain contact. Second, recent design efforts have created the *Enhanced Ground Proximity Warning System (EGPWS),* which augments the command display ("pull up") with a status display that provides a simplified two-dimensional electronic map of the terrain in front of the aircraft that presents a hazard (Phillips, 1997). This display represents more dangerous terrain (greater likelihood of potential contact) by more intense red color coding. Providing such status information can both create a greater level of trust in the command alerts that do occur (Pritchett, 2001) and provide some prediction of locations of ground hazards near the pathway ahead, thereby supporting the *principle of predictive aiding.*

Current developments in military fighter aircraft cockpits have considered a system called the Auto-GCAS (Automatic Ground Collision Avoidance System; Scott, 1999), which takes the command input of the GPWS system a step farther. The Auto-GCAS will automatically assume control of the aircraft, restore it to a wings level attitude, and fly up for the pilot, in order to avoid the ground collision that it (the automation) inferred would have otherwise occurred. The implementation of such automation features into display technology is an issue that will be addressed further in a later section of this chapter.

The Traffic Alert and Collision Avoidance System (TCAS)

The Traffic Alert and Collision Avoidance System is designed to warn pilots of air hazards. Although it was implemented more recently than the GPWS, TCAS has many parallels with it (Wickens, Mavor, Parasuraman, & McGee, 1998). Like GPWS, the mandated equipage of the TCAS in the aircraft cockpit followed the occurrence of disaster. In the case of TCAS, the disaster was in the form of a particularly lethal midair collision over San Diego in 1980 (Wiener, 1980). In this accident, as with most fatal aircraft accidents, multiple failures were responsible. However, a firm conclusion was drawn by accident investigators that, had the planes been equipped with status displays that depicted other aircraft in the nearby airspace and command displays that advised an appropriate evasive maneuver, the tragedy would have been avoided.

These two functions—status alerting and command maneuvering—are, in fact, incorporated into the modern TCAS, which has been designed with good sensitivity to some of the human factors issues discussed earlier (Chappell, 1990). A visual status display depicts nearby traffic that automated logic determines might be a potential conflict. At a low level of concern, this merely offers a "traffic advisory" indicated by a color change of the traffic symbol on the display. At a higher level of concern, the auditory alert "traffic traffic" is sounded. Finally, should automation infer that the likelihood of a collision is high, then two command displays (easily discriminable from the status display) clearly "advise" the pilot of the appropriate maneuver. This advice, called a *resolution advisory,* or *RA,* is embodied in the appearance of a range of vertical speeds to seek (depicted green) and to avoid (depicted red) located on the vertical speed indicator of the instrument panel [Fig. 5.1(a)] and in an auditory command, stating the vertical maneuver to accomplish (e.g., "climb, climb, climb"). In addition to the effective human factors approach of combining status information with (a discriminable) command display, the TCAS also adheres to another design feature that is important in warning systems by providing two "levels" of alert—a traffic advisory and a required resolution maneuver—rather than only a single all-or-none alarm (Pritchett, 2001; Sorkin, Kantowitz, & Kantowitz, 1988).

Still, despite the solid human factors guidelines that were imposed on the TCAS design at the outset, the system continues to present some human factors challenges. First, like the GPWS, TCAS also suffers from occasional "alarm false alarms" that signal resolution advisories when the pilot does not need to make an avoidance maneuver (Bliss et al., 1999). Second, there is a concern that pilots do not always adhere to TCAS resolution advisories, even when the false alarm rate is low (Ciemer et al., 1993). Finally, whenever the pilot *does* follow the TCAS resolution advisory, he or she is now executing a maneuver that was *not* requested by ATC. Hence, the resolution advisories commanded by TCAS to some extent intrude on the authority of ATC (Wickens, Mavor, Parasuraman, & McGee, 1998).

Conclusions: Cockpit Hazard Warning Systems

In spite of some concerns regarding the alarm false alarm issue and, in the case of TCAS, concerns regarding ATC communications and authority, both GPWS and TCAS have clearly improved aviation safety (Wickens, Mavor, Parasuraman, & McGee, 1998). Furthermore, both TCAS and GPWS are defined as "systems" rather than "displays," and hence there is flexibility in both systems as to how and where their information is displayed. In particular, whereas TCAS status displays were originally presented on a separate screen from the navigational display shown in Fig. 5.2, current designs can integrate TCAS information into the navigational display (along with EGPWS terrain information) in a manner that is consistent with the proximity compatibility principle. To the extent that such a display begins to incorporate progressively more air traffic information than just the most immediate threats, it has been described as a *cockpit display of traffic information* or *CDTI* (Johnson, Battiste, & Bochow, 1999; Kreifeldt, 1980; Olmos, Mundra, Cieplak, Domino, & Stassen, 1998; Shelden & Belcher, 1999). In the following sections, we address two design issues highlighted by, but not exclusive to, TCAS and GPWS—the human factors issues in the design of all alarm/alerting warning systems and the human factors issues involved in integrating displayed information.

Hazard Alerting: The Alarm Issue

Pilots (and humans in general) are not very effective in monitoring for infrequent and/or unexpected visual events; their vigilance declines over time, and as a result, those events may be detected late or not at all (Parasuraman, 1987; Warm, 1984). Such "events" include the unexpected loss of separation from terrain or other aircraft, as discussed earlier, as well as engine, instrument or onboard computer failures. Of course, the design solution to address these vigilance issues has been the development of a variety of alarms and alerting displays that have been incorporated into the modern aircraft. Such devices are designed to provide *salient* and even intrusive mechanisms to inform the pilot of the existence of the hazardous or abnormal condition (Patterson, 1990; Pritchett, 2001; Satchell, 1993; Stanton, 1994; Wickens, Gordon, & Liu, 1998).

There are, however, a number of critically important human factors concerns that must be considered with any alarm or warning device. First, designers must adopt a somewhat arbitrary criterion as to how serious the situation becomes before the alert is displayed (auditory or visual). Such a criterion should be set so that a serious condition is never "missed' by the system (an alert is not given when the condition exists); but the somewhat inevitable consequence of such a criterion setting is the occurrence of the alarm false alarm or "nuisance alarm" (Pritchett, 2001), as we noted earlier in the context of the GPWS and TCAS systems. Such events can be more serious than mere annoyances if they lead to pilot distrust in the alarm system (Parasuraman & Riley, 1997) such that true alarms are ignored. However, such distrust itself can be mitigated by alarm systems that are designed to convey graded levels of seriousness, as if the alarm system is informing the pilot as to its own level of uncertainty that the dangerous condition exists (Sorkin, Kantowitz, & Kantowitz, 1988). As we have noted in an earlier section, this is done with the TCAS. Also, it is possible to mitigate the mistrust caused by alarm false alarms by providing pilots with easy access to the "raw data" on which the alarm decision is based (Pritchett, 2001). This characteristic, for example, is implemented in the enhanced ground proximity warning system discussed earlier, in which pilots can see the terrain image ahead that triggered the alarm. It is also embodied in the status traffic display of the TCAS system.

A second human factors issue with alarms and alerts relates to their interpretability. What does an alarm mean when it signals? In the case of the EGPWS and TCAS, there is little problem in this regard, because the voice display is easy to interpret and because of the close association of the alerts with meaningful status displays. However, for a host of engine malfunctions, designated sometimes with arbitrary tones and sounds, this is a nontrivial issue (Patterson, 1990).

A third human factors issue is somewhat more subtle, but equally important, as software algorithms become more sophisticated and designers become more confident that their software not only can (a) detect the existence of a problem, but also can (b) diagnose the nature of the problem and (c) *recommend* the appropriate action to take to address the problem. (These three levels of knowledge were well represented in TCAS.) As automation moves beyond mere alerting to diagnosing and then recommending actions, the concern here is the danger that pilots may overtrust the diagnosis and recommendations and follow them without consulting the "raw data" on which the diagnosis or automation recommendation is based (Merlo, Wickens, & Yeh, 1999; Mosier, Skitka, Heers, & Burdick, 1998; Ockerman, 1999; Parasuraman & Riley, 1997). The pilot should only employ such a trusting strategy if the automation diagnoses and recommendations are themselves 100% reliable. But perfect reliability is a difficult goal for automation to achieve in areas of diagnosis and prediction in an uncertain world, just as it is unrealistic to assume 100% reliability from the human pilot. Hicks & DeBrito (1998) offer a good contrast between these levels of alarm automation in the context of engine failure alarms in the modern "glass cockpit" civil transport. Further discussion of these issues of overtrust is presented in a later section and in chapter 9. Pritchett (2001) provides an excellent review of human factors issues in cockpit alerting and alarm systems.

DISPLAY OVERLAY, INTEGRATION, AND CLUTTER

The previous section has considered three classes of spatial-geographical information that all pertain to the pilots' task of navigating to avoid hazards: flight paths on navigational displays, terrain on an enhanced GPWS, and other traffic depicted on a TCAS or CDTI. To this list, one may add weather information on a radar display (Lindholm, 1999) and, for combat aircraft, other geographically distributed threat information. The potential relevance to the pilot of so many different classes of spatial/geographical information, all often pertaining to the same or overlapping regions of space, presents a challenge for the cockpit display designer in addressing the trade-off between several different psychological factors, as we describe in detail in this section.

To introduce and simplify this discussion at the outset, consider a cockpit design in which each of these classes of spatial information is presented on its own dedicated display channel. (Such a design is quite possible, given that the different classes of information will come from very different sensor systems and may be rendered by different avionics manufacturers.) This solution has one fairly obvious cost, simply because of the amount of visual scanning required, as the pilot must look from display to display. Visual scanning is both effortful (Wickens, Vincow, Schopper, & Lincoln, 1997) and time consuming, and such a design will also keep the eyes focused inside the cockpit. Psychologists describe the extensive time required to select and access a channel to look at as a *cost of selective attention* (Wickens, 2000; Wickens & Hollands, 2000).

The opposite extreme solution is to employ software that will integrate all of these classes of information onto a single map or display panel, so that a pilot may, for example, see at a glance that an intended air route will guide her through bad weather or that a maneuver designed to avoid traffic, may bring the aircraft close to a mountaintop. Referring to our discussion of the proximity compatibility principle, we can see that such display integration will be useful to the extent that the navigational task requires *integration* (comparison between databases) as characteristic of the two examples. Indeed, experiments by O'Brien and Wickens (1997) using traffic and weather information and by Kroft and Wickens (2001) using traffic, weather, terrain, and air route information all revealed that pilots could make judgments about the joint implications of these different classes of information more rapidly and accurately if they were overlayed, than if they were displayed in separate panels. Examples of two such displays are shown in Fig. 5.3.

At the same time, however, such overlay can impose another cost, related to the *clutter* that results when too much graphical information is displayed within too small a spatial angle. Clutter here may be operationally defined as the number of marks (visually contrasting regions) within a given area or, more formally, as the amount of power at spatial frequencies above some arbitrary cutoff. As described by the proximity compotibility principle, such clutter will impose information-processing difficulty on the pilot to the extent that he or she needs to *focus* attention on one class of information, while ignoring or filtering other classes, which now may be physically overlaying the relevant class (Wickens & Hollands, 2000; Yeh & Wickens, 2001a). Such a need, for example, may be imposed by the questions: "Is there traffic ahead of me on the map?" "What is altitude?" Consistent with the proximity compatibility principle, this focused attention task may benefit more from spatially separated displays, a conclusion also supported by experimental research (Kroft & Wickens, 2001; O'Brien & Wickens, 1997).

A simplified representation of the trade-off between these two design solutions (integrated versus separated) is shown at the top of Fig. 5.4 in panels (a) and (b), respectively. The figure depicts two classes of information, for example air routes (the lines) and contour representations of either weather or terrain (the blobs). As we have noted, the integrated solution (a) is preferable to the extent that the task requires integration. The separated solution (b) will be more appropriate for focused attention tasks. Panels (c), (d), and (e) in the figure illustrate three alternative design approaches often used in practice, each of which raises human factors issues of their own. These design issues, in fact, apply to a much broader class of display design problems than simply the hazard database problem described here.

First, suppose the designer wishes to implement the separated display philosophy (b) to avoid clutter, but finds that there is simply no room on the panel to add the second display. Hence, the designer must place the two separated displays adjacently in the same space occupied by the integrated display. This solution, shown in (c), will lead to smaller displays, which, by halving the visual angle, can lead to problems in the legibility of the display (Kroft & Wickens, 2001). These legibility problems may be amplified if the display is to be read under vibrating, high glare, or low-illumination conditions (often characteristic of the cockpit). This "readability/resolution" factor will act to favor solution (a) over solution (c).

The design solution illustrated in the three panels of (d) is one that allows the pilot to exercise display options by using some control interface to temporarily "erase" entire classes of information (e.g., weather or all high-altitude aircraft). These erased classes can then be restored by similar control interface procedures. Thus, in the example shown in (d), the pilot can choose to represent airways or contours in the panel shown. Two approaches are possible with this solution. One is to allow either *or both* sources to occupy the same space, a technique known as *decluttering,* so named because the pilot can remove (declutter) either of the

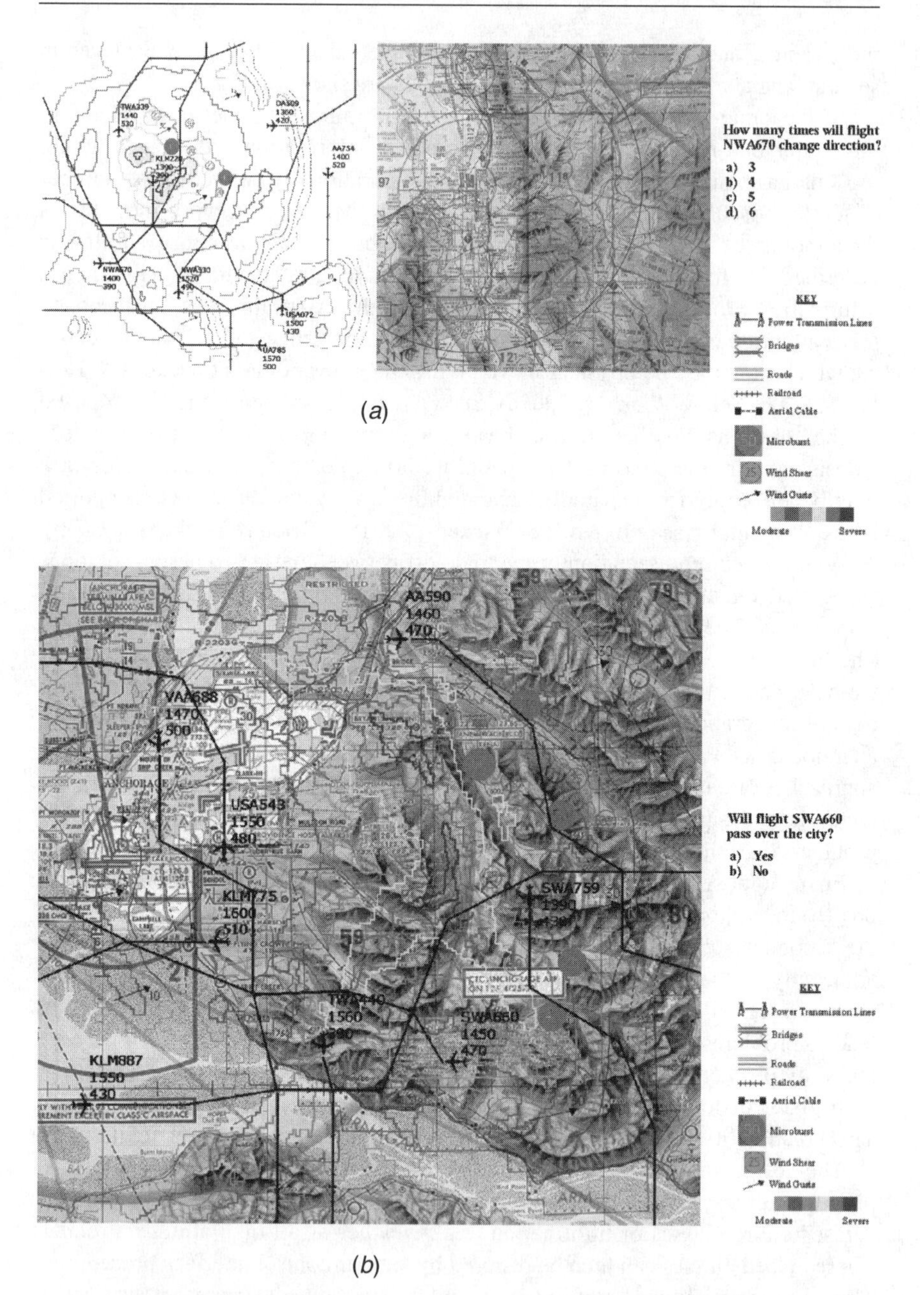

FIG. 5.3. Two examples of aviation display of multiple databases (traffic/weather and terrain): (a) Separated. (b) Overlayed. Both are presented within the same display area.

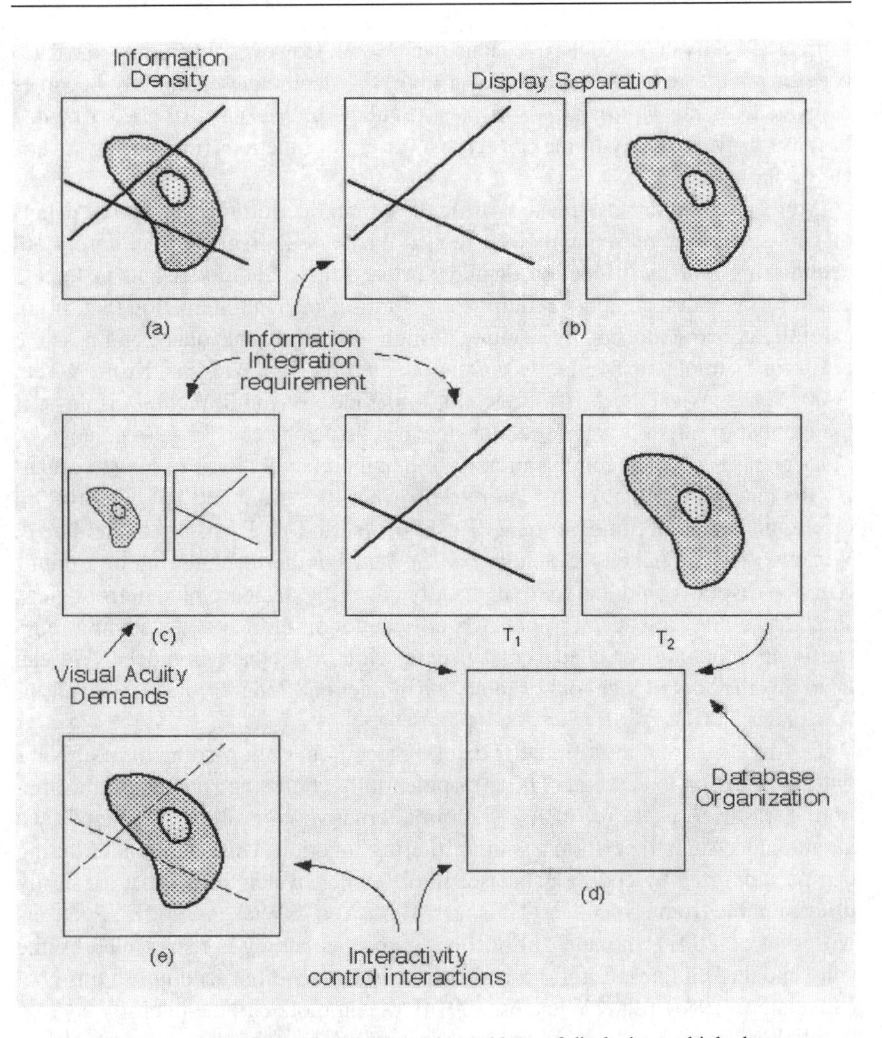

FIG. 5.4. Various display solutions to the problems of displaying multiple databases representing the same geographical area. The straight lines are a schematic representation of one database (e.g., traffic or air routes), the blob is a representation of another database (e.g., weather or terrain). The labels point to particular human factors concerns with particular solutions. (a) Overlaying databases. (b) Separation of databases with the same display resolution as (a). (c) Separation of databases within the same spatial area as (a), hence sacrificing display resolution. (d) Interactive decluttered or multifunction display. In the area or viewport below, each database can be called up as different "pages" of information at different time T_1 and T_2. (e) Highlighting or lowlighting of different databases. Highlighting could be interactively controlled.

sources (Mykityshyn, Kuchar, & Hansman, 1994). However, when only one database or another can be displayed at a time, this technique essentially becomes equivalent to the *multifunction display* (Seidler & Wickens, 1992), so named because a given display frame can serve a variety of functions (i.e., display a variety of classes of data).

Multifunction displays make it difficult to integrate information across panels that must be viewed sequentially (Yeh & Wickens, 2001a). Furthermore, both decluttering *and* multifunction displays impose three additional human factors costs. First, both techniques require some form of manual interaction that, if not carefully constructed, can be a source of high workload, time delay, and possible error, for example, inadvertantly pressing the wrong key (Wickens, Kroft, & Yeh, 2000; Yeh & Wickens, 2001a). Second, in the case of multifunction displays, if the number of possible sources of data that can be displayed is large (e.g., up to 40 radar pictures in some combat aircraft), the human factors issue of how to organize the database so that pilots can retrieve what they want and not get "lost" in electronic space becomes critical (Roske-Hofstrand & Paap, 1986; Seidler & Wickens, 1992; Wickens & Seidler, 1997). Third, either decluttering or a multifunction display could leave a dynamically changing database hidden from view, such that a pilot might fail to notice a critical condition that is developing (e.g., traffic approaching) or change that occurs while a database is hidden. We can label this the "out of sight out of mind" phenomenon (Podczerwinski, Wickens & Alexander, 2001).

Decluttering (and multifunction display) techniques allow some form of computer interaction to serve as a filter of potentially confusing sources of information. Turning to panel (e) in Fig. 5.4, an alternative is to allow the pilot's own attentional system to perform a similar filtering function. This attentional filtering can be supported by coding databases in different physical forms that are easily discriminable from each other (Treisman, 1986; Yeh & Wickens, 2001a; Podczerwinski et al. 2001). In panel (e) of this figure, this coding is represented by the solid and dashed lines. Careful use of color coding can often accomplish this goal (Silverstein, 1987; Yeh & Wickens, 2001a), as can monochrome intensity coding, highlighting, or other "foreground-background" techniques (Ververs & Wickens, 1998; Wickens, Kroft, & Yeh, 2000), although this monochrome coding is often restricted to two levels and care must be taken to assure that the "dim" level does not hurt legibility. Finally, intensity coding (or highlighting) may be coupled with control interactivity, allowing pilots to temporarily select certain databases for highlighting and leave others still visible, but "lowlighted" in the background. In this way, the lowlighted information is not lost from view, thereby addressing the "out of sight out of mind" problem for decluttering, but at the same time, it does not distract the focus of attention from the highlighted information class, if only the latter is required for task performance.

Of the various possible design solutions discussed and represented in Fig. 5.4, it is not apparent that any one will always be the best. Certainly some solutions such as those requiring display interactivity require more computer support (and therefore cost) than others. There are also a number of possible task and information factors that mediate the importance of certain information-processing principles and that will, therefore, favor some displays over others (Wickens, 2000). These factors and their mediating principles are shown in the text of Fig. 5.4. For example, in panel (a) we noted that denser displays will impose a greater cost on information overlay because there will be a greater likelihood that marks, pixels, or vectors for one class of information will overlap with those from a different class (Wickens, Kroft, & Yeh, 2000). Correspondingly, in panel (b), greater information integration requirements will impose a greater cost on database separation (O'Brien & Wickens, 1997; Wickens, Kroft, & Yeh, 2000), and greater acuity demands (i.e., smaller font, poorer reading conditions) will impose greater costs on the smaller displays in panel (c) (Kroft & Wickens, 2001). More dynamic information will impose a greater cost on the multifunction or decluttered displays because of the "out of sight out of mind" problem, and greater difficulty with keyboard interaction (for example, when wearing protective gloves or operating in vibrating environments) will impose greater costs on all interactive displays. Unfortunately, adequate human performance research has not yet revealed how all of these factors trade off and interact in a manner that will allow the designer to predict the best display for a particular circumstance with certainty. We know, for example, that when integration is required, solution (a) is superior to (b) and (c), suggesting that reducing scanning is more valuable than reducing the clutter of overlap (Wickens, 2000; Wickens, Kroft, & Yeh, 2000). However, the magnitude of this benefit or its dependency on the amount of spatial separation of the panels in (b) cannot be estimated with great confidence. But some progress continues to be made along these lines as researchers begin to develop computational models of display layout (Fisher, Coury, Tengs, & Duffy, 1989; Tullis, 1988; Wickens, 2000; Wickens et al., 1997).

PROCEDURAL DISPLAYS: TEXT, INSTRUMENTATION, CHECKLISTS

The previous sections have generally discussed dynamic displays that represent the state of the aircraft relative to features in the world, because these displays are most critical for supporting the high-priority tasks of aviating and navigating. However, pilots are also supported by a vast array of visual information that is far more routine and unchanging in nature, supporting aviation and navigation, as well as tasks related to communications and systems management. Many of these information sources are characterized by *checklists* and *instructions.* We will not discuss these displays in detail here and refer the reader instead to excellent readings by Degani and Wiener (1993, 1997) on the topic of checklists and procedures, as well as discussions by Wright (1998) and Orlady and Orlady (2000) on the human factors issues involved in writing instructions. However, we do devote some coverage to the critical issue of checklists, because of their great importance to flight safety and because of their relevance to some more traditional issues in display design. Checklists are also discussed briefly in chapter 7 from the perspective of cockpit task management.

Although the checklist formally serves as a memory aid to remind the pilot to follow critical procedures in either normal or emergency conditions (Degani & Wiener, 1993), two psychological processes often conspire to prevent the full or accurate completion of checklists (Hawkins, 1993). First, pilots, particularly those with high experience, may get in the habit of automatically following through the checklist without carefully using perception to insure that the state of the aircraft depicted in the checklist (e.g., "switch X is on") is in fact matched by the true physical state or "configuration" of the aircraft. Instead, the pilots' expectations that the state *should be* as represented in the text (and as encountered in scores of previous cycles through the checklist) dominates perception. In a process known as "top-down" perceptual processing (Wickens & Hollands, 2000), the state is seen as it is expected to be seen, not as it really is. Second, under time pressure or interruption, pilots may truncate a checklist before the last items are completed (time pressure) or may miss an item that occurred at the time of an interruption, resuming the checklist below the omitted item. Such an event apparently occurred prior to the crash of an airliner at Detroit Airport in 1987 as a result of a diversion of attention. Pilots missed a checklist item to set the flaps and slats on the wings (necessary for adequate lift on takeoff) (Degani & Wiener, 1993).

In response to these concerns about the human factors frailties of checklist following, various levels of automation have been considered, and many of these are embodied in the electronic checklist (Bresley, 1995; Boorman, 2001). At the simplest level, computer automation can sense whether or not an item has been addressed by the pilot by requiring the pilot to "touch" each item (or designate it with a cursor). Therefore onboard automation will know which items have not been designated and can highlight or flash these items if the pilot tries to move on to the next checklist item. Such a technique might well have called the pilots' attention to the unset flaps in the Detroit crash described. At a more complex level, the onboard automation could actually sense the *state* of all items on the checklist and assess if these states are appropriate. The checklist could then highlight those that are not in the appropriate state, saving the pilot a good deal of time by not requiring an examination of those items that are already correctly configured. Although these two levels of sophistication are already embodied in the electronic checklists of some advanced aircraft (e.g., the Boeing 777; Bresley, 1995), a third level has not been implemented. This is one in which the aircraft automation could not only sense the state of systems, but could actively *configure* those states that are sensed to be in the wrong state (e.g., set the flaps or turn a heating system on).

The three levels of automated checklist sophistication described illustrate the concept of levels of automation, discussed by Parasuraman and Byrne (chap. 9) and referred to briefly in our discussion of alarms. At this point, we will simply mention the potential human factors dangers of progressively higher automation levels. If the pilot begins to place too much reliance or trust on the automated features, he or she may fail to monitor the actual state of the aircraft and hence be unaware of a condition that the checklist automation does not itself perceive or configure correctly.

ADVANCED DISPLAY TECHNOLOGY: 3-D DISPLAYS

The turn of the century has brought tremendous advances in display technology in the cockpit; we have discussed examples such as traffic, weather, and sophisticated navigational displays. All of these displays are essentially two-dimensional, most showing only a "God's eye" plan view perspective of the earth (or spatial database in question). Yet, an argument can be made that flying is very much of a three-dimensional task. Both altitude and the integration of the vertical with lateral axes are information-processing concerns of the pilot, who may need to extract 3-D information from a display, such as: "If I turn here, will I still be able to climb over the weather?" Given that rapid advances of technology have supported the development of three-dimensional displays in other domains (such as data visualization or video games), accordingly it is important to consider the strengths and possible weaknesses of 3-D display technology for aviation. In the following section, we do so, focusing on how these strengths and weaknesses are modulated by psychological principles in the same way that such principles as scanning and clutter were seen to modulate the effectiveness of information overlay in an earlier section. In this context, we define the term "3-D display" to mean any display that uses perceptual depth or distance cues to create a three dimensional image. These may include cues capitalizing on the anatomical structure of the visual system (e.g., stereo; Patterson & Martin, 1992), as well as incorporating those "pictorial" cues that are properties of the 2-D image plane on which a 3-D view is projected (e.g., linear perspective, motion parallax, interposition). All such depth cues have varying degrees of effectiveness, some more and some less so, under different circumstances (Cutting & Vishton, 1995; Sedgwick, 1986; Wickens & Hollands, 2000; Wickens, Todd, & Seidler, 1989).

An important feature of any 3-D display is the *viewpoint* from which it presents a particular area of interest (e.g., the aircraft, or surrounding hazards). Although an infinite variety of such viewpoints can be created (Wickens, 1999), an important distinction for aviation displays is that between the *egocentric* or *immersed* and the *exocentric* viewpoint. These viewpoints are contrasted in Fig. 5.5. The immersed viewpoint in Fig. 5.5(a) displays the world as it would look

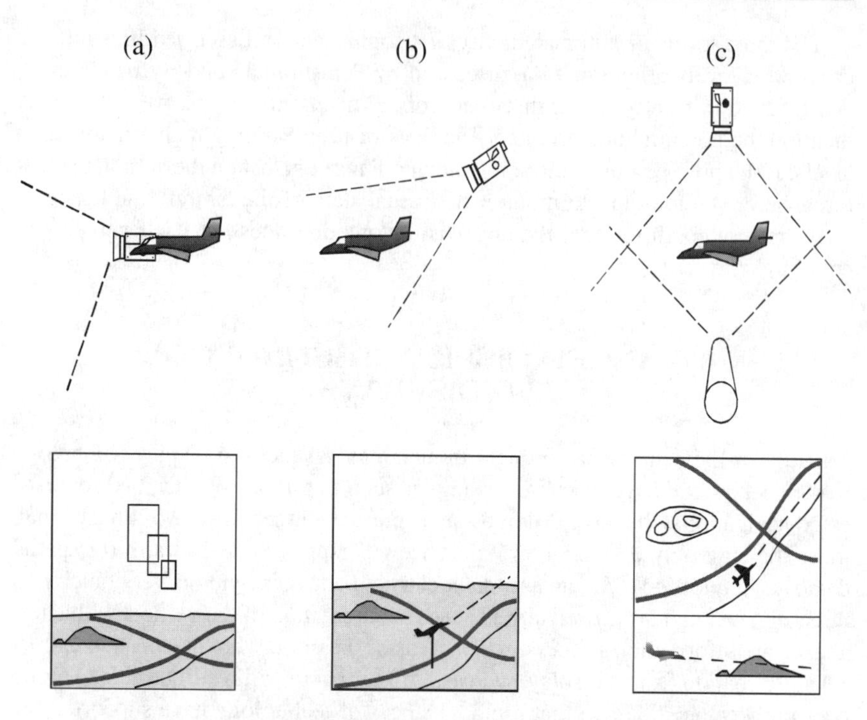

FIG. 5.5. Three different frames of reference with which flight path and hazard information can be presented to the pilot. Each panel represents the viewpoint of a camera or image generator, relative to the pilot. At the bottom is a schematic rendering of the view the pilot would see. (a) Immersed or egocentric view. The geometric field of view can be expanded as required to show a broader range of coverage. The three boxes represent a 3-D "tunnel" or flight path, turning downward. (b) Exocentric or "tethered" viewpoint. The view is rendered as if there is a tether linking the aircraft to the viewpoint at a constant distance behind and above. (c) Coplanar view, showing a top-down lateral view and a vertical profile view. In the current rendering, the profile is a side view. However, it could also be rendered as a view from behind. Adapted from Wickens, Liang, Prevett, and Olmos (1996). *Note.* From "Electronic Maps for Terminal Area Navigation," by C. D. Wickens, C. C. Liang, T. Prevett, and O. Olmas, 1996, *International Journal of Aviation Psychology, 6,* p. 244. Copyright 1996 by Lawrence Erlbaum Associates, Publishers. Reprinted with permission.

from the pilot's own location, looking forward. In contrast, as shown in Fig. 5.5(b), the exocentric viewpoint shows the view of the aircraft from outside. This viewpoint is typically positioned behind and above the aircraft, as if the viewpoint generator were created from a camera that is "tethered" to the back of the aircraft (Olmos, Wickens, & Chudy, 2000; Wickens & Prevett, 1995). Both of these viewpoints, in turn, can be contrasted with the 2-D plan view display, typical of the electronic map in most cockpits (e.g., Fig. 5.2 & 5.3). However, because 3-D viewpoints contain a spatial representation of altitude, whereas 2-D maps normally represent altitude by contour lines, color coding or digital indicators, a fair comparison of the strengths and weaknesses of 2-D versus 3-D displays requires that any 2-D display also represent altitude in analog form. This representation is typically done by coupling the 2-D map with a vertical or "profile" display, as shown in Fig. 5.5(c), creating what is sometimes called the *coplanar* display (or display suite; St. John, Cowen, Smallman, & Oonk (2001); Wickens, 1999). In the following sections, we discuss six important information-processing mechanisms that modulate the differences between these three fundamental display viewpoints, and then we show how these are relevant to different tasks that the pilot must perform.

Information-Processing Mechanisms

Most obviously, by presenting lateral and vertical information in two different display panels, the coplanar display imposes a (a) *cost of visual scanning,* which is not evident (or is less pronounced) when lateral and vertical information is integrated in a single "picture." (This integration is analogous to the advantage of spatial information integration, discussed in an earlier section.) It follows from the proximity compatibility principle that the scanning cost for the coplanar display would be amplified, to the extent that the task requires (b) *integration* of information in the lateral and vertical axes, as, for example, when the pilot performs a climbing turn. It also follows that a 3-D display presents a more realistic picture of three-dimensional features than that rendered by two orthogonal slices, thereby addressing the (c) *principle of pictorial realism* (Roscoe, 1968), discussed earlier.

However, 3-D displays can also impose certain costs, and some task requirements can neutralize their benefits. For example, if a task only requires that people consider or (d) *focus attention* on one plane of space at a time (e.g., two axes, forming the horizontal plane), then the benefits of 3-D integration are eliminated, and advantages of the less-integrated coplanar display may emerge, as suggested also by the proximity compatibility principle. Examples of tasks requiring such a focus would be purely lateral maneuvering with no climbs or descents or purely vertical maneuvers with no turns. To the extent that pilots think about and fly their vertical maneuvers separately from their lateral maneuvers, the advantages of the coplanar display will be enhanced. For example, the design of aircraft flight automation (to be discussed in a later section) is nicely categorized in terms of vertical navigation and lateral navigation. One enduring cost of nearly all 3-D displays is the (e) *ambiguity* with which the location of objects along the line-of-sight into the display (or depth axis) can be determined (Gregory, 1977; McGreevy & Ellis, 1986). One can see this in Fig. 5.5(b) and in more detail in Fig. 5.6, which presents examples of (a) landing and (b) traffic displays. In both examples, it is difficult to establish precisely how far away another aircraft is or

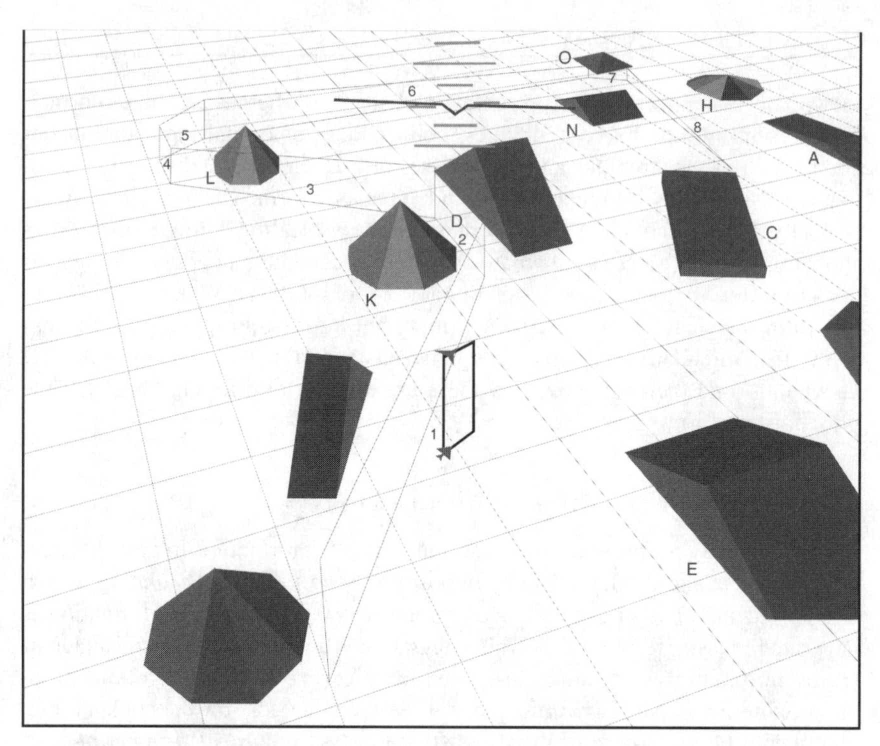

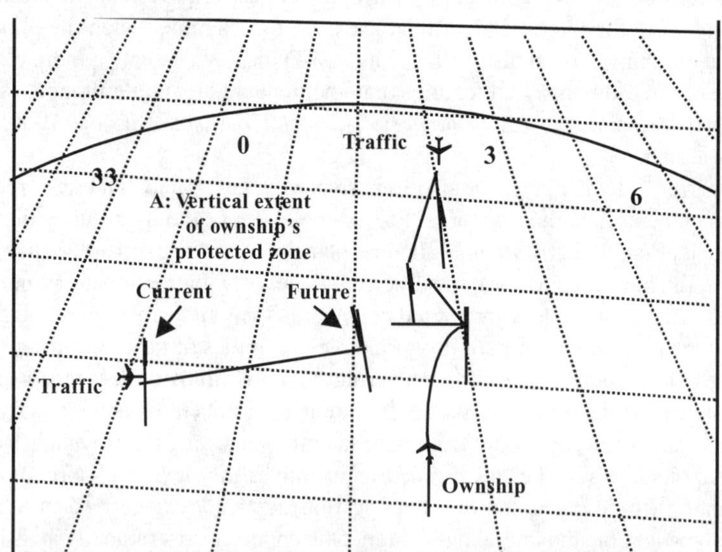

FIG. 5.6. Exocentric tethered display. (a) Display for landing in terrain (Wickens, Liang, Prevett, & Olmos, 1996. Copyright permission from Lawrence Erlbaum Associates). (b) Cockpit display of traffic information. *Note.* From "Electronic Maps for Terminal Area Navigation," by C. D. Wickens, C. C. Liang, T. Prevett, and O. Olmas, 1996, *International Journal of Aviation Psychology, 6,* p. 250. Copyright 1996 by Lawrence Erlbaum Associates, Publishers. Reprinted with permission.

which of two aircraft are closer. This ambiguity is greater with the tethered display [Fig. 5.5(b)] than with the immersed display (Fig. 5.5a) because ambiguity associated with the tethered display affects both where the pilot perceives her own aircraft to be and where she perceives the surrounding airspace elements to be, that is, a double ambiguity. Finally, when the more immersed perspective of a 3-D display is chosen [Fig. 5.5(a)], there is a cost associated with a (f) *keyhole* view of the world (Woods, 1984). The pilot simply is not presented with information to the side, above, below, and behind the aircraft, information that could be important for flight safety, particularly traffic on a conflict course from the side, or in hostile combat environments (Olmos et al., 2000). Note that the magnitude of this "keyhole cost" will be increased or decreased by reducing or expanding, respectively, the geometric field of view (GFOV) of the display.

The associated costs and benefits of the different display viewpoints are summarized in Table 5.1. In the following section, we show how a general categorization of aircraft tasks whose performance is mediated by the mechanisms in Table 5.1 has allowed the development of guidelines for the optimization of task-display combinations.

Aviation Tasks: The Task-Viewpoint Interactions

For flight path guidance, the immersed "pathway in the sky" or "highway in the sky" shown in Fig. 5.5(a) and 5.7 generally offers superior performance compared to both the coplanar counterpart and to the 3-D exocentric (tethered) display (Haskell & Wickens, 1993; Wickens & Prevett, 1995). When compared with the coplanar display, the benefit of the immersed pathway display results from the *integration* of lateral, vertical, and along-track (longitudinal) information regarding the

TABLE 5.1
Costs and Benefits of Different Display Perspectives

	3-D		*2-D*
	Immersed	*Tethered*	*Coplanar*
Cost of scanning	Low	Low	High[1]
Cost of cognitive integration across planes (Axis pairs)	Low	Low	High[1]
Principle of pictorial realism	Confirmed[2]	Confirmed[2]	Violated
Requirement of focused attention on an axis or plane (Axis pair)	Violated	Violated	Supported
Line of sight ambiguity	Cost[3]	Double cost[3]	
Keyhole view	Cost[4]		

[1]Increased with greater physical separation between lateral and vertical display panels.
[2]Less benefit at higher altitudes.
[3]Cost is decreased with more depth cues.
[4]Cost is decreased with larger geometric field of view.

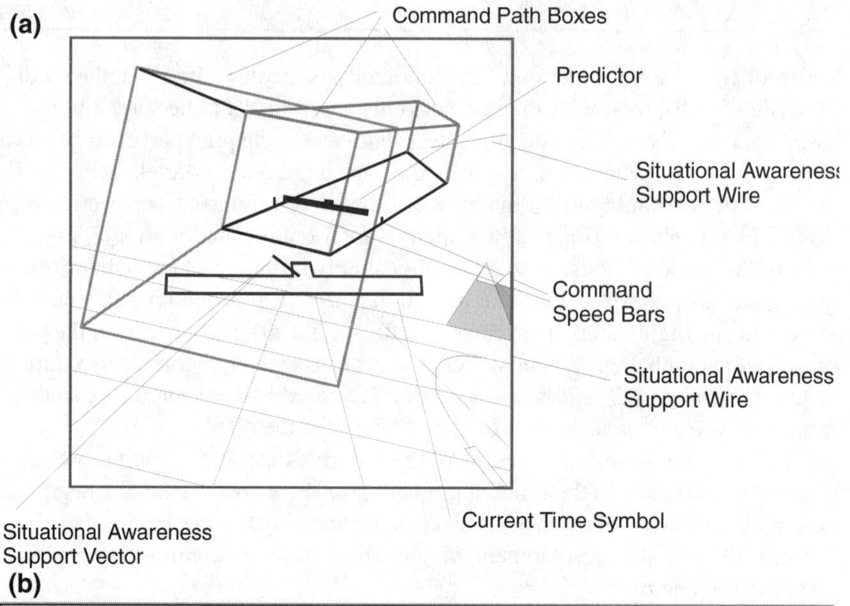

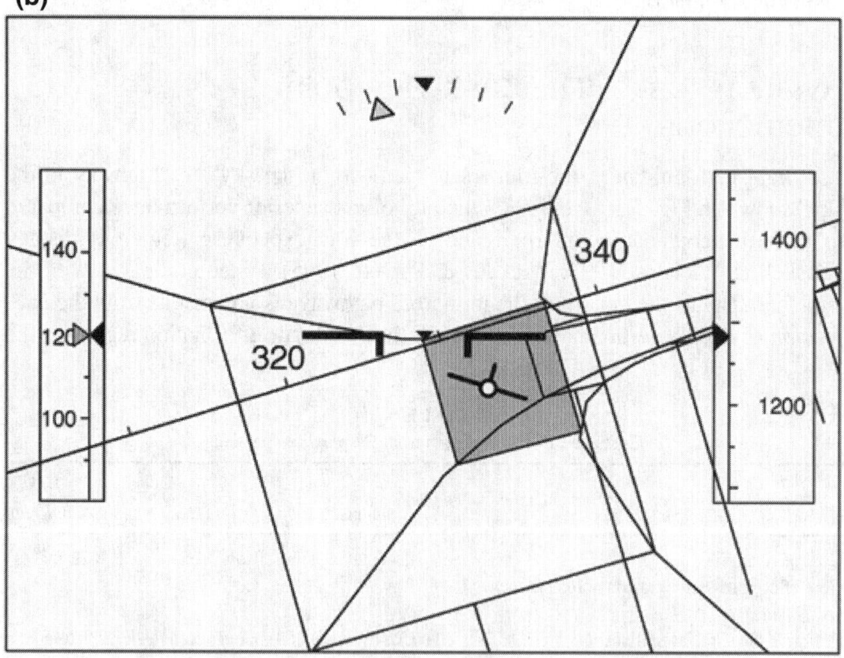

FIG. 5.7. Two examples of an Immersed Display or "pathway in the sky" display. (a) From Haskell and Wickens (1993). *Note.* From "Two and Three Dimensional Displays for Aviation," by I. D. Haskell and C. D. Wickens, 1993, *International Journal of Aviation Psychology, 3,* p. 95. Copyright 1993 by Lawrence Erlbaum Associates, Inc. Reprinted with permission. (b) From Theunissen (1997). Courtesy of University of Delft.

flight path and the aircraft's current and future positions. This integration supports the pilot's information-processing requirements for flight control, which also must integrate motion along the various axes. For example, given the characteristics of flight dynamics discussed earlier, a change in aircraft bank will produce both a change in lateral motion *and* a change in pitch, which in turn predicts a future (longitudinal) change in altitude. Consistant with the proximity compatibility principle, this cognitive integration task is better supported by the 3-D integrated display than by the separated panels of the coplanar suite (Haskell & Wickens, 1993).

When compared with the 3-D tethered display [Fig. 5.5(b) and 5.6)], the pathway in the sky display (Fig. 5.7) imposes a reduced cost of ambiguity, because the pilot's own current position is precisely perceived at the center of the display, whereas this location is perceived ambiguously in the tethered display. It is for these reasons that the flight path highway in the sky has received a great degree of attention in recent design (Beringer & Ball, 2001; Doherty & Wickens, 2001; Flohr & Huisman, 1997; Reising et al., 1998; Sachs, Dobler, & Hermle, 1998; Theunissen, 1997; von Viebahn, 1998). The pathway has been found to support good flight path tracking performance when contrasted with other displays (Fadden, Ververs, & Wickens, 2001; Haskell & Wickens, 1993; Wickens & Prevett, 1995) and has generally received favorable pilot comments.

When the pilot must make navigational judgments of the similarity, or the "match," between the image presented on the electronic map and the direct view of the world, the 3-D exocentric tethered display [Figs. 5.5(b) and 5.6(a)] appears to offer the best performance (Hickox & Wickens, 1999; Olmos, Liang, & Wickens, 1997; Schreiber et al., 1998). For example, these judgments might be required during "pilotage" or low-level visual navigation by reference to ground features (Williams, Hutchinson, & Wickens, 1996), or they might be required for confirmation of the identity between a display-rendered image and a ground target. The advantage of the exocentric over the coplanar display in this task is related both to its integration and to the principle of pictorial realism. Quite simply, the 3-D image looks more like its counterpart in the world than does the separated coplanar image. This perceptual similarity is greater as the pilot flies lower over the ground and hence sees the terrain from a more forward-looking perspective (Hickox & Wickens, 1999; Schreiber et al., 1998). The problem with the 3-D immersed display for these comparative judgments is that the small field of view can create the keyhole effect, which might hide critically important features that would assist image comparison and recognition. This task of obtaining an overall understanding of what a scene "looks like" is often required for tasks of situation or hazard awareness (Endsley, 1999; see also chap. 4). Hence, the exocentric 3-D display is often considered superior for many situation awareness tasks.

Many piloting tasks require more precise spatial judgments than those involved in general image recognition or comparison. For example, a pilot may need to know if her current trajectory is on a collision course with another aircraft, with a terrain feature in front of the landing strip, or toward an enemy radar coverage zone or a region of bad weather. In such tasks, errors of estimation of a few hundred feet (or a few degrees of heading) can be catastrophic. For these reasons, it is not surprising that the line-of-sight ambiguity problems inherent in the 3-D displays can be a concern, particularly the double ambiguity of the exocentric tether display. These limitations of 3-D displays, relative to coplanar counterparts, have been observed in tasks requiring location of both traffic and weather hazards (Boyer & Wickens, 1994; Merwin, O'Brien, & Wickens, 1997; Olmos, Wickens, & Chudy, 2000), as well as in terrain judgments (May, Campbell, & Wickens, 1996; Wickens, Liang, Prevett, & Olmos, 1996). It is for this reason in particular that 3-D displays may not be advisable for air traffic controllers, where precise, unambiguous judgments of position and trajectory are critical. Somewhat puzzling, however, is the fact that *coplanar* displays, which represent an undistorted view of the vertical axis, have never been developed for air traffic controllers, who rely instead on the uniplanar display, with only a digital representation of altitude.

3-D Displays: Conclusions and Solutions

The previous discussion has revealed a situation similar to that concerning the integration of spatial information; that is, there are several possible display solutions, and each is associated with costs and benefits for different tasks and environments, as these are mediated by a host of human information-processing capabilities and limitations. Given this situation, there are two general approaches to design that could be considered.

The first approach is to provide multiple viewpoints that can be either dedicated or pilot selectable and can be used for the appropriate task. A prototypical example of this approach is the T-NASA taxiway display shown in Fig. 5.8, designed to support aircraft ground operations in low visibility (Foyle et al., 1996; McCann, Foyle, Andre, & Battiste, 1996). This display suite contains an ego-referenced (immersed) pathway display as a head-up display—to aid the pilot in precise runway guidance—coupled with a head-down 3-D exocentric navigational display, shown to the right of the primary flight display, which will provide the pilot with a larger view representation that would support navigational awareness and planning. Different options within this approach involve giving the pilot a greater or lesser degree of choice in selecting the viewing parameters. An example with a high amount of such choice is to provide a standard 3-D exocentric view [Fig. 5.5(b)] with two controls; one control can "zoom in" the viewpoint to an immersed location [simulating a shortening of the tether length in Fig. 5.5b to zero, as in Fig. 5.5(a)], whereas the other control can "rotate" the exocentric viewpoint to an overhead perspective, thereby creating a 2-D plan view [Fig. 5.5(c)]. When this approach is taken it is important that pilots be made aware, through training and instruction, of the strengths and limitations of different viewpoints, as discussed in the previous two sections. The multiviewpoint solu-

FIG. 5.8. Display suite to support pilot taxi operations. The "T-NASA" display consisting of a HUD (top) and a head-down 3-D navigational display. These represent the immersed and exocentric perspectives, respectively. Courtesy of NASA Ames Research Center.) Note the conformal symbology on the HUD. Note also that the white translucent "wedge" in the bottom 3 D exocentric map display represents the region of space that is depicted in the upper head-up display, or HUD, creating visual momentum.

tion is being considered in the design of 3-D synthetic-vision systems display suites for terrain avoidance (Comstock, 2001).

The second design approach is to develop various display "supports" that can offset the liabilities of certain views, without themselves imposing additional penalties (Ellis, 1993; Olmos et al., 2000). As one example, display "posts" or droplines can be extended from each aircraft to the ground, as shown in Fig. 5.6(b), hence resolving some of the ambiguity of positional and altitude location that is inherent in an exocentric display (Ellis, McGreevy, & Hitchcock, 1987). As another example, ego-referenced displays can be supplemented with small exocentric or plan view inset maps of the full region of space surrounding ownship in order to reduce the keyhole effects (Olmos et al., 2000; Ruddle, Payne, & Jones, 1999).

Whenever two maps or viewpoints are used (as in either design approach 1 or when using an inset map), then principles of *visual momentum* should be implemented to allow the pilot to more easily transition between the two related spatial displays (Aretz, 1991; Olmos et al., 2000; Woods, 1984). The creation of visual momentum is an important display design feature that helps the pilot to understand how information presented in one display or display panel relates to information in another display of an overlapping or nearby region. For example, consider two display panels that must represent the large geographical area shown at the top of Fig. 5.9. The two display panels in Fig. 5.9(a) are shown to have visual momentum by their positions relative to each other. The two in Fig. 5.9(b) do not, since the display of the upper regim is located to the left, not above. Fig. 5.9(c) shows an effort to create some visual momentum between the two panels by highlighting their common boundary. When an immersed display with a forward-looking viewpoint is coupled with a more exocentric map display of a larger area, then visual momentum between them can be created by presenting a "wedge" on the exocentric map display, portraying the momentary field of view that the pilot

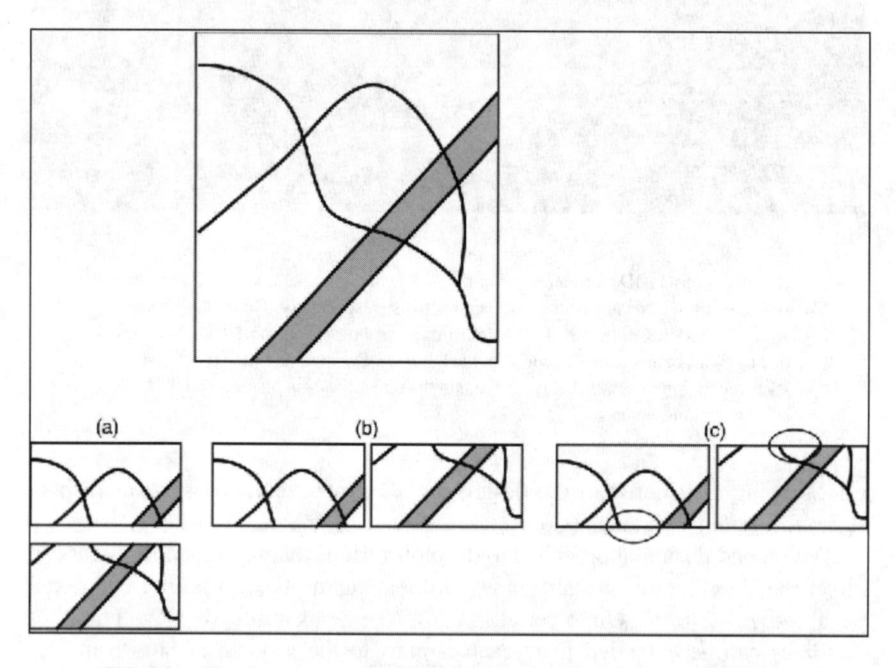

FIG. 5.9. Illustrates the concept of visual momentum in configuring two display panels of adjacent regions that are shown in the large figure at the top. (a) Visual momentum is achieved by relative positioning of the two display panels one above the other. (b) Visual momentum is absent because the two views of the north and south half of the space are now side-by-side. (c) Visual momentum is partially restored to (b) by highlighting the (oval) linkage between the two panels.

can see in the ego-referenced head-up display (Aretz, 1991). This feature enhances performance because it makes it considerably easier for the pilot to understand the common regions depicted by the two views. One example of this wedge is shown in the T-NASA display of Fig. 5.8.

It is not likely that adding any of these display supports to any single viewing perspective could make it superior to all other displays for all pilot tasks. Hence, some application of the first design approach (multiple views) is probably necessary to fully support the range of pilot tasks. The issue of whether these views should be selected by the pilot, selected by the designer, or selected in real time by automation is one that we do not address here (see Hammer, 1999). Nevertheless, the multiple-view design approach can be augmented by design supports related to the viewing perspectives chosen (design approach 2), and this augmentation can, in turn, reduce the total number of perspectives that may be required. The T-NASA display nicely illustrates this point.

ADVANCED TECHNOLOGY: SEE-THROUGH DISPLAYS: THE HUD AND THE HMD

Three-dimensional displays discussed in the previous section represent one relatively radical departure from conventional aviation displays using advanced technology. A second departure, somewhat less radical because it is already implemented in several aircraft, is the "see-through display," characteristic of the head-up display, or HUD, and the helmet-mounted display, or HMD.

The importance of see-through displays results from the fact that the eyeball (and its foveal vision) is probably the most critical resource that pilots have available for the spatial tasks involved in aviation. This resource must be employed for scanning the various displays described in the proceeding pages as well as scanning the airspace beyond the cockpit in visual meteorological conditions for traffic and other hazards. Much of this information, like the print on a map or the midair target, is of small visual angle (therefore imposing high visual acuity demands) and hence requires foveal vision. Based on the desire to keep the eyes looking forward out of the cockpit as much as possible, designers have developed head-up display technology, which, through clever optical techniques, can superimpose critical display imagery in focus on the outside view through the windscreen (Newman, 1995; Weintraub & Ensing, 1992; Wickens, Ververs, & Fadden, in press) (Fig. 5.8 & Fig. 5.10). A corresponding technique, imposing some of the same human factors concerns, is the head-mounted or helmet-mounted display (HMD) (Geiselman & Osgood, 1995; Melzer & Moffitt, 1997). Here the see-through display is mounted to the head rather than to the frame of the aircraft.

FIG. 5.10. Example of a head-up display conformal imagery (the horizon line and the runway overlay), along with nonconformal symbology (the digital readouts and round gauges). "Honeywell/ GEC-Marconi 2020 HUD." Courtesy of Honeywell Inc.

The Head-Up Display

The initial success of the HUD was realized in military tactical aviation, in which HUDs were designed to support the pilot in air-to-air combat, where even a fraction of a second with the pilot's head down in the cockpit could sacrifice a combat advantage. An important concept in the military HUD was a target reticle that could be superimposed over an enemy target in the far domain by appropriately pointing the aircraft (i.e., toward the target) and then superimposing the reticle on the target. Further developments in military HUDs have focused on providing assistance to the fighter pilot in recovering from unusual attitudes. To assist in this task, the HUD can offer command guidance cues of how to maneuver pitch and roll (Geiselman & Osgood, 1995), as well as status information regarding the location and orientation of the horizon. As discussed earlier, the Auto-GCAS is a HUD-mounted ground proximity warning system that can alert the military pilot on the need to "pull up" and displays the instant at which the autopilot assumes control in the terrain avoidance maneuver.

Both the target reticles and the horizon line illustrate the important HUD concept of *conformality,* that is, HUD imagery that directly overlays and, when the plane rotates, will move in synchrony with its real-world far-domain counterpart. Thus, conformal imagery is that which appears to "belong" to the far domain.

In the last decade of the 20th century, HUDs have gradually been incorporated into commercial, and subsequently into general aviation, aircraft (McClellan, 1999). In these civilian environments, as in the military environments, HUD benefits can be directly linked to at least two principles of visual attention. First, most obviously, the HUD will reduce the amount of visual scanning necessary to consult the display imagery and to monitor the far domain. Furthermore, because most HUDS are collimated (projected at or near optical infinity by the use of special lenses), they lessen the amount of reaccommodation that the eyes must accomplish, to focus on high-detail images in both the near and far domain (e.g., read a digital indicator and then scan for traffic). Second, to the extent that HUDs use conformal imagery, they can facilitate the integration of, or divided attention between, near and far domain information sources when the two domains are related. This is a direct application of the proximity compatibility principle discussed earlier. Two representations that are displayed in similar (overlapping) locations will help the pilot to mentally integrate the information presented by those two representations. One example of this conformal integration is air combat target cueing—the conformal superimposition of a target reticle on the actual target in the airspace beyond. Other examples of conformal imagery are the horizon line that overlays its true counterpart, the runway outline that overlays the true runway and helps in low visibility landings or taxi (see Fig. 5.8), traffic symbols that can overlay the location of potential traffic conflicts, or even a virtual "highway in the sky" that overlays a hypothetical 3-D path to be followed (Fadden et al. 2001). Some of these concepts are illustrated in Fig. 5.10.

In all of these cases of conformal imagery, the display designer is rendering an "augmented reality" (Milgram & Colquhoun, 1999) and is thereby facilitating the perceptual fusion or divided attention between the near and far domain (Levy, Foyle, & McCann, 1998). Basic research in human visual attention suggests that such conformal imagery creates a single, rigid, and integrated "object" linking the near and far domain in a way that facilitates divided attention between the two domains (Fadden et al., 2001; Kramer & Jacobson, 1991; Wickens, 1997; Wickens & Long, 1995). This facilitation lies at the core of *object-based* theories of visual attention (Driver & Baylis, 1989; Kahneman & Treisman, 1984).

HUDs have proven advantageous to pilots, who report favorably on their benefits (Leger, Aymeric, Audrezet, & Alba, 1999; McClellan, 1999), and more formal experimentation has revealed their benefits for many aspects of flight path tracking (Wickens & Long, 1995), as well as the detection of discrete events, such as traffic or an alert on the display (Fadden, et al., 2001; Martin-Emerson & Wickens, 1997; Ververs & Wickens, 1998; see also Fadden, Ververs, & Wickens, 1998; Fadden, Wickens, & Ververs, 2000; Wickens, Ververs, & Fadden, in press for summaries). However, competing against these advantages of reduced scanning and conformal image fusion is the prominent cost of *clutter* resulting when two images (the near and far domain) overlap in the same region of visual space. This clutter/overlap can be problematic for both conformal and nonconformal imagery, but is found to be considerably greater for nonconformal imagery (Martin-Emerson & Wickens, 1997; Fadden et al., 2001; Fadden et al., 2000; Wickens & Long, 1995). Such costs are quite analogous to those discussed earlier regarding display overlays and may result from either or both of two mechanisms. First, the clutter may hinder the speed of visual search to locate particular items of information, although it does not appear that this cost is much greater than when the two search field are separated (i.e., with a head-down presentation of the same information; Ververs & Wickens, 1998; Wickens & Hollands, 2000). Second, even after a target point of interest is found, the clutter of overlapping imagery may make the interpretation of that target more difficult, such as reading a highway sign through a dirty windshield or confirming the presence of a distant aircraft that may be partially hidden by the strokes of HUD imagery.

Research generally suggests that the costs of clutter with a HUD are less than the costs of scanning with a head-down display (and therefore the cost trade-off favors the HUD; Fadden et al., 1998, 2000, 2001). However, there is a more equal balance between these two costs (i.e., the HUD benefit is reduced), to the extent that HUD imagery is *unrelated* to that in the far domain (e.g., nonconformal). This balance may even shift toward a HUD cost and head-down benefit for the detection of very unexpected events. The classic example of such an event is the appearance of an aircraft on the runway on which a pilot is about to land. Evidence suggests that pilots are slower to detect such an unexpected event when looking through the HUD than when scanning between the windscreen and the head-down display (Fadden et al., 2001; Fischer, Haines, & Price, 1980; Hofer, Braune, Boucek, & Pfaff, 2000; Wickens & Long, 1995). Thus, designers of HUDs must resist the temptation to display large amounts of information on the HUD, which can produce dangerously masking clutter, obscuring and disrupting the detection of unexpected far-domain events and objects.

In addition to clutter concerns, another limitation of HUDs is that they tend to have a relatively narrow field of view, depicting a small region of forward space. Hence, if conformal imagery is to be used to represent information in the far domain, the HUD must present somewhat of a keyhole view of that domain, not unlike the constraints on the 3-D immersive display discussed earlier (Beringer & Ball, 2001). Furthermore, pilots can only benefit from image superimposition when they are looking forward. These are human factors issues that are directly addressed by the helmet-mounted display, which we discuss in the following section.

In closing our treatment of HUDs, we should note the possibility that HUDs may be joined with 3-D displays in the concept of a "synthetic image display" (Comstock, 2001), in which a highly realistic graphic depiction of terrain (or perhaps a photo image from an externally mounted camera) can be presented head down, while overlayed with important "HUD" information. Such a display has the advantage of avoiding image collimation issues and of providing 3-D terrain renderings that are much more realistic than those of the enhanced GPWS. However, the very compelling nature of such a head-down suite may ironically disrupt the very outside scanning that the original HUD was designed to support.

The Helmet-Mounted Display

The helmet-mounted display, or HMD, is analogous in some respects to the HUD in that the pilot can look at or through a displayed image to view the world beyond (Melzer & Moffitt, 1997). However, the see-through display is rigidly attached to the head, rather than to the airframe. In some recent applications, HMDs have been described as "wearable computers" (Ockerman, 1999). The HMD may be presented either monocularly (to one eye) or binocularly (to both). In most aviation applications, the binocular display must of necessity be translucent, so that the far domain can be seen through the HMD imagery. However, the monocular image can either occlude the far domain entirely or can be displayed on a translucent surface, so that the far domain can be seen through this image. The latter design is greatly preferable because a wider range of the far domain can be seen, and it reduces the costs of binocular rivalry (i.e., seeing one image or the other, but not both).

Although the HUD and HMD both present overlapping images to the pilot, there are some important functional differences between the two displays. Most importantly, HUD imagery can only be viewed when the pilot is looking more or less straight ahead because the HUD is rigidly mounted to the aircraft. In contrast, HMD imagery can be viewed independently of the head's orientation. This feature has two distinct benefits. First, it means that pilots can continue to see the HMD imagery when they scan the far domain beside and behind the aircraft heading vector (referred to as "off-boresight" scanning) or even when they look down into the cockpit. This feature is particularly advantageous for "eyes-out" viewing in the military cockpit (where the enemy can be very much of a threat behind; Geiselman & Osgood, 1995), or in low-altitude rotorcraft operations (when the hovering pilot must keep eyes out of the cockpit, in order to avoid contact with ground hazards; Hamilton, 1999; Hart, 1988; Haworth & Seery, 1992). The second advantageous feature of the HMD is that it enables conformal imagery or augmented reality to be employed across a much wider range of space than is possible with the narrow field of view of the cockpit-mounted HUD. This greater range then allows, for example, the pilot to look at an off-boresight axis target in order to place an HMD-displayed recticle on the target (Brickman, Hettinger, Haas, & Dennis, 1996) or to view a conformal horizon symbol as he looks to the side (Haworth & Seery, 1992).

These two benefits that the HMD provide for a greater range of superimposed viewing and conformal imagery are also offset by some competing costs. The first—the cost of clutter of overlapping imagery—is the same as that discussed in the HUD context: If HMD imagery is dense, it may hinder the focus of attention on the far-domain information which it overlays, and correspondingly, a busy rich-textured scene in the far domain can disrupt the readability of the HMD imagery (Yeh, Wickens, & Seagull, 1999). The second cost applies only to conformal HMD imagery, and this is the need to update or redraw the location of that imagery rapidly as the head is rotated so that the imagery always overlays its far-domain counterpart. Where such imagery is complex and the head movement is rapid, a lag in image updating can be quite disruptive to human perception and performance, can sometimes be disorienting, and can lead to a less than natural, more constricted amount of head movement (Seagull & Gopher, 1997).

Operational evaluations and comparisons between HMDs and head-down or vehicle-mounted head-up displays for similar tasks are fewer than comparisons between HUDs and their head-down counterparts discussed in the previous section. Although HUDs are common features in many cockpits, HMDs are only standard equipment in some military aircraft (the Apache helicopter) and will probably be considerably slower to be implemented in the cockpit of the civil fixed wing aircraft. Yet those few comparisons that have been carried out (Geiselman & Osgood, 1995; Yeh et al., 1999) suggest that, as with HUDs, the helmet-mounted display can provide very effective target and orientation cueing, so long as its potential costs are considered in design. However, a large range of psychological design issues in HMDs remain to be addressed (Wickens et al., in press).

AUTOMATION AND DISPLAYS

The issues in aviation displays cannot be discussed without reference to the tremendous implications of aviation automation, an issue that can be examined from two somewhat different perspectives. One perspective considers the display implications or display requirements that automation imposes or that automation changes. The other perspective addresses how automation (computer technology) has been harnessed to improve displays for the pilot or air traffic controller. We first consider the first of these perspectives and then discuss examples of the second perspective.

Display Implications of Aircraft Automation and Autopilots

A host of developments in autopilot technology has, in more advanced aircraft, eliminated the absolute need for the pilots' "inner loop" or attitude control of the aircraft. For example, a "heading select" mode of an autopilot allows the pilot to directly set a desired heading; the aircraft's computers will then execute the appropriate bank to yield a turn that will level out on the new heading (and sustain it in the face of disturbances) while automatically compensating for the downward pitch that results. If the pilot does not then need to *control* bank, then the pilot's need to *perceive* bank via the attitude indicator is also reduced; the instrument becomes less important. Indeed, in many modern aircraft, the *mode control panel* now allows direct setting of the desired heading, altitude, vertical speed, and airspeed, hence reducing the requirement of consulting many of the basic flight instruments shown in Fig. 5.1(a). (However, the advisability of monitoring these instruments still remains an issue in automation trust addressed in Parasuraman, 1987; see also chap. 9).

More sophisticated still is the *flight management system* (or FMS), which includes the automatic controls or autopilots along the lateral, vertical, and longitudinal axes as described earlier, along with a set of much more complex control algorithms and modes that can command and manage many different "4-D" trajectories over the ground (chap. 9; Sarter & Woods, 1997; Sherry & Polson, 1999). Although one might initially think that such autopilot technology would reduce the need for displayed information (i.e., because inner loop control of pitch, roll, and heading is not needed), there are two reasons why this clearly should *not* be the case. First, there remain circumstances in which the pilot may unexpectedly need or desire to *hand fly* the aircraft, either because of the pilot's own desires or out of necessity because of an unexpected failure in some aspect of the aircraft's dynamics, FMS, or navigational system. The information necessary for this inner loop control [i.e., the flight instruments of Fig. 5.1(a)], should always be maintained relatively available to the pilot, should these requirements develop.

Secondly, and somewhat more subtly, the *way* in which the logic of the FMS decides to execute various maneuvers (for example, minimizing fuel consumption or maximizing the benefit of tailwinds by selecting altitudes and climb/descent profiles) is not always apparent or intuitive to the pilot (Sarter & Woods, 1994, 1995, 1997; Sherry & Polson, 1999) As discussed in chapter 9, the pilot may be unaware of which one of the many different *modes* of automation operation is currently implemented by the FMS to achieve these maneuvers, and such a lack of awareness can be the source of hazardous and sometimes disastrous flight conditions (Dornheim, 1995). An example described earlier in this chapter is provided by the accident at Strassbourg, France, when pilots of one commercial aircraft apparently confused a flight path angle mode setting with a vertical speed mode setting, yielding a rate of descent much faster than that which was intended, crashing the aircraft as a result (Sarter & Woods, 1997). These human factors concerns related to FMS understanding and mode awareness require more effective attention-grabbing displays to signal the pilot when automation has changed the mode of vertical flight (Sklar & Sarter, 2000; Nikolic & Sarter, 2001), as well as intuitive pictorial displays of what vertical maneuvers are being carried out and are planned within the FMS logic. Two examples of displays to support vertical automation displays are shown in Fig. 5.11.

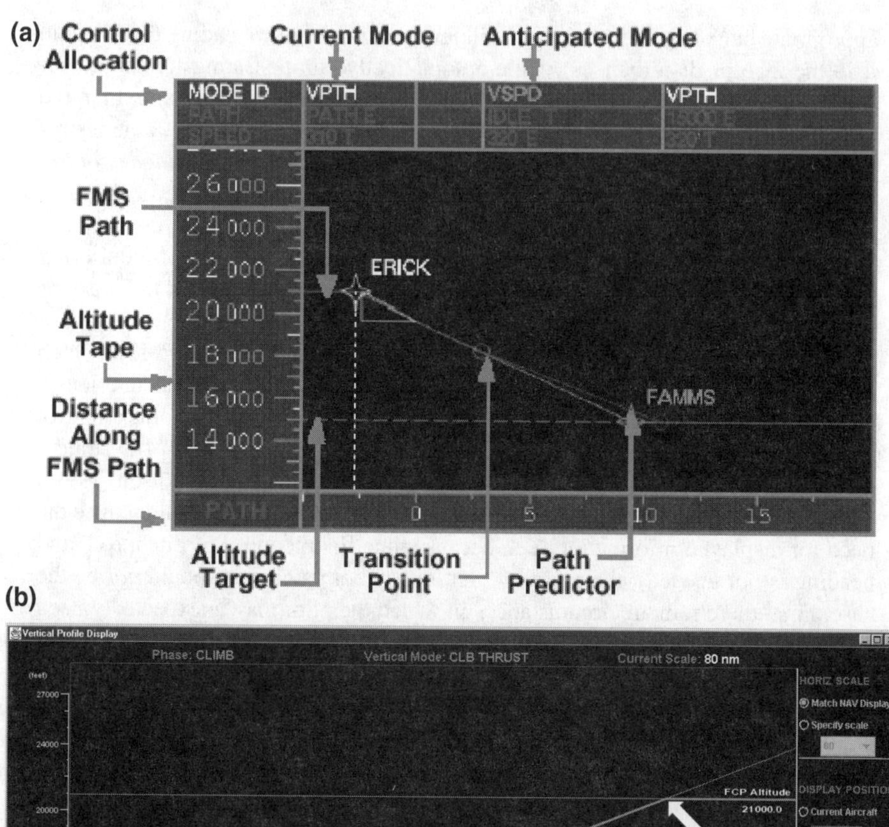

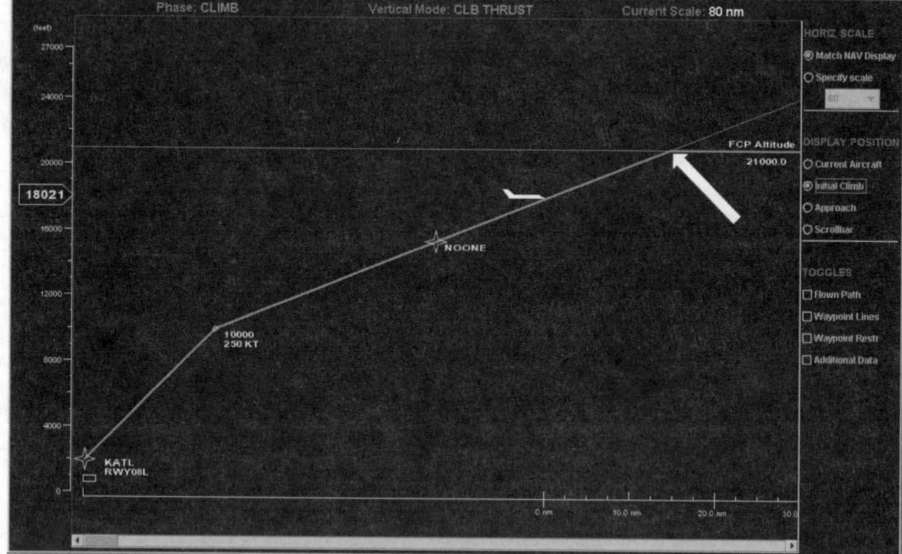

FIG. 5.11. Two examples of vertical profile displays: (a) From Vakil and Hansman, 1999, Courtesy of MIT. (b) From Gray, Chappell, Thurman, Palmer, and Mitchell, 2000.

Harnessing Automation for Improved Displays

Computer power can be, and has been, harnessed in many different ways to improve the job of the pilot or controller. In an earlier section, we addressed the role of automation to monitor for geographical/spatial hazards and to warn the pilot of their existence, a clearly valuable service. In the present section, we consider three other classes of display-related services that automation can provide: the role of an information host in the multifunction display, the role of a communicator in data link, and the role of an attention guider, or information advisor, in a variety of circumstances.

Multifunction Displays. Earlier we described the multifunction display (MFD) as a possible joint solution to the clutter problem caused by a single fixed display and to the scanning/resolution problem caused by multiple dedicated displays of a geographical area. In a broader (nongeographical) context, the multifunction display can serve as a host for vast quantities of information pertaining to such databases as emergency checklists, airport characteristics, or airplane system status. Rather than representing these on multiple pages of paper, in the multifunction display, only a single display viewport is employed, and this viewport can be used through some form of manual (or possibly vocal) interaction (Roske-Hofstrand & Paap, 1986; Seidler & Wickens, 1992) to call up the appropriate information "pages" at the right time. As we noted there, two costs were associated with this form of computer-based automation of information retrieval. First, if the databases themselves contain dynamic information, it is possible that important changes to a database could occur while it was hidden from view. This is the "out of sight out of mind" problem (Podczerwinski, Wickens, & Alexander 2001). For example, a weather database on an MFD could easily change in such a way as to increase the hazard during a time when it is out of view on an MFD page.

Second, the MFD invites challenges and problems in navigating through the "electronic space" or menu that represents the different options available. If the menu is simple and has perhaps only two levels, as recommended by human factors guidelines (Shneiderman, 1987), then these challenges are not serious. However, to the extent that large numbers of different pages are possible for display in a single viewport, then problems can arise. On the one hand, it may require a number of perceptual-motor operations (and resulting head-down time) for the pilot to "move" from one desired screen to another. Such travel time includes not only the actual time required to press the button (or otherwise implement the display change), but also the cognitive time required to decide which button to select (Seidler & Wickens, 1992; Wickens & Seidler, 1997; Yeh & Wickens, 2001a). On the other hand, deep and complex menus invite the possibility of "getting lost" or disoriented and not knowing how to find a particular page from one's current position in the database (Billingsley, 1982). This is an experience familiar to many users of computer databases with deep menu structures.

To avoid such problems, it is important that the structure of the MFD menu (and corresponding database) be designed to be consistent with the pilot's mental model of the knowledge required to carry out the tasks that are supported by the MFD database (Roske-Hofstrand & Paap, 1986; Seidler & Wickens, 1992) and should not be designed to include too many levels of depth. Furthermore, options should be made available to allow the pilot to rapidly and easily recover a known position (i.e., get to the "top" of the menu structure) if he or she becomes lost or disoriented (Seidler & Wickens, 1992).

Data Link and Digital Communications. The aviation community is embarked on a path to replace or augment traditional air-ground radio communications with a system commonly referred to as *digital data link,* by which the same communications information is now coded and conveyed electronically between ground and air in digital data packets (Kerns, 1999; Wickens, Mavor, Parasuraman, & McGee, 1998). Data link has major implications for the pilot's display, because the presence of electronically communicated information in the cockpit makes possible a wide variety of different formats for representing that information to the pilot (Hahn & Hansman, 1992). The evolving design approach for many data link displays has been that this uplinked information should appear in text format on a screen located somewhere on the flight deck (Kerns, 1999; Navarro & Sikorski, 1999). This format has the obvious advantage (over current radio-telephone communications) that it is semipermanent and hence will not be subject to the same sort of working memory forgetting (Loftus, Dark, & Williams, 1979) and confusion (Morrow, Lee, & Rodvold, 1993) that often disrupts voice communications. Furthermore, the permanent representation of a visual data link display means that the pilot need not interupt ongoing higher priority tasks to enter data link information (or write it down) before it is forgotten (Helleberg & Wickens 2001), an issue addressed in chapter 7. Yet, a visual text display has the possible cost of increasing pilot visual workload (head-down time) relative to an auditory display and of providing what may be very spatial information (e.g., way points and directional changes) in a noncompatible symbolic (text) format, thereby violating the *principle of pictorial realism.* Because data link information exists electronically in the cockpit, these two problems could be addressed, respectively, by a display format using synthetic voice (essentially mimicking the current radio system) or by a graphical format in which relevant information is painted onto an appropriate map display (e.g., a desired rerouting vector around bad weather) (Hahn & Hansman, 1992; Helleberg & Wickens, 2001). Indeed, there is no reason why any of these three formats must be used exclusively; a *redundant* combination of any of these may best address the pilot's need for error-free comprehension of such critical communications (Helleberg & Wickens, 2001).

Automated Attention Guidance. There are a variety of ways in which sophisticated automation can make inferences regarding a pilot's momentary information needs and then offer such information (Hammer, 1999; Pritchett, 2001). A simple example already encountered is the electronic checklist whereby automation infers that the pilot should be reminded about a checklist item not yet completed and does so by blinking or highlighting the remaining item (Boorman, 2001). As previously discussed, the modern TCAS system accomplishes a corresponding function for potential air conflicts. Advanced air traffic control systems, like the User Request Evaluation Tool (or URET; Wickens, Mavor, Parasuraman, & McGee, 1998), accomplish this function by examining programmed flight trajectories and calling the controller's attention (by highlighting and color coding) to paths of aircraft pairs that may lose separation in the future. Other forms of reminders (Herrmann, Brubaker, Yoder, Sheets, & Tio, 1999) or "pop-up menus" can alert the pilot if the computer infers that a pilot should be notified of a particular condition or reminded to take a particular action.

In general, such systems can be expected to be effective to the extent that the pilot's attention is overloaded or the pilot's memory system is sometimes fallible (and we know that this is the case) *and* to the extent that the automation inferences regarding the pilot's information needs are also correct. However, the latter is not always the case, and automation designers should be very wary of the consequences to pilot performance if computer inferences are sometimes incorrect (Wickens, Conejo, & Gempler, 1999; Wickens, Gempler, & Morphew, 2000); the consequences are pilot distrust, in which case the guidance offered by automation may be ignored even when it is correct. More subtle, but of equal concern, is a situation in which the pilot's (or controller's) attention becomes overly focused on the area or state highlighted by the automation, and as a consequence he or she is less effective in scanning other areas of space or considering other alternatives that are *not* called out by the automation (Mosier et al., 1998; Ockerman, 1999; Wickens et al., 1999; Yeh & Wickens, 2001b; Yeh, Wickens, & Seagull, 1999). An example of this was observed in an aircraft preflight inspection simulation carried out by Ockerman (1999). When a "wearable computer" directed the pilot's attention to particular locations where faults might be found, pilots became more vigilant at detecting such faults, but were more likely to *overlook* faults at unannounced locations. Similar results have been observed when automation has been used to guide attention to hostile ground targets (Yeh & Wickens, 2001b; Wickens et al., 1999; Yeh et al., 1999), to potential flight path conflicts (Wickens, Gempler, & Morphew, 2000), rotorcraft ground hazards (Davison & Wickens, 2001), or potential aircraft system failures (Mosier et al., 1998). Such findings of attentional narrowing should not themselves cause a rejection of "smart" attention directing automation; rather, however, they should highlight the need to carefully consider the consequences of such systems and quite possibly point to the implications for training the pilot as to the right level of attentional breadth that should accompany automated attention guidance. Such systems should always be accompanied by presentation of the raw data, both that to which attention is guided, and that to which it is not.

CONCLUSION

This chapter has barely scratched the surface regarding the many important issues that link the psychology of human perception and information processing to the human factors of aviation display design, and we have focused nearly exclusively on *visual* displays. But we have illustrated the importance of that linkage with examples of displays that are familiar, are novel, and are still on the drawing board. As both new display technology evolves and changes in airspace structure and procedures change the pilot's task, the nature of cockpit displays can be expected to change as well. Yet, the principles of human information processing that mediate between the display and the task will remain constant. Thus, we hope understanding of these principles and their applications to display design will support the designers' requirement to provide the pilot with continuous interpretable, reliable information.

REFERENCES

Aretz, A. J. (1991). The design of electronic map displays. *Human Factors, 33,* 85–101.

Bellenkes, A. H., Wickens, C. D., & Kramer, A. F. (1997). Visual scanning and pilot expertise: The role of attentional flexibility and mental model development. *Aviation, Space, and Environmental Medicine, 68,* 569–579.

Beringer, D. B., & Ball, J. (2001). General aviation pilot visual performance using conformal and non-conformal head-up and head-down highway-in-the-sky displays. *Proceedings of the 11th International Symposium on Aviation Psychology.* Columbus: The Ohio State University.

Beringer, D. B., Williges, R. C., & Roscoe, S. N. (1975). The transition of experienced pilots to a frequency-separated aircraft attitude display. *Human Factors, 17,* 401–414.

Billingsley, P. A. (1982). Navigation through hierarchical menu structures: Does it help to have a map? In *Proceedings of the 26th Annual Meeting of the Human Factors Society* (pp. 103–107). Santa Monica, CA: Human Factors Society.

Bliss, J. P., Freeland, M. J., & Millard, J. C. (1999). Alarm related incidents in aviation: A survey of the aviation safety reporting system database. In *Proceedings of the 43rd Annual Meeting of the Human Factors and Ergonomics Society* (pp. 6–10). Santa Monica, CA: Human Factors and Ergonomics Society.

Boff, K., & Lincoln, J. (1988). *Engineering data compendium.* Wright-Patterson Air Force Base, OH: Harry G. Armstrong Aerospace Medical Research Lab.

Boorman, D. (2001). Safety benefits of electronic checklists: An analysis of commercial transport accidents. *Proceedings of the 11th International Symposium on Aviation Psychology.* Columbus: The Ohio State University.

Boyer, B. S., & Wickens, C. D. (1994). *3D weather displays for aircraft cockpits* (Tech. Rep. No. ARL-94-11/NASA-94-4). Savoy: University of Illinois, Aviation Research Laboratory.

Bresley, B. (1995, April–June). 777 flight deck design. *Boeing Airliner,* 1–9.

Brickman, B. J., Hettinger, L. J., Haas, M. W., & Dennis, L. B. (1996). Designing the supercockpit. *Ergonomics in Design, 6*(2), 15–20.

Chappell, S. L. (1990). Pilot performance research for TCAS. *Managing the modern cockpit: Third human error avoidance techniques conference proceedings* (SAE 902357). Warrendale, PA: Society of Automotive Engineers.

Ciemer, C. E., Gabrarani, G. P., Sheridan, E. H., Stangel, S. C., Stead, R. P., & Tillotson, D. H. (1993). *The results of the TCAS II transition program (TTP).* Annapolis, MD: ARINC Research Corporation.

Clay, M. C. (1993). *Key cognitive issues in the design of electronic displays of instrument approach procedure charts* (DOT/FAA/RD-93/39). Washington, DC: Federal Aviation Administration.

Cohen, D., Otakeno, S., Previc, F. H., & Ercoline, W. R. (2001). Effect of "inside-out" and "outside-in" attitude displays on off-axis tracking in pilots and nonpilots. *Aviation, Space, and Environmental Medicine, 72*(3), 170–176.

Comstock, J. R., Jr. (2001). Can effective synthetic vision system displays be implemented on limited size display spaces? *Proceedings of the 11th International Symposium on Aviation Psychology.* Columbus: The Ohio State University.

Cutting, J. E., & Vishton, P. M. (1995). Perceiving layout and knowing distances: The integration, relative potency, and contextual use of different information about depth. In W. Epstein & S. Rogers (Eds.), *Perception of space and motion* (pp. 69–117). San Diego, CA: Academic Press.

Davison, H. J., & Wickens, C. D. (2001). Rotorcraft hazard cueing: The effects on attention and trust. *Proceedings of the 11th International Symposium on Aviation Psychology.* Columbus: The Ohio State University.

Degani, A., & Wiener, E. L. (1993). Cockpit checklists: Concepts, design and use. *Human Factors, 35,* 345–360.

Degani, A., & Wiener, E. L. (1997). Philosophy, policies, procedures and practice: The four P's of flight deck operation. In N. Johnston, N. McDonald, & R. Fuller (Eds.), *Aviation psychology in practice* (pp. 44–67). Brookfield, VT: Ashgate.

Delzell, S., & Battiste, V. (1993). Navigational demands of low-level helicopter flight. In R. S. Jensen and D. Meumlister (Eds.), *Proceedings of the 7th International Symposium on Aviation Psychology* (pp. 838–842). Columbus: Ohio State University.

Doherty, S. M., & Wickens, C. D. (2001). Effects of preview, prediction, frame of reference, and display gain in tunnel-in-the-sky displays. *Proceedings of the 11th International Symposium on Aviation Psychology.* Columbus, OH: The Ohio State University.

Dornheim, M. A. (1995, January 30). Dramatic incidents highlight mode problems in cockpit. *Aviation Week & Space Technology,* 57–59.

Driver, J. S., & Baylis, G. C. (1989). Movement and visual attention: The spotlight metaphor breaks down. *Journal of Experimental Psychology: Human Perception and Performance, 15,* 448–456.

Ellis, S. R. (Ed.). (1993). *Pictorial communication in visual and real environments* (2nd ed.). London: Taylor & Francis.

Ellis, S. R., McGreevy, M. W., & Hitchcock, R. J. (1987). Perspective traffic display format and airline pilot traffic avoidance. *Human Factors, 29,* 371–382.

Endsley, M. R. (1999). Situation awareness in aviation systems. In D. J. Garland, J. A. Wise, & V. D. Hopkin (Eds.), *Handbook of aviation human factors* (pp. 257–276). Mahwah, NJ: Lawrence Erlbaum Associates.

Fadden, S., Ververs, P., & Wickens, C. D. (1998). Costs and benefits of head-up display use: A meta-analytic approach. In *Proceedings of the 42nd Annual Meeting of the Human Factors and Ergonomics Society* (pp. 16–20). Santa Monica, CA: Human Factors and Ergonomics Society.

Fadden, S., Ververs, P. M., & Wickens, C. D. (2001). Pathway HUDS: Are they viable? *Human Factors, 43,* 173–193.

Fadden, S., Wickens, C. D., & Ververs, P. M. (2000). Costs and benefits of head-up displays: An attention perspective and a meta analysis (Paper # 2000–01–5542). *2000 World Aviation Congress.* Warrendale, PA: Society of Automotive Engineers.

Fischer, E., Haines, R. F., & Price, T. A. (1980). *Cognitive issues in head-up displays* (NASA Technical Paper 1711). Moffett Field, CA: NASA Ames Research Center.

Fisher, D. L., Coury, B. G., Tengs, T. O., & Duffy, S. A. (1989). Minimizing the time to search visual displays: The role of highlighting. *Human Factors, 31,* 167–182.

Fitts, P., Jones, R. E., & Milton, E. (1950). Eye movements of aircraft pilots during instrument landing approaches. *Aeronautical Engineering Review, 9,* 24–29.

Flohr, E., & Huisman, H. (1997). Perspective primary flight displays in the 4D ATM environment. In *Proceedings of the Ninth International Symposium on Aviation Psychology* (pp. 882–886). Columbus: The Ohio State University.

Foyle, D. C., Andre, A. D., McCann, R. S., Wenzel, E. M., Begault, D. R., & Battiste, V. (1996). Taxiway Navigation and Situation Awareness (T-NASA) System: Problem, design philosophy, and description of an integrated display suite for low-visibility airport surface operations. *SAE Transactions: Journal of Aerospace, 105,* 1411–1418.

Fradden, D. M., Braune, R., & Wiedemann, J. (1993). Spatial displays as a means to increase pilot situational awareness. In S. R. Ellis (Ed.), *Pictorial communication in virtual and real environments* (pp. 172–181). Bristol, PA: Taylor & Francis.

Geiselman, E. E., & Osgood, R. K. (1995). Head vs. aircraft oriented air-to-air target location symbology using a helmet-mounted display. In R. J. Lewandowski, W. Stephens, & L. A. Haworth (Eds.), *Proceedings—SPIE Vol. 2465* (pp. 214–225). Bellingham, WA: SPIE—International Society for Optical Engineering.

Gray, W. M., Chappell, A. R., Thurman, D. A., Palmer, M. T., & Mitchell, C. M. (2000). The VProf tutor: Teaching MD-11 vertical profile navigation using GT-ITACS. In *IEA 2000/HFES 2000 Congress Proceedings* (pp. 239–242). San Diego, CA: Human Factors and Ergonomics Society.

Gregory, R. L. (1977). *Eye and brain.* London: Weidenfeld & Nicolson.

Hahn, E. C., & Hansman, R. J. (1992). Experimental studies on the effect of the effect of automation and pilot situational awareness in the data ink ATC environment. *SAE AEROTECH Conference and Exposition.* Warrendale, PA: Society of Automotive Engineers.

Hamilton, B. E. (1999). Helicopter human factors. In D. J. Garland, J. A. Wise, & V. D. Hopkin (Eds.), *Handbook of aviation human factors* (pp. 405–428). Mahwah, NJ: Lawrence Erlbaum Associates.

Hammer, J. M. (1999). Human factors of functionality and intelligent avionics. In D. J. Garland, J. A. Wise, & V. D. Hopkin (Eds.), *Handbook of aviation human factors* (pp. 549–566). Mahwah, NJ: Lawrence Erlbaum Associates.

Harris, R. L., & Christhilf, D. M. (1980). What do pilots see in displays? In *Proceedings of the 24th Annual Meeting of the Human Factors Society* (pp. 22–26). Santa Monica, CA: Human Factors Society.

Hart, S. G. (1988). Helicopter human factors. In E. L. Wiener & D. C. Nagel (Eds.), *Human factors in flight* (pp. 591–638). San Diego, CA: Academic Press.

Haskell, I. D., & Wickens, C. D. (1993). Two- and three-dimensional displays for aviation: A theoretical and empirical comparison. *International Journal of Aviation Psychology, 3,* 87–109.

Hawkins, F. H. (Ed.). (1993). *Human factors in flight* (2nd ed.). Aldershot, England: Ashgate.

Haworth, L. A., & Seery, R. E. (1992). *Rotorcraft helmet mounted display symbology research* (SAE Technical Paper Series 921977). Warrendale, PA: Society of Automotive Engineers.

Helleberg, J., & Wickens, C. D. (2001). Effects of data link modality on pilot attention and communication effectiveness. *Proceedings of the 11th International Symposium on Aviation Psychology.* Columbus: The Ohio State University.

Hendrick, H. W. (1998). *Good ergonomics is good ergonomics.* Santa Monica, CA: Human Factors and Ergonomics Society.

Herrmann, D., Brubaker, B., Yoder, C., Sheets, V., & Tio, A. (1999). Devices that remind. In F. T. Durso (Ed.), *Handbook of applied cognition* (pp. 377–407). New York: Wiley.

Hickox, J. C., & Wickens, C. D. (1999). Effects of elevation angle disparity, complexity, and feature type on relating out-of-cockpit field of view to an electronic cartographic map. *Journal of Experimental Psychology: Applied, 5*(3), 284–301.

Hicks, M., & DeBrito, G. (1998). Civil aircraft warning systems: Who's calling the shots? In G. Boy, C. Graeber, & J. Robert (Eds.), *International Conference on Human-Computer Interaction in Aeuronautics* (pp. 205–212). Montreal, Canada: Ecale Polytechnique de Montreal.

Hofer, E. F., Braune, R. J., Boucek, G. P., & Pfaff, T. A. (2000). *Attention switching between near and far domains: An exploratory study of pilots' attention switching with head-up and head-down tac-*

tical displays in simulated flight operations (D6–36668). Seattle, WA: Boeing Commercial Airplane Group.

Hofer, E. F., & Wickens, C. D. (1997). Part-mission simulation evaluation of issues associated with electronic approach chart displays. In *Proceedings of the Ninth International Symposium on Aviation Psychology* (pp. 347–352). Columbus: The Ohio State University.

Jensen, R. S. (1981). Prediction and quickening in perspective flight displays for curved landing approaches. *Human Factors, 23,* 355–363.

Johnson, S. L., & Roscoe, S. N. (1972). What moves, the airplane or the world? *Human Factors, 14,* 107–129.

Johnson, W. W., Battiste, V., & Bochow, S. H. (1999). A cockpit display designed to enable limited flight deck separation responsibility. *Proceedings of the 1999 World Aviation Conference.* Anaheim, CA.

Kahneman, D., & Treisman, A. (1984). Changing views of attention and automaticity. In R. Parasuraman & R. Davies (Ed.), *Varieties of attention* (pp. 29–61). New York: Academic Press.

Kerns, K. (1999). Human factors in air traffic control/flight deck integration: Implications of data-link simulation research. In D. J. Garland, J. A. Wise, & V. D. Hopkin (Eds.), *Handbook of aviation human factors* (pp. 519–546). Mahwah, NJ: Lawrence Erlbaum Associates.

Kirwan, B., & Ainsworth, L. K. (1992). *A guide to task analysis.* London: Taylor & Francis.

Kovalenko, P. A. (1991). Psychological aspects of pilot spatial orientation. *ICAO Journal, 46,* 18–23.

Kramer, A. F., & Jacobson, A. (1991). Perceptual organization and focused attention: The role of objects and proximity in visual processing. *Perception and Psychophysics, 50,* 267–284.

Kreifeldt, J. G. (1980). Cockpit displayed traffic information and distributed management in air traffic control. *Human Factors, 22,* 671–691.

Kroft, P. D., & Wickens, C. D. (2001). Integrating aviation databases: Effects of scanning, clutter, resolution, and interactivity. *Proceedings of the 11th International Symposium on Aviation Psychology.* Columbus: The Ohio State University.

Leger, A., Aymeric, B., Audrezet, H., & Alba, P. (1999). Human factor issues associated with HUD-based hybrid landing systems: SEXTANT's experience. In *Proceedings of the World Aviation Congress* (1999–01–5512). Warrendale, PA: Society of Automotive Engineers.

Levy, J. L., Foyle, D. C., & McCann, R. S. (1998). Performance benefits with scene-linked HUD symbology: An attentional phenomenon? In *Proceedings of the 42nd Annual Meeting of the Human Factors and Ergonomics Society* (pp. 11–15). Santa Monica, CA: Human Factors and Ergonomics Society.

Lindholm, J. M. (1999). Weather information presentation. In D. J. Garland, J. A. Wise, & V. D. Hopkin (Eds.), *Handbook of aviation human factors* (pp. 567–590). Mahwah, NJ: Lawrence Erlbaum Associates.

Loftus, G., Dark, V., & Williams, D. (1979). Short-term memory factors in ground controller/pilot communication. *Human Factors, 21,* 169–181.

Martin-Emerson, R., & Wickens, C. D. (1997). Superimposition, symbology, visual attention, and the head-up display. *Human Factors, 39,* 581–601.

May, P. A., Campbell, M., & Wickens, C. D. (1996). Perspective displays for air traffic control: Display of terrain and weather. *Air Traffic Control Quarterly, 3*(1), 1–17.

McCann, R. S., Foyle, D. C., Andre, A. D., & Battiste, V. (1996). Advanced navigation aids in the flight deck: Effects on ground taxi performance under low-visibility conditions. *SAE Transactions: Journal of Aerospace, 105,* 1419–1430.

McClellan, J. M. (1999, May). Heads up. *Flying,* 68–73.

McGreevy, M. W., & Ellis, S. R. (1986). The effect of perspective geometry on judged direction in spatial information instruments. *Human Factors, 28,* 439–456.

Melzer, J. E., & Moffitt, K. (Eds.) (1997). *Head mounted displays: Designing for the user.* New York: McGraw-Hill.

Merlo, J. L., Wickens, C. D., & Yeh, M. (1999). *Effect of reliability on cue effectiveness and display signaling* (Technical Report ARL-99-4/FED-LAB-99-3). Savoy: University of Illinois, Aviation Research Lab.

Merryman, R. F. K., & Cacioppo, A. J. (1997). The optokinetic cervical reflex in pilots of high-performance aircraft. *Aviation, Space, and Environmental Medicine, 68,* 479–487.

Merwin, D., O'Brien J. V., & Wickens, C. D. (1997). Perspective and coplanar representation of air traffic: Implications for conflict and weather avoidance. *Proceedings of the Ninth International Symposium on Aviation Psychology* (pp. 362–367). Columbus: The Ohio State University.

Merwin, D. H., & Wickens, C. D. (1996). *Evaluation of perspective and coplanar cockpit displays of traffic information to support hazard awareness in free flight* (Technical Report ARL-96-5/NASA-96-1). Savoy: University of Illinois, Aviation Research Laboratory.

Milgram, P., & Colquhoun, H. (1999). A taxonomy of real and virtual world display integration. In Y. Ohta & H. Tamura (Eds.), *Mixed reality—Merging real and virtual worlds.* Tokyo: Ohmsha and Berlin, Germany: Springer-Verlag.

Morphew, E. M., & Wickens, C. D. (1998). Pilot performance and workload using traffic displays to support free flight. In *Proceedings of the 42nd Annual Meeting of the Human Factors and Ergonomics Society* (pp. 52–56). Santa Monica, CA: Human Factors and Ergonomics Society.

Morrow, D., Lee, A., & Rodvold, M. (1993). Analysis of problems in routine controller-pilot communication. *International Journal of Aviation Psychology, 3,* 285–302.

Mosier, K. L., Skitka, L. J., Heers, S., & Burdick, M. (1998). Automation bias: Decision making and performance in high-tech cockpits. *International Journal of Aviation Psychology 8,* 47–63.

Mykityshyn, M. G., Kuchar, J. K., & Hansman, R. J. (1994). Experimental study of electronically based instrument approach plates. *International Journal of Aviation Psychology, 4,* 141–166.

Navarro, C., & Sikorski, S. (1999). Datalink communication in flight deck operations: A synthesis of recent studies. *International Journal of Aviation Psychology, 9,* 361–376.

Newman, R. L. (1995). *Head-up displays: Designing the way ahead.* Brookfield, VT: Avebury.

Nikolic, M. I., and Sarter, N. B. (2001). Peripheral visual feedback: A powerful means of supporting effective attention allocation in event-driven, data rich environments. *Human Factors, 43,* 30–38.

O'Brien, J. V., & Wickens, C. D. (1997). Free flight cockpit displays of traffic and weather: Effects of dimensionality and data base integration. *Proceedings of the 41st Annual Meeting of the Human Factors and Ergonomics Society* (pp. 18–22). Santa Monica, CA: Human Factors and Ergonomics Society.

Ockerman, J. J. (1999). Over-reliance issues with task guidance systems. *Proceedings of the 43rd Annual Meeting of the Human Factors and Ergonomics Society* (pp. 1192–1196). Santa Monica, CA: Human Factors and Ergonomics Society.

Olmos, O., Liang, C. C., & Wickens, C. D. (1997). Electronic map evaluation in simulated visual meteorological conditions. *International Journal of Aviation Psychology, 7,* 37–66.

Olmos, O., Mundra, A., Cieplak, J. J., Domino, D. A., & Stassen, H. P. (1998). Evaluation of near-term applications for ADS-B/CDTI implementation. *Proceedings of SAE/AIAA World Aviation Congress.* Warrendale, PA: Society for Automotive Engineers.

Olmos, O., Wickens, C. D., & Chudy, A. (2000). Tactical displays for combat awareness: An examination of dimensionality and frame of reference concepts and the application of cognitive engineering. *International Journal of Aviation Psychology, 10*(3), 247–271.

Oman, C. M., Kendra, A. J., Hayashi, M., Stearns, M. J., & Burki-Cohen, J. (2001). Vertical navigation displays: Pilot performance and workload during simulated constant angle of descent GPS approaches. *International Journal of Aviation Psychology, 11*(1), 15–31.

Orlady, H., & Orlady, L. M. (2000). *Human factors in multicrew flight operations.* Brookfield, VT: Ashgate.

Parasuraman, R. (1987). Human-computer monitoring. *Human Factors, 29,* 695–706.

Parasuraman, R., & Riley, V. (1997). Humans and automation: Use, misuse, disuse, abuse. *Human Factors, 39,* 230–253.

Patterson, R. D. (1990). Auditory warning sounds in the work environment. *Philosophical Transactions of the Royal Society of London, 327,* 485–492.

Patterson, R., & Martin, W. L. (1992). Human stereopsis. *Human Factors, 34,* 669–692.

Phillips, E. H. (1997, April 21). FAA may mandate enhanced GPWS. *Aviation Week & Space Technology,* 22–23.

Podczerwinski, E. S., & Wickens, C. D., & Alexander, A. (2001). *Exploring the "out of sight, out of mind" phenomenon in dynamic settings across electronic map displays* (ARL-01-8/NASA-01-4). Savoy: University of Illinois, Aviation Research Laboratory.

Previc, F. H. (1998). The neuropsychology of 3-D space. *Psychological Bulletin, 124,* 123–164.

Previc, F. H. (2000, March/April). Neuropsychological guidelines for aircraft control stations. *IEEE Engineering in Medicine and Biology,* 81–88.

Previc, F. H., & Ercoline, W. R. (1999). The "outside-in" attitude display concept revisited. *International Journal of Aviation Psychology, 9,* 377–401.

Pritchett, A. (2001). Reviewing the role of cockpit alerting systems. *Human Factors and Aerospace Safety, 1,* 5–38.

Reising, J., Liggett, K., Kustra, T. W., Snow, M. P., Hartsock, D. C., & Barry, T. P. (1998). Evaluation of pathway symbology used to land from curved approaches. *Proceedings of the 42nd Annual Meeting of the Human Factors and Ergonomics Society* (pp. 1–5). Santa Monica, CA: Human Factors and Ergonomics Society.

Roscoe, S. N. (1968). Airborne displays for flight and navigation. *Human Factors, 10,* 321–332.

Roscoe, S. N. (1997). Horizon Control reversals and the grave yard spiral. *CSERIAC GATEWAY, VII,* 1–4.

Roscoe, S. N., & Williges, R. C. (1975). Motion relationships in aircraft attitude and guidance displays: A flight experiment. *Human Factors, 17,* 374–387.

Roske-Hofstrand, R. J., & Paap, K. R. (1986). Cognitive networks as a guide to menu organization: An application in the automated cockpit. *Ergonomics, 29,* 1301–1311.

Ruddle, R. A., Payne, S. T., & Jones, D. M. (1999). The effects of maps on navigation and search strategies in very large scale virtual environments. *Journal of Experimental Psychology: Applied, 5,* 54–75.

St. John, M., Cowen, M. B., Smallman, H. S., & Oonk, H. M. (2001). The use of 2-D & 3-D displays for understanding vs. relative positioning tasks. *Human Factors, 43,* 1, 79–98.

Sachs, G., Dobler, K., & Hermle, P. (1998). Synthetic vision flight tests for curved approach and landing. *Proceedings of the 27th Digital Avionics Systems Conference.* New York: Institute of Electrical and Electronics Engineers.

Salvendy, G. (Ed.). (1997). *Handbook of human factors and ergonomics.* New York: Wiley.

Sanders, M. S., & McCormick, E. J. (1993). *Human factors in engineering and design* (7th ed.). New York: McGraw-Hill.

Sarter, N. B., & Woods, D. D. (1994). Pilot interaction with cockpit automation II: An experimental study of pilots' model and awareness of the flight management system. *International Journal of Aviation Psychology, 4,* 1–28.

Sarter, N. B., & Woods, D. D. (1995). How in the world did we ever get into that mode? Mode error and awareness in supervisory control. *Human Factors, 37,* 5–19.

Sarter, N. B., & Woods, D. D. (1997). Team play with a powerful and independent agent: Operational experiences and automation surprises on the Airbus A-320. *Human Factors, 39,* 553–569.

Satchell, P. M. (1993). *Cockpit monitoring and alerting systems.* Brookfield, VT: Ashgate.

Schreiber, B. T., Wickens, C. D., Renner, G. J., Alton, J., & Hickox, J. C. (1998). Navigational checking using 3D maps: The influence of elevation angle, azimuth, and foreshortening. *Human Factors, 40,* 209–223.

Scott, W. B. (1999, February 1). Automatic GCAS: You can't fly any lower. *Aviation Week & Space Technology,* 76–80.

Seagull, F. J., & Gopher, D. (1997). Training head movement in visual scanning: An embedded approach to the development of piloting skills with helmet-mounted displays. *Journal of Experimental Psychology: Applied, 3*(3), 163–180.

Sedgwick, H. (1986). Depth perception. In K. Boff, L. Kaufman, & J. Thomas (Eds.), *Handbook of perception and human performance* (Vol. 1, Chap. 21). New York: Wiley.

Seidler, K., & Wickens, C. D. (1992). Distance and organization in multifunction displays. *Human Factors, 34,* 555–569.

Shelden, S., & Belcher, S. (1999). Cockpit traffic displays of tomorrow. *Ergonomics in Design, 7*(3), 4–9.

Sherry, L., & Polson, P. G. (1999). Shared models of flight management system vertical guidance. *International Journal of Aviation Psychology, 9,* 139–154.

Shneiderman, B. (1987). *Designing the user interface: Strategies for effective human-computer interaction.* Reading, MA: Addison-Wesley.

Silverstein, L. D. (1987). Human factors for color display systems: Concepts, methods, and research. In H. J. Durrett (Ed.), *Color and the computer* (pp. 27–61). New York: Academic Press.

Sklar, A. E., & Sarter, N. B. (1999). Good vibrations: Tactile feedback in support of attention allocation and human-automation coordination in event-driven domains. *Human Factors, 41,* 543–552.

Smith, D. R., Cacioppo, A. J., & Hinman, G. E. (1997). Aviation spatial orientation in relationship to head position, altitude interpretation, and control. *Aviation, Space, and Environmental Medicine, 68,* 472–478.

Sorkin, R. D., Kantowitz, B. H., & Kantowitz, S. C. (1988). Likelihood alarm displays. *Human Factors, 30,* 445–460.

Stanton, N. (Ed.). (1994). *Human factors in alarm design.* Bristol, PA: Taylor & Francis.

Steenblik, J. W., (1989, December). Alaska airlines H6S. *Airline Pilot,* 10–14.

Theunissen, E. (1997). *Integrated design of a man-machine interface for 4-D navigation.* Doctoral dissertation, Faculty of Electrical Engineering, Delft University of Technology. Delft, The Netherlands.

Treisman, A. (1986). Properties, parts, and objects. In K. R. Boff, L. Kaufman & J. P. Thomas (Eds.), *Handbook of perception and human performance* (Vol. 2, pp. 35-1–35-70). New York: Wiley.

Tullis, T. S. (1988). A system for evaluating screen formats: Research and application. In H. R. Hartson & D. Hix (Eds.), *Advanced in human-computer interaction* (pp. 214–286). Norwood, NJ: Ablex.

Vakil, S., & Hansman, R. J. (1999). Approaches to mitigating complexity-driven issues in commercial autoflight systems. In *Third Workshop on Human Error, Safety, and System Development (HESSD'99).* Leige, Belgium.

van Breda, L. (1999). *Anticipatory behavior in supervisory control.* Delft, The Netherlands: Delft University Press.

Ververs, P. M., & Wickens, C. D. (1998). Head-up displays: Effects of clutter, symbology intensity, and display location on pilot performance. *International Journal of Aviation Psychology, 8,* 377–403.

von Viebahn, H. (1998). The 4-D display. *Proceedings of the 17th Digital Avionics Systems Conference.* New York: Institute of Electrical and Electronics Engineers.

Warm, J. S. (1984). *Sustained attention in human performance.* London: Wiley.

Weintraub, D. J., & Ensing, M. J. (1992). *Human factors issues in head-up display design: The book of HUD* (CSERIAC State of the Art Report 92–2). Wright-Patterson Air Force Base, OH: Crew System Ergonomics Information Analysis Center.

Wickens, C. D. (1986). The effects of control dynamics on performance. In K. R. Boff, L. Kaufman, & J. P. Thomas (Eds.), *Handbook of Perception and Performance* (Vol. 2, pp. 39–1–39–60). New York: Wiley.

Wickens, C. D. (1997). Attentional issues in head-up displays. In D. Harris (Ed.), *Engineering psychology and cognitive ergonomics: Transportation systems (Vol. 1,* pp. 3–21). Aldershot, England: Ashgate.

Wickens, C. D. (1999). Frames of reference for navigation. In D. Gopher & A. Koriat (Eds.), *Attention and performance* (Vol. 16, pp. 113–144). Orlando, FL: Academic Press.

Wickens, C. D. (2000) Human factors in vector map design: The importance of task-display dependence. *Journal of Navigation, 53,* 1–14.

Wickens, C. D., & Carswell, C. M. (1995). The proximity compatibility principle: Its psychological foundation and its relevance to display design. *Human Factors, 37,* 473–494.

Wickens, C. D., Conejo, R., & Gempler, K. (1999). Unreliable automated attention cueing for air-ground targeting and traffic maneuvering. *Proceedings of the 43rd Meeting of the Human Factors and Ergonomics Society* (pp. 21–25). Santa Monica, CA: Human Factors and Ergonomics Society.

Wickens, C. D., Gempler, K., & Morphew, M. E. (2000). Workload and reliability of predictor displays in aircraft traffic avoidance. *Transportation Human Factors Journal, 2*(2), 99–126.

Wickens, C. D., Gordon, S., & Liu, Y. (1998). *An introduction to human factors engineering.* New York: Addison Wesley Longman.

Wickens, C. D., & Hollands, J. G. (2000). *Engineering psychology and human performance* (3rd ed.). Upper Saddle River, NJ: Prentice Hall.

Wickens, C. D., Kroft, P., & Yeh, M. (2000). Database overlay in electronic map design. In *Proceedings of the IEA 2000/HFES 2000 Congress* (pp. 3-451–3-454). Santa Monica, CA: Human Factors and Ergonomics Society.

Wickens, C. D., Liang, C. C., Prevett, T., & Olmos, O. (1996). Electronic maps for terminal area navigation: Effects of frame of reference and dimensionality. *International Journal of Aviation Psychology, 6,* 241–271.

Wickens, C. D., & Long, J. (1995). Object- vs. spaced-based models of visual attention: Implications for the design of head-up displays. *Journal of Experimental Psychology: Applied, 1,* 179–194.

Wickens, C. D., Mavor, A. S., Parasuraman, R., & McGee, J. P. (Eds.). (1998). *The future of air traffic control: Human operators and automation.* Washington, DC: National Academy Press.

Wickens, C. D., & Prevett, T. (1995). Exploring the dimensions of egocentricity in aircraft navigation displays. *Journal of Experimental Psychology: Applied, 1,* 110–135.

Wickens, C. D., & Seidler, K. S. (1997). Information access in a dual-task context: Testing a model of optimal strategy selection. *Journal of Experimental Psychology: Applied, 3,* 196–215.

Wickens, C. D., Todd, S., & Seidler, K. (1989). *Three-dimensional displays: Perception, implementation, and applications* (Technical Report ARL-89-11/CSERIAC-89-1). Savoy: University of Illinois, Aviation Research Laboratory (also CSERIAC SOAR 89–001). Wright-Patterson Air Force Base, OH: Crew System Ergonomics Information Analysis Center, December).

Wickens, C. D., Ververs, P. M., & Fadden, S. (in press). Head up displays. In D. Harris (Ed.), *Human factors for civil flight deck design.*

Wickens, C. D., Vincow, M. A., Schopper, A. W., & Lincoln, J. E. (1997). *Computational models of human performance in the design and layout of controls and displays* (CSERIAC SOAR Report 97–22). Wright-Patterson Air Force Base, OH: Crew System Ergonomics Information Analysis Center.

Wiener, E. L. (1977). Controlled flight into terrain accidents: System-induced errors. *Human Factors, 19,* 171–181.

Wiener, E. L. (1980). Mid air collisions. *Human Factors, 22,* 521–534.

Williams, H., Hutchinson, S., & Wickens, C. D. (1996). A comparison of methods for promoting geographic knowledge in simulated aircraft navigation. *Human Factors, 38,* 50–64.

Woods, D. D. (1984). Visual momentum: A concept to improve the cognitive coupling of person and computer. *International Journal of Man-Machine Studies, 21,* 229–244.

Wright, P. (1998). Printed instructions: Can research make a difference? In H. Zwaga, T. Boersema, & H. Hoonout (Eds.), *Visual information for everyday use: Design and research perspectives* (pp. 45–66). London: Taylor & Francis.

Yeh, M., & Wickens, C. D. (2001a). Attentional filtering in the design of electronic map displays: A comparison of color-coding, intensity coding, and decluttering techniques. *Human Factors, 43,* 543–562.

Yeh, M., & Wickens, C. D. (2001b). Explicit and implicit display signaling in augmented reality: The effects of cue reliability, image realism, and interactivity on attention allocation and trust calibration. *Human Factors, 43,* 355–365.

Yeh, M., Wickens, C. D., & Seagull, F. J. (1999). Target cueing in visual search: The effects of conformality and display location on the allocation of visual attention). *Human Factors, 41,* 524–542.

ACKNOWLEDGMENTS

The authors would like to acknowledge the support of Grant NAG 2–1120 from NASA Ames Research Center. Many of the ideas expressed with this chapter were developed under the support of this grant. Dr. David Foyle was the technical/scientific monitor. Sandra Hart was also responsible for monitoring grants that supported much of the research reported here.

6

Aeronautical Decision Making: Metaphors, Models, and Methods

David O'Hare
University of Otago

Few people have an extensive knowledge of the landmarks of unmanned flight. It is the history of manned flight that captures the imagination. The individuals who first "slipped the bonds" of gravity are now household names, with the Wright brothers and Yuri Gagarin foremost in the public knowledge of the history of flight. One reason for this could be that the drama of early flight came from the very real risks to which these early proponents were exposed. Human risk takers exert a powerful fascination on the human mind, as they exemplify the constant struggle between the power of the human mind and the untamed forces of nature. As aviation has progressed from a high-risk endeavor at the beginning of the century to a highly predictable and safe transportation system at the end of the same century, the way in which we view the human actors involved has changed correspondingly. The next section presents some of the early views of the pilot as gambler, daredevil, ace, and romantic. In the following sections I will discuss a variety of metaphors that have guided the scientific analysis of human performance in aviation with a particular emphasis on the pilot as aeronautical decision maker.

GAMBLERS, DAREDEVILS, ACES, AND ROMANTICS

Many people, before and after the demonstration of sustained powered flight by the Wright brothers in 1903, gambled their lives on a belief that their contraptions would temporarily defeat the forces of gravity. Gwynn-Jones (1981, p. 7) refers to these early pioneers as "magnificent gamblers": "The gamblers of the air were cast in a different mould. They were the long-shot punters who placed their bets when the odds were a hundred to one against. And the stakes were the highest of all . . . their lives!" History remembers the victors—those who won their gambles, such as Harriet Quimby, first woman across the English Channel in 1912, and Charles Lindbergh, who conquered the Atlantic alone in 1927.

The early gamblers were initially thought of in the same tradition as sportsmen and women, but with the advent of aerial warfare in the First World War, the image changed to that of "the flying ace, an airborne knight armed with a machine gun who jousted in the sky" (Wohl, 1994, p. 203). One of these was the French ace Roland Garros who first solved the problem of how to fire a machine gun straight ahead and through the propeller arc. Another great ace, Max Immelmann, died when his machine gun malfunctioned and a bullet shattered one of his propeller blades. These wartime aces tended to be outstanding sportsmen and athletes with a preference for high-risk sports such as mountain climbing.

With the end of the war, aviators began to take on the role of explorers and romantics undertaking hitherto unimaginable journeys, such as Lindbergh's trans-Atlantic crossing and Amy Johnson's flight from England to Australia in just 19 days. As Wohl (1994, p. 1) puts it, "The miracle of flight, once achieved, opened vistas of conquests over nature that excited people's imagination." These feats of conquest laid the basis for the subsequent development of aviation as a means of commercial transport. For this to occur, however, there had to be significant improvements in reliability and safety. By the 1930s, air transport had started to become an accepted form of travel in the United States. Pilots, at least in the public imagination, now became captains of the sky—uniformed, reassuring and "in command."

The effects of these metaphors on practical matters such as selection and training (see chap. 10) can be seen in one example. English (1996) describes the qualities sought in Canadian pilots around the time of the First World War: "Clear-headed, keen young men . . . sturdy physique . . . with a measure of recklessness thrown in" (p. 23). Evidence of interests in motoring or riding horses was considered proof of the required recklessness! The U.S. Army Signal Corps also described their pilots in equestrian terms as a "twentieth century cavalry officer mounted on Pegasus" (Henmon, 1919, cited in Koonce, 1984, p. 500). Such concepts have little relevance to pilot selection today, as we have discarded the early metaphors of pilots as gamblers, daredevils, aces, and romantics. In the remainder of this chapter I propose to review five contemporary metaphors of pilot performance and examine the models of aeronautical decision making that each has generated.

WHAT IS AERONAUTICAL DECISION MAKING (ADM)?

Decision making is conventionally characterized as the act of choosing between alternatives under conditions of uncertainty. Decision making is thus different from the simple act of choice (e.g., choosing which of 20 flavors of ice cream to purchase) by virtue of the presence of uncertainty. The factoring in of both the value or utility of an outcome and its subjective likelihood of occurrence began in the Renaissance (Bernstein, 1996) and now forms the normative standard of rational decision making. Questions about aeronautical decision making have often arisen in the aftermath of a crash.

For example, the first fatal U.S. commercial airline crash in almost two years on June 2, 1999, raised immediate concerns about the crew's decision making in landing their American Airlines MD-80 into the midst of a severe thunderstorm at Little Rock, Arkansas. Without knowing the results of the detailed accident investigation, there appear to be considerable similarities with a previous crash of a Delta Air Lines L-1011 Tristar at Dallas–Fort Worth in August 1985 (National Transportation Safety Board [NTSB], 1986). This flight penetrated a heavy storm cloud containing lightning. The wind shear conditions caused the indicated airspeed to fluctuate by − 44kts to + 20kts. There was heavy rain and severe turbulence. The airplane struck the ground about a mile short of the runway, killing most of those on board.

The "probable causes" of the crash were held to be "the lack of definitive, real-time wind shear hazard information" and "the flightcrew's decision to initiate and continue the approach into a cumulonimbus cloud which they observed to contain visible lightning" (p.i.). The National Transportation Safety Board (NTSB) report devoted much attention to pilot decision making, quoting at length from a National Aeronautics and Space Administration (NASA) memorandum (NASA, 1975, cited in NTSB, 1986):

> . . . a pilot must first seek and acquire information from whatever sources are available. He must then make some determination regarding the quantity, and the quality, of the information. . . . Having determined that he has enough information, and that it is reasonably reliable, the pilot must then process these data in predetermined ways in order to reach a wise decision from a limited number of alternatives. . . . A large part of this process involves the pilot's judgment of probabilities; he is attempting to make wise decisions, often in the face of uncertainty. (p. 69)

This summary of pilot decision making very much reflects the normative characterization of decision making in which the pilot's essential task is to choose the best option, bearing in mind the possible outcomes associated with each choice. This reflects the dominant metaphor that has guided thinking about the aeronautical decision maker—the essentially vigilant, but somewhat prone to error, processor of information. Since this report was published, other metaphors have appeared that have challenged and shaped our thinking about ADM. The following sections will outline these different metaphors and the associated theoretical and empirical work that has characterized aeronautical decision making in the past two decades or so. An excellent introduction to the decision making field can be found in Wickens and Holland (2000).

THE DECISION MAKER AS FAULTY COMPUTER

In the early 1970s, psychologists Tversky and Kahneman published a review of some simple experiments that showed that people's judgments under uncertainty were largely based on several heuristic processes (Tversky & Kahneman, 1974). A heuristic is a mental "shortcut" involving psychological processes such as assessing the similarity of one event to another or the ease with which an example can be brought to mind, rather than reasoning with probabilities. This led to widespread interest in the apparent fallibility of human judgment summarized in an edited collection of 35 chapters published in 1982 with the title *Judgment Under Uncertainty: Heuristics and Biases.* The "heuristics and biases" literature was based on a paradigm that provided participants with two solutions to a problem. The correct solution was derived from a normative analysis of the situation, and the incorrect response was based on a heuristic process. By choosing the incorrect response, participants were shown to have followed the heuristic rather than the normative approach. The message that was taken from these studies, and which has been hugely influential in many applied fields such as economics and business, was that humans use judgmental heuristics and thereby reach largely incorrect conclusions. However, because the paradigm does not include a third possibility—a correct answer reached by heuristic processes, the studies do not in fact show that heuristic reasoning normally leads to incorrect outcomes (Lopes, 1991). More-recent work has emphasized the positive utility of heuristic reasoning that can be both fast and accurate (Gigerenzer, Todd, & The ABC Research Group, 1999).

The message that "human incompetence was . . . a fact like gravity" (Lopes, 1991, p. 67) has, however, been popularized in numerous volumes with titles like *Irrationality: The Enemy Within* (Sutherland, 1992) and *Inevitable Illusions* (Piattelli-Palmarini, 1994). For many, the heuristics and biases literature developed from Tversky and Kahneman's innovative studies confirms the second-rate status of human information processing in comparison to the normative models of classical decision theory. According to this view, people have difficulty making judgments under uncertainty because of faulty routines in processing the necessary data. Clearly, the antidote to such a problem would be to train people how to use the correct routines for processing information. Although there have been few actual heuristics and biases studies in aviation, there have been a number of attempts to apply corrective training.

Wickens and Flach (1988) outlined a general decision-making model for aeronautical decision making with explicit reference to the possible operation of various heuristics and biases (see Fig. 6.1). The model outlines a sequence of information-processing activities from cue sampling through situation assessment, option generation, choice, and action. Sources of bias due to the use of heuristic processes are presumed to affect particular stages in this process. For example, framing operates at the point where a decision or choice is formulated. The confirmation bias affects cue sampling and so forth.

Wickens and Flach focus their attention on the decision point where the pilot must choose between several options. Wickens and Flach use the example of a

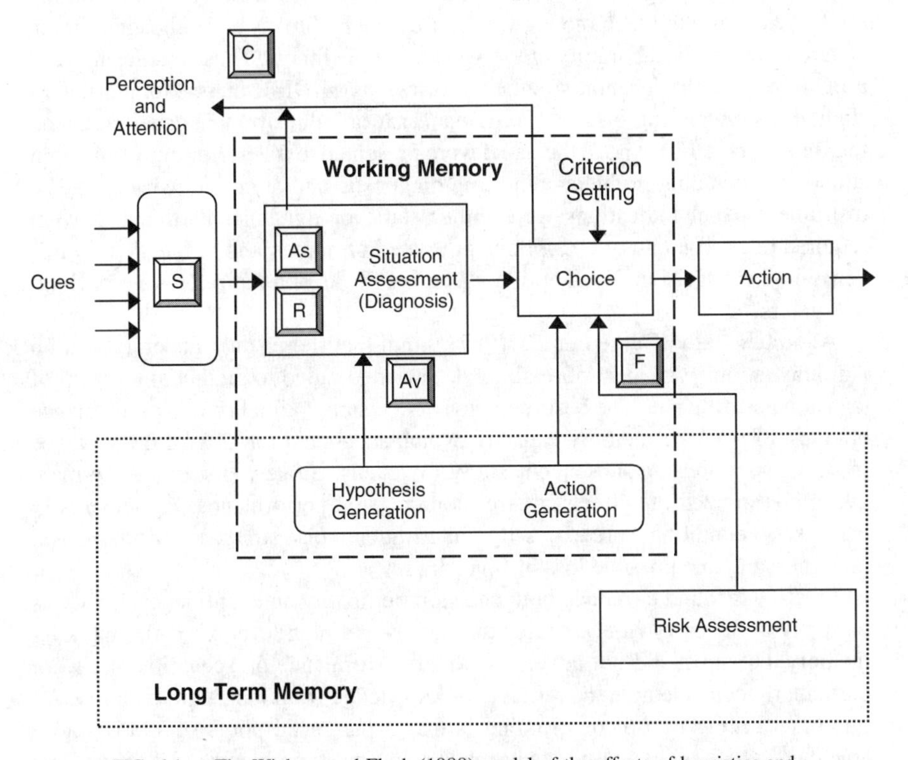

FIG. 6.1. The Wickens and Flach (1988) model of the effects of heuristics and biases on information processing in ADM. Information flows from cue sampling at the left to selection of an action at the right. The suggested influence of the various heuristics and biases is indicated by the following letters: S = Salience bias; C = Confirmation bias; As = "As If" heuristic; R = Representativeness heuristic; Av = Availability heuristic, and F = Framing. *Note.* From *Human Factors in Aviation* (Figure 5.2), by E. Wiener and D. Nagel, 1988, San Diego, CA: Academic Press. Copyright 1988 by Academic Press. Adapted with permission.

pilot whose fuel gauges read empty considering a forced landing decision. Each option (e.g., continue on, land immediately) can be described in terms of possible outcomes (e.g., reach airport, disastrous landing) and associated probabilities (e.g., likelihood of having enough fuel). Following normative theory, the pilot's task is clear: "the decision maker should choose the course of action with the most favorable expected outcome—the highest expected utility" (p. 133). Expected utility is derived by considering each option in turn and multiplying the value of each potential outcome by the likelihood of it occurring. Their description emphasizes the prescriptive nature of this representation and discusses how various heuristics and biases might derail the process. As Wickens and Flach noted, however, "their actual investigation in an aviation context has not been carried out" (p. 127).

An empirical program of research into pilot decision making, based initially on the Wickens and Flach model, was carried out by Stokes and colleagues at the Aviation Research Laboratory at the University of Illinois. These studies utilized a PC-based "flight decision simulator" known as MIDIS (microcomputer-based flight decision training system). Participants saw a full instrument panel, based on the Beech Sport 180. Above the panel were presented text descriptions of in-flight situations, including problems requiring diagnostic and/or corrective action. The instrument panel indications were either static or dynamic. Participants were required to choose one of six options presented on screen and to rate their confidence in their decision. Each option was rated for its optimality by several flight instructors.

An initial study (Barnett et al., 1987) found that scores on a Federal Aviation Administration (FAA) test of textbook knowledge failed to predict any aspect of decision performance. The best predictor of decision optimality was performance on a test of working memory capacity, which accounted for 22% of the variance in decision scores. A subsequent study (Wickens, Stokes, Barnett, & Hyman, 1988) investigated the effects of stress on decision performance. Participants in the "stress" condition were exposed to intermittent noise stress, a secondary task, time pressure, and possible loss of financial reward.

Stress was found to reduce both decision optimality and confidence. This was particularly the case with scenarios that involved a high level of spatial working memory, but stress did not adversely affect performance on scenarios that were particularly dependent on the retrieval of knowledge from long-term memory. An important implication of this finding is that experienced pilots should be more able than novice pilots to find ready-made solutions to problems in LTM and thus be less susceptible to the effects of stress than novice pilots forced to solve problems in real-time using fragile resources such as working memory. Some support for this view was provided by Barnett's (1989) study of expert versus novice instrument-rated pilots. More recent work in this area (e.g., Stokes, Belger, & Zhang, 1990; Stokes, Kemper, & Marsh, 1992) is discussed in a later section, Enquiring Expert.

A direct examination of the effects of the salience heuristic (see *S* in Fig. 6.1) on decision making was reported by Stokes et al. (1990). The task used was "a dynamic time-limited CRT based decision task like (the) dynamic scenarios in the MIDIS simulation" (p. 71). Participants fired "missiles" at an array of up to five target aircraft moving down the screen. Some targets were made more salient by use of blue or red coloring. Some participants were exposed to stress in the form of 90dBa white noise throughout the experiment. Results indicated that participants were more likely to attack the salient targets and that this tendency was significantly increased in the noise condition.

Mosier, Skitka, Heers, and Burdick (1998) have suggested that "automated systems introduce opportunities for new decision making heuristics and associated biases" (p. 48; see also chap. 9). According to Mosier et al., automated systems introduce new, highly salient cues that supplant the pilot's traditional reliance on assessing patterns or combinations of cues (see the work of Bellenkes, Wickens, & Kramer, 1997 described later). Reduced situational assessment and overreliance on cues provided by automated systems was dubbed *automation bias* by the authors.

Empirical evidence was provided by a study of 25 airline pilots (all qualified on glass cockpit aircraft such as the B-767 or MD-11) who "flew" two simulated sectors on a part-task simulator. This simulator consisted of two touchscreen Silicon Graphics monitors displaying primary flight displays, navigation and communication information, electronic checklists, engine and systems information, and flight management systems (FMS) information. There were three flight-related automation failures (e.g., a commanded heading change was incorrectly executed) and a false automated engine-fire warning occurred during a missed-approach procedure.

Results showed that the three failures were missed by 44%, 48%, and 71% of the pilots, respectively. Surprisingly, these error rates were positively correlated with flight experience (i.e., greater experience was associated with more errors in detecting the automation failures). The authors suggest that this may either have been due to the experienced pilots usually flying as captains, with the cross-checking role normally delegated to first-officers, or that experience with highly reliable automated systems breeds complacency. Every single pilot responded to the false engine-fire warning by shutting down the engine. Interestingly, 67% reported seeing at least one additional cue, such as a master warning light, which was not actually present. The authors suggest that this "false memory mechanism" may, in part, account for the automation bias in that pilots recall "a pattern of cues that was consistent with what should have been present . . . errors are not traced back to a failure to cross-check automated cues" (Mosier et al., 1998, pp. 60–61). More-recent research with nonpilot participants (Skitka, Mosier, & Burdick, 1999) has shown that performance in a cockpitlike setting was improved in the presence of a perfectly performing automated monitoring aid. However, when the aid was inaccurate, participants in the nonautomated condition performed better.

Overall, the effects of the Tversky and Kahneman (1974) work on decision heuristics and biases have been highly significant in a number of fields. Surprisingly, this has not been the case in research on aeronautical decision making. The model proposed by Wickens and Flach (1988) has not stimulated much research on the role of heuristics and biases in aeronautical decision making outside the Aviation Research Laboratory where the MIDIS work was conducted.

Nevertheless, the metaphor of "human as faulty computer" has led to the development of a number of procedures for reducing the potential influence of heuristics and biases on ADM. These will be discussed in the following section on the human decision maker as "rational calculator."

THE DECISION MAKER AS RATIONAL CALCULATOR

The normative models of classic decision theory also form the backdrop to this metaphor. However, in place of a focus on heuristics and biases, these models and their associated research (some of it actually conducted in aviation!) have emphasized the rational nature of human decision making. In recent times, the nature of what might be considered rational has changed from that which is strictly in accordance with normative theory to encompass processes that are, on the whole, generally adaptive for the decision maker.

The Wickens and Flach (1988) model outlined earlier is based on the assumption that the pilot's task is to process information concerning the probable outcomes of each option and to combine this information with the estimated values and costs associated with each outcome in order to arrive at a subjective estimate of the utility or worth of each of the available options. Some empirical work has investigated the ways in which pilots arrive at these estimated values and the relationship between pilot characteristics and these decision preferences.

Flathers, Giffin, and Rockwell (1982) tested 30 instrument-rated pilots on a sequence of pencil-and-paper scenarios. The last scenario required the pilot to make a diversion decision in response to an alternator failure midway through the flight. At this point, the pilot was asked to rank-order 16 possible airports from "most preferable" to "least preferable" in terms of diversion. Each airport was described in terms of four characteristics: Air Traffic Control (ATC) facilities, weather, time from present position, and instrument approach facilities.

Multiple-regression techniques were used to estimate weights for each attribute for each pilot. Multiple regression finds a set of weights for the predictor variables (e.g., ATC facilities) that provides the best prediction of the actual value of the predicted variable (the diversion decision). In this case, over 90% of the variance in pilots' rankings could be accounted for by a simple linear combination of attributes. Differences in the weightings given to the certain characteristics were found to exist between civil and military trained pilots for ATC facilities, between

Airline Transport Pilot Licence (ATPL) holders and other pilots (Private Pilot Licence and Commercial Pilot Licence holders) for weather, and marginally, for time from present position. The most significant difference was between professional (civil and military) pilots and those flying for business or pleasure on the weight given to the ATC facilities attribute. There were no differences as a function of total flight hours.

More recently, Driskill et al. (1997) extended this line of research to identifying the worth functions for weather and terrain variables in light aircraft cross-country flying over three different kinds of terrain (water, nonmountainous, and mountainous). An initial set of scenarios was developed to reflect the four most common factors associated with "Visual Flight Rules (VFR) into Instrument Meteorological Conditions (IMC)" accidents—visibility, terrain, ceiling, and precipitation. The initial pool of scenarios was examined by expert pilots and meteorologists, and implausible scenarios were removed and replaced. The final set of 81 scenarios was judged by 152 pilots varying widely in age, experience, and certification. Pilots were asked to rate their comfort level about completing a flight under the conditions described in the scenario.

The weights for each characteristic and for each interaction between two or more characteristics were identified by multiple-regression and then subject to a clustering technique to identify groups with similar emphases. This resulted in the identification of six different worth functions. In contrast to the Flathers et al. (1982) results, the findings of this study showed that the worth functions were dominated by interactions between characteristics. In other words, no single characteristic such as ceiling or visibility alone provided a good prediction of the scenario's rating, but the interaction between these variables was often a significant predictor. For example, a low ceiling combined with poor visibility was significantly related to pilots' ratings of comfort level for a scenario. The finding of Flathers et al. (1982) that worth functions varied as a result of professional versus business/pleasure reasons for flying was supported. In terms of pilot characteristics, the amount of hours flown in the past 90 days was often associated with differences in attribute weightings between the judgment groups.

Whereas the previous section focused primarily on the potentially distorting effects of heuristics and biases on the assessment of probability in decision making, these studies illustrate the complexity of the second part of rational judgment—the calculation of value or worth of an option. Such judgments vary according to circumstance and according to some particulars of pilot training and reasons for flying.

One potentially concerning issue highlighted by Driskill et al. (1997) is that the presence of significant interaction effects indicates that the judgment model that pilots seem to be using is a compensatory one in which levels of one characteristic can be traded off against levels of another. For example, in purchasing an automobile one might reasonably trade off fuel economy for off-road capability. Although in general, compensatory strategies make better use of available information, in aviation trading off ceiling for visibility might not always be such a good idea. Setting an absolute minimum value for a certain characteristic is known as a noncompensatory strategy. The decision maker rejects all alternatives that fall below the minimum cutoff. This is common practice in aviation, for example, with the minimum descent altitude (MDA) for an approach. This simplifies a complex decision situation by reducing the cognitive effort involved (Payne, Bettman, & Johnson, 1993). It may be that training techniques such as the Personal Minimums program for general aviation (described later), which attempt to create an individualized set of nontradeable minimum values for characteristics involved in decision making, are valuable for precisely this reason.

The Vigilant Decision Maker

The view that the key to the quality of decision making lies in the process used by the decision maker is the basic assumption of the classic approach to decision making. Janis and Mann (1977), in their model of the vigilant information processor, articulated seven criteria for good decision making including careful weighing of risks and benefits, searching for new information, and reevaluating options. Their main interest, however, was in showing how stress—often generated by the decision making itself—might adversely affect the requirements for vigilant information processing. The combination of stress and the requirements for vigilant information processing lead to five coping patterns. O'Hare (1992) has described these patterns in the context of a framework model of ADM.

A modified version of the ARTFUL model of ADM is illustrated in Fig. 6.2. Based on a review of the ADM and related literature in cognitive engineering, six identifiable components of decision making can be identified. Goal setting is at the apex of the model, with reciprocal connections to the processes of situational awareness (SA; see chap. 4) and planning. Linking these two is a process of risk assessment, which is central to Janis and Mann's (1977) theory. In the theory, the risk associated with the current goal is continually assessed. If the level of risk rises or appears to be becoming unacceptable, the decision maker determines if time is available to generate new goals. If no time is available, the theory predicts that the outcome will be "hypervigilance," involving an impulsive act or panic. If time is available, then possible alternative goals are appraised and evaluated for risk.

Although Janis and Mann's (1977) model offers testable propositions and has been explained in the context of ADM by O'Hare (1992), it appears to have led to no direct empirical investigations in the ADM context. However, Fischer, Orasanu, and Wich (1995) had professional pilots sort 22 written scenarios taken from the Aviation Safety Reporting System (ASRS) database into groups involving similar kinds of decisions. Multivariate analyses (multidimensional scaling and hierarchical clustering) led to the conclusion that there were two dimensions used to sort the decisions. The first was the amount of risk associated with the

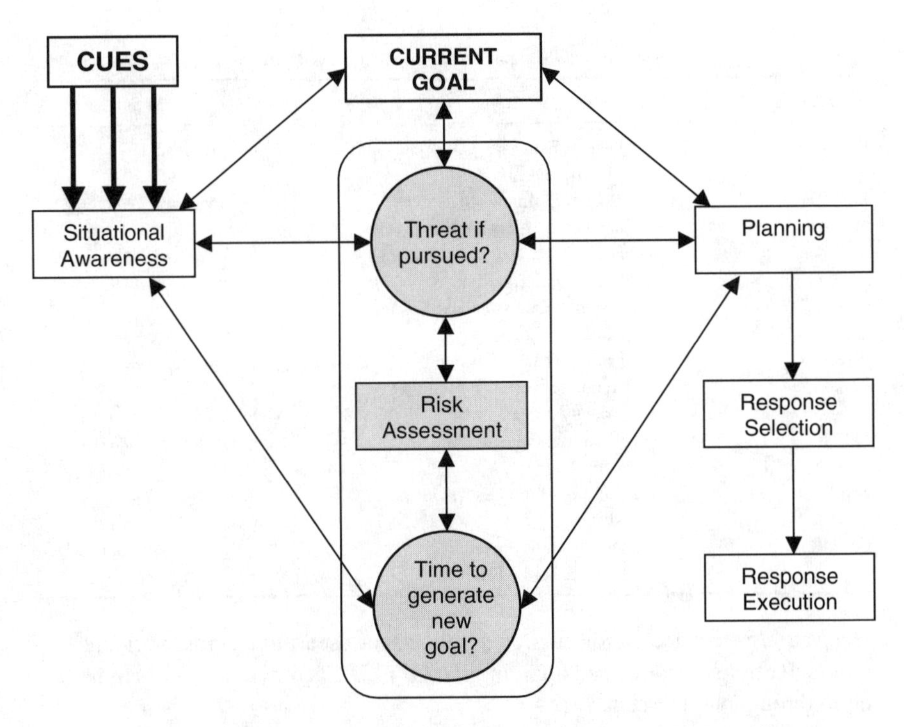

FIG. 6.2. An outline of the components and processes of the ARTFUL decision-making model of ADM proposed by O'Hare (1992). The model suggests that a current goal will only be altered if the pilot's situational awareness indicates a need for change and there is time to select an alternative, less-risky goal.

decision, and the second was the amount of time available to make the decision. These are the two key distinctions outlined in O'Hare's (1992) and Orasanu's (1995) models. Jensen's (1995) model of ADM is in a similar mold, describing eight processes involved in ADM: problem vigilance, problem recognition, problem diagnosis, alternative identification, risk assessment, motivation, decision, and action. Jensen's recommendation that such a model be used to develop prescriptive training approaches in ADM is reflected in a number of programs, mostly airline based, which provide acronyms for systematically ordering the processes of vigilant decision making.

Training Rational Decision Making

The best-known approach is the DECIDE model (Benner, 1975) adapted for use in aviation by Clarke (1986). A number of decision acronyms are listed in Table 6.1. The aim of these techniques is to foster a systematic or vigilant approach to decision making that should be less affected by the heuristics and biases illustrated in Fig. 6.1. A typical training approach is described by Jensen (1995). Participants

TABLE 6.1.
A selection of decision acronyms used in ADM training

Acronym	*Steps*	*Reference*
DECIDE	Detect, Estimate, Choose, Identify, Do, Evaluate	Benner, 1975
DESIDE	Detect, Estimate, Set safety objectives, Identify, Do, Evaluate	Murray, 1997
FOR-DEC	Facts, Options, Risks & benefits, Decision, Execution, Check	Hormann, 1995
PASS	Problem identification, Acquire information, Survey strategy, Select strategy	Maher, 1989
SOAR	Situation, Options, Act, Repeat	Oldaker, 1996
SHOR	Stimuli, Hypotheses, Options, Response	Wohl, 1981
QPIDR	Questioning, Promoting Ideas, Decide, Review	Prince and Salas, 1993

received a 5-hour lecture/discussion seminar that used three actual accident scenarios. Participants practiced applying the DECIDE acronym at every significant point throughout the scenarios.

Following the training session, participants made a short simulated VFR familiarization flight in a Frasca 141 simulator, which was configured to represent a typical single-engine general aviation aircraft such as the PA-28-181. Participants then undertook a flight that involved up to three problems presented at 10-minute intervals. Five of the participants had taken the DECIDE model training and five others served as control subjects. The experimental design, involving multiple problems, meant that each problem was faced by different numbers of participants depending on their response to earlier problems. In addition, only seven pilots in total completed the evaluation flight. These difficulties make the results hard to interpret, although the fact that none of the experimental participants "crashed" whereas all the controls did, is suggestive that a positive training effect could be achieved. Unfortunately, there appear to have been no other attempts to empirically evaluate the effectiveness of these decision-structuring techniques.

In addition to lack of empirical validation, the attempt to train decision making to be more systematic and analytical suffers from a more serious flaw. As the nature of expert decision making becomes better understood (see The Decision Maker as Enquiring Expert), it has become apparent that slow, analytical decision making is a characteristic of novice, rather than expert, decision makers (O'Hare, 1993; Wiggins & O'Hare, 1995).

A more-promising approach has been developed by Jensen and colleagues (see Jensen, 1995) as part of the FAA's Back to Basics safety program. The aim was to promote a conservative approach to risk management by pilots not constrained by organizational Standard Operating Procedures (SOPs) and the like. The key element of the program is the creation of a set of minimums for various aspects of aircraft operation, covering the pilot, the aircraft, the environment, and external pressures. Participants can watch specially prepared video segments or work on a CD-ROM program designed to highlight the various risks that can affect flight safety. The participants then develop their own set of *personal minimums* "defined as an individualized set of decision criteria (standards) to which the pilot is committed as an aid to preflight decisions" (Jensen, 1995, p. 279). These minimums can be more conservative than the regulatory minimums and should reflect each pilot's own values, experience, risk preferences, and so forth. Participants are encouraged to have their minimums reviewed by other pilots and instructors.

The program has been well received by pilots (Jensen, Guilkey, & Hunter, 1998), with 85% of participants rating the program as "helpful" or "extremely helpful" and 82% indicating that they would definitely use the program. Studies of the effectiveness of the personal minimums program in changing pilot risk management and behaviors are obviously required.

Effects of Technology on Rational Decision Making

The past few decades have seen the development of a number of technologies designed to provide pilots with better information regarding their position (e.g., Global Positioning System [GPS]) or the position of other aircraft (e.g., Traffic Collision Avoidance System [TCAS]). Enhanced graphical weather displays and flight-planning aids have also been developed although not yet implemented in the cockpit (Lindholm, 1995). A full discussion of the human factors aspects of these various technologies is clearly beyond the scope of the present chapter. Beringer (1999) provides a good overview of many of these developments in the general aviation (GA) context. In this section, we will briefly discuss the effects of such technologies on the basic processes of rational decision making as described earlier.

Satellite-based navigation systems such as GPS have become widely used in aviation. Because of the low cost of the receivers, GPS has been taken up on a significant scale in GA. Unfortunately, the GPS systems in current use leave much to be desired in terms of their basic ergonomics (Heron & Nendick, 1999). Concerns about the potential effects of GPS on pilot decision making have been raised (e.g., O'Hare & St. George, 1994). For example, pilots might be more likely to fly "VFR on top" (i.e., above a cloud layer in VFR conditions) or to take off in marginal weather if they have GPS. There is plenty of anecdotal evidence to support these suggestions, but well-designed empirical studies or detailed investigations of air crash databases are required to establish the nature and dimensions of the effects (if any) on ADM.

Lee (1991) compared the decision-making process of aircrews receiving conventional ATC methods of providing information about the possible presence of windshear or microbursts with a real-time graphical display. Eighteen airline crews were tested in the Advanced Concept Flight Simulator (ACFS) at NASA Ames. At this time, the ACFS provided a computer-generated external scene and five CRTs displaying flight instrumentation. Some crews also had a CRT display of simulated ground-based Doppler radar precipitation returns. The results showed that crews using the visual display responded more quickly (equating to 3 nm further out and 700 ft higher) to a microburst event on approach. Data from intracrew communication showed that the crews receiving conventional ATC warnings were preoccupied with a single cue—surface winds, whereas crews using the visual display discussed a wider range of cues.

The effects of visual displays of weather information on pilot decision making have also been investigated by Lind, Dershowitz, and Bussolari (1994). The study looked at GA pilot decision making based on Graphical Weather Service (GWS) images, which would be transferred by data link technology. Aircraft would have to be equipped with a Control and Display Unit (CDU) for displaying the uplifted images. In their initial study, 20 instrument-rated pilots were given five hypothetical flights in an Instrument Flight Rules (IFR)-equipped C172. Each flight involved three decision points. At the decision point participants were asked: "What will you do now?" Participants could search for information using the GWS or by other means or make a decision regarding the flight. The GWS images were presented on an Apple Macintosh computer monitor. Various GWS images, based on actual recorded radar precipitation images, were available for each decision point.

GWS significantly increased pilots' confidence in their ability to assess the weather and reduced the number of requests to ATC for weather information. There is some evidence from the data obtained in this study of a less-conservative approach to decision making by the participants using GWS. For example, 50% of the non-GWS participants rejected one of the flights, whereas all the GWS equipped participants took off. In another scenario, 40% of non-GWS participants made precautionary landings, whereas none of the GWS participants did so. Although these decisions cannot be easily characterized as "good" or "bad," clearly the potential effects of GWS and related technologies on pilot decision making processes deserve further scrutiny.

The potential for graphical decision aids to affect decision-making strategies is illustrated in a study by Layton, Smith, and McCoy (1994). This study used a graphical flight-planning tool to study the effects of different design features on pilot decision making. The Flight-Planning Test Bed involved two side-by-side displays consisting of a map display showing weather information, navigational fixes, and jet routes and a spreadsheet display showing flight information for four alternative routes. In a mixed design, 30 airline pilots used one of three versions of the system to select a flight route for each of four scenarios. In the simplest version of the flight-planning system, additional routes had to be suggested by the pilot; in the intermediate version, the system could generate alternative routes if requested; in the most advanced version, route deviations were automatically suggested in order to avoid turbulence, precipitation, and so forth.

Various adverse effects were noted. For example, the large amount of data available led to some pilots overlooking some basic cues (e.g., forecast winds). Computer-generated suggestions seemed to carry undue weight in at least one case, with 40% of pilots accepting a "poor plan" in the most automated condition (see chap. 9). This interesting study raises a number of important questions that must be considered in designing systems to support human decision making in aviation and illustrates just how brittle decision making strategies can be. The authors usefully point out that exhortations such as to "avoid excessive automation" and "keep the operator in the loop" are overly simplistic. The specific effects of system features on decision making should be closely scrutinized in each case. However, as Gopher and Kimchi (1989) have pointed out, it is important to adopt a more considered, theoretically driven approach to researching the effects of changes in technology on decision making rather than simply chasing each new technological development as it comes along.

THE ADAPTIVE DECISION MAKER

A number of new approaches to decision making have emerged in response to evidence that the normative decision models (such as expected-utility theory) fail to provide an adequate descriptive framework for predicting decision behavior. These developments have had relatively little impact on the study of ADM as yet but may be expected to do so as ADM research becomes more connected to its theoretical foundations. The term *adaptive decision maker* was coined by Payne et al. (1993) as they developed a theoretical framework to account for the observed behavior of decision makers.

The theory of the adaptive decision maker proposes that decision makers adapt their strategies according to the features of a situation based on a desire to trade off accuracy and cognitive effort. The use of simple heuristics might be perfectly satisfactory for resolving inconsequential decision problems, such as what to eat for dinner, whereas complex, analytical strategies might be preferred for dealing with major decisions that involve high levels of personal accountability. Individuals generally show intelligent, adaptive use of different strategies according to circumstances.

In aviation, intelligent flexibility in the use of decision strategies is also required. A "one size fits all" approach will not explain the variety of different decision strategies used by pilots. In some cases, the desire to avoid a bad decision may be more potent than the desire to make an optimal choice, as this description of pilot decision making makes clear: "He had to take decisions that, even if not ideal, were not basically wrong. He tended to solve problems, not by weighing factors, but by choosing the dominant feature, and acting as if it alone existed" (Anderson, 1975, p. 42). Wiggins and Henley (1997) using a simulated decision task found that inexperienced flight instructors (median instructing experience = 215 hr) were significantly less likely to send students on a solo cross-country flight than experienced instructors (median instructing experience = 1,500 hr). Those who decided against sending the students examined less information and focused more on the meteorological information category. The authors suggest that these instructors might have been following a low effort/low risk strategy in arriving at their decisions.

There are several models in the decision-making literature that are consistent with this description. One is Montgomery's (1993) dominance model, which assumes that the decision maker utilizes various strategies to structure the problem so that one alternative clearly dominates the others. Any decision rule, such as elimination-by-aspects (Tversky, 1972) may be utilized. The emphasis is thus on the decision maker's structuring or representation of the problem. If the problem can be structured so that one alternative clearly stands out from the rest of the field, then the decision itself becomes straightforward. The investigation of this model in the ADM context would seem warranted.

Another model that emphasizes the decision maker's representation of the problem is prospect theory (Kahneman & Tversky, 1979). The central feature of prospect theory is that the decision maker structures the decision problem in terms of potential gains and losses from some reference point rather than in terms of anticipated end states, as would be expected in classical utility theory. An important empirical observation is that people are generally risk averse when considering the prospects of a gain, preferring a certain gain over a risky gain of the same or even greater magnitude. However, when the prospect of a loss is involved, people generally become risk seeking, preferring the chance of a greater loss over a certain loss of equal or somewhat lower amount. What determines whether a prospect is seen as a gain or loss is the reference point. For example, a pay increase of $1,000 might be seen as a gain of $1,000 if the reference point is current pay or a loss of $1,000 if the reference point was a promised $2,000 pay rise.

The principles of prospect theory have been utilized in a study of ADM by O'Hare and Smitheram (1995). Twenty-four active pilots were asked to indicate which of two statements describing the gain framework and loss framework respectively best characterized the way in which they would normally view problems in cross-country flight. The majority (79.2%) selected the gain framework that described the certain gains achieved by diverting a flight away from deteriorating weather. Of course, this could simply be considered the more socially desirable viewpoint to express publicly. This interpretation is supported by the lack of relationship between this initial frame preference and decision making in a subsequent simulated VFR flight.

The main part of the study involved a simulated flight where all the information (weather descriptions, route maps, etc.) could be accessed on the computer via a series of menus. At the decision point, participants were informed that destination weather had deteriorated, and they had to decide whether to continue or abandon the flight. Half the participants received information in a loss frame emphasizing the time already spent on the flight. The other half received the same information in a gains framework emphasizing the benefits (e.g., added flight time logged) accrued. Overall, the decision to continue the flight was evenly balanced with 54% continuing and 46% diverting. However, there was a significant effect of the framing of the information. Those who viewed the choice in terms of gains were much more risk averse, as predicted by prospect theory, with only 25% continuing the flight. Those viewing the choice from the loss perspective were inclined towards risk seeking with 67% electing to continue.

The empirical confirmation of prospect theory predictions in an ADM task is significant. Anecdotal evidence (e.g., Hobbs, 1997, p. 10) suggests that these effects do indeed occur in aviation: "We sometimes see accidents where a pilot under pressure chooses to take a gamble that the serious outcome won't happen, rather than taking the safe but inconvenient option." Replication and further extension of the O'Hare and Smitheram (1995) work would be highly desirable. Useful theoretical predictions could equally be derived from other models in this area such as the adaptive decision-making model (Payne et al., 1993) and the dominance structure model (Montgomery, 1993).

THE DECISION MAKER AS CHARACTER DEFECTIVE

One of the most widely known approaches to pilot decision making, particularly in the field of general aviation (GA), has involved an analysis of the underlying attitudinal bases of decisions. This originated in an exercise designed to characterize "the thought patterns which cause pilots to exhibit . . . 'irrational pilot judgment' " (Berlin et al., 1982, p. 9). This exercise was carried out at the Aviation Research Center of Embry-Riddle Aeronautical University in late 1979. The exercise was designed very much as an applied effort with the aim of developing instructional materials for targeting pilot judgment and decision making. The following paragraph from Berlin et al. (1982) describes the approach:

> It was postulated that if these thought patterns could be identified, then pilots could be trained to recognize them in their own thinking, and to apply corrective actions. Little prior research was found in which such thought patterns were described. Thus, it was found appropriate to consult experts in the psychological and sociological sciences to obtain informed opinions on the nature of such hazardous thought patterns. This resulted in the identification of five thought patterns and the assigning of descriptive names for these thoughts. (p. 9)

Given that little research had been done into the attitudinal bases of pilot decision making, it might have been more appropriate to initiate a program of basic research to empirically identify whether decision making is, in fact, influenced by general character traits, and if so, to empirically determine the nature of those influences. Unfortunately, the applied focus on delivering a training product necessitated that this essential step be passed over as quickly as possible. The result was the description of five hazardous attitudes (antiauthority, invulnerability, macho, impulsivity, resignation) and the development of a self-assessment questionnaire to "reveal" the presence of these in a pilot's thinking.

A small-scale evaluation study was conducted at Embry-Riddle using 81 young (mean age = 19 years) student pilots. The training included study of the student manual and three training flights covering judgment and decision making. The training flights involved instruction in the application of concepts such as the "poor judgment chain" and "avoidance of hazardous thoughts." The training was followed by a written test of knowledge of the materials and an observation flight during which the pilot was rated for judgment over 20 situations. For example, the instructor set up an unstabilized approach, and the student was expected to demonstrate good judgment by executing a go-around rather than attempting to land.

Results showed that the experimental participants performed significantly better on both the written test and in the observation flight than controls. The authors note some limitations to the results by virtue of the age of the participants and the nature of the training and testing, which were both carried out over a 1-month period. It should also be noted that the judgment training might have been effective through raising participants' general awareness of the decision-making process rather than through specific knowledge of the hazardous attitudes used as a basis for most of the exercises.

Two further evaluation studies were forthcoming. The first, at two Canadian flying schools, involved 50 student pilots divided into a control group and an experimental group who received modified Embry-Riddle materials. During the observation flights, 18 judgment situations were assessed by four observers. Again, the results showed a significant improvement in the rated judgment of the experimental participants (Buch & Diehl, 1984). The second, conducted at an Australian flying school, involved 20 participants divided into one experimental and two control groups. Participants studied the Embry-Riddle materials and then took a written test and an observation flight covering nine activities. The experimental participants performed slightly better on the observation flight than the controls, although there were few differences on the written tests (Telfer & Ashman, 1986).

Attempts to put the theoretical basis of the hazardous thought model on a sounder footing were reported by Lester and Bombaci (1984) and Lester and Connolly (1987). The plan was to relate self-assessed propensity toward each of the five hazardous thoughts to selected personality characteristics, such as locus-of-control. Overall, these data showed that ratings of the five patterns were highly intercorrelated (mean $r = 0.39$ on a sample of 152) with only three patterns (invulnerability, impulsiveness, and macho) endorsed by significant numbers of respondents. The dominance of these three response patterns was more recently noted by Ives (1993). Principal components analysis is commonly used to provide insight into the factors underlying a pattern of intercorrelations between variables. I carried out a principal components analysis of the data presented by Lester and Connolly (1987) and found that only one factor, accounting for 52% of the variance, was present. Four of the five scales loaded very highly ($> .7$) on this factor, with the fifth (invulnerability) loading moderately highly (.56) on this factor. This single factor was responsible for between 50% and 60% of the variance in each scale, except for invulnerability, where the variance accounted for was 31%. This confirms the findings of Lubner and Markowitz (1991), who also found that only one factor was needed to account for the intercorrelations between the hazardous thoughts items.

Remarkably, no relationship has been found between the impulsivity hazardous thoughts pattern and the impulsivity scale of a standard personality questionnaire (Lester & Bombaci, 1984). There was also no relationship between the hazardous thought ratings and self-reported accident-involvement (Lester & Connolly, 1987). Other studies (e.g., Platenius & Wilde, 1989) have also failed to find any relationship between reported accident histories and measures of impulsiveness, invulnerability or machismo.

When reading an accident report describing some particularly outlandish piece of foolishness, it becomes fairly easy to endorse the view that pilots do exhibit poor judgment for no apparent reason. Whether the same pilots consistently exhibit the same kinds of poor judgment has never been empirically established, however, so the existence of a character flaw is unproven. The validity of the five hazardous thoughts has not been clearly established, and indeed the structure of responses as revealed by principal components analysis indicates that only one underlying factor may actually be involved. The potential role of underlying dispositional factors in pilot judgment and decision making remains controversial in any case (Besco, 1994). On the positive side, as noted previously, several studies have shown that structured programs of decision training can lead to immediate improvements in observed judgments in flight, at least with young, relatively inexperienced pilots. Whether the same is true for older groups and whether the effects are sustainable has not been demonstrated.

There is now almost universal agreement on the basic structure of personality in terms of the "Big Five" dimensions (Hampson, 1999). These are extraversion, agreeableness, conscientiousness, emotional stability, and openness or intellect. There is evidence for the utility of at least some of these five dimensions in predicting job performance in a variety of areas (Barrick & Mount, 1991). The underlying dispositional nature of performance tendencies in aviation might be more amenable to good theory-based empirical investigation using the Big Five framework. For example, Arthur and Graziano (1996), using two different personality measures and two different samples of participants, found a significant relationship between the personality trait of conscientiousness and self-reported vehicle accident involvement. Conscientiousness has also been found to be a valid predictor of other job-related criteria (Barrick & Mount, 1991) and might well be related to decision style and performance in the aviation domain.

In the aviation domain, Lubner, Hwoschinsky, and Hellman (1997) have recently developed self-report measures of decision-making styles (the DMS) and risk profile (the PIRIP). The measures were shown to have sound psychometric properties (e.g., good internal consistency), and preliminary data on a small sample (n = 46) indicated that all but one of the subscales showed modest correlations with self-reported involvement in occurrences. This research offers the prospect of establishing valid scales of individual differences that may actually relate to decision making in the cockpit.

THE DECISION MAKER AS ENQUIRING EXPERT

In recent years, the field of decision making has been revitalized by a new approach referred to as the "naturalistic decision making" (NDM) approach. Instead of examining departures from the normative models of decision making, NDM research focuses on the skills and capabilities displayed by expert decision makers in action. As the term *naturalistic* implies, the field is concerned with expert decision makers in real-world tasks such as business, military command, medicine, and aviation. The NDM approach includes a range of models and methods. The most well known is the Recognition-Primed Decision (RPD) model of Klein (1989). An overview of the NDM approach to aeronautical decision making is provided by Kaempf and Klein (1994).

The crucial difference between some of the models of decision making reviewed previously and the NDM models is that the normative models of classical decision theory (e.g., subjective expected utility theory) assume a well-defined problem, often set up in terms of explicit probabilities and payoffs, whereas the NDM models assume that the problems that people normally confront are ill defined. The key decision-making skill required by experts is therefore that of assessing the problem rather than in deliberating between options. This task is made more difficult by the presence of other features of natural environments such as time pressure, dynamic tasks, questionable data, and organizational constraints.

Orasanu (1990) reported a study of the decision making of 10 experienced airline B737 crews in a simulated scenario involving two critical decisions (a missed approach at the original destination followed by a diversion decision). The crews were divided equally into two groups on the basis of ratings of their performance in handling the two critical decisions by experienced pilot trainers. No reliability data were reported for these ratings. All crew communication was recorded and coded. Although the small numbers precluded statistical analysis, on the missed-approach decision there appeared to be more statements indicative of situational awareness and more statements or commands by the captains in high-performance crews regarding possible future actions. In other words, the better-performing crews exhibited greater sensitivity to the current cues, and the captains were more proactive. Very similar findings were reported in a study of 34 military crews dealing with an in-flight communication problem (Jentsch, Sellin-Wolters, Bowers, & Salas, 1995).

On the diversion decision, there were differences between the crews in terms of information requests and communication with the ground, but not in terms of situational awareness statements. The crew communications were subsequently examined in much greater detail. There was evidence that crews considered multiple options (mean = 3.2 out of a possible 8) and used an elimination-by-aspects strategy for choosing the alternate. Overall, the study confirms the importance of problem discussion and elaboration in ADM but does not rule out the importance of option consideration by means of conventional choice strategies as described in previous sections.

More recently, Orasanu (1993, 1995) has described a taxonomy of decision types and put forward a decision process model. The decision taxonomy (see Table 6.2) is defined by two dimensions: cue clarity and response options available. Cues can be unambiguous or ambiguous, in which case additional effort is required in diagnosis. Response options can be divided into three categories: a single prescribed response, a choice from several response options, or no prescribed response. Examples of each of the six combinations of cue clarity × response options are shown in the table. The simplest situation (unambiguous cue/single prescribed response) is on the upper left, with the most complex (ambiguous cues/no prescribed responses) on the lower right. Orasanu, Dismukes, and Fischer (1993) used this taxonomy to reinterpret the B737 crew decision-making data from Orasanu (1990). The taxonomy provides insight into the information-processing requirements and associated error forms generated by the different types of decisions.

TABLE 6.2.
Orasanu's taxonomy of decision types

	Retrieve Response	*Select or Schedule Response*	*Create Response*
Clear Cues	Simple Go/ No-Go (e.g., missed approach)	Choice (e.g., shut down engine with oil leak?)	Procedural Management (e.g., coping with sick or disruptive passenger)
Ambiguous Cues	Condition-Action (e.g., stop runaway stabilizer trim)	Scheduling (e.g., completing manual gear/flap extension in time)	Creative Problem Solving (e.g., loss of control of all flight surfaces)

This taxonomy represents a useful basis for designing research and training in ADM. It is closely related to Rasmussen's (1983) work on defining three levels of control of skilled actions (the Skill-Rule-Knowledge framework). An interesting question raised by Rasmussen (1986) concerns the problems of transitioning between these decision-making modes. There is certainly anecdotal evidence in aviation of pilots persevering with an unsuitable skill or rule-based solution to a problem when the situation clearly requires a switch to a more analytical, knowledge-based level of problem solving. According to Rasmussen (1986), this occurs because most situations provide familiar cues that can be interpreted with frequently used procedures. Thus the path of least effort involves trying the standard procedure over again rather than reasoning from first principles. Orasanu (1995) presents a decision process model (see Fig. 6.3), which combines the taxonomy of decision types with the kind of framework described by O'Hare (1992).

By specifying expected relationships between variables, models of this type provide a useful foundation for empirical research. For example, the models of

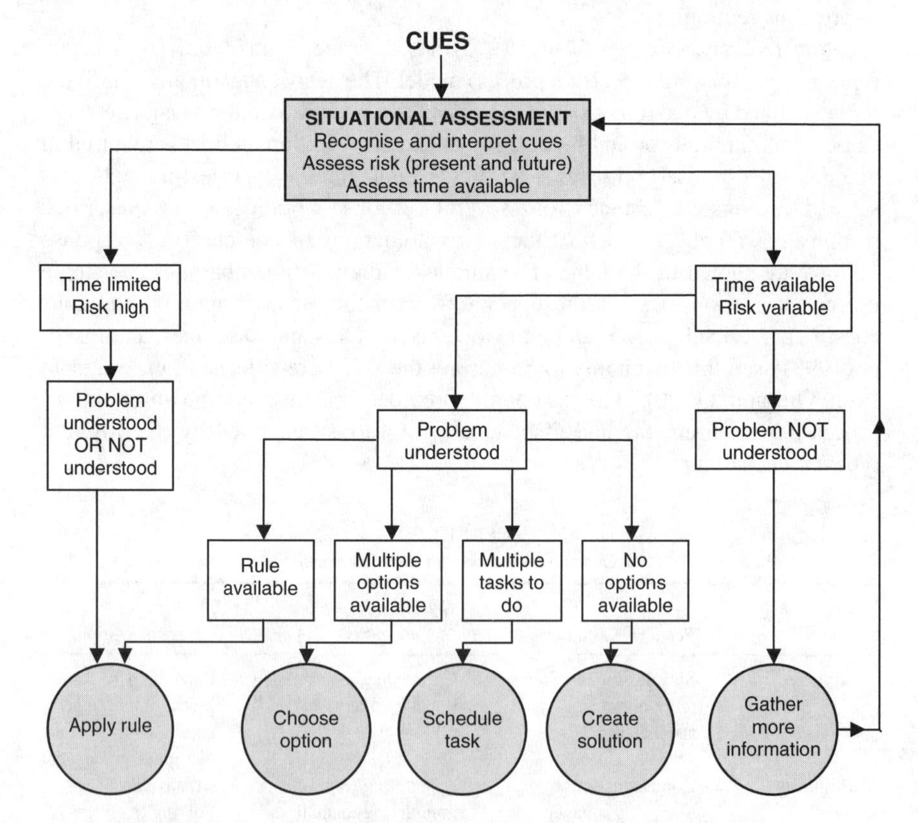

FIG. 6.3. An outline of Orasanu's (1995) Decision Process Model. There are two phases: Situational Assessment and Response Selection. The response possibilities include applying a rule, scheduling tasks, and creative problem solving.

ADM described by O'Hare (1992) and Orasanu (1995) both suggest that it will be easier to continue with an existing course of action than to change to a new one, as this requires continual monitoring, planning, and risk assessment. As such, the models predict that errors due to continuing on with an unprofitable course of action will be more prevalent than errors due to prematurely switching to other courses of action. We are currently examining air crash records for evidence that this is the case. Unfortunately, such research as there is in ADM has more often been of an ad hoc or a theoretical nature (O'Hare, 1992).

Expertise in ADM

The NDM approach emphasizes the ability of knowledgeable experts to retrieve ready-made solutions to problems from memory. In terms of Orasanu's (1993) taxonomy, expertise in a domain should have the effect of increasing the clarity of the critical cues and strengthening the association between critical cues and prepared responses in memory. This underlies the ability of the expert to quickly and accurately recognize the appropriate way of responding to a situation (Klein, 1989). In combination, the effect is to simplify the decision task by moving away from the lower right of the matrix in Table 6.2 toward the upper left. Tasks such as those in the upper left of the matrix put less emphasis on fragile resources like working memory and should therefore be less vulnerable to the effects of stressors.

Stokes, Belger, and Zhang (1990) investigated the effects of a stressor on performance in a simulated flight using an upgraded MIDIS simulator described earlier and performance on a battery of cognitive tests known as SPARTANS (Simple Portable Aviation-Relevant Task-battery and Answer-scoring System), which included measures of a range of information-processing abilities such as digit span (the number of digits that can be held in working memory) and logical reasoning (reading a series of syllogisms and deciding on the validity of the conclusions). These widely used measures have high reliability and stability. In addition, two tests of domain knowledge were given. One was an FAA book test of knowledge and the other was a test of recall for ATC messages. The participants were 16 private pilots (mean age = 23 yrs, mean flight hours = 117) and 10 instrument-rated flight instructors (mean age = 43 yrs, mean flight hours = 1,089). Stress was again manipulated through a combination of noise, time pressure, financial incentive, and a secondary task. The secondary task was a visually presented Sternberg memory scanning task that consisted of deciding whether a series of verbally presented letters were or were not part of a previously learned sequence. The design was within subjects, so all participants completed a MIDIS flight with and without stress.

As expected, working memory resources were significantly affected by stress in both the inexperienced and experienced groups. There were no significant differences between the groups in performance on any of the other cognitive measures. This is consistent with other research (Ericsson & Lehmann, 1996), which has shown that experts are not superior to novices in terms of general memory or other cognitive abilities. Neither stress nor expertise affected performance on the domain-specific tests. For example, both groups scored around the 55% level on the test of book knowledge. In terms of decision-making optimality (as rated by expert judges) on the MIDIS task, there was a significant interaction between stress and expertise for dynamic problems. The inexperienced group showed a significant decrease in decision quality under stressful conditions, whereas the experienced pilots were not affected by the stress manipulation.

Stokes, Kemper, and Marsh (1992) replicated the previous study with another modified version of MIDIS. The previous study, rather surprisingly, failed to show any overall difference between expert and novice decision-making quality. The authors attributed this to the use of multichoice response options, which may have artificially inflated the inexperienced pilots' performance. The MIDIS task was therefore modified to require participants to generate their own options. In addition, participants were asked to input the nature of the cues used to analyze the problem and to list the response options considered. Thirteen inexperienced and 11 experienced pilots participated. The stress manipulation was restricted to an on-screen stopwatch counting the time spent on each scenario.

The results confirmed the previous finding of no difference in cognitive performance between the groups. However, a significant difference on domain-specific performance on the ATC recall task was found in this experiment. A significant difference in decision optimality was evident between the experienced and inexperienced groups. The experienced pilots identified significantly more relevant cues and also generated significantly more response alternatives than inexperienced pilots. Expert pilots were found to actually choose the first alternative generated 71% of the time, compared to 53% for the inexperienced group.

Although the ratio of cases to variables was well below that recommended for multiple regression (Tabachnik & Fidell, 1996), various regression analyses were conducted. However, confidence in the findings is strengthened by the replication of previous results that age and total flight hours are not predictive of decision performance. Once again, training background, as reflected in type of pilot certificate held (i.e., PPL, CPL, or ATPL) was predictive. With pilot certificate removed, the next most significant predictor was number of relevant cues identified.

The findings of these recent MIDIS studies are highly consistent with the NDM view of decision-making expertise. Expert pilots seem to utilize a long-term memory strategy based on the identification of situationally relevant cues. Their performance appears to be more resistant to stress effects. They are more likely to generate an appropriate action as the first alternative considered than are novices. However, as in Orasanu's (1990) study, there is evidence that experts are perfectly able to generate multiple response alternatives. Enthusiasm for the NDM model should not, therefore, be allowed to completely occlude some of the findings of previous research into factors that drive the alternative comparison process.

Differences between domain experts and novices have been explored in a variety of areas such as chess and bridge (Charness, 1991), physics (Larkin, McDermott, Simon, & Simon, 1980), livestock judging (Phelps & Shanteau, 1978) and submarine search (Kirschenbaum, 1992). Several studies of expert-novice differences have been reported in aviation. For example, Bellenkes, Wickens, and Kramer (1997) compared visual scanning differences between 12 novice and 12 expert pilots in order to determine how expert pilots differ from novice pilots in terms of attention allocation and flexibility. Participants wearing a head-mounted eye tracker flew a series of maneuvers in an instrument flight simulator. A number of differences in scanning the instruments were observed, with experts making shorter, more frequent scans of each instrument whereas novices took fewer, longer looks.

The authors interpret these data to indicate that experts have a more refined scan pattern based on a better understanding of the interrelationships between the variables displayed. Although decision performance was not measured in this study, the findings of Stokes et al. (1992) and Bellenkes et al. (1997) might be jointly interpreted as suggesting that the basis for expert superiority in cue recognition and decision optimality has its basis in attention management and visual scanning patterns. O'Hare (1997) has shown that elite soaring pilots show superior performance on a dynamic test of attention management to matched controls. Gopher, Weil, and Bareket (1994) have shown that attention control skills can be developed with practice on a computer game and can transfer to actual flight performance in a jet aircraft trainer.

McKinney (1993) looked at the role of experience in the decision-making ability of military pilots leading multiship formations of fighter aircraft. The data examined consisted of 195 USAF Class A (i.e., one involving major damage and/or serious injury or death) fighter mishap reports. To avoid confounding flight leadership and mishap type, only mishaps that were caused by mechanical malfunction were examined. The decision-making effectiveness of the flight leads was rated by two experienced judges. Contrary to expectations, the decision making of the elite lead pilots was rated as less effective than other pilots in the formation. There was no difference in the decision effectiveness of experienced pilots (defined as $>$ 500 hours on type) and inexperienced pilots ($<$ 500 hours on type).

The effects of overall experience on the decision performance of general aviation (GA) pilots was measured by Wiggins and O'Hare (1995). Forty qualified pilots responded to six computer-presented scenarios involving simulated VFR cross-country flights. Participants could access any information required via an on-screen menu covering topographical maps, meteorological information, aerodrome specifications, and so forth. Maps and airport diagrams were scanned in and could be viewed in the appropriate section. The decision made at the decision point (i.e., continue the flight, divert to alternate, or return to departure point) was recorded along with latency and confidence ratings. All information requested was recorded and patterns of information search examined.

Few differences in information acquisition or decision making were evident when pilots were classified into groups based on total flight hours. However, a number of differences emerged when pilots were grouped on the basis of cross-country flight experience. The expert group (mean cross-country hours = 3,640) differed from the remaining groups (intermediate and novice) in terms of decision making (experts were more likely to continue the flight and to decide more quickly) and information search patterns (experts acquired less information, were less likely to revisit information, and were more confident). The conclusion to be drawn seems to be that expertise is related to specific experience with a task rather than overall or global experience. This conclusion is supported by evidence that inexperienced, but specifically trained, pilots can outperform far more experienced pilots in the performance of a key flight task.

In one study, for example, the highly experienced instructor pilots (> 2000 hrs flight time) and inexperienced graduates (200 hours total flight time) of a 15-hour course on Nap-of-the-Earth (NOE) flying flew six short NOE tasks in a UH-1H helicopter (Farrell & Fineberg, 1976). Accuracy of performance and ability to find the designated landing zones were recorded. The highly experienced pilots performed no better than the inexperienced pilots with 15 hours of NOE training. Once again, it appears that overall level of experience is not predictive of performance with respect to a particular task. This has important implications for the provision of training. If performance in general, and decision making in particular, is developed through exposure to specific task contexts, then pilots at all levels need access to training built around tasks rather than around general knowledge or procedures, as is often the case, particularly in general aviation.

Several recent studies have focused on identifying the characteristics of the "expert" pilot. For example, Jensen, Chubb, Adrion-Kochan, Kirkbride, and Fisher (1995) looked at the qualities of pilots flying high-performance GA aircraft such as the Cessna P-210 or Piper Malibu. Using structured interviews and observing performance in flight simulator exercises, a list of qualities of the "expert" pilot was drawn up. This included self-confidence, motivation to learn, superior ability to focus one's attention, being skeptical about "normal" aircraft functioning, and constantly makes contingency plans. In general, the authors suggest that the key differences between the expert and the merely competent are to be found in the domain of cognitive (e.g., attention management, judgment, and decision making), rather than stick-and-rudder skills. They argue that most competent pilots possess the necessary skills to control and maneuver an aircraft, but fewer reliably exhibit the cognitive qualities outlined.

In a recent study (Guilkey et al., 1998), the Ohio research team have looked at the relationship between aspects of GA flight experience and performance in dealing with an in-flight event in a Frasca 142 simulator. Both think-aloud data and performance measures were analyzed. Twenty-two instrument-rated pilots flew the scenario, which involved a 500RPM drop due to a propeller governor malfunction. Responses varied, with some pilots continuing and others diverting. Although 64% did not obtain any current weather for any airport, confidence in the decision was high (approximately an 8 on a 10-point scale).

Inspection of the intercorrelations between the various measures of experience assumed to contribute to expertise (e.g., recency, variety) and performance in the simulation revealed a small number of significant relationships at the $p < .05$ level. Given that 84 correlations were computed, a more-conservative approach would have been to set the significance level to $p < .01$, in which case no significant relationships were present. All but one of the marginally significant (i.e., $p < .05$) relationships between the variables involved the simple measure of number of flight hours. From a qualitative analysis of the verbal protocol data, the authors suggest that three distinct decision-making strategies can be observed.

One group of pilots was found to make immediate decisions with little attempt to seek out further information. The second group exhibited a degree of tunnel vision, or we might call it route myopia, by attending only to information about airports on the flight plan, ignoring more suitable alternatives elsewhere. The third group made an initial assessment of the severity of the problem and then moved into a stepwise or progressive search for further information. This latter group seemed to consist of the more flight-experienced pilots. These findings are therefore at variance with previous research described earlier (e.g., Flathers et al., 1982; McKinney, 1993; Wiggins & O'Hare, 1995), where no relationship between decision-making performance and hours of flight experience was evident. However, there are a number of differences between the tasks used (e.g., performance in simulated flight versus review of accident records), the type of simulation involved, and the presence of time pressure that make comparisons between these studies difficult.

In summary, the preponderance of empirical evidence indicates that expertise in aviation is more a function of experience with a particular task rather than with overall levels of flight experience or age, although these may show some relationship with decision-making strategy. This is consistent with the emphasis in NDM on domain experience rather than with general characteristics or problem-solving skills. ADM might therefore be classified as a cognitive skill (Anderson, 1982), and the principles of cognitive skill acquisition could be applied to the development and maintenance of ADM (O'Hare, 1997; Wiggins & O'Hare, 1995).

A practical application of cognitive skill acquisition principles to the development of a skill-based CD-ROM for training weather-related decision making has been described by Wiggins (1999) and Wiggins, O'Hare, Guilkey, and Jensen (1997). This tool, developed as part of an FAA program to improve ADM training, emphasizes the importance of recognizing the critical cues for action in weather situations. The Weatherwise program provides the user with experience in judging and responding to a variety of critical weather cues before applying this knowledge in the context of a simulated flight task.

Another approach to training decision making, which combines the NDM concern with pattern recognition with the classical concern with uncertainty and risk, has been suggested by Cohen, Freeman, and Thompson (1995). This approach is based on the Recognition/Metacognition or R/M model. This assumes that the initial basis for decision making is recognition, but that this is supplemented by metacognitive processes such as critiquing and correcting plans based on initial assessments. The authors report a successful application of this approach in training decision making in naval officers.

THE DECISION MAKER AS ORGANIZATIONAL COG

Several of the metaphors discussed previously focus on purely internal shortcomings—either in character or in information processing—that might lead to inadequate decision making. More recent work in the decision-making literature such as that on adaptive decision making (Payne, et al., 1993) or naturalistic decision making (Zsambok & Klein, 1996) has emphasized the interrelationship between decision maker and the context of the decision. An early discussion of contextual factors on ADM was provided by Mosier-O'Neill (1989). There has been a strong movement in the field of aviation safety away from individualistic explanations of human performance towards more complex sociotechnical theories that consider different levels within a system (Reason, 1997).

The work of Manchester University professor James Reason has become so well known in aviation that it is often simply referred to as the "Reason model" of human error (Reason, 1990). The Reason model emphasizes the contribution of preexisting or latent circumstances in accident generation. Decisions taken by equipment designers or boardroom members can have considerable downstream effects on working conditions, which may be instrumental in the genesis of an accident. The infamous Braniff "fast buck" program, whereby the airline offered passengers a dollar if their flight did not arrive within 15 minutes of schedule (Nance, 1986) would be a good example of a latent condition that, in combination with other factors, might eventually manifest itself in an accident.

Reason (1990) has distinguished between several varieties of human error at the individual level, such as slips, mistakes, and violations. Reason (1997) considers these individual mechanisms within the work context. The key question here is whether the work context provides good procedures for the situation, poor or inadequate procedures, or no procedures at all. The combination of work context (good rules, bad rules and no rules) and individual performance defines the possible outcomes, as shown in Fig. 6.4.

The individual decision-making process consists of deciding whether there is an accepted procedure for the situation and then deciding whether, and how, to follow that procedure. This analysis is at a different level from that of the models

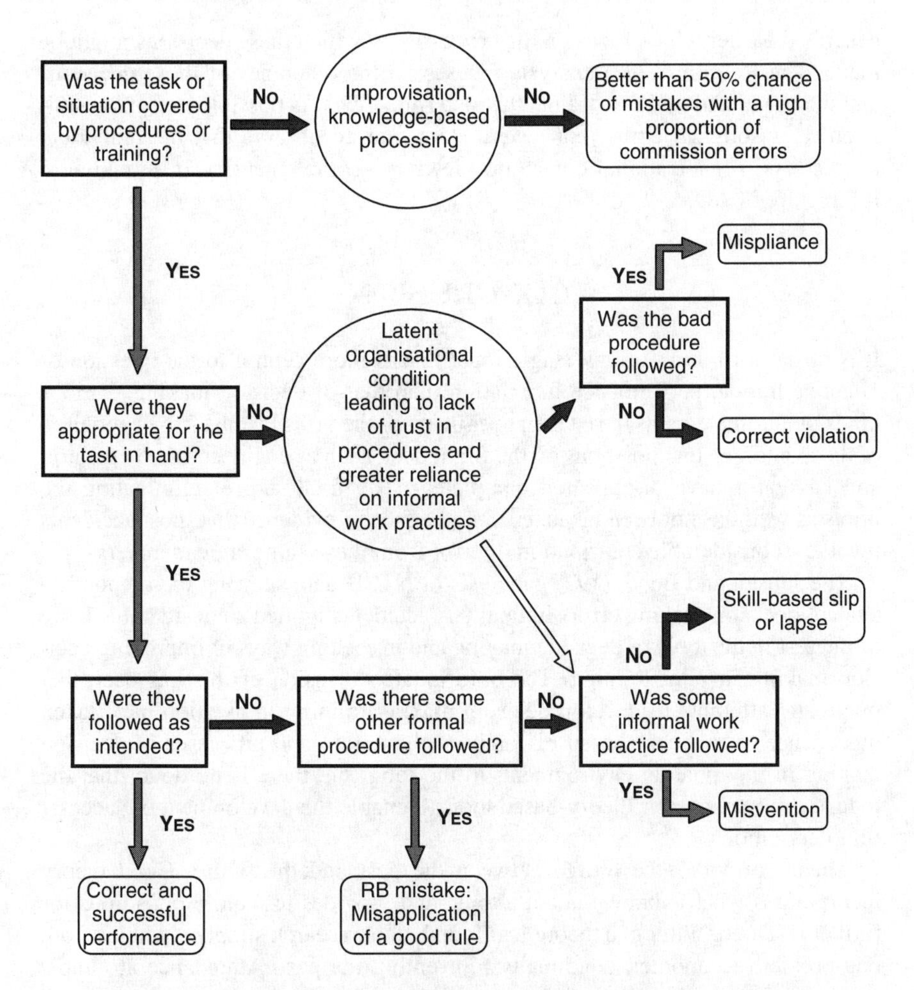

FIG. 6.4. Reason's (1997) model of decision making within the work-related context. The path directly down the left-hand side shows that if the individual correctly applies the appropriate procedures for the task at hand, the outcome should be successful. There are a variety of paths to unsafe performance. When the operator is forced to improvise at a knowledge-based level, the probability of failure is extremely high. The remaining pathways lead to errors at the rule-based (e.g., misapplying a rule) and skill-based (e.g., slip or lapse) levels. *Note.* From *Managing the Risks of Organizational Accidents* (Figure 4.10), by J. Reason, 1977, Aldershot, England: Ashgate Publishing Limited. Copyright 1977 by James Reason. Adapted with permission.

described earlier which focus on the processes by which these decisions might be made. The Reason (1997) analysis focuses on the outcomes of those decisions and the organizational constraints that determine what is possible in a given situation. It is somewhat trite, but nevertheless true, to observe that the individual process and organizational constraint views are complementary in providing a full picture of ADM.

CONCLUSION

It is difficult to think of any single topic that is more central to the question of effective human performance in aviation than that of decision making. Various kinds of automated aids have taken over many of the tasks that used to be required of the aviator, so that positions on the flight deck such as engineer, radio operator, and navigator have disappeared completely. The final step of eliminating the pilots as well has not been taken, even in the face of evidence that most accidents involve a considerable contribution of error from these same crew members.

The Jensen and Benel (1977) analysis of NTSB data showing the preponderance of decision-making errors in fatal GA accidents ignited a considerable body of interest in the topic of decision making and in seeking ways of improving decision making in pilot training. The unfortunate downside of this has been that pragmatic attempts to develop decision-making training in aviation have taken precedence over theory-based research on the nature and processes of decision making in the aviation environment. In the long run, there is no doubt that the industry requires better theory-based tools to enable the development of successful interventions.

Theory provides the bridge between the past and the future. Good theory incorporates what is known about a topic and provides testable propositions for further research. Without a theoretical foundation, research simply stumbles from one problem to another, continually reinventing the past. Attendance at almost any conference will likely provide examples where researchers have "discovered" a problem and embarked on a study involving currently fashionable tools and measures with no apparent knowledge of previous research on similar problems. Good theory allows us to learn and profit from the past by accumulating knowledge in a useable form.

For a topic of such central importance, there has been a remarkable lack of coverage in the collections of overview chapters that have appeared in recent years. For example, Wiener and Nagel's (1988) early landmark volume contains chapters on information processing, workload, error, and fatigue but not on decision making (although some coverage of ADM is provided in the chapter on information processing). A more recent substantial collection (Garland, Wise, & Hopkin, 1999), which describes itself as "a comprehensive source covering all current applications of human factors to aviation systems and operations" remarkably contains but a single index entry on the topic of decision making in 668 pages of text!

The reasons for this apparent neglect can be adduced from an examination of the current research literature in aviation psychology. The *International Journal of Aviation Psychology* (IJAP) can lay claim to being the flagship of published empirical research in the field. Although high-quality research in aviation psychology can be found in many other journals, IJAP provides the largest single pool of research evidence in the field. When the articles published in the first 5-year span of IJAP were categorized and classified (O'Hare & Lawrence, 2000) the three leading topics were training, workload, and displays, accounting for 32% of all the articles published in IJAP. Decision making was 15th on the list.

Nevertheless, the research reviewed in previous sections provides reason for cautious optimism concerning the future development of knowledge about ADM and the potential to translate that knowledge and understanding into tools to aid and develop ADM in aviators. The most important consideration for both researchers and funding providers is to consolidate their efforts into research and development programs that have a good theoretical foundation.

There are a number of theoretical frameworks that have been developed to account for decision making in other domains whose applicability to the aviation environment remains untested or undeveloped. It is reasonable to assume that theoretical principles that account for decision making in other contexts (e.g., framing) may well account for some aspects of ADM also. However, a note of caution is appropriate because as proponents of the NDM framework have pointed out, many studies of decision making have been conducted with inexpert college students considering problems in isolation. It cannot be assumed that the findings will generalize to expert performance in the aviation domain. There are, however, theoretical frameworks specifically elaborated within ADM that could provide useful guidance for future research. The evidence suggests that a pragmatic ad hoc quick-fix approach to developing practical applications is rarely successful. Solutions based on theoretical understanding are likely to be slower and more costly to develop but offer more long-term promise.

At the time of writing the "enquiring expert" and "organizational cog" metaphors are clearly in the ascendancy. The NDM framework has generated valuable insights into ADM, but there remain aspects of ADM that fit better into other frameworks (Guilkey et al., 1998). Both naturalistic studies of real-world decision making and controlled experimentation can be useful in developing and refining the theoretical basis of decision making (Azar, 1999; Cialdini, Cacioppo, Bassett, & Miller, 1978). No one camp has a monopoly on knowledge, and more progress is likely to be made by critically pursuing the interplay between problem and theory than by adopting a "methodology in search of a problem" approach.

The only sure thing about the future is that it will not be identical to the present. There are likely to be significant technological changes in GA as well as in commercial aviation in the near future (Beringer, 1999). The introduction of technologies designed to provide enhanced information for collision avoidance, navigation, weather evaluation, and so forth will raise increasingly pressing questions concerning ADM. Because such changes tend to occur in a piecemeal and fragmented way, it becomes particularly difficult to predict the overall effects on pilot performance. There is already some evidence that the provision of additional cockpit information, especially that in highly salient graphic form (e.g., weather displays, GPS-driven moving maps) may significantly alter decision making based on integrating patterns of cues. As Hancock, Smith, Scallen, Briggs, and Knecht (1995, p. 771) have observed, "Information is not some panacea which inevitably and automatically reduces the opportunity for failure. In this respect, information serves as both problem and answer."

The field of aeronautical decision making is likely to become increasingly central to future research on human performance in the technologically driven field of aviation. The utility of the field will be increased by stronger links with basic theory and greater attention to methodological rigor in the design of research. The field of ADM can provide an excellent example of O'Hare and Roscoe's (1990, p. ix) dictum that "research is truly basic to the degree that its findings are generalizable to a wide-range of real-world applications."

REFERENCES

Anderson, E. W. (1975). *Man the aviator.* London: Priory Press.

Anderson, J. R. (1982). Acquisition of cognitive skill. *Psychological Review, 89,* 369–406.

Arthur, W., & Graziano, W. G. (1996). The five-factor model, conscientiousness, and driving accident involvement. *Journal of Personality, 64,* 593–618.

Azar, B. (1999, May). Decision researchers split, but prolific. *APA Monitor,* 14.

Barnett, B. J. (1989). Information-processing components and knowledge representations: An individual differences approach to modeling pilot judgment. In *Proceedings of the Human Factors Society 33rd Annual Meeting* (pp. 878–882). Santa Monica, CA: Human Factors Society.

Barnett, B., Stokes, A., Wickens, C. D., Davis, T., Rosenblum, R., & Hyman, F. (1987). A componential analysis of pilot decision making. In *Proceedings of the Human Factors Society 31st Annual Meeting* (pp. 842–846). Santa Monica, CA: Human Factors Society.

Barrick, M. R., & Mount, M. K. (1991). The big five personality dimensions and job performance: A meta-analysis. *Personnel Psychology, 44,* 1–26.

Bellenkes, A. H., Wickens, C. D., & Kramer, A. F. (1997). Visual scanning and pilot expertise: The role of attentional flexibility and mental model development. *Aviation, Space, and Environmental Medicine, 68,* 569–579.

Benner, L. (1975). D.E.C.I.D.E in hazardous materials emergencies. *Fire Journal, 69*(4), 13–18.

Beringer, D. (1999). Innovative trends in general aviation: Promises and problems. In D. O'Hare (Ed.), *Human performance in general aviation* (pp. 225–261). Aldershot, England: Ashgate.

Berlin, J. I., Gruber, E. V., Holmes, C. W., Jensen, P. K., Lau, J. R., Mills, J. W., & O'Kane, J. M. (1982). *Pilot judgment training and evaluation volume 1* (Report No. DOT/FAA/CT-82/56-I). Atlantic City Airport, NJ: Federal Aviation Administration Technical Center.

Bernstein, P. L. (1996). *Against the gods: The remarkable story of risk.* New York: Wiley.

Besco, R. O. (1994, January). Pilot personality testing and the emperor's new clothes. *Ergonomics in Design,* 24–29.

Buch, G., & Diehl, A. (1984). An investigation of the effectiveness of pilot judgment training. *Human Factors, 26,* 557–564.

Charness, N. (1991). Expertise in chess: The balance between knowledge and search. In K. A. Ericsson & J. Smith (Eds.), *Toward a general theory of expertise* (pp. 39–63). New York: Cambridge University Press.

Cialdini, R. B., Cacioppo, J. T., Bassett, R., & Miller, J. A. (1978). Low-ball procedure for producing compliance: Commitment then cost. *Journal of Personality and Social Psychology, 36,* 463–476.

Clarke, R. (1986). *A new approach to training pilots in aeronautical decision making.* Frederick, MD: AOPA Air Safety Foundation.

Cohen, M. S., Freeman, J., & Thompson, B. (1995). Training metacognitive skills for decision making. In R. S. Jensen & L. A. Rakovan (Eds.), *Proceedings of the Eighth International Symposium on Aviation Psychology* (pp. 789–794). Columbus: The Ohio State University.

Driskill, W. E., Weissmuller, J. J., Quebe, J., Hand, D. K., Dittmar, M. J., & Hunter, D. R. (1997). *The use of weather information in aeronautical decision making* (DOT/FAA/AAM-97/3). Washington, DC: Federal Aviation Administration.

English, A. D. (1996). *The cream of the crop: Canadian aircrew, 1939–1945.* Montreal/Kingston, Canada: McGill-Queens University Press.

Ericsson, K. A., & Lehmann, A. C. (1996). Expert and exceptional performance: Evidence of maximal adaptation to task constraints. *Annual Review of Psychology, 47,* 273–305.

Farrell, J. P., & Fineberg, M. L. (1976). Specialized training versus experience in helicopter navigation at extremely low altitudes. *Human Factors, 18,* 305–308.

Fischer, U., Orasanu, J., & Wich, M. (1995). Expert pilots' perceptions of problem situations. In R. S. Jensen & L. A. Rakovan (Eds.), *Proceedings of the Eighth International Symposium on Aviation Psychology* (pp. 777–782). Columbus: The Ohio State University.

Flathers, G. W., Giffin, W. C., & Rockwell, T. (1982). A study of decision-making behaviour of aircraft pilots deviating from a planned flight. *Aviation, Space, and Environmental Medicine, 53,* 958–963.

Garland, D. J., Wise, J. A., & Hopkin, V. D. (Eds.). (1999). *Handbook of aviation human factors.* Mahwah, NJ: Lawrence Erlbaum Associates.

Gigerenzer, G., Todd, P. M., & The ABC Research Group. (1999). *Simple heuristics that make us smart.* New York: Oxford University Press.

Gopher, D., Weil, M., & Bareket, T. (1994). Transfer of skill from a computer game trainer to flight. *Human Factors, 36,* 387–405.

Gopher, D., & Kimchi, R. (1989). Engineering psychology. *Annual Review of Psychology, 40,* 431–455.

Guilkey, J. E., Jensen, R. S., Tigner, R. B., Wollard, J., Fournier, D., & Hunter, D. R. (1998). *Intervention strategies for aeronautical decision making* (Report DOT/FAA/AM-99/XX). Washington, DC: Office of Aviation Medicine.

Gwynn-Jones, T. (1981). *Aviation's magnificent gamblers.* London: Orbis.

Hampson, S. (1999). State of the art: Personality. *The Psychologist, 12,* 284–288.

Hancock, P. A., Smith, K., Scallen, S., Briggs, A., & Knecht, W. (1995). Shared decision-making in the national airspace system: Flight deck applications. In R. S. Jensen & L. A. Rakovan (Eds.), *Proceedings of the Eighth International Symposium on Aviation Psychology* (pp. 769–772). Columbus: The Ohio State University.

Heron, R., & Nendick, M. (1999). Lost in space; warning, warning, satellite navigation. In D. O'Hare (Ed.), *Human performance in general aviation* (pp. 193–224). Aldershot, England: Ashgate.

Hobbs, A. (1997, July). Decision making for helicopter pilots. *Asia-Pacific Air Safety,* 8–12.

Hormann, H.-J. (1995). FOR-DEC—A prescriptive model for aeronautical decision making. In R. Fuller, N. Johnston, & N. McDonald (Eds.), *Human factors in aviation operations* (pp. 17–23). Aldershot, England: Ashgate.

Ives, J. R. (1993). Personality and hazardous judgment patterns within a student civil aviation population. *Journal of Aviation/Aerospace Education and Research, 3,* 20–27.

Janis, I. L., & Mann, L. (1977). *Decision making: A psychological analysis of conflict, choice, and commitment.* New York: Free Press.

Jensen, R. S. (1995). *Pilot judgment and crew resource management.* Aldershot, England: Avebury.

Jensen, R. S., & Benel, R. (1977). *Judgment evaluation and instruction in civil pilot training* (FAA Report No. FAA-Rd-78–24). Savoy: University of Illinois, Aviation Research Laboratory.

Jensen, R. S., Chubb, G. P., Adrion-Kochan, J., Kirkbride, L. A., & Fisher, J. (1995). Aeronautical decision making in general aviation: New intervention strategies. In R. Fuller, N. Johnston, & N. McDonald (Eds.), *Human factors in aviation operations* (pp. 5–10). Aldershot, England: Ashgate.

Jensen, R. S., Guilkey, J. E., & Hunter, D. R. (1998). *An evaluation of pilot acceptance of the Personal Minimums training program for risk management* (Report DOT/FAA/AM-98/6). Washington, DC: Office of Aviation Medicine.

Jentsch, F. G., Sellin-Wolters, S., Bowers, C. A., & Salas, E. (1995). Crew coordination behaviors as predictors of problem detection and decision making times. In *Proceedings of the Human Factors and Ergonomics Society 39th Annual Meeting* (pp. 1350–1354). Santa Monica, CA: Human Factors and Ergonomics Society.

Kaempf, G. L., & Klein, G. (1994). Aeronautical decision making: The next generation. In N. Johnston, N. McDonald, & R. Fuller (Eds.), *Aviation psychology in practice* (pp. 223–254). Aldershot, England: Ashgate.

Kahneman, D., and Tversky, A. (1979). Prospect theory: An analysis of decision making under risk. *Econometrica, 47,* 263–291.

Kirschenbaum, S. S. (1992). Influence of expertise on information-gathering strategies. *Journal of Applied Psychology, 77,* 343–352.

Klein, G. A. (1989). Recognition-primed decisions. In W. B. Rouse (Ed.), *Advances in man-machine systems research* (Vol. 5, pp. 47–92). Greenwich, CT: JAI Press.

Koonce, J. M. (1984). A brief history of aviation psychology. *Human Factors, 26,* 499–508.

Larkin, J., McDermott, J., Simon, D. P., & Simon, H. A. (1980). Expert and novice performance in solving physics problems. *Science, 208,* 1335–1342.

Layton, C., Smith, P. J., & McCoy, C. E. (1994). Design of a cooperative problem-solving system for en-route flight planning: an empirical evaluation. *Human Factors, 36,* 94–119.

Lee, A. T. (1991). Aircrew decision-making behavior in hazardous weather avoidance. *Aviation, Space, and Environmental Medicine, 62,* 158–161.

Lester, L. F., & Bombaci, D. H. (1984). The relationship between personality and irrational judgment in civil pilots. *Human Factors, 26,* 565–572.

Lester, L. F., & Connolly, T. J. (1987). The measurement of hazardous thought patterns and their relationship to pilot personality. In R. S. Jensen (Ed.), *Proceedings of the Fourth International Symposium on Aviation Psychology* (pp. 286–292). Columbus: The Ohio State University.

Lind, A. T., Dershowitz, A., & Bussolari, S. R. (1994). *The influence of data link-provided graphical weather on pilot decision making* (Report No. DOT/FAA/RD-94/9). Springfield, VA: National Technical Information Service.

Lindholm, T. A. (1995). Advanced aviation weather graphics—information content, display concepts, functionality, and user needs. In R. S. Jensen & L. A. Rakovan (Eds.), *Proceedings of the Eighth International Symposium on Aviation Psychology* (pp. 839–844). Columbus: The Ohio State University.

Lopes, L. (1991). The rhetoric of irrationality. *Theory and Psychology, 1,* 65–82.

Lubner, M., & Markowitz, J. (1991). Towards the validation of the five hazardous thoughts measure. In R. S. Jensen & L. A. Rakovan (Eds.), *Proceedings of the Sixth International Symposium on Aviation Psychology* (pp. 1049–1054). Columbus: The Ohio State University.

Lubner, M., Hwoschinsky, P., & Hellman, F. (1997). A proposed measure of expert aeronautical decision making for training programs. In R. Jensen (Ed.), *Proceedings of the Ninth International Symposium on Aviation Psychology* (pp. 733–738). Columbus: The Ohio State University.

Maher, J. (1989). Beyond CRM to decisional heuristics: An airline generated model to examine accidents and incidents caused by crew errors in deciding. In R. S. Jensen (Ed.), *Proceedings of the Fifth International Symposium on Aviation Psychology* (pp. 439–444). Columbus: The Ohio State University.

McKinney, E. H. (1993). Flight leads and crisis decision-making. *Aviation, Space, and Environmental Medicine, 64,* 359–362.

Montgomery, H. (1993). The search for a dominance structure in decision making: Examining the evidence. In G. Klein, J. Orasanu., R. Calderwood, & C. E. Zsambok (Eds.), *Decision making in action: Models and methods* (pp. 182–187). Norwood, NJ: Ablex.

Mosier, K. L., Skitka, L. J., Heers, S., & Burdick, M. (1998). Automation bias: Decision making and performance in high-tech cockpit. *International Journal of Aviation Psychology, 8,* 47–63.

Mosier-O'Neill, K. L. (1989). A contextual analysis of pilot decision making. In R. S. Jensen (Ed.), *Proceedings of the Fifth International Symposium on Aviation Psychology* (pp. 371–376). Columbus: The Ohio State University.

Murray, S. R. (1997). Deliberate decision making by aircraft pilots: A simple reminder to avoid decision making under panic. *International Journal of Aviation Psychology, 7,* 83–100.

Nance, J. (1986). *Blind trust: How deregulation has jeopardized airline safety and what you can do about it.* New York: Morrow.

National Transportation Safety Board. (1986). *Aircraft accident report—Delta Air Lines, Inc., Lockheed L-1011-385-1, N726DA, Dallas/Fort Worth International Airport, Texas, August 2, 1985* (Report No. NTSB/AAR-86/05). Springfield, VA: National Technical Information Service.

O'Hare, D. (1992). The "artful" decision maker: A framework model for aeronautical decision making. *International Journal of Aviation Psychology, 2,* 175–191.

O'Hare, D. (1993). Expertise in aeronautical decision making: A cognitive skill analysis. In R. S. Jensen (Ed.), *Proceedings of the Seventh International Symposium on Aviation Psychology* (pp. 252–256). Columbus: The Ohio State University.

O'Hare, D. (1997). Cognitive ability determinants of elite pilot performance. *Human Factors, 39,* 540–552.

O'Hare, D., & Lawrence, B. (2000). The shape of aviation psychology: A review of articles published in the first five years of *The International Journal of Aviation Psychology. International Journal of Aviation Psychology, 10,* 1–11.

O'Hare, D., & Roscoe, S. N. (1990). *Flightdeck performance: The human factor.* Ames: Iowa State University Press.

O'Hare, D., and Smitheram, T. (1995). "Pressing on" into deteriorating conditions: An application of behavioral decision theory to pilot decision making. *International Journal of Aviation Psychology, 5,* 351–370.

O'Hare, D., & St. George, R. (1994). GPS—(Pre)cautionary tales. *Airways, 7,* 12–15.

Oldaker, I. (1995, January). *Pilot decision making—an alternative to judgment training.* Paper presented at the XXIV Organisation Scientific et Technique International du Vol a Voile (OSTIV) Congress, Omarama, New Zealand.

Orasanu, J. M. (1990). *Shared mental models and crew decision making.* (Cognitive Science Laboratory Report 46). Princeton, NJ: Princeton University.

Orasanu, J. (1993, September). Lessons learned from research on expert decision making on the flight deck. *ICAO Journal,* 20–22

Orasanu, J. (1995). Training for aviation decision making: The naturalistic decision making perspective. In *Proceedings of the Human Factors and Ergonomics Society 39th Annual Meeting* (pp. 1258–1262). Santa Monica, CA: Human Factors and Ergonomics Society.

Orasanu, J., Dismukes, R. K., & Fischer, U. (1993). Decision errors in the cockpit. In *Proceedings of the Human Factors and Ergonomics Society 37th Annual Meeting* (pp. 363–367). Santa Monica, CA: Human Factors and Ergonomics Society.

Payne, J. W., Bettman, J. R., & Johnson, E. J. (1993). *The adaptive decision maker.* New York: Cambridge University Press.

Phelps, R. H., & Shanteau, J. (1978). Livestock judges: How much information can an expert use? *Organizational Behavior and Human Performance, 21,* 209–219.

Piatelli-Palmarini, M. (1994). *Inevitable illusions: How mistakes of reason rule our minds.* New York: Wiley.

Platenius, P. H., & Wilde, G. J. S. (1989). Personal characteristics related to accident histories of Canadian pilots. *Aviation, Space, and Environmental Medicine, 60,* 42–45.

Prince, C., & Salas, E. (1993). Training and research for teamwork in the military aircrew. In E. L. Wiener, B. G. Kanki, & R. L. Helmreich (Eds.), *Cockpit resource management* (pp. 337–366). San Diego, CA: Academic Press.

Rasmussen, J. (1983). Skills, rules, and knowledge: Signals, signs, and symbols, and other distinctions in human performance models. *IEEE Transactions on Systems, Man, and Cybernetics, SMC-15,* 234–243.

Rasmussen, J. (1986). *Information processing and human-machine interaction: An approach to cognitive engineering.* Amsterdam: Elsevier.

Reason, J. (1990). *Human error.* New York: Cambridge University Press.

Reason, J. (1997). *Managing the risks of organizational accidents.* Aldershot, England: Ashgate.

Skitka, L. J., Mosier, K. L., & Burdick, M. (1999). Does automation bias decision-making? *International Journal of Human-Computer Studies, 51,* 991–1006.

Stokes, A., Belger, A. & Zhang, K. (1990). *Investigation of factors comprising a model of pilot decision making: Part II. Anxiety and cognitive strategies in expert and novice aviators* (Technical Report ARL-90-8/SCEEE-90-2). Savoy: University of Illinois, Aviation Research Laboratory.

Stokes, A. F., Kemper, K. L., & Marsh, R. (1992). *Time-stressed flight decision making: A study of expert and novice aviators* (Technical Report ARL-93-1/INEL-93-1). Savoy: University of Illinois, Aviation Research Laboratory.

Sutherland, S. (1992). *Irrationality: The enemy within.* London: Constable.

Tabachnik, B. G., and Fidell, L. S. (1996). *Using multivariate statistics* (3rd ed). New York: HarperCollins.

Telfer, R., & Ashman, A. (1986). *Pilot judgment training—An Australian validation study.* Newcastle, Australia: University of Newcastle.

Tversky, A. (1972). Elimination by aspects: A theory of choice. *Psychological Review, 79,* 281–299.

Tversky, A., & Kahneman, D. (1974). Judgments under uncertainty: Heuristics and biases. *Science, 185,* 1124–1131.

Wickens, C. D., & Flach, J. M. (1988). Information processing. In E. Wiener & D. Nagel (Eds.), *Human factors in aviation* (pp. 111–155). San Diego, CA: Academic Press.

Wickens, C. D., & Hollands, J. G. (2000). *Engineering psychology and human performance* (3rd ed.). Upper Saddle River, NJ: Prentice Hall.

Wickens, C. D., Stokes, A. F., Barnett, B., & Hyman, F. (1988). Stress and pilot judgment: An empirical study using MIDIS, a microcomputer-based simulation. In *Proceedings of the Human Factors Society 32nd Annual Meeting* (pp. 173–177). Santa Monica, CA: Human Factors Society.

Wiener, E., & Nagel, D. (Eds.). (1988). *Human factors in aviation.* San Diego, CA: Academic Press.

Wiggins, M. (1999). The development of computer-assisted learning (CAL) systems for general aviation. In D. O'Hare (Ed.), *Human performance in general aviation* (pp. 153–172). Aldershot, England: Ashgate.

Wiggins, M., & Henley, I. (1997). A computer-based analysis of expert and novice flight instructor preflight decision making. *International Journal of Aviation Psychology, 7,* 365–379.

Wiggins, M., & O'Hare, D. (1995). Expertise in aeronautical weather-related decision making: A cross-sectional analysis of general aviation pilots. *Journal of Experimental Psychology: Applied, 1,* 305–320.

Wiggins, M., O'Hare, D., Guilkey, J., & Jensen, R. (1997). The design and development of a computer-based pilot judgement tutoring system for weather-related decisions. In R. S. Jensen & L. A. Rakovan (Eds.), *Proceedings of the Ninth International Symposium on Aviation Psychology* (pp. 761–766). Columbus: The Ohio State University.

Wohl, J. G. (1981). Force management decision requirements for air force tactical command and control. *IEEE Transactions on Systems, Man, and Cybernetics, SMC-11,* 618–639.
Wohl, R. (1994). *A passion for wings: Aviation and the western imagination 1908–1918.* New Haven, CT: Yale University Press.
Zsambok, C. E., & Klein, G. A. (Eds.) (1996). *Naturalistic decision making.* Mahwah, NJ: Lawrence Erlbaum Associates.

ACKNOWLEDGMENTS

I am grateful for the helpful comments of David Hunter, the editors, and two anonymous reviewers on the first draft of this chapter. I am grateful to Douglas Owen for preparing the figures.

7

Pilot Actions and Tasks: Selections, Execution, and Control

Christopher D. Wickens
University of Illinois at Urbana-Champaign

In most circumstances, a pilot's task involves a continuous stream of activities. Many of these activities are overt and easily observable, such as movement of the flight control sticks, communications with air traffic control, or manipulating switches. Others are much more covert and less observable, such as planning, diagnosing or monitoring. A skilled pilot will selectively *choose* which tasks and actions to perform at the appropriate time, knowing which tasks to emphasize and which ones to ignore when workload is high (Adams, Tenney, & Pew, 1995; Funk, 1991; Orasanu & Fischer, 1997). This skilled pilot will also execute those actions smoothly and appropriately, the most important of which is *control* of the aircraft. In this chapter, we will first focus on the choice of actions and tasks, and then describe the execution of the most important of those tasks—those involved in flight control and navigation.

TASK CHOICE AND TASK MANAGEMENT

Much of task selection in aviation is *procedural.* Tasks are carried out at certain predesignated times or sequences during a flight. For example, the landing gear must be lowered prior to landing and raised after takeoff; particular communications must be initiated when the pilot is instructed or when the aircraft transitions between ATC sectors in the airspace. With repeated practice, the pilot builds up what are called "schemata" of the appropriate procedures and actions to be performed at the appropriate times. The airlines provide great emphasis on training pilots in what they call "standard operating procedures" (Degani & Wiener, 1993; Hawkins, 1993; Orlady & Orlady, 1999), capturing many of these sequences of tasks or actions. Correspondingly, there is great emphasis on "proceduralizing" most aspects of air traffic control through standard communications protocols, standardized departure and arrival routes, and so forth (Wickens, Mavor, & McGee, 1997).

Although extensive training, practice, and rehearsal may be necessary to ensure that procedures are carried out fluently, such training is never entirely sufficient to ensure safe flight because the appropriate sequence of procedures will not have been learned well by the student pilot and may occasionally be forgotten or missed even by the skilled pilot. Furthermore, not all aspects of aviation can be proceduralized, for example, when unexpected circumstances develop, such as an in-flight engine failure. Under such circumstances, even emergency procedures must be applied flexibly and with care, and many of those procedures may not apply at all, as in the case of the crash of United Airlines Flight 232 at Sioux City, Iowa, for which there were simply no procedures available to cope with the total hydraulics failure (Predmore, 1991). There are two alternative approaches to address these limitations of procedures, which we shall discuss. On the one hand, checklists can provide external support to help ensure that procedures are not forgotten. On the other hand, training can guarantee more optimal forms of flexible *task management* when a standard order of procedures (defined by training or by a checklist) cannot be applied.

Checklists (Degani & Wiener, 1993) provide a useful, although not foolproof, means of following the appropriate script of actions. Some of the perceptual and display aspects of checklists were described in chapter 5. Different airlines and aviation organizations vary considerably in the precise philosophy with which checklists are performed, the roles of two pilots in carrying out the checklists, and the roles of the spoken word in acknowledging that each item or task is, in fact, accomplished. Such checklists are particularly vulnerable to two sorts of cognitive errors. First, checklist items may be missed altogether. This could occur under circumstances of time pressure, so that a checklist is prematurely halted before completion, and hence the final items are not accomplished. This situation has led Degani and Wiener (1993) to advocate that "killer" items (those causing possible disaster if missed) should be placed earlier in a checklist. Items may also be missed when the pilot's attention is diverted from the checklist by an "interrupting task" (as is discussed later) and then returned to the checklist at an item beyond the point at which it left. This error apparently occurred prior to the crash of a Northwest Airlines plane at Detroit in 1987, when the pilots needed to divert their attention to deal with an unexpected takeoff runway change midway through the checklist. When they resumed the checklist, their attention returned apparently to a step *after* the critical "flaps set" item. The flaps did not get set to the appropriate configuration for takeoff, and a crash occurred after takeoff when the aircraft did not achieve adequate lift.

The second cognitive error in checklists may result when checklists are followed rapidly without full attention being given to the underlying state of the system to be checked. It is as if the pilot may be relying on top-down perceptual processing to partially "see" the item in the state it is supposed to be (as dictated by the checklist), rather than carefully inspecting it visually to ensure that it is physically in that state. This top-down processing bias would be particularly likely to occur if the checklist has been performed numerous times before, and the particular item (e.g., "switch "on") has always been in that position (but is now "off"). An excellent source of information on other human factors issues in checklist design can be found in Degani and Wiener (1993).

Cockpit Task Management

It might be ideal if all of a pilot's tasks could be proceduralized; in practice, this is not possible. There are simply too many unexpected events that can occur in the airspace, as well as numerous situations when two (or more) tasks compete for attention, and the pilot must then choose which to perform and which to defer, if they cannot be accomplished in parallel (see chap. 2). The issue of *cockpit task management* (CTM) Chou, Madhavan, & Funk, 1996; Dismukes, 2001; Funk, 1991), or *strategic workload management* (Hart & Wickens, 1990), describes the characteristics of such strategies that are employed by the pilot to choose which tasks to perform, which to delay or "shed," and which may interrupt or "preempt" other ongoing tasks (see also Adams, Tenney, & Pew, 1995). An understanding of these findings is provided in the following pages.

Optimal Scheduling. In order to provide some framework for the study of CTM, it is necessary to try to establish both some general rules for the *optimal* scheduling of tasks, as well as to derive some general conclusions regarding the extent to which this optimal scheduling is followed and the characteristics of tasks that may cause a departure from the optimal. A key aspect in establishing a gold standard for optimal task management is the concept of *task importance* (see also chap. 2). For example, Raby and Wickens (1994) had a group of pilots rank-order 19 tasks to be carried out in a landing simulation in terms of their importance or relevance to flight safety; from these, they were able to group the tasks into those of high, medium, and low priority. A related approach, to adopt the standard hierarchy of "aviate—navigate—communicate—systems management (ANCS)," is introduced in chapter 5 (Schutte & Trujillo, 1996). According to such a hierarchy, when a conflict exists between two tasks, the task toward the top of the hierarchy should generally dominate the one below and be performed first rather than deferred. This prioritization scheme underlies some of the key principles that NASA proposes should be taught in any CRM program (see Sherman, chap. 13).

There is little doubt that air safety is compromised by poor cockpit task management when more important tasks are allowed to be preempted or superceded by those of lesser importance. Such a conclusion comes from several sources. For example, the crash of Eastern Airlines Flight 401 into the Florida Everglades in 1972 resulted because pilots became fixated on a landing gear problem (systems management), failing to heed a loss of altitude (aviate) that resulted from an accidentally disengaged autopilot (Wiener, 1977). Indeed, Chou et al. (1996) have documented that 23% of 324 National Traffic Safety Board (NTSB)-reported accidents during the period (1960 to 1989) had poor CTM as one underlying cause. Those authors also identified a large number of Aviation Safety Reporting System (ASRS) incident reports describing poor CTM. Dismukes (2001) also examined a large number of ASRS incidents related to poor tasks management, documenting the sorts of task management errors that occurred most frequently. A strong case can also be made that examples of controlled flight into terrain (CFIT) accidents also result from poor CTM, (chap. 2; see also Shappell & Wiegmann, 1997; Wiener, 1977), because these accidents by definition involve an inappropriate "shedding" of the task of altitude monitoring, one that can be associated with the two highest priority tasks (aviate or navigate). Finally, a careful analysis of simulated fault management carried out by Orasanu and Fischer (1997) revealed that differences between those aircrews who handled unexpected failures properly and those who did not could be largely attributed to differences in task management. For example, better crews were found to offload lower priority tasks, whereas more poorly performing crews did not.

It is the case, however, that accidents only reflect the behavior of a minority of pilots, and that the "poorly performing crews" in studies such as that carried out by Orasanu and Fischer (1997) are also typically a minority of the general population. Are there other more general "rules" of CTM that can be used to characterize the pilot population as a whole? A small handful of studies have attempted to reveal such characteristics. For example, Damos (1997), Dismukes (2001), and Latorella (1996) have all studied task "interruptions," in which the performance of certain ongoing tasks is terminated by the arrival of a new task (Reason, 1990). Analyzing a series of videotaped Line-Oriented Flight Training (LOFT) scenarios, Damos found that ATC communications (a task third on the ANCS hierarchy) would often interrupt or preempt a pilot's involvement with navigational planning tasks, which may be considered a second-level task, a characteristic that can be described as nonoptimal task management. In contrast, however, pilots were fairly effective at protecting the hand flying of the aircraft (Aviate: first priority) from being interrupted. Latorella (1996) and Dismukes (2001) have also both noted the preemptive nature of ATC communications into a pilot's ongoing task, an issue that will be addressed further.

Schutte and Trujillo (1996) attempted to categorize the task management strategies of a group of 16 pilots, each flying a simulation with a "confederate" copilot and each encountering a series of unexpected systems failures (e.g., a fuel leak). They found that the crews who performed best were those whose task management strategies adhered to a "perceived importance" strategy, in which tasks at any level of the ANCS hierarchy could jump to the top if they were perceived momentarily to be of greatest importance. That is, a certain flexibility (as opposed to rigidity) of task management served them well. However, these most effective pilots also adhered to the ANCS hierarchy for routine task management, when emergencies did not occur. They were more successful than crews who either adhered to a procedural task strategy or a strategy in which any new activity preempted ongoing activities.

Raby and Wickens (1994) analyzed the CTM behavior of pilots on a routine, simulated approach in which workload was sometimes unexpectedly elevated to higher levels (imposed by ATC asking for an expedited approach). They found that most pilots generally did adhere to an optimal task management strategy when workload increased, by "shedding" those tasks that had been rated to be of lower importance. However, pilots did sometimes allow flight control to be preempted by lower priority tasks, and the researchers also found that pilots' task management planning strategies were not terribly sophisticated (see also Moray, Dessouky, Kijowski, & Adapathya, 1991). Pilots did not, for example, alter the scheduling of higher priority tasks as workload increased. A further observation made by Raby and Wickens (1995) was that the better performing pilots tended to switch attention between tasks more frequently than those pilots who performed less well, and Laudeman and Palmer (1995) and Raby and Wickens (1994) both found that better performing pilots were more optimal in *when* they performed the higher priority activities, tending to accomplish these earlier.

Preemptive Activities. In characterizing task management, it is important to establish if there are certain characteristics of tasks and activities, unrelated to their overall priorities, that tend to either interrupt or preempt other activities or, conversely, tend to protect an ongoing activity from being interrupted by others (Latorella, 1998). One such characteristic appears to be the modality (auditory vs. visual) of the perceptual information involved in the activity. Indeed, it appears that the somewhat interrupting nature of ATC communications, described earlier, may be a result of the fact that such activity is typically auditory in nature. Conversely, the *failure* of pilots to sometimes notice important, automation commanded "mode transitions" on a flight management system (an event relevant to the task of navigation monitoring; Sklar & Sarter, 1999) can also be described by the *less-preemptive* nature of the visual events. A review of literature on dual task interference carried out by Wickens and Liu (1988) revealed that discrete auditory tasks have a tendency to disrupt or interrupt ongoing (and continuous) visual tasks, whereas discrete visual tasks do not, a characteristic that supports the use of the auditory modality for highest priority warnings and alarms (Stanton, 1994). Furthermore, Latorella (1998) has argued that ongoing visual tasks, like reading a checklist or planning from a visual display, are somewhat more "interruptible" than ongoing auditory tasks, such as understanding a communications string from ATC. This difference results because the visual tasks usually leave a visual reminder of where the task was interrupted (e.g., the print of the last checked item on the checklist), which can be returned to after the interruption, whereas auditory tasks do not have such a "permanent" reminder. Hence, the pilot is more reluctant to interrupt the ongoing auditory task, because this point of departure must be remembered and cannot be seen.

Such a distinction between auditory and visual interruptions is quite important for considering the implications of pilot-controller digital *data link* for replacing or augmenting the current voice-dependent ATC communications link (see chap. 5; Navarro & Sikorski, 1999). This is because the typical conceptions of data link involve presenting a visual display of ATC-generated communications to replace the conventional "auditory display" from the ATC radio. Will such a visual input of the same communications information that had previously been presented auditorily reduce the interruptions of ongoing activity that the auditory communications tends to produce (Damos, 1997)? Simulations carried out by Latorella (1998) and Helleberg and Wickens (2001) revealed that this was indeed the case. Pilots tended less to interrupt ongoing tasks when ATC information was presented visually than auditorily. This finding suggests that a visual communications display may be less likely than voice communications to disrupt the optimal task prioritization hierarchy. The pilot with a visual data link display thus has more flexibility to manage the communications task—to deal with it later if a higher priority task is ongoing (Navarro & Sikorski, 1999).

Conclusions. The data suggest that, although most pilots are reasonably effective at appropriate task management, selecting higher priority tasks over lower priority ones, nonoptimal allocation sometimes does occur. This fact reinforces the importance of including CTM training within the broader context of both CRM training (how the pilot manages his or her *own* resources to deal with the tasks imposed) and aeronautical decision-making (ADM) training (how the pilot *decides* what tasks to perform and what ones to postpone or defer). Despite such a recommendation, it does not appear that CTM instruction is formally or consistently implemented in most pilot training programs (Dismukes, 2001). The data also suggest that design considerations related to the modality of task information and the salience of that information within a modality can also affect, positively or negatively, the appropriate choice of which tasks to perform.

Finally, automation can sometimes be employed to guide appropriate task management. The simple auditory warning system provides such an example. Indeed, the ground proximity warning system (GPWS) can be thought of as an auditory reminder to the pilot to switch attention to the task of altitude monitoring (see chap. 5). And the choice of the auditory modality to implement most critical warnings is based on the inherent preemptiveness of that channel (Stanton, 1994). More elaborate forms of automated task reminders are also being proposed and implemented. One key item is the automated checklist (Bresley, 1995; Palmer & Degani, 1991; see also chap. 5), but other more sophisticated systems are proposed to monitor the pilots' ongoing pilot activities and tasks and to remind them of what needs to be done contingent on the momentary circumstances (Funk & Braune, 1997) or present them with task-relevant information given those circumstances (Hammer, 1999), thereby implicitly advising them to shift task activity. Although such systems have obvious benefits in addressing task management issues, their ultimate effectiveness will depend on the *reliability* of automation's inference as to what should be performed, an issue dealt with elsewhere (see chap. 9).

Concurrent Activities

The previous discussion has generally dealt with situations in which the concern has been on what tasks are dropped, interrupted, or resumed, that is, the *sequential* or serial aspects of multiple task performance. However, it is certainly the case that pilots are able to "time-share" many tasks (i.e., to perform two tasks in parallel or concurrently). For example, flying the aircraft can be accomplished concurrently with communications, perhaps explaining Damos' (1997) finding that pilots did *not* allow flying (aviate) to be interrupted by ATC communications. That is, because the pilots could continue to fly while concurrently listening to ATC, the question of interruption was not relevant. The issue of concurrent task activities is addressed by theories of multiple task performance (Damos, 1991; Wickens & Hollands, 2000). Such theories highlight four critical characteristics that determine the extent to which two or more activities *can* be accomplished concurrently (hence, mitigating the need for task management interruption and scheduling): (a) attention resource demand, (b) resource allocation, (c) multiple resources, and (d) task similarity. These characteristics will be discussed briefly.

Resource Demand. The ability to perform two tasks concurrently will be improved to the extent that one or both demand less attentional resources for their performance or impose less mental workload (chap. 4). In the extreme, tasks that demand almost no resources are said to be "automated" (Damos, 1991; Schneider, 1985). Such differences in resource demand can be found between tasks, as well as between pilots performing a given task. As an example of between-task differences, an important difference between the flight dynamics of different aircraft is in terms of their resource demands, and as a consequence, the extent to which other tasks can be performed concurrently (see next section; see also chap. 8). As an example of between-pilot differences, we may contrast performance of the novice, who requires more resources to perform a given task, with that of the expert, who requires fewer (see chap. 14). Damos (1978), for example, found differences between students and flight instructors in their ability to time-share flight tasks. Bellenkes, Wickens, and Kramer (1997) found that flight instructors could extract information from the attitude indicator in about half the time as student pilots, availing the instructors with many more attention resources to check other flight instruments.

Resource Allocation. In addition to how many resources a task demands (its mental workload), the concurrent performance of two (or more) tasks will also be determined by the strategy of allocating resources between tasks (Gopher, 1992; Gopher, Weil, & Bareket, 1994; Tsang, Velazquez, & Vidulich, 1996; Tsang & Wickens, 1988), that is, which task gets relatively more resources and which gets fewer. This might, for example, describe the relative frequency with which the pilot visually samples the outside world (for traffic monitoring) and the instrument panel (for aviating and navigating; Wickens, Xu, Helleberg, & Marsh, 2001). This factor of resource allocation is in many respects closely tied to that of cockpit task management discussed in the previous section. However, CTM tends to refer more to the management of time, when tasks are performed in an all-or-none fashion (i.e., task shedding or preemption). In contrast, resource allocation characterizes the circumstances in which both tasks are being "performed" at some level, but the level of cognitive effort invested in the task is modulated. For flight control tasks, this effort can often be inferred from the amount of control activity the pilot imposes (Wickens & Gopher, 1977). For visual tasks, it may be related to the allocation of visual attention measured via scanning (see the following discussion).

Not surprisingly, differences between novice and expert pilots can also be seen in the skill of appropriately allocating resources to tasks and subtasks (Bellenkes et al., 1997). For example, Gopher et al. (1994) found that the skill in such dynamic resource allocation was a very important one for tactical fighter pilots, and that those who were trained on such a skill with a generic video game were more likely to be selected to pursue fighter pilot training in the Israeli Air Force.

Multiple Resources. The multiple resources demanded by concurrent tasks (Sarno & Wickens, 1995; Wickens, 1991; Wickens & Hollands, 2000; Wickens, 2002) is an important influence on the effectiveness of their time-sharing. For example, although discrete auditory tasks are more likely to preempt ongoing visual ones, as described earlier, it is also the case that it is easier for the pilot to time-share an auditory task with a visual one than it is to carry out two concurrent visual tasks or two concurrent auditory ones, as if the two modalities require separate processing resources (Wickens, Sandry, & Vidulich, 1983). This is particularly true if the two visual tasks are widely separated in space (see chap. 5) or if the two auditory tasks tend to mask each other. Corresponding claims can be made for the benefits of voice input over manual input for discrete tasks that must be carried out concurrently with the pilot's continuous manual control of the aircraft, as if the voice and manual responses rely on separate resources that can be deployed concurrently (Farrell, Cain, Sarker, & Ebbers, 1999; Sarno & Wickens, 1995).

A model of multiple resources developed by Wickens (Wickens, 1991; Wickens & Hollands, 2000) proposes that there are actually four dimensions of the human information-processing system that define separate resources. The efficiency of concurrent performance of two tasks will depend on the extent to which the two require different levels along the four dimensions. These four dimensions are as follows:

1. Auditory versus visual perception as discussed above.
2. Focal versus ambient vision (Leibowitz, 1987; Previc, 2000). Focal vision is used for detecting and recognizing detail, whereas ambient vision is used for processing motion, often in peripheral vision, as the pilot must do in low-level flight (see chap. 2).
3. Spatial versus verbal or linguistic processing. Spatial processing is used in *perceiving* motion and spatial locations, *remembering* spatial information such as maps, and in *manipulating* the hands continuously, as in such tasks as pointing or tracking. Verbal tasks involve the *perception* of the printed or spoken word, the *remembering* of speech (as when rehearsing an ATC instruction), and in actual *response* speaking.
4. Perceptual/cognitive processes versus responding. This dimension distinguishes those perceptual and memory processes such as monitoring, reading, listening, rehearsing, or planning from those action-oriented processes such as speaking or manipulating controls with the hands.

Simply put, to the extent that any two tasks demand common levels along any of the these four dimensions, they will be less likely to be time-shared efficiently and more likely to disrupt each other's performance (given, of course, that resource demand is held constant, as discussed earlier; Sarno & Wickens, 1995). In the extreme, when resource competition is excessive, concurent performance becomes impossible and the rules of sequential task management apply.

Similarity and Confusion. A fourth influence on the effectiveness of concurrent ask performance is related to the similarity of the specific information or material processed in the two tasks, an issue that we highlighted in the display principle of discriminability (chap. 5). For example, a pilot will be more likely to confuse voice messages from air traffic control and from a cockpit voice synthesizer, if both messages emanate from the same source (e.g., both ears of the headphones), both have the same voice quality (e.g., both male or both female), and both deal with the same class of information (e.g., both are presenting numerical information). Reducing the similarity of the two messages in any of these ways will increase the likelihood that they can both be perceived or retained correctly and in parallel (i.e., successful time-sharing).

Conclusions. Collectively, then, the picture of how pilots deal with multiple competing tasks is a complex one. Through cockpit task management they may choose to perform some and ignore, abandon, or postpone others. If the collective resource demands of two (or more) tasks are low enough, pilots may choose to perform them concurrently and if they do, their success will be determined by how low those demands are, by the resource allocation policy between tasks (providing more resources to those tasks that are more important), by the extent to which different resources are employed, and by the extent to which different, nonconfusable material is processed in the tasks at hand.

Collectively, all of these factors present a sufficiently complex picture that it is difficult to predict the precise nature of multiple task performance in the cockpit. Nevertheless, some models have tried to harness at least the resource demand and multiple resource components in order to predict how changes in cockpit technology might influence the overall capacity to perform (Groce & Boucek, 1987; North & Riley, 1989; Sarno & Wickens 1995). The interested reader should consult Sarno and Wickens (1995) for a more detailed review and description of those models.

Visual Scanning, Task Management, and Concurrent Task Performance. As we noted previously, when the pilot must attend to two widely separated sources of visual information for two tasks, it becomes difficult to accomplish both tasks concurrently. On the one hand, this characteristic provides the rationale for using auditory input (e.g., voice synthesis or spatial tones) in an otherwise heavily visual environment of the cockpit, for superimposing images with head-up displays, or for integrating the information on the display (see chap. 5). On the other hand, it is this separation characteristic that often allows the analyst to diagnose the task that is being performed at the moment by assessing the momentary direction of visual scan, even if that task has no observable components. For example, Wickens, Xu, Helleberg, and Marsh (2001) measured the pilots' distribution of eye fixations between the outside world (for air traffic monitoring), a cockpit display of traffic information (for maneuver planning), and the instrument panel (for aviating). By characterizing the proportion of time that the eye spent examining any of these three sources, they were able to ascertain that pilots did indeed adhere to some aspects of the ANCS task hierarchy, fixating on the instrument panel (aviate) far more frequently than on the collective sources of navigational information (the outside world and the traffic display). The investigators' use of visual scanning as a measure of task involvement also allowed them to assess the frequency with which the pilot switched the focus of attention between navigational planning, traffic monitoring, and flight control tasks, even though the former two tasks were not directly observable from the pilot's hands or vocal activity.

We now shift the focus of our attention from the choice of what activities to perform in high workload environments and the capability to perform them concurrently to a discussion of the execution of those activities that are in the highest priority tasks confronting the pilot in the ANCS hierarchy—the control of the aircraft.

AVIATION CONTROL

The pilot's control of the aircraft's attitude and position in 3-D space serves both tasks at the top of the ANCS hierarchy: aviating and navigating. In this section, we consider the human information-processing demands of flight control from a relatively intuitive perspective. In chapter 8, characteristics of flight control and flight dynamics are addressed from a more formal mathematical/engineering perspective, with an emphasis on the computational modeling of the human pilot. For those lacking the formal engineering knowledge of control theory, the present section can provide a useful background for the following chapter.

The pilot is part of the closed-loop tracking system as shown in Fig. 7.1. The "closed-loop" description of the figure is evident from the feedback loop, by which the "output" of the aircraft state at the right is fed back to provide a displayed input for the pilot to control. Such a system is sometimes called a *negative* feedback system because the output of the system will be designed to be in the opposite direction (negating) from the error. Thus, if the pilot is too far to the left of the desired flight path (an error), he will correct to the right. (Note that most fixed wing aircraft are themselves negative feedback or "self-correcting" systems, even without the pilot in the control loop. If the pilot takes her hands off the yoke, most aircraft will inherently seek a wings level attitude.)

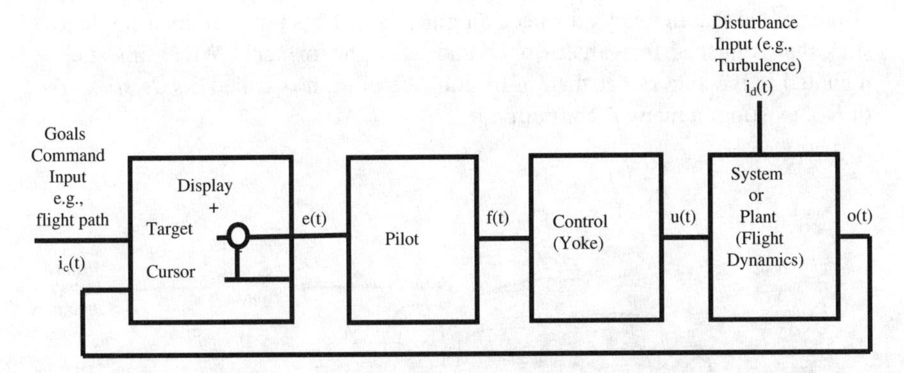

FIG. 7.1. The tracking loop. The figure depicts how a pilot, shown in the middle box, perceives an error that changes over time [$e(t)$], responds to this with a force on the control [$f(t)$] that is normally exerted in the opposite direction of the error. The changed position of the control, $u(t)$ that results from the force, is delivered to the aircraft via its control surfaces, and this produces a changed flight path output of the aircraft, $o(t)$. The aircraft output may also be driven by disturbances such as turbulence, $i_d\ (t)$. Both the new flight path and some representation of the commanded (or desired) flight path are perceived by the pilot in a display. The differences between the two is the error, $e(t)$, which is where our description of the tracking loop begin. In some descriptions, the display of aircraft output is called the cursor (controlled by the human), whereas the command input is called the target.

When flying, the pilot has navigational *goals* (shown in Fig. 7.1 as command inputs), such as "level off at flight level 110"; "turn to heading 250"), which in nonautomated aircraft are sought by exerting continuous manual control (usually of the *yoke*). The challenge to the pilot is that the navigational goals are *hierarchical,* as shown pictorially in Fig. 7.2(a) and more formally in Fig. 7.2(b). That is, in the current generation of nonautomated aircraft, in order to obtain higher level goals (i.e., obtain a position over the ground in space, specified by *xyz* coordinates), the pilot must seek intermediate goals (i.e., obtain a heading or vertical speed pointing toward that point) and to achieve these intermediate goals (e.g., a heading), the pilot must change the orientation or "attitude" of the aircraft (e.g., bank), thus turning until the required heading is reached.

As shown in Fig. 7.2(b), two tracking loops are represented simultaneously, one for lateral tracking and one for vertical tracking. In the left column of the figure are the hierarchical *goals* to be met for both dimensions, arrayed from top (high level) to bottom (low level). Each goal in the left column is associated with a particular corresponding *action* in the center column. However, in most nonautomated aircraft (or in "hand flying" the automated ones), these actions cannot be directly commanded, but are accomplished by manipulating the *yoke* shown at the bottom, whose lateral motion affects the ailerons (and thus the bank of the aircraft) and whose fore-aft motion affects the elevators (and thus the pitch of the aircraft). These two primary controls are often coupled with movement of the throttle, to effect airspeed and with the rudder pedals to coordinate turns. It should also be noted that in many advanced aircraft, the yoke is replaced by a single joystick that is pushed fore-aft for pitch and left-right for bank. When this stick is mounted to the side rather than in front of the pilot, it is called a *side stick controller,* existing in many Airbus aircraft.

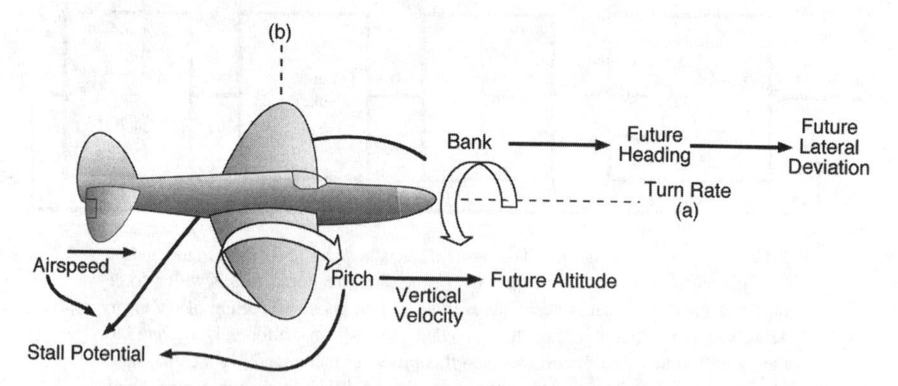

FIG. 7.2. Flight dynamics. (a) A schematic representation of the relation between the axes of rotation (pitch and bank) and translation of the aircraft. The white arrows show the current control of attitude (pitch and bank); the black arrows show how these influence the future state. Coupled with airspeed, pitch and bank also affect the potential for stalling the aircraft.

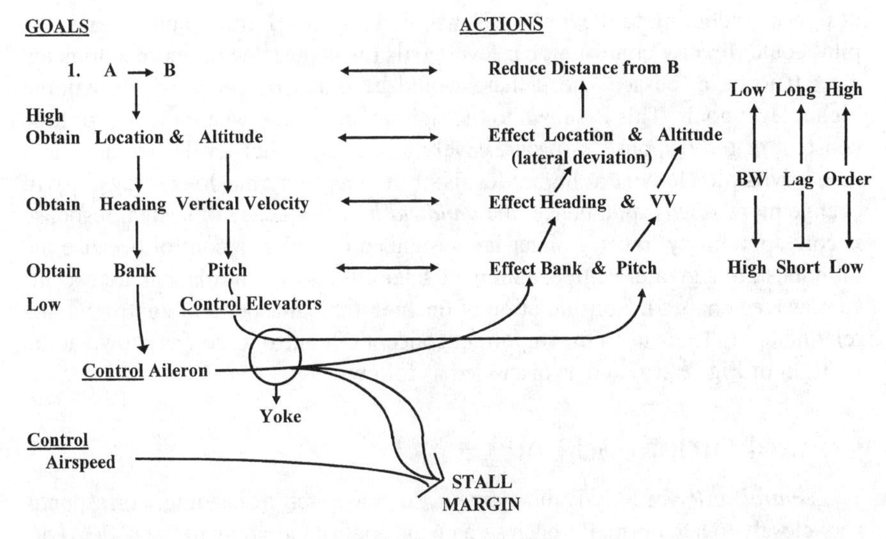

FIG. 7.2. (b) Hierarchical control. The relation between the vertical and lateral navigational *goals* sought by a pilot (left column), and the *actions* required to achieve these goals (center column). The pilot is assumed to be at location A, with the goal of getting to location B. Each goal on the left is accomplished by achieving the goal underneath it, which in turn is accomplished by carrying out the associated action in the center column. The direct physical action required is movement of the control yoke depicted at the bottom, coupled with adjustment of the throttle to control airspeed. Shown in the right columns are three properties that covary with the level of the hierarchy. Shown at the bottom is a second goal (2), preventing stalling of the aircraft.

In understanding flight control, it is important to realize that in seeking the navigational goals, shown in the left column, by implementing the actions in the center column, the pilot must also always keep in mind the critical limiting values of pitch, bank, and airspeed that prevent the aircraft from *stalling,* as shown at the bottom of the figure. Too high a pitch, too low an airspeed, or too large of a bank angle can all destroy the lift of the wings, causing the aircraft to stall. How close these three parameters are to the stalling state can be described as the "stall margin." Failure to monitor the stall margin is neglect of the highest task in the pilot's ANCS priority hierarchy—to aviate—as described in the first part of this chapter. If the plane does not remain in the sky, it cannot navigate!

As we have noted, in the standard nonautomated aircraft, the pilot does not directly track the higher level goals. Instead, the pilot must obtain these goals by tracking lower level goals, in order to make them satisfy the quest for higher level actions and goals (e.g., affect heading to obtain location, affect vertical velocity to obtain an altitude). The primary reason for this hierarchical control loop is that this was the fundamental way in which fixed wing aircraft were designed—adhering to the laws of Newtonian physics. Furthermore, even if the pilot could directly control higher level goals (as is possible on more automated aircraft to be discussed later), there would be much *longer lags* in obtaining higher level goals. This is shown to the right in the figure, where the *lag,* or time constant of the response, is progressively longer at higher levels. To the extent that a dynamic element at higher levels of the hierarchy has longer lags, it will change more slowly and, hence, the *bandwidth* or frequency of system response is correspondingly lower. Longer lag systems are harder to control because the pilot must, to a greater degree, anticipate later effects of controls that are executed now (see chap. 5). Consideration of the hierarchical aspects of control and its relation to lag leads us to the important concept of control order, as shown at the far right of Fig. 7.2(b) and as discussed as follows.

Control Order and Lag

The *control order* of a dynamic element, such as pitch or heading, corresponds very closely to lag. Formally, each step up the control hierarchy in Fig. 7.2(b) corresponds to an increase in control order by one. Also, each order of control corresponds to one *time integral* in the calculus of integral equations (Wickens, 1986). Thus, a constant position input to a first-order dynamic element will produce a constant *rate of change* of the output. For example, a constant lateral deflection of the control yoke (Y) will cause a constant *rate of change* of the Bank angle (B), which we see in Fig. 7.2(b). This is a first-order control, sometimes called velocity control. That is, if $B(t)$ is the bank angle as a function of time, then:

$$B(t) = \int Y(t)dt$$

The *second-order* response is equivalent to the heading (H) of the aircraft as controlled by the yoke. This is because a given bank angle (B) causes a given *rate of change* in heading. (The steeper the bank, the faster heading changes and the tighter is the turn.)

$$H(t) = \int B(t)dt \text{ or } H(t) = \iint Y(t)dt$$

Because each integration causes a lag, there is a greater lag between the Yoke deflection and the heading (second order), than between the yoke deflection and the bank (first order). Because a constant input to a second-order system leads to a constant rate of change of velocity—a constant acceleration—second-order systems are sometimes called acceleration control systems.

The *third-order* response to the yoke turn is equivalent to the lateral deviation (D) of the aircraft off some predefined flight path (e.g., the final approach path to a runway). This is because a given heading causes a *change* in the lateral deviation. Thus,

$$D(t) = \int H(t)dt \quad D(t) = \iint B(t)dt \text{ or } D(t) = \iiint Y(t)dt$$

These relations are shown in Fig. 7.3(a) (bottom) and are juxtaposed to a similar representation of the driving task with which some readers may be more familiar. Similar relations characterize vertical flight, except that there are only three levels of the hierarchy, not four. That is, as shown in Fig. 7.2(b), yoke fore-aft position (= elevator position) affects pitch (= vertical velocity), and pitch affects altitude. Thus, the relation between fore-aft yoke position and altitude is a second-order one:

$$Alt(t) = \iint Y(t)dt$$

In Fig. 7.3(b) we show the time response of a first-order velocity (bank or pitch), second-order acceleration (heading or altitude), and third-order (lateral position) system to an abrupt and then constant deflection of the yoke. (In control theory this is sometimes called the "step response".) The increasing lag of system response with increasing order is very much in evidence and is highlighted at the bottom of the figure. Thus, it will take a short time for bank (first order) to reach a desired state following a yoke input, but a longer time for heading (second order) to reach the desired state, and a still longer time for the third-order lateral deviation variable to reach a desired state. Not shown in the figure is the response of a 0 order, or position control system, in which the output would correspond, in position, to the solid line input. The mouse, which moves a cursor around a display screen, is an example of a 0 order control system. A new position of the mouse produces a new position of the cursor.

The important point here is that when pilots try to track systems with long lags (e.g., higher order) they must *anticipate* error ahead of the lags so that a control input delivered now will affect the plane's flight in the future, when it is needed to correct the anticipated error. Anticipation is cognitively difficult, and the source of high resource demands and mental workload (Wickens, 1986; Wickens, Gempler, & Morphew, 2000; Wickens & Hollands, 2000) and is sometimes impossible if lags are extremely long. If people do not adequately anticipate, the aircraft will at best lag behind its desired position and at worst can go unstable in its control. This latter phenomenon is known as *closed-loop instability* or *pilot-induced oscillations* (PIO) and might produce an altitude profile as shown in Fig. 7.4. (As we noted in chap. 5, the principle of predictive aiding is designed to help the pilot tracking a lagged system.)

It is important to note here that higher order control is not the only source of increasing lags (and therefore increasing anticipation demand) in flight controls. Larger aircraft with greater inertia will have longer lags than smaller aircraft on corresponding variables with the same order. It will, for example, take a 747 longer to change its heading than a light Cessna or F-16. Mechanical or computational delays can also produce lags, although these are generally much shorter than the lags associated with aircraft control and higher inertia.

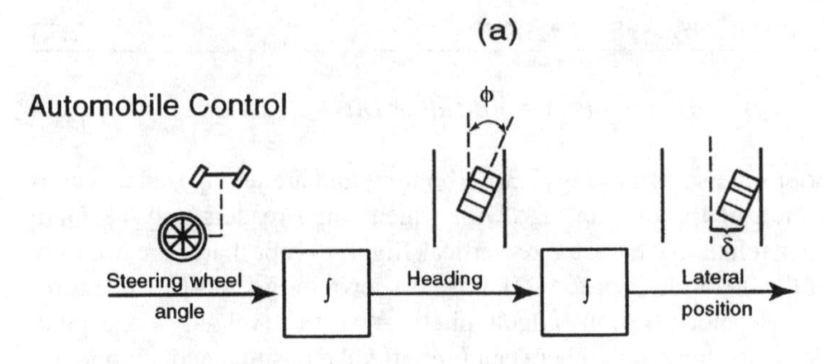

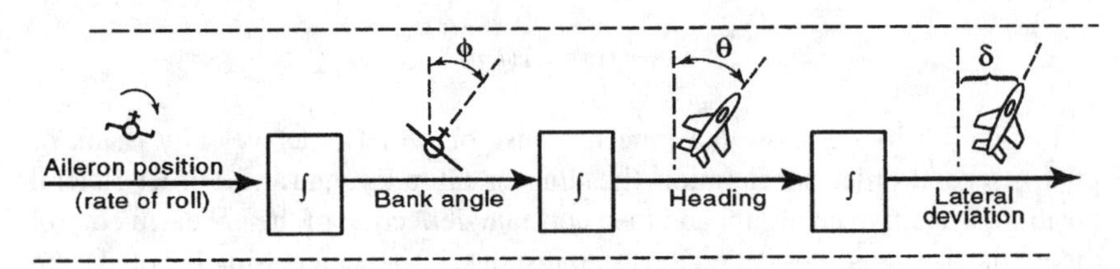

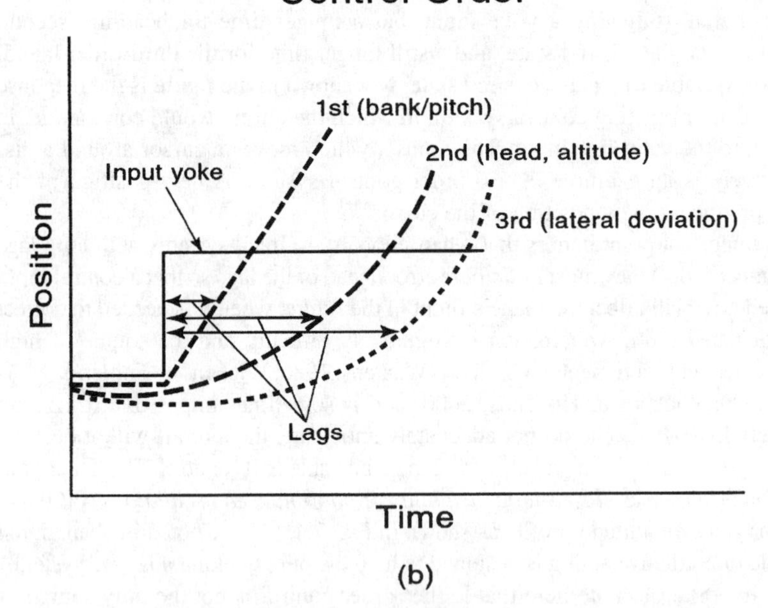

FIG. 7.3. Control order. (a) Control order relations in controlling a car (top) and an aircraft (bottom) in the lateral axis. In response to the control action on the left, each state variable on the right is a progressively higher control order. (b) Illustrates the increasing lag with progressively higher order variables, in response to a step deflection of the control yoke.

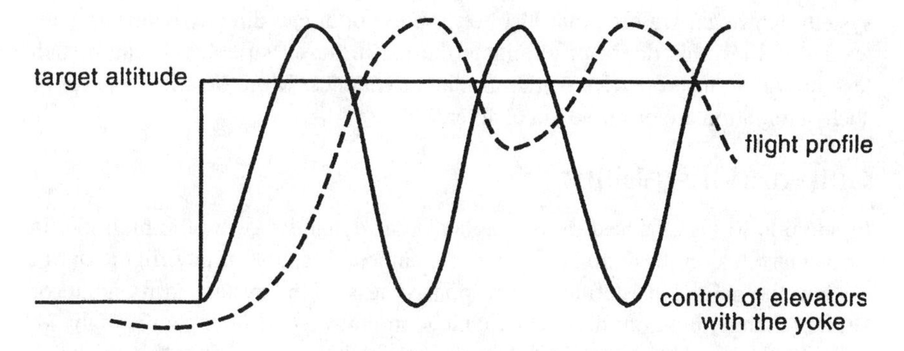

FIG. 7.4. Pilot induced oscillations in the vertical axis.

Anticipation. Given the control order relations inherent in Fig. 7.2, and shown in Fig. 7.3, it is possible for the pilot to use displays of lower order variables in order to predict or anticipate changes in higher order variables and thus reduce cognitive load. Thus the attitude indicator, or artificial horizon, discussed in chapter 5 presents both bank and pitch information and predicts the values of heading and altitude, respectively. The vertical velocity indicator also predicts altitude. The directional gyro (compass rose) depicts heading, and predicts the lateral deviations. As might be expected, such anticipation forms a critical component of flying skill. Consistent with this view, in a study of flight instrument scanning, Bellenkes et al. (1997) found that experienced flight instructors spent significantly more time fixating on the vertical speed indicator (a predictor of future altitude) than did novice student pilots.

Design solutions. To help pilots cope with the lags inherent in the higher order dynamics of aviation control, two sorts of solutions have been imposed: predictive displays and automation. Predictive displays are those that offer explicit prediction of where the future state of the variables will be, so that the pilot is not required to derive or estimate these from other instruments. As noted in chapter 5, these predictions must also be coupled with preview of where the aircraft should be (i.e., the commanded flight path, such as the "pathway in the sky").

Automation involves more complex computer-mediated flight controls or autopilots, which enable the pilot to directly control higher level variables. Many advanced aircraft, for example, have a *mode control panel,* which allows a pilot to directly set in a desired heading or vertical velocity (rather than using the yoke to achieve this through higher order control). This is equivalent to directly "crossing the bridges" in Fig. 7.2(b) at higher levels of the hierarchy; for example, if the goal is to obtain a heading, the pilot just "dials in" a heading on the mode control panel, and the aircraft autopilot will adjust ailerons and bank until the new heading is obtained. This makes heading control essentially a 0 order, position control system. However, we note that just because the pilot can directly control higher level variables, this does not eliminate the lag in those variables, because such lags are inherent properties of the airplane dynamics. More detailed aspects of such automation are discussed in chapter 9.

Gain and Instability

In addition to lags, caused by the higher order dynamics or by the high inertia (mass) of the aircraft, there is a second fundamental feature of any flight control system, the gain. This defines the responsiveness of the system to its input, or D0/DI (change in output divided by change in input). When the gain is high, we say the system is "responsive"; when the gain is low, we say it is "sluggish." A high-gain aircraft, such as an F-16 or Cessna, is highly maneuverable. A given deflection of the yoke will cause a fairly rapid roll or pitch rate. A low-gain aircraft, such as a Boeing 747 or a C5a Transport, is less maneuverable. The same angle of control deflection will lead to a slower roll or pitch rate (Fig. 7.5).

Just as we can talk of the gain of an aircraft system (its responsiveness) as the relationship of $u(t)$ to $o(t)$ in Fig. 7.1 [$o(t)/u(t)$], so we can also speak of the gain of the *pilot,* as characterizing how large a control response the pilot will deliver to the yoke when perceiving a given error or discrepancy between actual and desired variable (the relationship of $e(t)$ to $u(t)$ in Fig. 7.1 [$u(t)/e(t)$]). Generally, pilots adaptively *compensate* their pilot gain to the gain of the aircraft flight dynamics, so that higher gains of the aircraft lead the pilot to adopt lower gains (smaller yoke deflections in response to perceived errors; McRuer & Jex, 1967). An important concept in flight dynamics is the *combined* gain of the pilot and aircraft

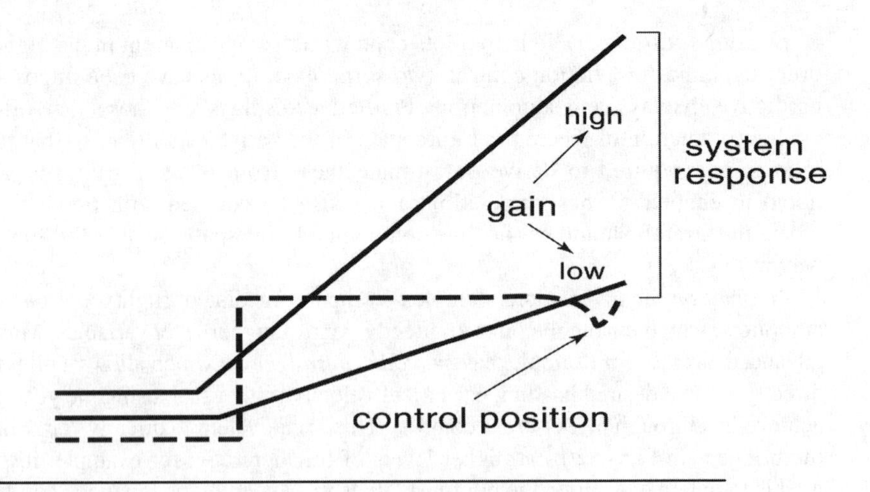

FIG. 7.5. The response of two first-order systems, differing in gain, to the step deflection of the control stick.

together, that is, the size of a change in aircraft state ($o(t)$ in Fig. 7.1) in response to an error perceived by the pilot ($e(t)$). This combined gain is sometimes referred to as the open-loop gain (Wickens, 1986; Wickens & Hollands, 2000).

A high open-loop gain in combination with the lags (which are imposed by higher control order and high inertia) are the two components that are together responsible for *closed-loop instability.* Control theory principles can predict the particular high values of gain *and* lag together that will produce unstable oscillatory flight control performance (PIOs) shown in Fig. 7.4 (see also chap. 8). Furthermore, psychological models of human performance can predict when the aircraft lags are so great that pilot anticipation becomes too difficult and, as a result, makes the airplane unflyable (McRuer, 1980). These issues become extremely important as new aircraft are certified and as experimental prototypes are flight tested (National Research Council, 1997). These issues of predictive models of pilot-in-the-loop flight control are discussed in more detail in chapter 8 and in Wickens and Hollands (2000).

Simplifications

The previous discussion has simplified the presentation of flight dynamics in certain ways. Most importantly, in the actual aircraft there is considerable *cross-coupling* between the three axes (lateral vertical, longitudinal), as shown in Fig. 7.6. For example, banking the aircraft will cause it to pitch downward, and any downward pitch will cause it to increase airspeed, just as any upward pitch will cause it to lose airspeed. Correspondingly, any increase in airspeed of a nonlevel aircraft will cause an increase in its upward or downward vertical velocity. Changes in airspeed can have somewhat unpredictable effects on the gain of both vertical and lateral control. Finally, the control order relations for pitch and bank shown in Fig. 7.3 only hold through a restricted range of values. For example, a constant yoke displacement will only lead the aircraft to pitch down (or up) at a constant rate for a limited time. An additional simplification in this presentation is the absence of rudders. These are generally used to make turns in a *coordinated* fashion, such that the centrifugal force during the turn is exerted directly parallel to the tail (at right angles to the wings).

Positive Feedback

The previous section has described the negative feedback closed tracking loop and has emphasized lag as the major source of difficulty within that context, imposing high workload and possibly leading to negative feedback instability. In contrast, there is another source of difficulty in some aircraft, particularly helicopters or "rotorcraft," and some unmanned air vehicles (UAVs) related to positive feedback. To provide a context for understanding positive feedback systems, recall that in a negative feedback system, the correction is issued in the opposite direction of the error (i.e., "negating" the error). As we noted, in many aircraft, this negative feedback will occur even if the pilot is removed from the control loop in Fig. 7.1.

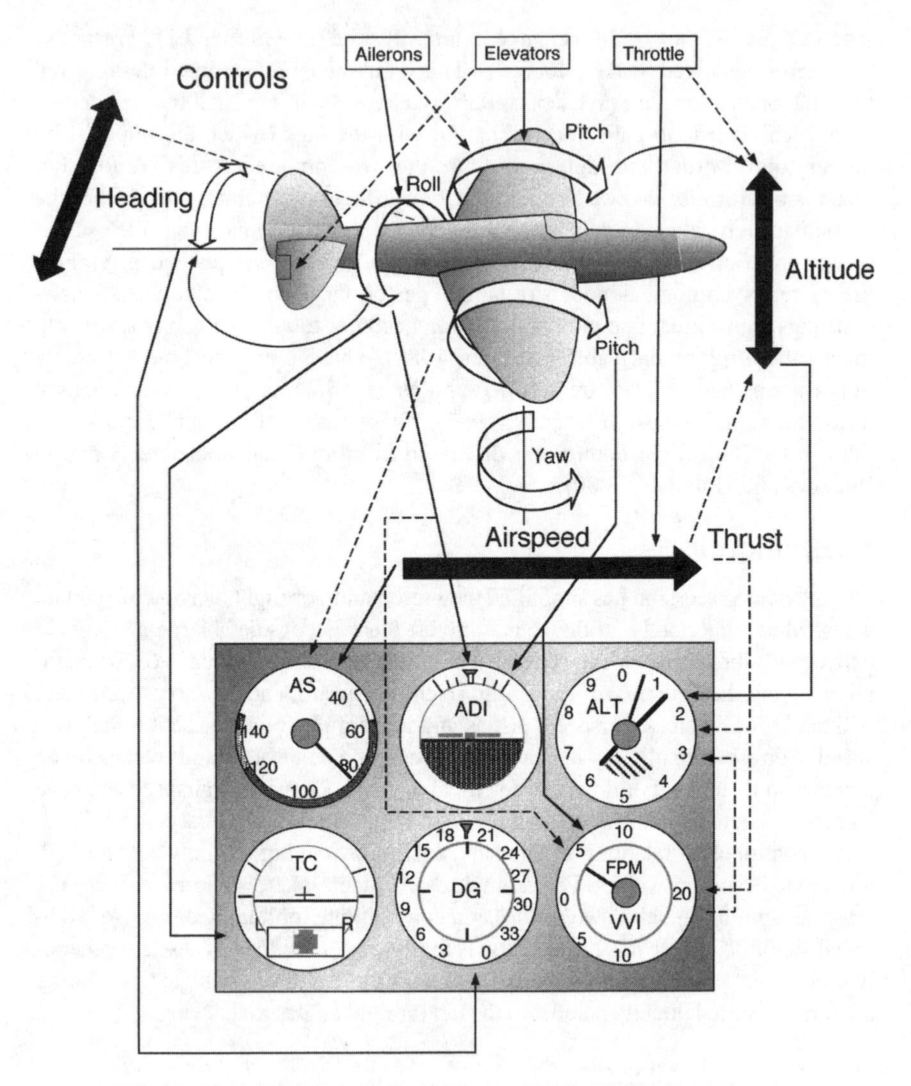

FIG. 7.6. Schematic representation of the causal influences in flight dynamics that underlie variation in flight instrument readings. Control inputs from the yoke and the throttle are shown at the top. The solid thin lines represent sources of influence within a primary flight axes (lateral, vertical, longitudinal). The dashed lines represent cross-coupling influence between axes. *Note.* From "Visual Scanning and Pilot Expertise: The Role of Attentional Flexibility and Mental Model Development," by A. H. Bellenkes, C. D. Wickens, and A. F. Kramer, 1997, *Aviation, Space, and Environmental Medicine, 68,* p. 570. Copyright 1997 by the Aerospace Medical Association. Reprinted with permission.

In contrast, in a system with *positive feedback,* an error will become larger, rather than diminishing. This situation might characterize a boat with heavy ice high on the mast or indeed any pole that is balanced. Once it begins to fall, it will fall progressively faster, unless corrected. Many helicopters during hover are characterized by such positive feedback. If left uncontrolled, a small "tilt" will magnify (Hart, 1988), creating what may be described as *positive feedback instability.* Such aircraft dynamics are extremely demanding of pilot visual attention, for the simple reason that the pilot cannot ignore any existing error, because it will magnify itself if left unattended. Continuous monitoring and corrections are required. Such systems are natural candidates for "stability augmentation" automation, as is being imposed in many advanced helicopters.

Adaptive Flight Control

One feature of many aircraft is that the particular nature of the flight dynamics—the gain, the amount of lag along certain axes at certain frequencies, or even the degree of positive versus negative feedback—may change with conditions of flight. For example, gain may change as airspeed is increased. Lag will increase as ice builds up on the wings, and helicopters will change their flight dynamics from inherently stable negative feedback systems to less-stable positive feedback systems as they transition from forward flight to hover. The tilt rotor aircraft shows large changes in dynamics from forward flight to vertical flight mode. How effectively the pilot adapts to these changes is characterized by the study of *adaptive manual control* (Kelley, 1968; Young, 1969).

The general finding is that pilots adapt relatively smoothly, and sometimes even unconsciously, to these changes, as long as they are gradual and/or anticipated (for example, the lower responsiveness or gain of an aircraft as its airspeed decreases). However, a critical aspect of adaptive manual control relates to the pilots' ability to adapt to unexpected and sudden changes, for example, a gradual buildup of ice is suddenly removed, heavy baggage shifts position in flight, a stability augmentation device suddenly fails, or the elevator or stabilizer suddenly is jammed in a fixed position. Evidence suggests that, even when the postfailure dynamics are controllable, it still may take 5 to 7 seconds before the pilot has successfully adapted to the new dynamics (Wickens, 2001; Wickens & Kessel, 1981). During this adaptive period, if control is exercised on the basis of the old prefailure dynamics (so the pilot actions are now inappropriate for the changed flight dynamics), there is obviously a possibility of an uncontrollable aircraft, leading to a crash. Such a situation points to the need for extensive training at transition points, where fixed changes in dynamics can be expected. It also suggests the desirability of feedback systems to warn pilots of any unanticipated changes in dynamics that might result from factors such as icing (Sarter & Schroeder, 2001).

Flight Control and Task Management

As we have noted, controlling the loop depicted in Figs. 7.1 and 7.2 may be considered either an *aviate* task or one of *navigating,* depending on the extent to which attention is directed toward lower level goals at the bottom of the hierarchy (attitude and speed control to prevent stall) or higher level ones. Thus, the distinction between aviation and navigation is a fuzzy one. Three sources of information can, however, be employed to infer a pilot's direction of attention between these goals and, hence, their task management strategies. As noted earlier in this chapter, visual attention can be monitored through eye scanning techniques to estimate the extent to which, for example, the attitude indicator is fixated (suggesting attention to aviation) rather than instruments depicting higher level navigational information, such as the compass or altimeter (Bellenkes et al., 1997; Wickens et al., 2001). Second, large errors or deviations in some flight parameters can sometimes suggest "neglect" of the axis in question. For example, a large altitude error would suggest vertical neglect. Third, the measurement of *control activity* (often estimated by the mean absolute control stick velocity) provides a fairly direct estimate of the degree of attention a pilot is allocating to the vertical or lateral axis of control (see also Wickens & Gopher, 1977, and chap. 14).

CONCLUSION

Despite the fact that automation is effectively "lowering the order" of control in many aircraft, an understanding of the nature of flight dynamics and how they impose limits on human information processing remains of critical importance. Autopilots may fail, or fail to perform as anticipated, requiring the pilot to hand fly the aircraft, and for a long time to come, students will continue to learn flight dynamics of the nonautomated aircraft. In this chapter, we have provided a fairly intuitive and psychological description of the nature of this flight control. The following chapter describes many of these same issues from the more formal and rigorous perspective of control engineering.

REFERENCES

Adams, M. J., Tenney, Y. J., & Pew, R. W. (1995). Situation awareness and the cognitive management of complex systems. *Human Factors, 37,* 85–104.

Bellenkes, A. H., Wickens, C. D., & Kramer, A. F. (1997). Visual scanning and pilot expertise: The role of attentional flexibility and mental model development. *Aviation, Space, and Environmental Medicine, 68,* 569–579.

Bresley, B. (1995, Apr.–Jun.). 777 flight deck design. *Airliner,* 1–10.

Chou, C., Madhavan, D., & Funk, K. (1996). Studies of cockpit task management errors. *International Journal of Aviation Psychology, 6,* 307–320.

Damos, D. L. (1978). Residual attention as a predictor of pilot performance. *Human Factors, 20,* 435–440.

Damos, D. L. (Ed.), (1991). *Multiple-task performance.* London: Taylor & Francis.

Damos, D. L. (1997). Using interruptions to identify task prioritization in Part 121 air carrier operations. *Proceedings of the Ninth International Symposium on Aviation Psychology* (pp. 871–876). Columbus: The Ohio State University.

Degani, A., & Wiener, E. L. (1993). Cockpit checklists: Concepts, design and use. *Human Factors, 35,* 345–359.

Dismukes, K. (2001). The challenge of managing interruptions, distractions, and deferred tasks. *Proceedings of the 11th International Symposium on Aviation Psychology.* Columbus: The Ohio State University.

Farrell, P. S. E., Cain, B., Sarkar, P., & Ebbers, R. (1999). Direct voice versus manual input technology for aircraft CDU operations. In *Proceedings of the Tenth International Symposium on Aviation Psychology* (pp. 210–215). Columbus: The Ohio State University.

Funk, K. H., II. (1991). Cockpit task management: Preliminary definitions, normative theory, error taxonomy, and design recommendations. *International Journal of Aviation Psychology, 1,* 271–286.

Funk, K. H., II, & Braune, R. (1997). Expanding the functionality of existing airframe systems monitors: The agenda manager. *Proceedings of the Ninth International Symposium on Aviation Psychology* (pp. 887–892). Columbus: The Ohio State University.

Gopher, D. (1992). The skill of attentional control: Acquisition and execution of attention strategies. In S. Kornblum & D. E. Meyer (Eds.), *Attention and performance, XIV* (pp. 299–322). Cambridge, MA: MIT Press.

Gopher, D., Weil, M., & Bareket, T. (1994). Transfer of skill from a computer game trainer to flight. *Human Factors, 36,* 387–405.

Groce, J. L, & Boucek, G. P., Jr. (1987). *Air transport crew tasking in an ATC data link environment* (SAE Technical Paper Series 871764). Warrendale, PA: Society of Automotive Engineers.

Hammer, J. M. (1999). Human factors of functionality and intelligent avionics. In D. J. Garland, J. A. Wise, and V. D. Hopkin (Eds.), *Handbook of human factors in aviation* (pp. 549–565). Mahwah, NJ: Lawrence Erlbaum Associates.

Hart, S. G. (1988). Helicopter human factors. In E. Weiner & D. Nagel (Eds.), *Human factors in aviation* (pp. 591–638). San Diego, CA: Academic Press.

Hart, S. G., & Wickens, C. D. (1990). Workload assessment and prediction. In H. R. Booher (Ed.), *MANPRINT: An approach to systems integration* (pp. 257–296). New York: Van Nostrand Reinhold.

Haskell, I. D., & Wickens, C. D. (1993). Two- and three-dimensional displays for aviation: A theoretical and empirical comparison. International Journal of Aviation Psychology, 3, 87–109.

Hawkins, F. H. (1993). *Human factors in flight.* Brookfield, VT: Ashgate.

Helleberg, J., & Wickens, C. D. (2001). Effects of data link modality on pilot attention and communication effectiveness. *Proceedings of the 11th International Symposium on Aviation Psychology.* Columbus: The Ohio State University.

Kelley, C. R. (1968). *Manual and automatic control.* New York: Wiley.

Kerns, K. (1999). Human factors in ATC-flight deck integration: Implications of datalink research. In D. J. Garland, J. A. Wise, & V. D. Hopkin (Eds.), *Handbook of aviation in human factors* (pp. 519–546). Mahwah, NJ: Lawrence Erlbaum Associates.

Latorella, K. A. (1996). Investigating interruptions: An example from the flightdeck. *Proceedings of the 40th Annual Meeting of the Human Factors and Ergonomics Society* (pp. 249–253). Santa Monica, CA: Human Factors and Ergonomics Society.

Latorella, K. (1998). Effects of modality on interrupted flight deck performance: Implications for data link. *Proceedings of the 42nd Annual Meeting of the Human Factors and Ergonomics Society* (pp. 87–91). Santa Monica, CA: Human Factors and Ergonomics Society.

Laudeman, I. V., & Palmer, E. A. (1995). Quantitative measurement of observed workload in the analysis of aircrew performance. *International Journal of Aviation Psychology, 5,* 187–198.

Liebowitz, H. (1987). The human senses in flight. In E. Weiner & D. Nagel (Eds.), *Human factors in aviation* (pp. 83–110). San Diego, CA: Academic Press.

McRuer, D. T. (1980). Human dynamics in man-machine systems. *Automatica, 16,* 237–253.

McRuer, D. T., & Jex, H. R. (1967). A review of quasi-linear pilot models. *IEEE Transactions on Human Factors in Electronics, 8,* 231.

Moray, N., Dessouky, M. I., Kijowski, B. A., & Adapathya, R. (1991). Strategic behavior, workload, and performance in task scheduling. *Human Factors, 33,* 607–629.

National Research Council. (1997). *Aviation safety and pilot control.* Washington, DC: National Academy Press.

Navarro, C., & Sikorski, S. (1999). Datalink communication in flight deck operations: A synthesis of recent studies. *International Journal of Aviation Psychology, 9,* 361–376.

North, R. A., & Riley, V. A. (1989). W/INDEX: A predictive model of operator workload. In G. R. McMillan, D. Beevis, E. Salas, M. H. Strub, R. Sutton, & L. Van Breda (Eds.), *Applications of human performance models to system design* (Vol. 1, pp. 81–89). New York: Plenum Press.

Orasanu, J., & Fischer, U. (1997). Finding decisions in natural environments: The view from the cockpit. In C. E. Zsambok & G. Klein (Eds.), *Naturalistic decision making* (pp. 343–357). Mahwah, NJ: Lawrence Erlbaum Associates.

Orlady, H., & Orlady, L. M. (1999). *Human factors in multicrew flight operations.* Brookfield, VT: Ashgate.

Palmer, E., & Degani, A. (1991). Electronic checklists: Evaluation of two levels of automation. *Proceedings of the Sixth International Symposium on Aviation Psychology* (pp. 178–183). Columbus: The Ohio State University.

Predmore, S. C. (1991). Micro-coding of cockpit communications in accident investigation: Crew coordination in United 811 and United 232. *Proceedings of the Sixth International Symposium on Aviation Psychology* (pp. 350–355). Columbus: The Ohio State University.

Previc, F. H. (2000, Mar.–Apr.). Neuropsychological guidelines for aircraft control stations. *IEEE Engineering in Medicine and Biology,* 81–88.

Raby, M., & Wickens, C. D. (1994). Strategic workload management and decision biases in aviation. *International Journal of Aviation Psychology, 4,* 211–240.

Reason, J. (1990). *Human error.* New York: Cambridge University Press.

Sarno, K. J., and Wickens, C. D. (1995). The role of multiple resources in predicting time-sharing efficiency: An evaluation of three workload models in a multiple task setting. *International Journal of Aviation Psychology, 5,* 107–130.

Sarter, N., & Schroeder, B. (2001). Supporting decision-making and action selection under time pressure and uncertainty: The case of inflight icing. *Human factors, 43,* 573–583.

Schneider, W. (1985). Training high performance skills. *Human Factors, 27,* 285–300.

Schutte, P. C., & Trujillo, A. C. (1996). Flight crew task management in non-normal situations. In *Proceedings of the 40th Annual Meeting of the Human Factors and Ergonomics Society* (pp. 244–248). Santa Monica, CA: Human Factors and Ergonomics Society.

Shappell, S. A., & Wiegmann, D. (1997). Why would an experienced aviator fly a perfectly good aircraft into the ground? In *Proceedings of the Ninth International Symposium on Aviation Psychology* (pp. 26–32). Columbus: The Ohio State University.

Sklar, A., & Sarter, N. B. (1999). Good vibrations: Tactile feedback in support of attention allocation and human-automation coordination in event-driven domains. *Human Factors, 41,* 543–552.

Stanton, N. (1994). *Human factors of alarm design.* London: Taylor & Francis.

Tsang, P. S., Velazquez, V. L., & Vidulich, M. A. (1996). Viability of resource theories in explaining time-sharing performance. *Acta Psychologica, 91,* 175–206.

Tsang, P. S., & Wickens, C. D. (1988). The structural constraints and strategic control of resource allocation. *Human Performance, 1,* 45–72.

Wickens, C. D. (1986). The effects of control dynamics on performance. In K. R. Boff, L. Kaufman, & J. P. Thomas (Eds.), *Handbook of perception and performance* (Vol. 2, pp. 39–1–39–60). New York: Wiley.

Wickens, C. D. (1991). Processing resources and attention. In D. Damos (Ed.), *Multiple-task performance* (pp. 3–34). London: Taylor & Francis.

Wickens, C. D. (2001). Keynote address: Attention to safety and the psychology of surprise. *Proceedings of the 11th International Symposium on Aviation Psychology.* Columbus: The Ohio State University.

Wickens, C. D. (2002). Multiple Resources. *Theocretical Issues in Ergonomic Science.*

Wickens, C. D., Gempler, K., & Morphew, M. E. (2000). Workload and reliability of predictor displays in aircraft traffic avoidance. *Transportation Human Factors Journal, 2*(2), 99–126.

Wickens, C. D., & Gopher, D. (1977). Control theory measures of tracking as indices of attention allocation strategies. *Human Factors, 19,* 249–366.

Wickens, C. D., & Hollands, J. G. (2000). *Engineering psychology and human performance* (3rd ed.). Upper Saddle River, NJ: Prentice Hall.

Wickens, C. D., & Kessel, C. (1981). Failure detection in dynamic systems. In J. Rasmussen and W. Rouse (Eds.), Human detection and diagnosis of system failures. New York: Plenum Press.

Wickens, C. D., & Liu, Y. (1988). Codes and modalities in multiple resources: A success and a qualification. *Human Factors, 30,* 599–616.

Wickens, C. D., Mavor, A. S., & McGee, J. P. (Eds.) (1997). *Flight to the future: Human factors in air traffic control.* Washington, DC: National Academy Press.

Wickens, C. D., Sandry, D., & Vidulich, M. (1983). Compatibility and resource competition between modalities of input, output, and central processing. *Human Factors, 25,* 227–248.

Wickens, C. D., Xu, X., Helleberg, J. R., & Marsh, R. (2001). Pilot visual workload and task management in freeflight: A model of visual scanning. *Proceedings of the 11th International Symposium on Aviation Psychology.* Columbus: The Ohio State University.

Wiener, E. L. (1977). Controlled flight into terrain accidents: System-induced errors. *Human Factors 19,* 171–181.

Young, L. R. (1969). On adaptive manual control. *Ergonomics, 12,* 635–675.

8

Pilot Control

Ronald A. Hess
University of California

THE AIRCRAFT CONTROL PROBLEM

The majority of aircraft now being flown and those planned for the near future will involve active pilot control for at least a part of their intended mission. *Control* means maintaining the aircraft along a desired trajectory in three-dimensional space. To accomplish this task, the pilot must exert control over the six degrees of freedom that the aircraft possesses. The six degrees of freedom are the six independent coordinates required to uniquely specify the position of any point in the aircraft with respect to the origin of a reference frame in the earth. These six are the three coordinates needed to specify the location of the center of mass of the aircraft and three angular rotations necessary to specify the orientation of the aircraft, all with respect to the reference frame fixed in the earth. These coordinates are shown in Fig. 8.1.

An aircraft is a dynamic system whose motion in three-dimensional space is described by differential equations, obtained by application of Newton's second law of motion. This law describes how the linear and angular accelerations of the aircraft depend on the applied forces and moments. A force is most easily described as a push or pull whereas a moment is a force applied through a distance or moment arm. Linear acceleration describes the rate of change of the

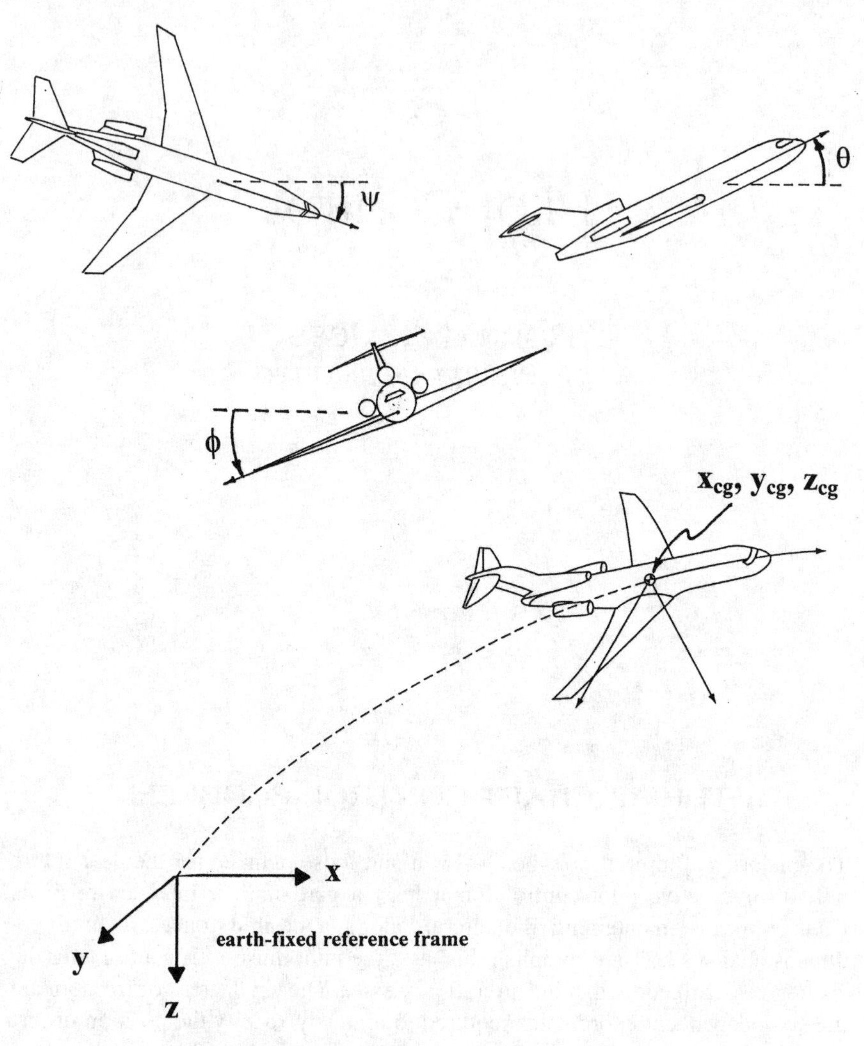

FIG. 8.1. The six degrees of freedom of an aircraft. The three values x_{cg}, y_{cg}, z_{cg} represent the coordinates of the aircraft center of gravity relative to a reference frame fixed in the earth. The remaining degrees of freedom (ψ, θ, φ) describe the angular orientation of the aircraft relative to the earth-fixed frame.

linear velocity of the object's center of mass, and angular acceleration describes the rate of change of angular velocity. The latter describes the rate at which the angular orientation of the aircraft is changing. The forces and the moments that an aircraft experiences can be categorized as aerodynamic, propulsive, and gravitational, with the first two being created by the aircraft's motion through the atmosphere. The pilot modulates the aerodynamic and propulsive forces on the aircraft and thus controls the aircraft's degrees of freedom and its movement in three-dimensional space. The aerodynamic forces are modulated through the deflection of aerodynamic control surfaces, such as the elevator, ailerons, and rudder. The propulsive forces are modulated through the control of fuel flow to the engine. See Fig. 8.2. The pilot effects these changes through manipulators in the cockpit, referred to as *inceptors*. In a typical cockpit, these inceptors consist of a column or wheel for elevator and aileron movement, foot pedals for rudder movement, and throttles to command fuel flow to the engine(s).

Aircraft control is usually described in terms of stabilization, regulation against disturbances, and maneuvering. Stabilization refers to maintaining the aircraft in an equilibrium state in which all the aerodynamic, propulsive, and gravitational forces and moments acting on the aircraft sum to zero. An aircraft in an equilibrium condition will be in steady, wings level flight at constant speed. Depending on the aircraft's physical configuration, the flight condition (altitude

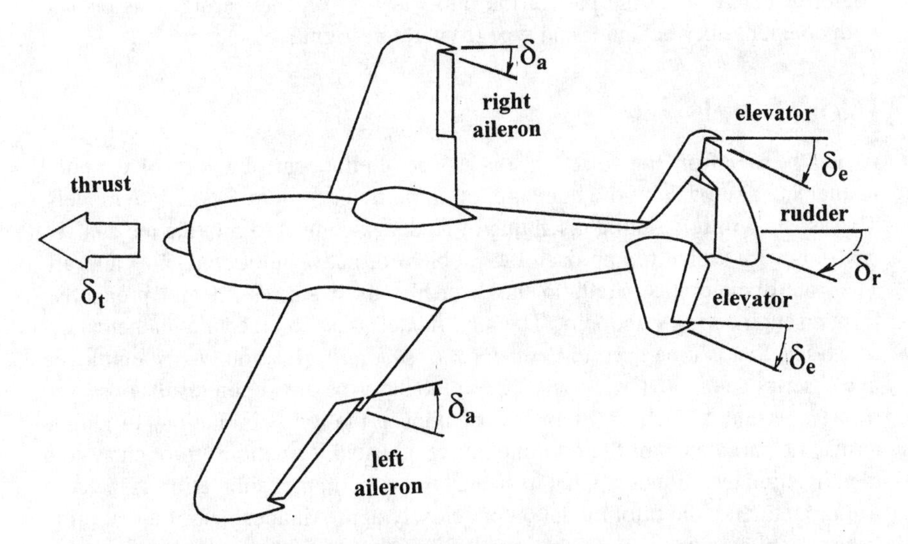

FIG. 8.2. Examples of aerodynamic and propulsive control effectors for an aircraft. In many aircraft, particularly general-aviation vehicles, the pilot directly controls the effectors from the cockpit. In high-performance aircraft such as fighters, a flight-control computer also contributes to the effector motion.

and Mach number), and the existence or absence of some form of artificial stability augmentation, the pilot's task of stabilization can range from easy to extremely difficult. The relative ease or difficulty of stabilization is one important part of the aircraft's *handling qualities*. In the early days of flight, stabilization itself constituted the lion's share of the pilot's control activity. For example, the Wright Flyer was an inherently unstable aircraft that required continuous pilot control inputs to maintain equilibrium flight.

Disturbance regulation refers to countering the effects of atmospheric turbulence on equilibrium flight. It is quite possible for an aircraft to be easily stabilized by the pilot in still air, but difficult to stabilize in turbulent air. The difficulty of pilot control in turbulence again depends on the aircraft configuration, the flight condition, and the existence or absence of artificial stability augmentation.

Maneuvering defines situations in which (a) the equilibrium condition of the aircraft is being changed, for example, when an aircraft is changing equilibrium altitude and Mach number, or (b) large, rapid changes in the position and orientation of the aircraft are occurring, for example, in air combat maneuvers or in aerobatics.

Stabilization and disturbance regulation tasks encompass a type of control activity in which the pilot is behaving much like an inanimate feedback control device (often referred to as *compensatory* behavior), employing little or nothing in the way of preprogrammed responses (often referred to as *precognitive* behavior). Maneuvering flight, on the other hand, can entail either compensatory or precognitive behavior on the part of the pilot. Type (a) maneuvering is associated with compensatory behavior and type (b) with precognitive behavior.

Pilot Models

As will be seen from the outset, the discussion of pilot control will revolve around mathematical models of the human pilot, in particular, *control–theoretic* models. That is, we will rely on the discipline of feedback control to provide the analytical tools with which to approach the problem of active pilot control of aircraft. The control–theoretic paradigm has been the most successful and productive from an engineering standpoint. The pilot models thus obtained are mathematical constructs that can be used in *descriptive* or *predictive* fashion. When employed in a descriptive mode, the models are used to describe or explain results observed from experiment, such as flight test or pilot-in-the-loop simulation. In simple terms, the parameters of the pilot model are adjusted, sometimes through system identification techniques, so that in an analysis or computer simulation of the control task at hand, the pilot model output closely approximates that of a real pilot. When employed in a predictive mode, the pilot models are used to predict pilot/aircraft performance and handling qualities in the absence of corroborative experimental data. Models used in descriptive fashion help the engineer in understanding human pilot behavior in various tasks, whereas models used in predictive fashion help the engineer in the design and/or improvement of the aircraft and its subsystems.

Chapter Overview

The following section will begin with a review of the feedback paradigm and how control–theoretic models of continuous pilot behavior can be viewed in this framework. Next, some control–theoretic models will be reviewed and their utility discussed. An example of a pilot/aircraft analysis follows (referred to as pilot/vehicle analysis herein), using one of the modeling approaches that have been developed. Next follows a discussion of automation and its impact on pilot control. The chapter ends with concluding remarks and future research directions in the discipline.

THE FEEDBACK PARADIGM IN PILOT CONTROL

Mathematical Preliminaries and Inner-Loop Pilot Control

Differential Equations. Fig. 8.3 is a simplified representation of the flight control elements typically found in a modern aircraft. Here, a pilot is attempting to control the aircraft's pitch attitude, that is, the angle the aircraft's longitudinal axis makes with the horizontal (one of the three rotational degrees of freedom of Fig. 8.1). To accomplish this control task, visual and possibly *vestibular cues* regarding the aircraft's motion are being employed, where vestibular cues refer to whole-body motion cues. All of these cues are continually processed by the pilot to produce inputs to the cockpit inceptor, for example, control wheel or column. In most modern aircraft, the pilot's inceptor input is electronically transmitted to the *flight control computer,* with the resulting control system referred to as *fly-by-wire.* The flight control computer is typically a digital device that also receives information from various onboard sensors. These latter devices, like the pilot, sense aircraft motion such as aircraft pitch attitude and its rate of change. Using a programmed *flight control law,* the flight control computer operates on the inputs from the pilot and sensors and produces a command to an actuator, typically an electrohydraulic device that moves an aerodynamic surface or control effector. The motion of this effector changes the aerodynamic forces on the aircraft and thus influences the motion of the aircraft.

Because the pilot's cockpit inceptors are responsible for changes in the aerodynamic and propulsive forces and moments on the aircraft, the relationships between the pilot's inceptor inputs and the resultant motion of the aircraft are also

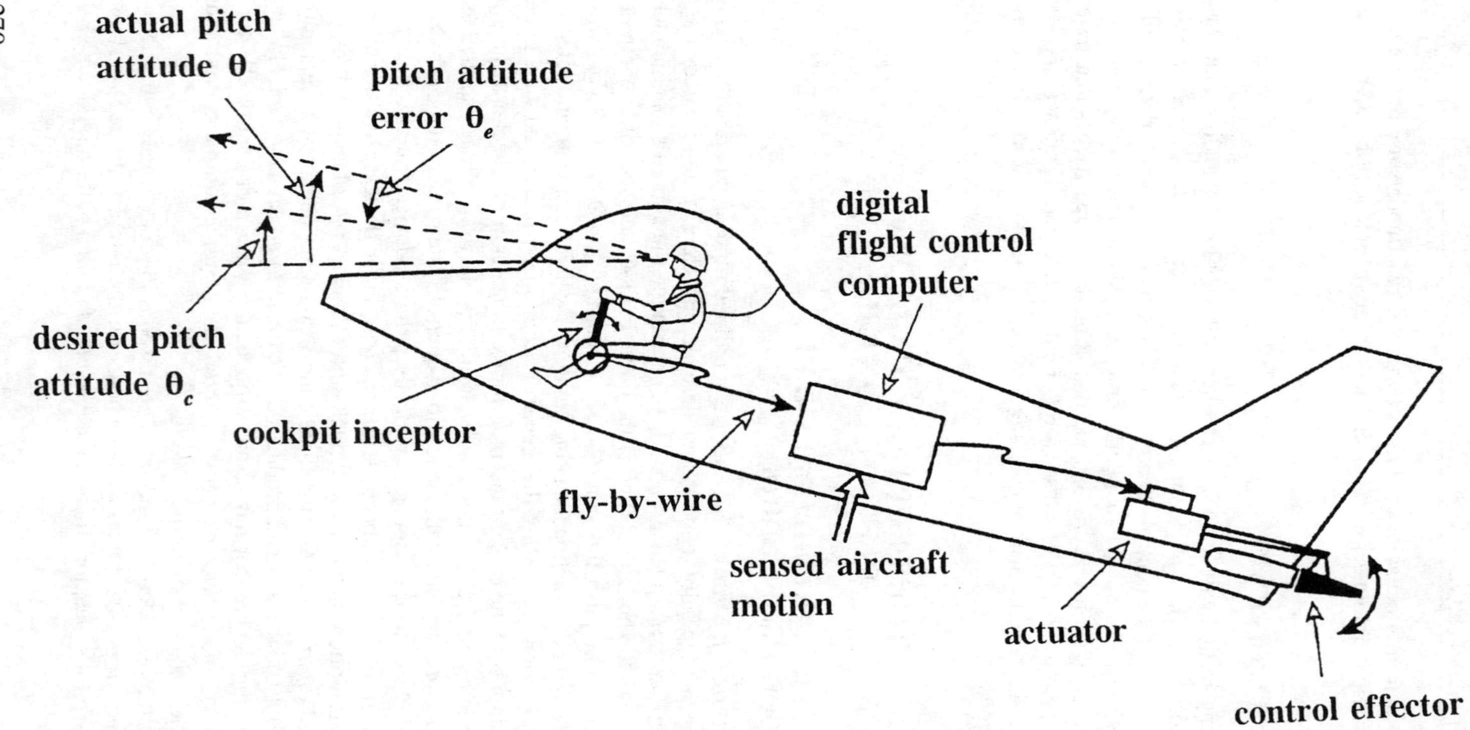

FIG. 8.3. The primary elements of a modern fly-by-wire flight control system. The pilot controls the vehicle by means of electrical commands to the flight-control computer and actuators for the aerodynamic and propulsive control effectors. *Note.* From *Aviation Safety and Pilot Control—Understanding and Preventing Unfavorable Pilot-Vehicle Interactions* (p. 119), by Committee on the Effects of Aircraft-Pilot Coupling on Flight Safety, 1997, Washington, DC: National Academy Press. Copyright 1997 by the National Academy of Sciences. Reprinted with permission.

described by differential equations. As an example, consider the control of aircraft pitch attitude exemplified in Fig. 8.3. We can examine three "stereotypical" types of differential equations that might describe the relationship between cockpit inceptor input, $\delta(t)$, and the pitch attitude of the aircraft, $\theta(t)$. These different relations can be created through the use of artificial stabilization provided by the flight control computer of Fig. 8.3.

The simplest relationship isn't really a differential equation at all but merely an equivalence given by

$$\theta(t) = K\delta(t) \tag{1}$$

In the parlance of aircraft stability and control engineers, Equation 1 describes an *attitude command* system in which inceptor input $\delta(t)$ creates proportional vehicle attitude changes $\theta(t)$, with $1/K$ being the constant of proportionality. Although basic aircraft response characteristics rarely, if ever, are as simple as that given in Equation 1, the use of artificial feedback stabilization can create an aircraft response that closely approximates this response type. Such systems are common in large transport aircraft.

A second type of differential equation is one given by

$$\frac{d\theta(t)}{dt} = \dot{\theta}(t) = K\delta(t) \tag{2}$$

Here the rate of change of $\theta(t)$ [denoted by the time derivative $d\theta(t)/dt = \dot{\theta}(t)$] is created by inceptor input $\delta(t)$. Equation 2 describes an *attitude-rate command system,* or more simply a rate-command system. Such systems are common in high-performance aircraft such as fighters.

A third type of differential equation is one given by

$$\frac{d^2\theta(t)}{dt^2} = \ddot{\theta}(t) = K\delta(t) \tag{3}$$

Equation 3 describes an *attitude-acceleration command system,* or simply an acceleration command system. Such systems are rarely found on aircraft because, as will be seen, they are difficult for the pilot to control; in other words, the handling qualities are poor.

The Laplace Transform. The Laplace transform is a mathematical tool that allows control system designers to analyze dynamic systems such as aircraft in an efficient manner. As the name implies, the Laplace operation transforms differential equations into algebraic equations, the latter of which are considerably easier to solve than differential equations. The Laplace transform of a time signal such as $\theta(t)$ is denoted $\Theta(s)$ and is formally defined as

$$\theta(s) = \int_0^\infty \theta(t)e^{-st}\,dt \tag{4}$$

The conversion of differential equations into algebraic ones by the Laplace transform can be simplified by the following convenient rule: Differentiation (integration) with respect to time is transformed into multiplication (division) by the Laplace variable s. For example, Equations 1 through 3 become, respectively

$$\theta(s) = K\delta(s) \tag{5}$$

$$s\theta(s) = K\delta(s) \tag{6}$$

$$s^2\theta(s) = K\delta(s) \tag{7}$$

The transformation using the Laplace transform allows the algebraic representation of more complicated differential equations than that given by Equations 1 to 3. For example, consider the differential relation

$$\ddot{\theta}(t) + 3\dot{\theta}(t) + 2\theta(t) = 6\dot{\delta}(t) + \delta(t) \tag{8}$$

In the Laplace domain, Equation 8 becomes

$$\begin{gathered} s^2\theta(s) + 3s\theta(s) + 2\theta(s) = 6s\delta(s) + \delta(s) \\ \text{or} \\ (s^2 + 3s + 2)\theta(s) = (6s + 1)\delta(s) \end{gathered} \tag{9}$$

Algebraic relationships such as those of Equations 5 to 7 and 9 lead to the introduction of the *transfer function,* defined as the Laplace transform of the system output, divided by that of the input (with zero initial conditions). In Equations 5 to 7 and 9, this leads to

$$\begin{gathered} \frac{\theta}{\delta}(s) = K \\ \frac{\theta}{\delta}(s) = \frac{K}{s} \\ \frac{\theta}{\delta}(s) = \frac{K}{s^2} \\ \frac{\theta(s)}{\delta(s)} = \frac{6s + 1}{s^2 + 3s + 2} \end{gathered} \tag{10}$$

Relations like Equation 10 completely summarize in the Laplace domain how the dynamic system in question produces an output or response $\theta(t)$ to an input or control $\delta(t)$ in the time domain. As in Equation 10, transfer functions are almost invariably expressed as ratios of polynomials in the Laplace variable s.

A final transform element needs to be introduced for the sake of completeness, and that is one that deals with the transform of a function of time that has been delayed by some time interval τ. Handling such delays is an important part of modeling the pilot as a control system element because human responses to stimuli are always accompanied by a time delay. The Laplace transform of a function of time, say $\theta(t)$, that has been delayed by τ secs is merely $e^{-\tau s} \cdot \theta(s)$, that is, the term $e^{-\tau s}$ multiplies the Laplace transform of the undelayed time function.

Block Diagrams and Feedback Systems. Although the pictorial representation of Fig. 8.3 is a useful as a heuristic device, a more efficient representation can be given by *block diagrams* such as Fig. 8.4. In diagrams such as these, blocks represent dynamic elements, that is, those that produce continuous outputs in response to continuous inputs. All system elements have been represented by their transfer functions, each of which will be represented as a ratio of polynomials in the Laplace variable *s*, with the possible inclusion of time delay factors $e^{-\tau s}$. The inputs and outputs are represented by the directed line segments entering and leaving the blocks. The various inputs and outputs in Fig. 8.4 are referred to as "signals" and are transformed functions of time; for example, the control effector angular motion is $\delta_e(t)$ in the time domain and $\delta_e(s)$ in the Laplace domain. Signals can be added or subtracted, as shown in Fig. 8.4, just as they are in a physical feedback system. The circular elements containing a cross are referred to as summing junctions, with the accompanying "+" or "−" sign indicating that the signal is being added or subtracted at the junction. Note in Fig. 8.4 that the pilot has been represented as a control system element described by the transfer function $Y_p(s)$. Justification for this representation will be given in a later section.

The system of Fig. 8.4 is referred to a *feedback system,* in that signals are being sensed, fed back, compared with other signals at the summing junctions, and then used as inputs to elements such as the pilot transfer function $Y_p(s)$. The

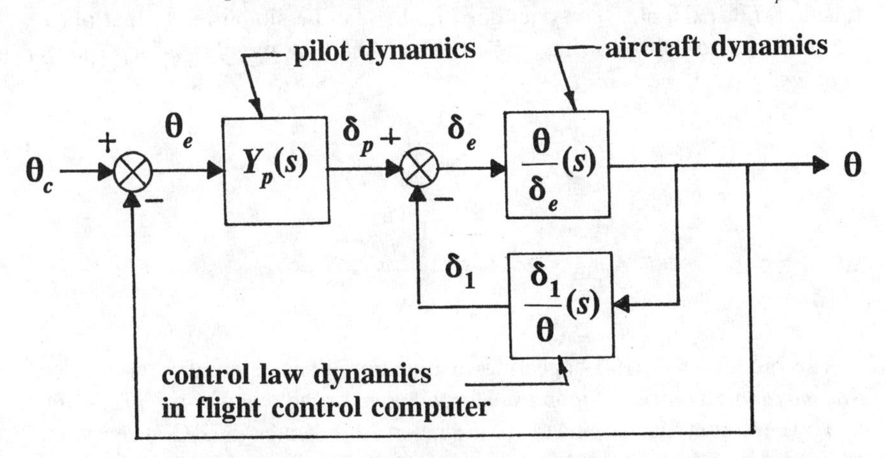

FIG. 8.4. A block diagram representation of a pilot/vehicle system. System elements, including the pilot, are represented by transfer functions. The control law dynamics can represent the action of a flight control computer contributing to the stability and control of the aircraft.

physical nature of the signals evident in Fig. 8.4 can vary considerably. For example, the aircraft's pitch attitude θ is shown as an input to the block labeled "control law dynamics" and also shown as an input to the summing junction to be compared with θ_c. In the former case, the pitch attitude is an angular orientation with units of radians and would typically be sensed by an electromechanical transducer such as an attitude gyroscope with an electrical output δ_1 with units of volts. This electrical output is then compared with the output of the pilot's inceptor (not shown), also assumed to have units of volts. The resulting voltage is sent to the actuator (also not shown) that drives the elevator and produces θ as completely described by the transfer function θ/δ_e. The pitch attitude sensing and comparison shown in the outer loop would be accomplished visually by the pilot. The term *closed-loop feedback system* is often applied to systems such as those in Fig. 8.4 to emphasize the fact that feedback loops are in operation. That is, they have been "closed" through the action of sensing devices, either inanimate or animate.

With the introduction of the Laplace transform, a simple algebra can be associated with the block diagrams; namely, the Laplace transform of the output of a dynamic element can be obtained by multiplying the Laplace transform of the input by the transfer function of the element. For example, in Fig. 8.4,

$$\theta(s) = \delta_e(s) \cdot \frac{\theta}{\delta_e}(s) \tag{11}$$

One advantage of the block diagram and transfer function presentation is that complicated, interconnected systems possessing the variety of feedback loops typical of those found in a modern aircraft can be simplified to a single transfer function. For example, the system of Fig. 8.4 can be simplified to that of Fig. 8.5, and then to the following single transfer function, the *closed-loop transfer function.*

$$\frac{\theta}{\theta_c}(s) = \frac{Y_p(s)G(s)}{1 + Y_p(s)G(s)} \tag{12}$$

$$\text{with} \quad G(s) = \frac{\dfrac{\theta}{\delta_e}(s)}{1 + \dfrac{\theta}{\delta_e}(s) \cdot \dfrac{\delta_1}{\theta}(s)} \tag{13}$$

The product $Y_p(s)G(s)$ appearing in Equation 12 is referred to as the *loop transmission,* the *forward-loop transfer function,* or the *open-loop transfer function* by control system engineers. As Equation 12 indicates, $Y_p(s)G(s)$ determines the form of the final, closed-loop transfer function. Equation 12 is a mathematical description of how the actions of the human pilot [now described by $Y_p(s)$] determine the dynamic behavior of the complete pilot/vehicle system (now described by θ/θ_c).

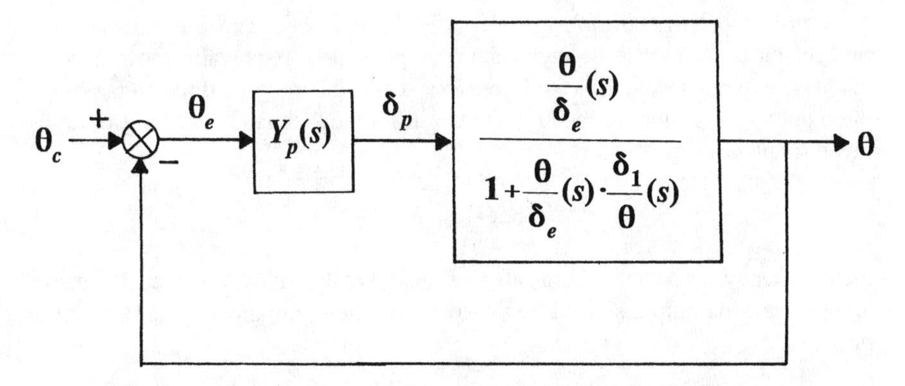

FIG. 8.5. A simplification of the block diagram of Fig. 8.4. Block diagram algebra is used to the represent the dynamics of the inner-feedback loop of Fig. 8.4 by a single transfer function.

Although the transfer function representation of a system is of paramount importance in the analysis of a pilot as an element in a control system, the performance of the pilot/vehicle system as a function of time is of final interest (e.g, $\theta(t)$ as opposed to $\theta(s)$). Once again, the Laplace transform is of value, because $\theta(t)$ can always be obtained from $\theta(s)$ through a straightforward inversion process.

In addition to obtaining continuous functions of time to describe the performance of a pilot/vehicle system, single-performance metrics or performance *norms* are often employed. These norms are typically chosen so that they can be calculated analytically and measured experimentally. One such metric is the *root mean square* (RMS) value of a signal. For example, assume one is interested in how well a pilot is able to control the pitch attitude $\theta(t)$ of an aircraft as it encounters a patch of atmospheric turbulence. The average value of $\theta(t)$ may be of little use, as large excursions in $\theta(t)$ can occur while still having an average value of zero over the time interval of the turbulence encounter. However, the RMS value can provide information regarding the average "strength" of a particular function of time. To obtain the RMS value of a signal, one squares the signal, then integrates this squared value over some time period, say the length of time that the aircraft is exposed to the turbulence patch. The resulting number is then divided by the length of the time interval, and the square root is taken. Alternately, the *mean square value* of a signal can be chosen as a metric. In this case, the square root operation is simply omitted. For the example just cited, assume that the pitch attitude deviation of the aircraft under piloted control can be approximated by $\theta(t) = 10\sin(2t)deg$. Now the average value of this time function calculated over an integral number of cycles of the sine wave is exactly zero. However, the RMS value is 7.07 deg. This latter figure obviously conveys the "strength" of the pitch attitude deviation much more effectively than does the average value.

Graphical Representation of Transfer Functions. The transfer functions of dynamic systems can be conveniently represented graphically by frequency response diagrams, sometimes referred to as Bode diagrams or Bode plots. When the input to a system described by a linear differential equation is a sinusoid of a given frequency, say,

$$\text{input} = A \sin(\omega t) \tag{14}$$

then, in steady state, the output is also a sinusoid at the same frequency. In general, however, the output sinusoid will have a different amplitude B and exhibit a phase lag, γ,

$$\text{output} = B(\omega) \cdot \sin\left[\omega t + \gamma(\omega)\right] \tag{15}$$

As Equation 15 indicates, both B and γ will be functions of the frequency of the input. The frequency response diagram of the transfer functions consists of two plots: One plot is $20 \cdot \log|B(\omega)|$ versus $\log(\omega)$ and one plot is $\gamma(\omega)$ versus $\log(\omega)$. The quantity $20 \cdot \log|B(\omega)|$ is referred to as the magnitude of $B(\omega)$ in *decibels* (dB). Fig. 8.6 shows the frequency response diagrams for the transfer functions from Equation 10 (with $K = 1$). The frequency response diagram may seem to provide limited information, as it is defined only for a single type of input. However, many types of inputs and responses that occur in piloted aircraft control, even random-appearing signals, can be broken down into component frequencies through Fourier analysis. Because the systems in question are assumed to be linear, in other words, described by linear differential equations, the system response to a sum of sinusoids is merely the sum of the responses to the individual sinusoids that make up the input. This property is referred to as *superposition* and is one of the fundamental advantages of linear systems analysis.

One parameter that can conveniently summarize the performance capabilities of a feedback structure in a pilot/vehicle system and that can be obtained from the frequency response diagram of the closed-loop system is the system *bandwidth* (ω_{BW}). The bandwidth is stated in terms of a frequency (rad/sec or cycles/sec) and defines the frequency beyond which the system output or response to a sinusoidal input will no longer follow the input with acceptable fidelity. Referring to Equations 14 and 15, this lack of fidelity means that beyond an input frequency $\omega = \omega_{BW}$, $B(\omega_{BW})$ will be substantially different than A and $\gamma(\omega_{BW})$ will be substantially different than zero. The bandwidth can be obtained from the frequency response diagram of the closed-loop transfer function through the following simple definition: The bandwidth of a closed-loop system is the lowest frequency at which the magnitude portion of the Bode diagram is 3 dB below the 0 dB line. Fig. 8.7 shows such the bandwidth of a hypothetical closed-loop feedback system. For example, Fig. 8.7 might represent the frequency response diagram of the closed-loop transfer function θ/θ_c given in Equation 12.

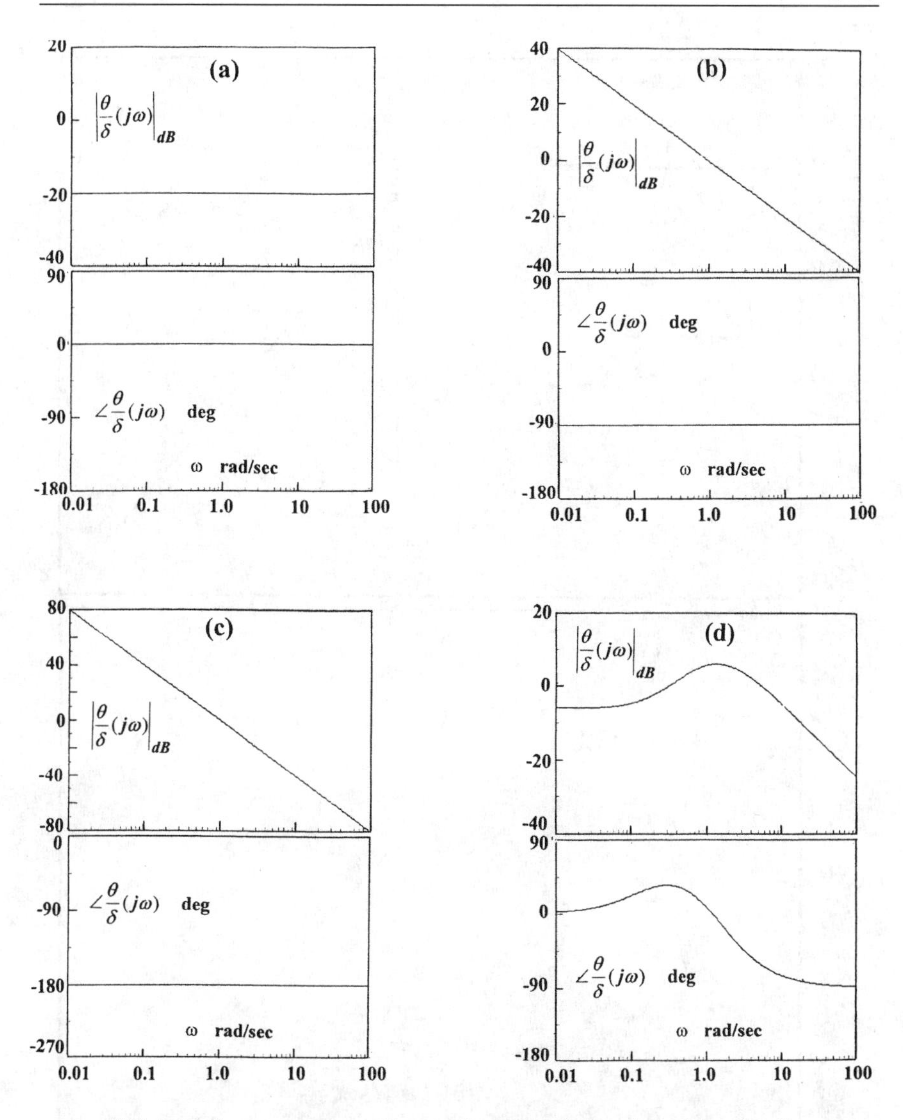

FIG. 8.6. Frequency response diagrams for the transfer functions of Equation 10. Parts (a) to (c) represent the three, simple, "stereotypical" transfer functions, respectively, and part (d) represents the more-complex transfer function.

Given the importance of the open-loop transfer function on its closed-loop counterpart, two questions may arise: First, are there "optimum" or "desirable" open-loop transfer function characteristics as interpreted on the frequency response diagram, and second, can the human pilot be expected to create these characteristics by adopting an appropriate transfer function, $Y_p(s)$? The answer to both questions is "yes." The optimum or desirable characteristics are shown in

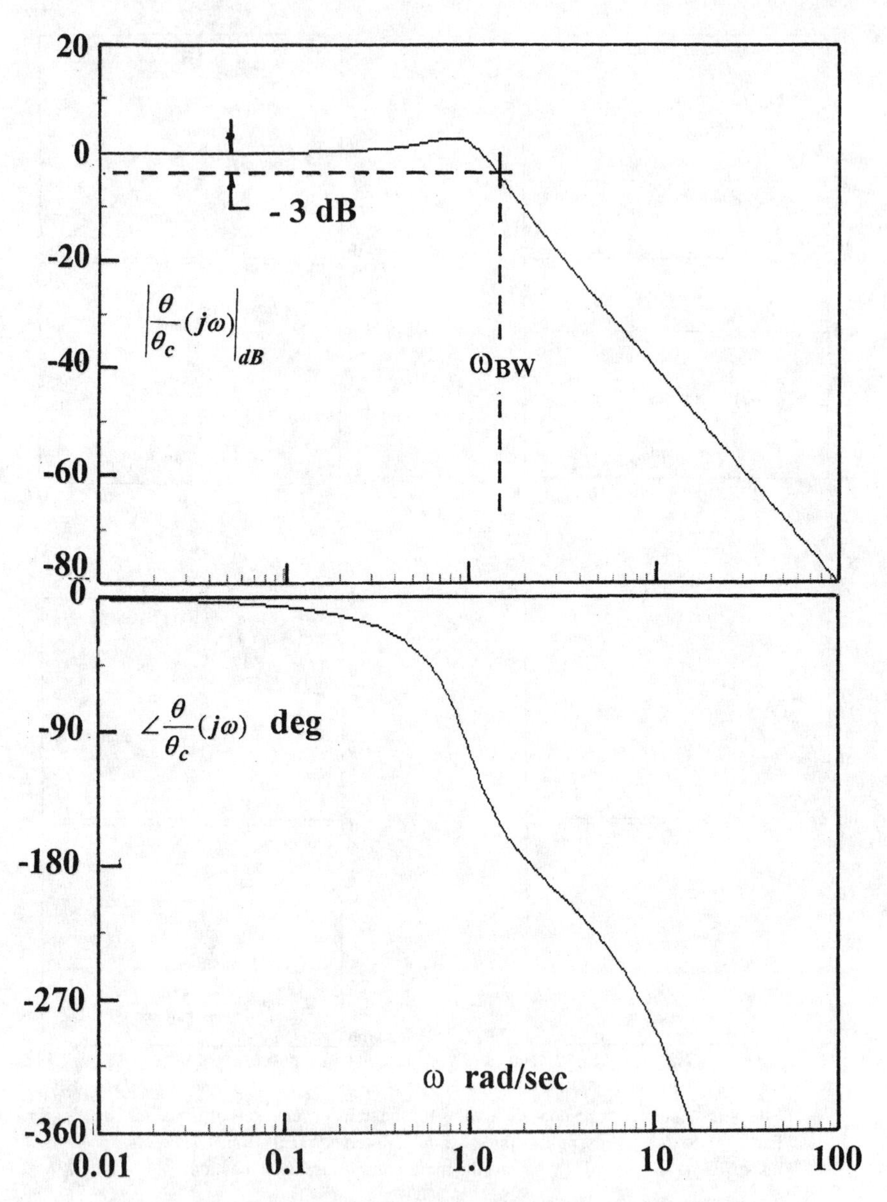

FIG. 8.7. The frequency response diagram for a closed-loop transfer function indicating the definition of the system bandwidth ω_{BW}. The closed-loop transfer function is identified as θ/θ_c using the nomenclature of Figs. 8.4 and 8.5.

Fig. 8.8. In Fig. 8.8, the frequency at which the magnitude of the open-loop transfer function is 0 dB is called the *crossover frequency* ω_c.

When the frequency response of the open-loop transfer function of a pilot/vehicle system appears as in Fig. 8.8, it can be shown that excellent command following and disturbance regulation characteristics will be exhibited (Nise, 2000). In addition, it can be shown that

$$\omega_{BW} \approx \omega_c \tag{16}$$

That is, the closed-loop bandwidth is approximately equal to the crossover frequency of the open-loop transfer function. The larger the open-loop crossover frequency, the larger the closed-loop bandwidth, and the larger will be the class of inputs that can be followed and disturbances that can be regulated by the pilot/vehicle system. The question now is: What limits the crossover frequency? The answer lies in control system *stability.*

Stability as applied to closed-loop systems requires that bounded inputs produce bounded outputs. This means that any system input that does not become arbitrarily large in amplitude must produce outputs that also do not become arbitrarily large. The presence or absence of closed-loop stability can be obtained by examination of the frequency response diagram of the open-loop system. All that is required is that when the phase lag first equals -180 deg, the magnitude must be less that 0 dB. We can now answer the question posed in the previous paragraph regarding an upper limit on the crossover frequency. It is the time delay inherent in human pilot dynamics that effectively places an upper limit on the attainable crossover frequency of any open-loop pilot/vehicle transfer function because the time delay transform $e^{-\tau s}$ adds phase lag as a linear function of log(ω).

Comparing the frequency response diagrams of Fig. 8.7 and 8.8, one sees a fundamental difference. The frequency response diagram for a closed-loop system will be approximately unity (0 dB magnitude and 0 degrees phase lag) over a frequency range extending from zero to near the system bandwidth. The frequency-response diagram for an "optimum" open-loop transfer function, however, will typically exhibit very large amplitude at low frequency $\omega << \omega_c$ and very small amplitude at high frequency $\omega >> \omega_c$ with the region around crossover being linear with a slope of -20 dB/decade. A decade is defined as the distance along the abscissa of the diagram separating any two frequencies differing by a factor of 10.

Outer-Loop Pilot Control

Sequential Loop Closures in Manual Control. The aerodynamic control surfaces shown in Fig. 8.2 are used primarily to control the attitude of the aircraft. The three angles that define the attitude are: the pitch attitude $\theta(t)$ and defined as the angle the fuselage axis makes with the horizontal; the roll-attitude $\phi(t)$, defined as the angle a line perpendicular to the aircraft's plane of symmetry makes with the

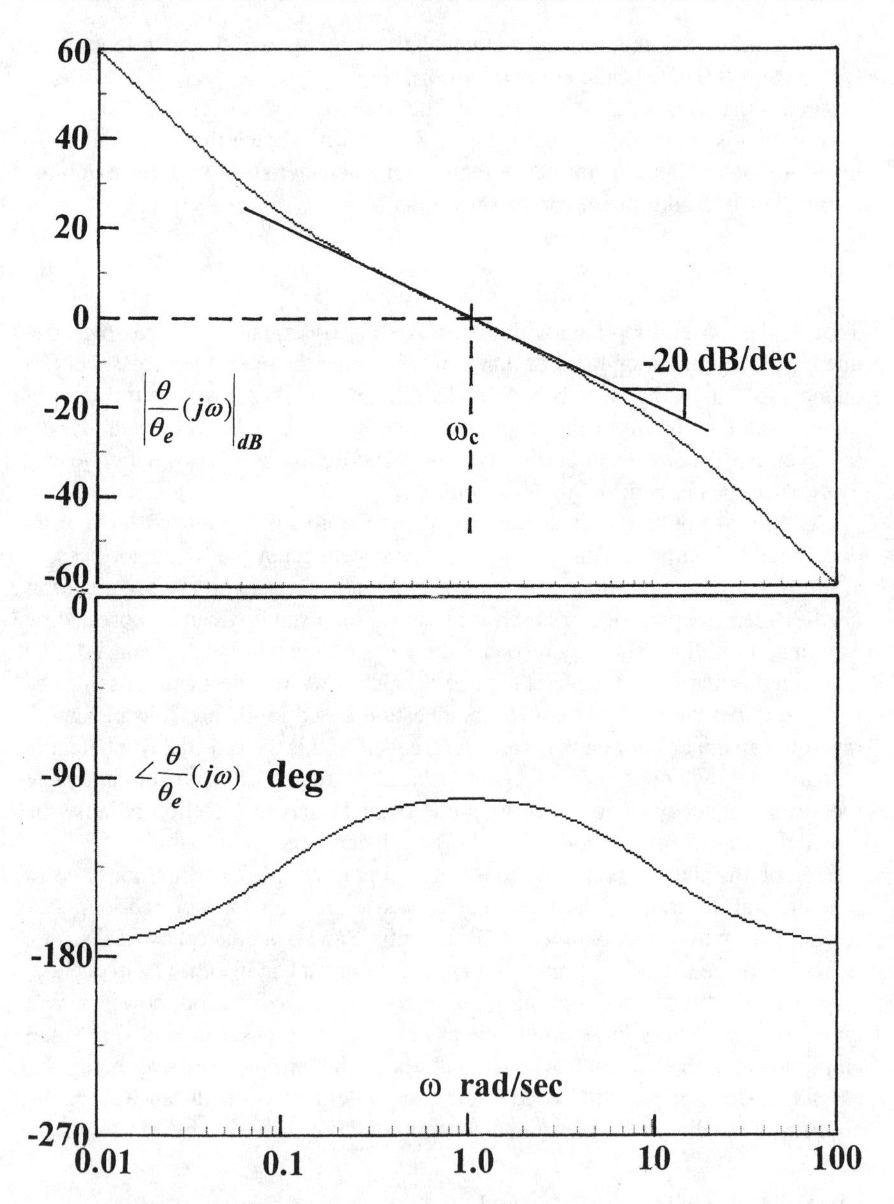

FIG. 8.8. The frequency response diagram for an "optimum" or "desirable" open-loop transfer function indicating the definition of the open-loop crossover frequency ω_c. The open-loop transfer function is identified as θ/θ_e using the nomenclature of Figs. 8.4 and 8.5.

horizontal; and finally the heading angle, $\psi(t)$, defined as the angle the fuselage axis makes with a particular earth-fixed direction, say due north. Simplifying somewhat, the elevator is used to control $\theta(t)$, the ailerons to control $\phi(t)$, and the coordinated use of the rudder and ailerons to control $\psi(t)$. But as stated at the outset of this chapter, the pilot must control not only the attitudes of the aircraft, but also the remaining three degrees of freedom, that is, the position of the aircraft center of mass relative to an earth-fixed reference frame. In terms of the control–theoretic approach discussed to this point, control of the aircraft's position defines maneuvering flight and is best analyzed as a multiloop feedback structure.

Consider the maneuver performed when a pilot is attempting to change the direction of flight and constrain the aircraft's ground track to a straight line. This maneuver occurs in a landing approach when the pilot is maneuvering from the "cross-wind" leg to the "final approach" leg. Fig. 8.9 diagrams the task. A block

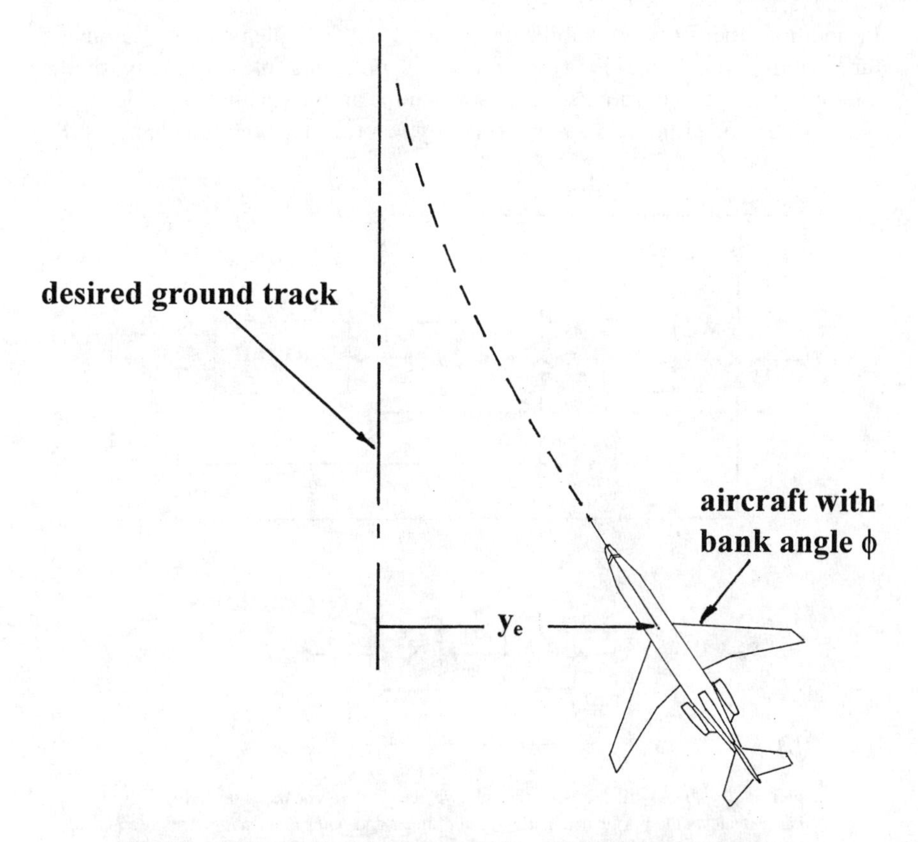

FIG. 8.9. A pilot task involving the control of more than a single aircraft response variable. The pilot is attempting to change the ground track of the aircraft. Two pilot feedback loops are employed involving bank angle ϕ (inner-loop control) and lateral displacement from the desired ground track y (outer-loop control).

diagram representation of a pilot/vehicle feedback structure that would accomplish the task is given in Fig. 8.10(a).

Note in Fig. 8.10(a) that the aircraft now has two outputs, the roll-attitude ϕ and the lateral coordinate of the aircraft center of mass, y. Here, the earth-fixed reference frame is assumed to have an axis coincident with the desired ground track. The pilot's inceptor is now commanding aileron angle δ_a. The aircraft block is also assumed to contain any artificial stabilization feedback loops, just as the transfer function θ/δ_p in Fig. 8.5 contains the feedback loop with δ_1/θ in Fig. 8.4. For example, a roll-rate command system may be in evidence in which

$$\dot{\phi}(t) = K\,\delta_a(t) \quad or \quad s\phi(s) = K\,\delta_a(s) \tag{17}$$

In addition, the artificial stabilization may also be implementing "automatic turn coordination" in which the rudder is deflected proportionally to the ailerons to effect a coordinated turn in which the aircraft exhibits no sideslip. In Fig. 8.10(a), the pilot is now represented by two transfer function elements, Y_{p_y}

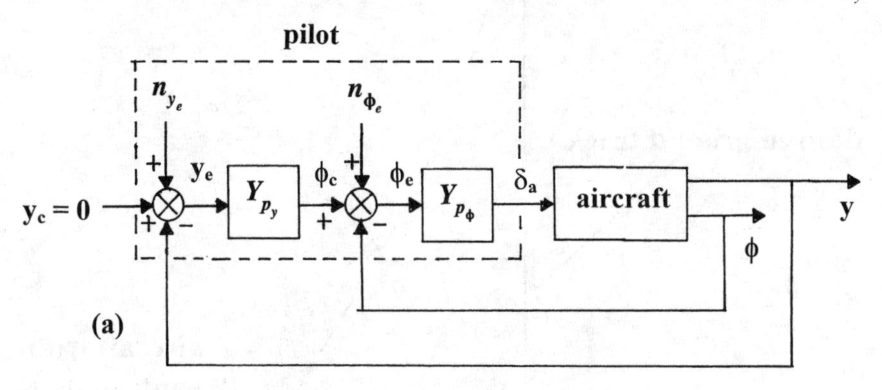

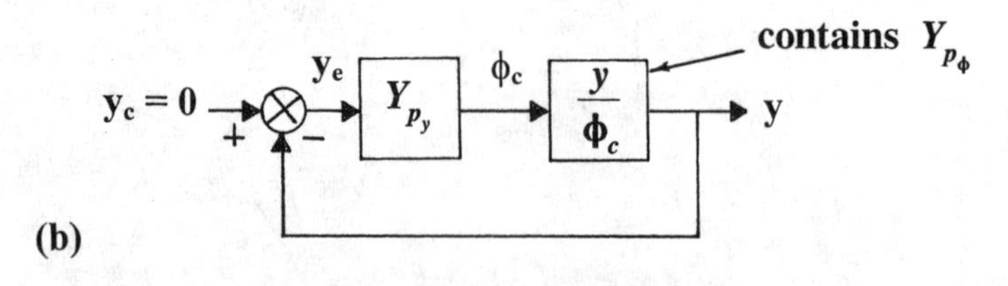

FIG. 8.10. Block diagrams of the pilot/vehicle system for the task of Fig. 8.9. Part (a) shows the pilot participating in both inner and outer loops and represented by the transfer functions Y_{p_y} and Y_{p_ϕ}. The signals n_{y_e} and n_{ϕ_e} are remnant signals modeling task interference occurring as the result of the pilot controlling two aircraft response variables. Part (b) is a simplification of part (a) obtained using block diagram algebra.

and Y_{p_ϕ}, operating in two feedback loops, one referred to as an inner loop and the second referred to as an outer loop. The inner-loop closure requires visually sensed ϕ and the outer-loop closure requires visually sensed *y*. The output of the element Y_{p_y} in the outer loop is the roll-attitude command ϕ_c for the inner loop, a signal internally generated by the pilot and not existing as a measurable entity. The inputs n_{y_e} and n_{ϕ_e} indicated at the two summing junctions will be addressed in the following section. The feedback structure of Fig. 8.10(a) is referred to as a multiloop structure. Fig. 8.10(a) can be simplified, just as was Fig. 8.4. The simplified diagram is shown in Fig. 8.10(b). Here, part of the pilot's dynamics have been subsumed into the block labeled y/ϕ_c. Figure 8.10(b) now resembles a single-loop feedback system. For satisfactory performance, Y_{p_y} should also exhibit the desirable open-loop frequency response characteristics of Fig. 8.8. However, bandwidth considerations require that the crossover frequency of the new open-loop transfer function be considerably lower than the bandwidth of the closed, inner-loop feedback system ϕ/ϕ_c that includes Y_{p_ϕ}, for example,

$$\omega_{cy} \approx \frac{\omega_{BW_\phi}}{n} \approx \frac{\omega_{c\phi}}{n}$$
$$n \approx 2 \rightarrow 3 \tag{18}$$

Thus, the forms of Y_{p_y} and Y_{p_ϕ} are constrained by the dictates of control system design principles for inanimate systems. In the design of inanimate controllers, the procedure just described is referred to as design by *sequential loop closure* (Maciejowski, 1989).

Task Interference. The human pilot is capable of closing only a small number of inner and outer control loops with any degree of precision. In a control–theoretic sense, this limitation is handled by considering the human pilot as a "single-channel processor" that shares attention between competing tasks or control loops. Such a modeling approach suggests introducing switches in the model such that control loops are activated individually, with the switching logic dependent on the loop structure and the signal magnitudes in any loop. Although attractive from the standpoint of placing obvious limits on the number of tasks that can be simultaneously addressed by the pilot, such a modeling procedure suffers from the fact that the description of the human pilot is now nonlinear in a mathematical sense. Nonlinearities unfortunately eliminate the transfer function approach to system analysis, along with many of the analytical tools associated with linear control systems design.

An alternative to switching models can be offered in which random "noise" or *remnant* signals are injected into the control loops that involve the human pilot. This approach models human channel capacity by corrupting the signals circulating in the various control loops in which the human is active. These injected noise signals are indicted by n_{y_e} and n_{ϕ_e} in Fig. 8.10(a). For simplicity, filtered "white" noise is usually used in these applications, where white noise can be described as possessing equal power or strength at all frequencies. In addition, the strength of the noise is often assumed to scale with the mean square value of the signal to which it is being added and with the number of loops being closed by the pilot. In Fig. 8.10(a), these signals are the two "error" signals y_e and ϕ_e. The term *error signals* is used because if y_e and ϕ_e are nonzero, the outputs y and ϕ are not equal to the command values y_c and ϕ_c. This assumption of scaling the strength of the noise signal with the mean square value of the signal to which it is added can lead to another type of instability in pilot models, referred to as *remnant-induced instability.* Although obviously undesirable from a control system standpoint, this instability can effectively model the human's limitations in simultaneously controlling multiple tasks or control loops.

Historical Perspectives

The preceding discussion allows us to provide a brief historical perspective of the mathematical modeling of the human controller and pilot. The discipline of what is now known as *manual control* began in the latter part of World War II, when engineers and engineering psychologists were seeking to improve the performance of human gunners by attempting to model the dynamics of the gunner much as one would that of an inanimate servomechanism (Tustin, 1947), that is, to employ the transfer-function representation just described. If such models could be provided, better means of aiding the operator could be obtained. The approach of these early researchers proved to be sound, and their methodology was extended in succeeding decades to modeling the human pilot in many continuous control tasks.

The concept and utility of the "human pilot transfer function" was firmly established in 1965 with the publication of the results of a study sponsored by the U.S. Air Force (McRuer, Graham, Krendel, & Reisener, 1965). In this report, the basic tenets describing how the human adapts to different vehicle dynamics were presented. A later treatment (McRuer & Krendel, 1974) summarized the state of the art in deriving models of human pilot dynamic behavior. Figure 8.11 shows the form of the "pilot describing function" discussed. The term *describing function* is used in lieu of transfer function to emphasize the fact that the early experiments proved that only a portion of the human controller's output could be obtained by a linear operation on a perceived input as would be obtained with a transfer function. The means of accommodating this modeling deficiency was through the introduction of a random *remnant* signal as shown in Fig. 8.11 and briefly described in the preceding section. As opposed to the remnant previously introduced, however, the remnant of Fig. 8.11 does not model task interference

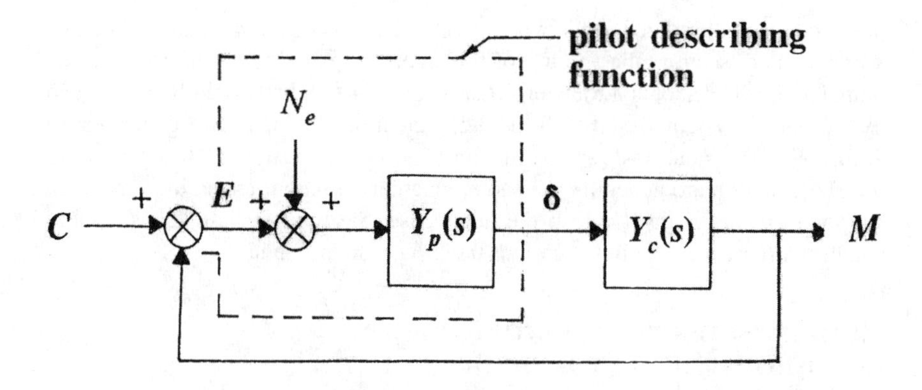

FIG. 8.11. A describing function representation of the pilot in a simple control task. The signal N_e represents remnant modeling nonlinear and/or time varying pilot behavior in the task. The combination of the transfer function Y_p and remnant N_e defines the describing function representation of the pilot.

but rather accounts for any inherent nonlinearities, time variations, or "noisy" visual perception in single-loop human control behavior. Because the measured remnant was indeed random and not amenable to description by a deterministic function of time (and thus a Laplace transform), it was represented by a power spectral density (PSD). The PSD representation is essentially a measure of the "power" in any signal expressed as a continuous function of frequency. The PSDs of the remnant signal were found to coalesce best if the remnant signal itself was assumed to be injected into the perceived visual input signal, denoted E in Fig. 8.11. This led some researchers to associate the remnant with a noise-injection process in the human visual system itself, for example, Levison, Kleinman, and Baron (1969).

The human describing function models were obtained from measurements of human operators in continuous, laboratory tracking tasks in which the input or command signal was random or random-appearing in nature. Such inputs were considered essential in manual tracking tasks because they most closely approximated the inputs that the human would face in control tasks of practical interest. The early researchers also found that inputs that were not random appearing could induce a type of human control behavior that was distinctly different from that exhibited when tracking random inputs. This behavior was characterized by human control actions that appeared well rehearsed, nearly automatic and with little or no apparent time delay in execution. This was the *precognitive* behavior described previously.

There exist, of course, human control tasks of engineering interest that do not involve random inputs or commands. Nonetheless, the transfer function model of the human is still utilized, often without consideration of the remnant. As will be seen in a later section, the validity of this approach is based on the fact that the human controller adopts a transfer function similar to what would be prescribed by a control system designer faced with creating an inanimate controller to accomplish the same task as the human. The characteristics of this controller transfer function are generally the same, whether a random-appearing input or a set-point change is involved. In the latter case, the remnant is of considerably smaller importance than in the case of tracking random inputs.

Human Sensors, Control Effectors, and Information Processors

The individual transfer function models of pilot behavior discussed in previous sections have considered only a single input that was tacitly assumed to be visual in nature. However, other sensory inputs are possible, indeed likely. The sensory capabilities of the pilot called on for the continuous tasks encountered in modern aircraft are *visual, vestibular,* and *proprioceptive* in nature. The visual organs are, of course, the eyes. In a control–theoretic sense, it is usually assumed that the human visual system can sense errors and possibly error rates between the visual information that is actually sensed and that indicating a nominal or desired condition. The vestibular organs are primarily those of the inner ear, *otoliths,* and *semicircular canals* that can sense linear and angular accelerations (Proctor & Proctor, 1997, see also chap. 3). The proprioceptive organs are those that sense mechanical displacement of the muscles and joints and consist of mechanorecpetors in the joints that respond to changes in the angles of the joints, muscle spindles in the muscle groups in arms and legs that respond to stretch, and cutaneous mechanorecptors that respond to touch. In the majority of control theoretic models, only visual and vestibular inputs are considered. One important exception to this is the *structural pilot model* to be discussed.

Human control effectors are primarily the hands, feet, and limbs that drive the cockpit inceptors such as control sticks and pedals. With the advent of voice recognition systems in the cockpit, the vocal chords might be added to this list. However, voice inputs are typically employed to make discrete commands rather than the continuous ones associated with the pilot models to be discussed herein.

Information processing for piloting tasks occurs in higher levels of the central nervous system, whose activity has been described as that of a *central nervous processor* (CNP) (Phillips, 2000). The elements of the CNP consist of (a) a cranial sensory array that receives information from the sensory systems outlined in the previous paragraph, (b) the brain/midbrain that performs a large amount of higher level data processing, (c) the brainstem/cerebellum that continues the data processing, and finally (d) the spinal cord that serves to transfer sensory information to the brain and to transfer motor commands to the effectors.

With the forgoing discussions as background, we can now begin a general discussion of some specific control–theoretic approaches to modeling pilot control behavior. The approaches covered are not exhaustive, but have been selected based on (a) their success in previous applications, and (b) the ability of the introductory material of the preceding sections to adequately prepare the reader for the discussions.

CONTROL–THEORETIC MODELS OF PILOT CONTROL BEHAVIOR

Isomorphic Models

Isomorphic models refer to those models of pilot control behavior in which some effort has been directed toward explicitly modeling the dynamics of the human sensory and control effector systems. Two pertinent examples will be given here: (a) the crossover and precision models and (b) the structural model.

The Crossover and Precision Models. The *crossover model* itself is not an isomorphic model. However, its central importance in manual control theory and its relationship to the precision model, which is an isomorphic representation, justifies its inclusion in this section. The crossover model is based on the experimentally verified fact that, in the region of the crossover frequency, the open-loop transfer function of a pilot/vehicle system, for example, that of Fig. 8.4, can be approximated as

$$Y_p Y_c(s) \approx \frac{\omega_c}{s} e^{-\tau_e s} \tag{19}$$

where

$Y_p(s)$ is the pilot transfer function,
$Y_c(s)$ is the appropriate transfer function for the vehicle, $\frac{\theta}{\delta_p}(s)$ in Fig. 8.4,
ω_c is the crossover frequency, rad/sec, and
τ_e is an "effective" time delay, sec.

The τ_e represents the cumulative effect of pure time delays in the human sensing, central processing, and control-effecting systems and includes the low-frequency effects of higher frequency dynamics such as those of the pilot's neuromuscular system as well as high-frequency response modes of the vehicle.

If one were to make a frequency response diagram of the right-hand side of Equation 19, it would appear as in Fig. 8.12, which closely approximates Fig. 8.8. The primary difference between the two plots is the increasingly negative phase angle apparent in Fig. 8.12 attributable to the effective time delay just defined.

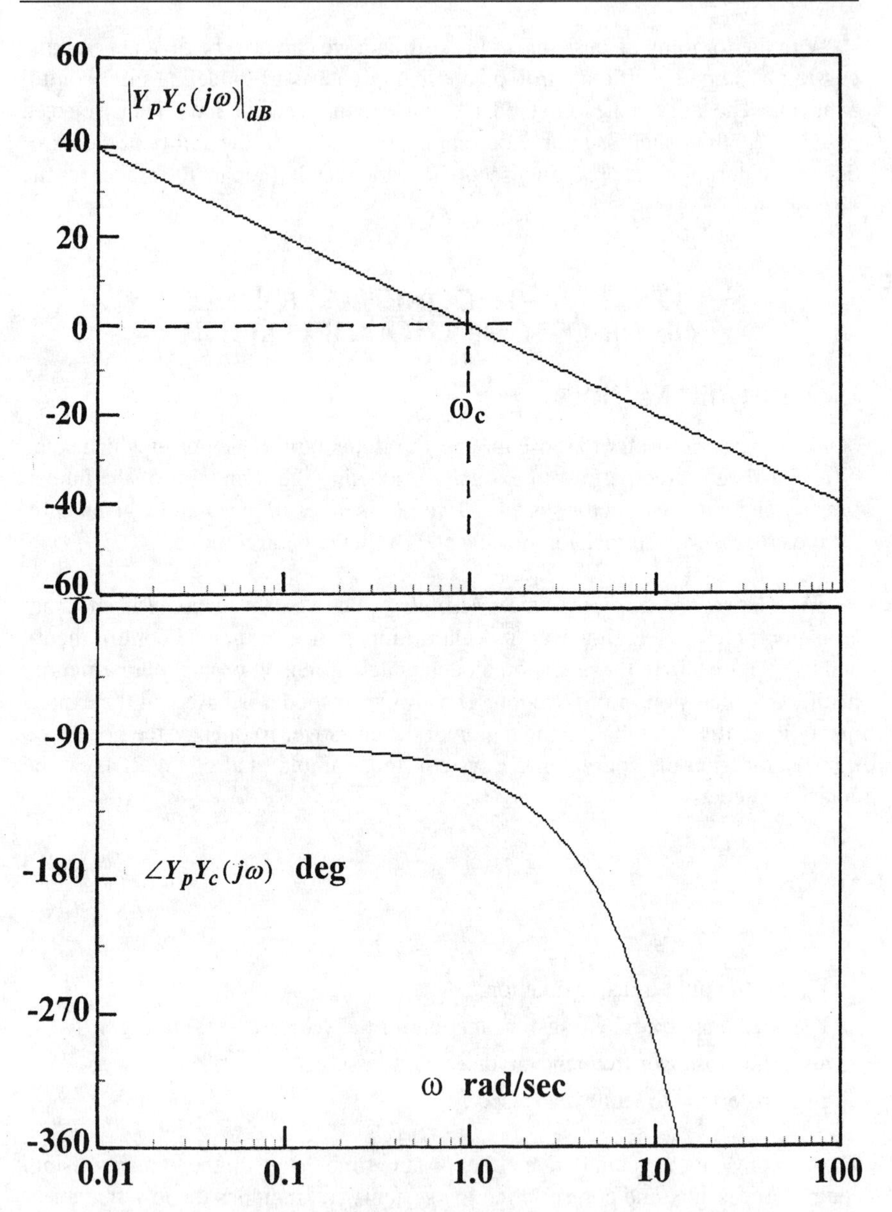

FIG. 8.12. A frequency response diagram of the crossover model of the pilot/vehicle open-loop transfer function given by Equation 19. Although plotted over a frequency range of four decades, the model's validity extends only over a limited frequency range about the crossover frequency ω_c.

Equation 19 and the frequency response diagrams of Figs. 8.8 and 8.12 imply that the human adopts dynamic characteristics (or control compensation) similar to that which would be specified by a control system engineer attempting to design an inanimate controller to serve in the place of the human. The primary limitations of the human compensation are (a) a limitation on the crossover frequency ω_c (and by implication the closed-loop bandwidth ω_{BW}) that can be obtained while still achieving stability of the closed-loop system and (b) a limitation on the form of the compensation required to achieve the ω_c/s-like characteristics evident in Equation 19.

Typically, this latter limitation arises when factors of the form $(T_L s + 1)^n$ are required in the numerator of $Y_p(s)$. Recalling that multiplying by the Laplace variable s in the Laplace domain is equivalent to differentiation in the time domain, the appearance of terms such as $(T_L s + 1)^n$ in the numerator of the pilot transfer functions implies the pilot sensing or generation of higher order time derivatives of the error signal in the control loop in question. For example, with $n = 1$, the numerator term $(T_L s + 1) \cdot e(s)$ translates in the time domain to $T_L[de(t)/dt] + e(t)$. With $n = 2$, the numerator term becomes $(T_L s^2 + 2T_L s + 1) \cdot e(s)$ with a time-domain interpretation of $T_L[d^2e(t)/dt]^2 + 2T_L[de(t)/dt] + e(t)$. Human control compensation corresponding to $n = 1$ has been shown to be difficult for the human, in other words, to involve considerable workload and poor handling qualities. This was the reason for stating that vehicle dynamics similar to that given in Equation 3 are usually avoided in flight control systems because these dynamics require pilot compensation with $n = 1$. Pilot compensation with $n = 2$ implies sensing the first and second derivatives or first- and second-order time differentiation of a visual signal. Such sensing/generation of a quality sufficient for continuous control has been shown to be extremely difficult for the human pilot. Indeed, rarely is the human capable of creating this compensation for $n \geq 2$.

Note that Equation 19 does not describe the pilot transfer function per se, but rather the product of the pilot and vehicle transfer functions. Initially, this may seem to present a significant limitation of the crossover model. However, as pointed out earlier, from a control system standpoint, it is the characteristics of the open-loop transfer function, the combined pilot/vehicle system, that determines closed-loop performance and stability. McRuer and Krendel (1974) and Hess (1997a) provide information on how the parameters of the crossover model can be estimated to provide a predictive tool for estimating pilot/vehicle dynamics.

A more detailed representation of the pilot transfer function that displays some isomorphism can be presented in the form of the *precision model.* The precision model is a higher order transfer function representation of human pilot dynamics. For example, the numerator polynomial is of second order in s, and the denominator polynomial is of fifth order in s. Figure 8.13, from Hess (1997a) demonstrates that the precision model can provide close matches to pilot transfer functions obtained from experiment. Here, data from laboratory tracking tasks with unstable vehicle dynamics were used. The fit of Fig. 8.13 is noteworthy because it demonstrates the fidelity of the neuromuscular system modeling employed in for-

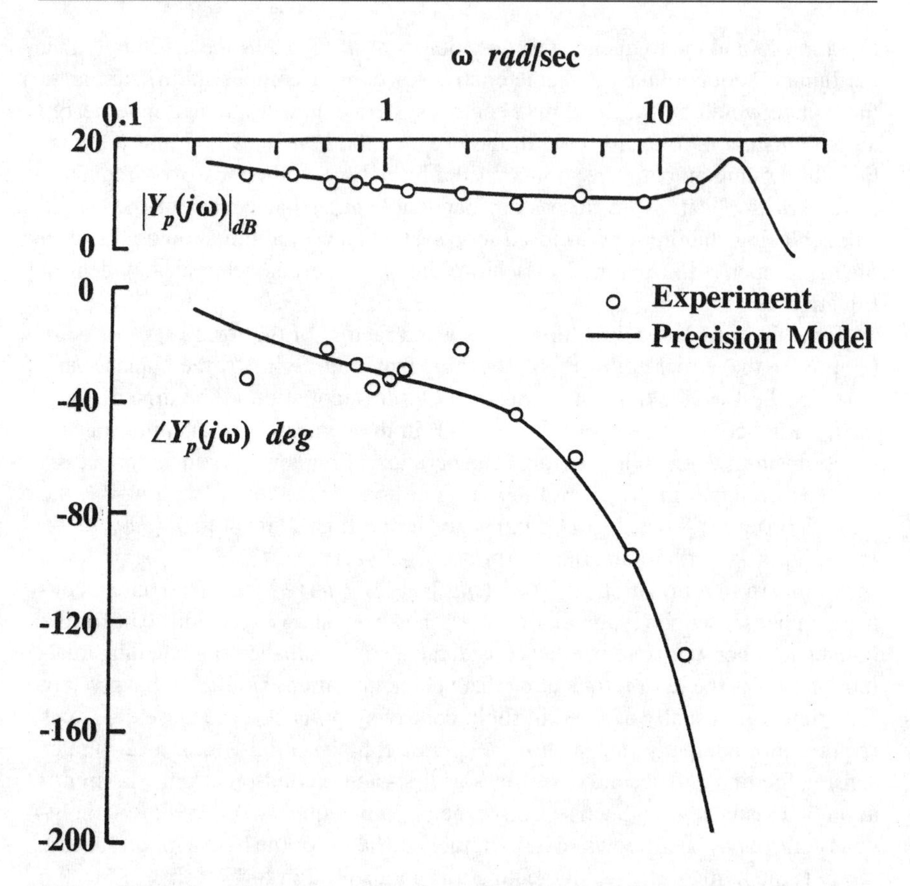

FIG. 8.13. A frequency response diagram indicating the ability of the precision model of the pilot to match experimental results obtained from a laboratory tracking task.

mulating the precision model. This is an example of the precision model being used in descriptive fashion. Although vestibular feedback has not been explicitly considered in the model, the model parameters can be adjusted to match pilot transfer functions obtained when such feedback is present.

The Structural Model. Hess (1997b) offers a *structural model* of the human pilot. In this model, specific feedback paths are used to describe visual, vestibular, and proprioceptive activity, although the dynamics of the respective sensors are ignored. In addition, Hess offers rules for selecting the model parameters so that the model may be used in predictive fashion. The model is shown in Fig. 8.14. Note that although the pilot model produces a single output *M* based on a single input *C,* there exist multiple internal feedback paths in the structure. Of particular importance in both the philosophy and implementation of the model are the dynamics of the proprioceptive element denoted Y_{PF} in Fig. 8.14. Hess

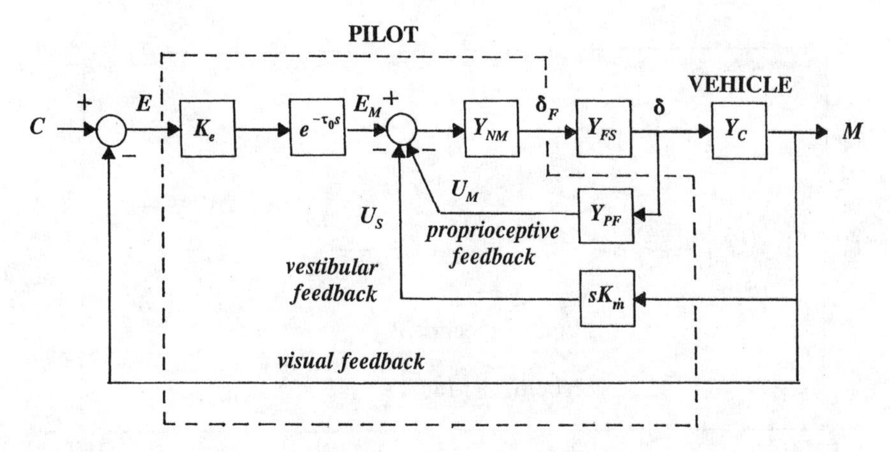

FIG. 8.14. A block diagram representation of the structural model of the pilot. The model incorporates visual, vestibular, and proprioceptive feedback signals.

hypothesizes that the primary compensation capabilities of the pilot, those capabilities that allow the dictates of Equation 19 to be met, are obtained from the proprioceptive feedback activity modeled by $Y_{PF}(s)$. The elements of the model are presented by individual transfer functions defined as:

K_e	A pilot "gain" used to establish the desired crossover frequency
$e^{-\tau_0 s}$	A time delay to model human response delays to stimuli
Y_{NM}	A simplified model of the neuromuscular system of the particular limb responsible for inceptor inputs
Y_{FS}	A model of the cockpit inceptor
Y_c	A model of the vehicle
Y_{PF}	An adjustable proprioceptive feedback element
$s K_{\dot{m}}$	A gain ($K_{\dot{m}}$)-derivative (s) combination to model vestibular feedback

Figure 8.15 demonstrates the ability of the model to match the frequency response data of the open-loop pilot/vehicle transfer function of an aircraft roll-control task where the data was obtained from flight test (Mitchell, Aponso, & Klyde, 1992). This fit is noteworthy because with the exception of the gain K_e, the model parameters were chosen in a priori fashion. This is an example of using the structural model in descriptive fashion. An efficient, computer-aided design tool is available for generating the structural model of the pilot (Zeyada & Hess, 1998).

A Biophysical Model. Van Paassen (1994) presents a detailed *biophysical model* of the human pilot, with particular emphasis on the dynamics of the pilot's neuromuscular system. The impetus behind the development of this model was the introduction of fly-by-wire systems in modern aircraft with cockpit control

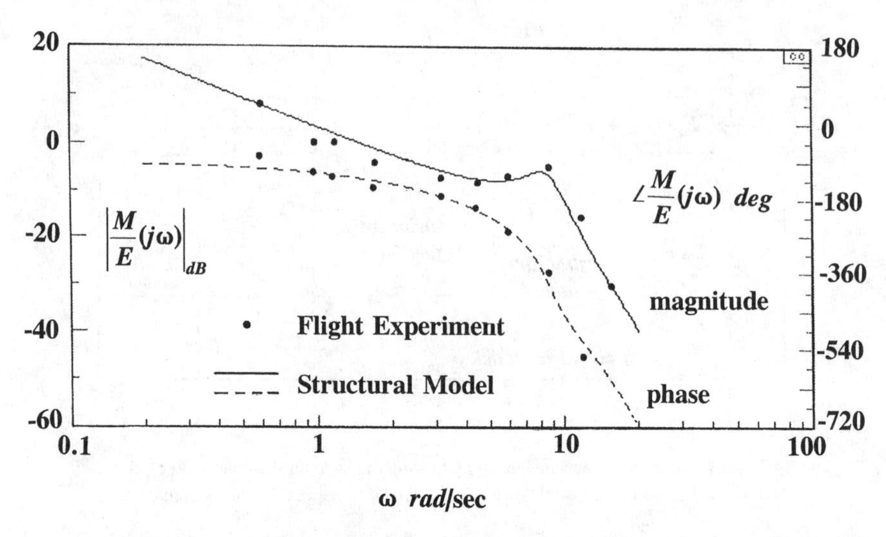

FIG. 8.15. A frequency response diagram indicating the ability of the Structural Model of the pilot to match experimental results obtained from a flight-test roll-control task. As opposed to Fig. 8.13, the diagram represents the open-loop pilot/vehicle transfer function, not just that of the pilot alone.

inceptors whose dynamics could be tailored by the flight-control engineer. This tailoring was possible because the inceptor and the pilot were no longer connected to the aircraft's control effectors with large cables, pulleys, and other mechanical linkages. Because of this design freedom, a number of transport and military aircraft with fly-by-wire flight control systems now eschew the normal center stick for small side-stick inceptors. All this leads to the fact that the interaction of the pilot's neuromuscular system with these inceptors can become an important part of the handling qualities and even the stability of the pilot/vehicle system (Hess, 1998; Johnston & Aponso, 1988). Figure 8.16 shows the configuration of van Paassen's biophysical model for a side-stick controller. It must be emphasized that the simplified nature of Fig. 8.16 masks all the detail in the neuromus-

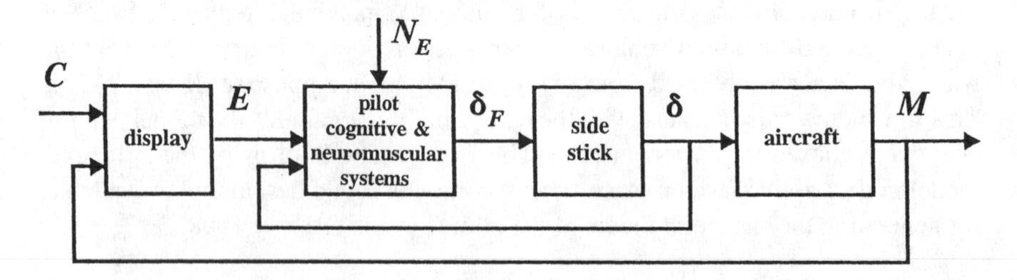

FIG. 8.16. A block diagram representation of a biophysical model of the pilot in a feedback system. Detailed submodels are contained within the block labeled "Pilot Cognitive & Neuromuscular Systems."

cular system of the model. Van Paassen's representation includes (a) separate modeling of the inceptor; (b) an explicit and detailed neuromuscular model combined with a model of control (compensation) behavior of the pilot; and (c) a set of representative, characteristic parameter values to make quantitative predictions of the behavior of the manipulator and neuromuscular system.

The characteristics just enumerated are also exhibited by the structural model just described. However, the biophysical model involves a much more detailed description of the human's neuromuscular dynamics. Fig. 8.17 shows the pilot/vehicle structure of Fig. 8.16, now configured so that identification of various elements can be made on the basis of experiment. Figure 8.18, taken from van Paassen's thesis, compares a series of these measurements with model-generated results. The vehicle transfer function was representative of the roll-attitude response of an aircraft to aileron inceptor inputs. The fits shown are noteworthy, given the level of detail found in van Paassen's model. This again is an example of a pilot model used in descriptive fashion.

A Biodynamic Model. The pilots of modern aircraft can be subjected to accelerating or vibrating environments that can adversely affect their performance as control system elements. For example, the helicopter pilot may be subjected to whole-body vibrations caused by the dynamics of the main rotor (Glusman, Landis, & Dabundo, 1986). Pilots of large transport aircraft can be subjected to oscillatory accelerations in the cockpit attributable to the flexible nature of the aircraft structure (Chan et al., 1992). High-performance fighter aircraft can produce rolling accelerations that induce whole-body pilot movement relative to the cockpit and control inceptors (Hess, 1998; Hohne, 1999). These potential problems point to a need for a *biodynamic model* of the human pilot, one that can incorporate the effects of an accelerating/vibrating environment on the pilot's control capabilities. The complexity of this modeling problem is evident in the

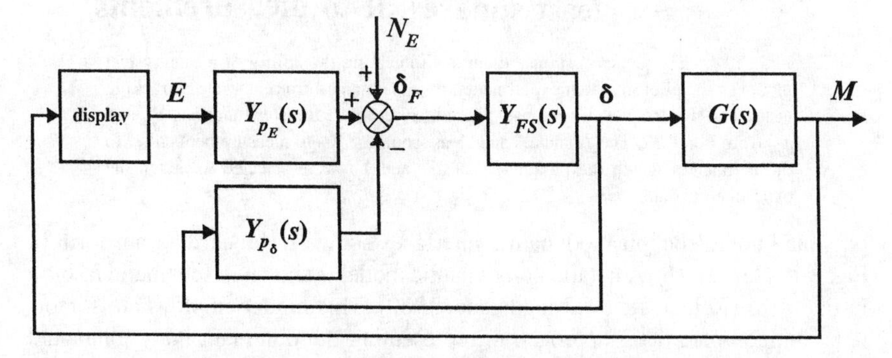

FIG. 8.17. A detailed block-diagram representation of the biophysical model of Fig. 8.16 showing individual transfer function elements. The transfer functions Y_{p_E} and Y_{p_δ} operate on visual and proprioceptive signals, respectively.

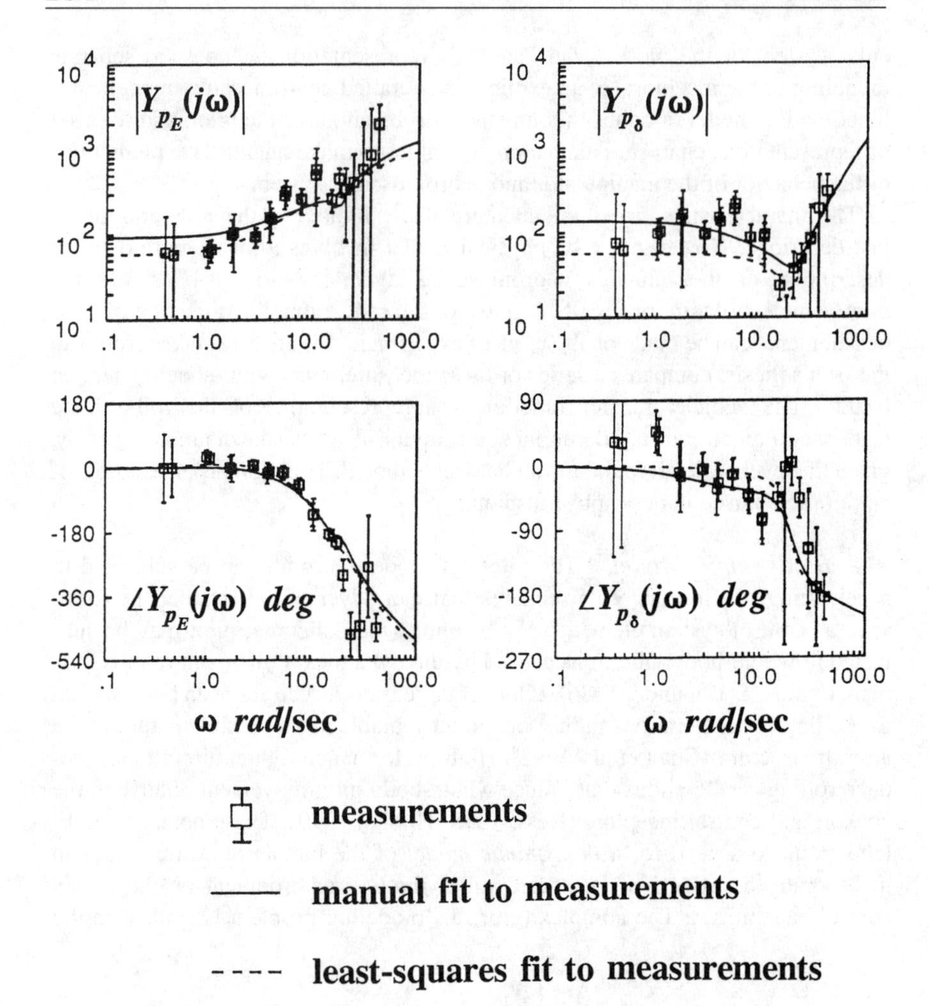

FIG. 8.18. Frequency response diagrams indicating the ability of a biophysical model of the pilot to match experimental results obtained from laboratory tracking tasks. The two frequency response diagrams refer to the transfer functions Y_{p_E} and Y_{p_δ} from Fig. 8.17. The "manual" and "least-squares" fits to measurements refer to the manner in which the parameters of Y_{p_E} and Y_{p_δ} were selected to match the experimental data.

possible human/cockpit biodynamic interfaces shown in block diagram form in Fig. 8.19 (Jex, 1971). A detailed biodynamic model is shown in schematic fashion in Fig. 8.20 (Reidel, Jex, & Magdaleno, 1980). This model employs an isomorphic, lumped-parameter approach in representing the dominant body joints and resulting modes of motion.

Biodynamic models differ somewhat from those that have been described in that they are not typically used as closed-loop compensatory controllers. Rather,

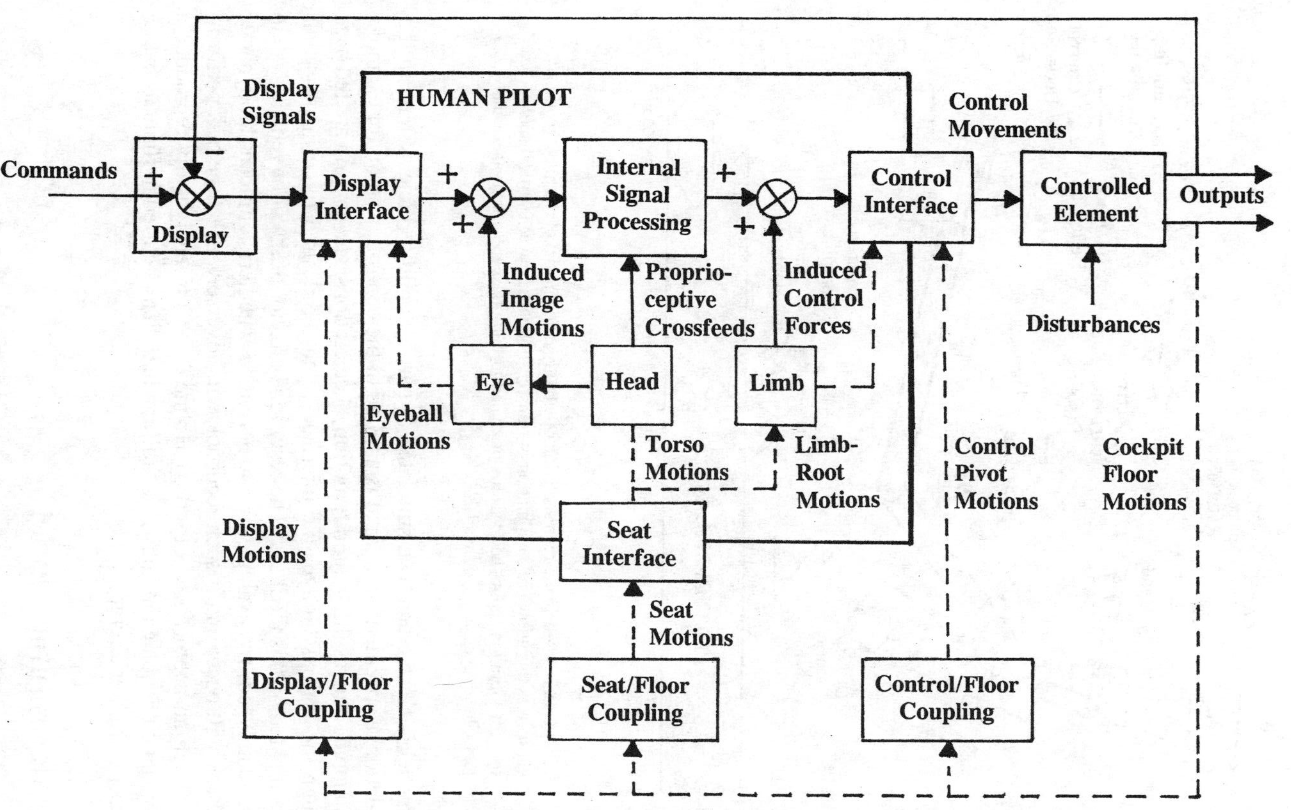

FIG. 8.19. Biodynamic interfaces that can influence pilot control. The top of the figure represents the feedback control loop that would exist in a nonvibrating, nonmoving environment. The bottom of the figure represents those additional elements necessary to model the biodynamic environment.

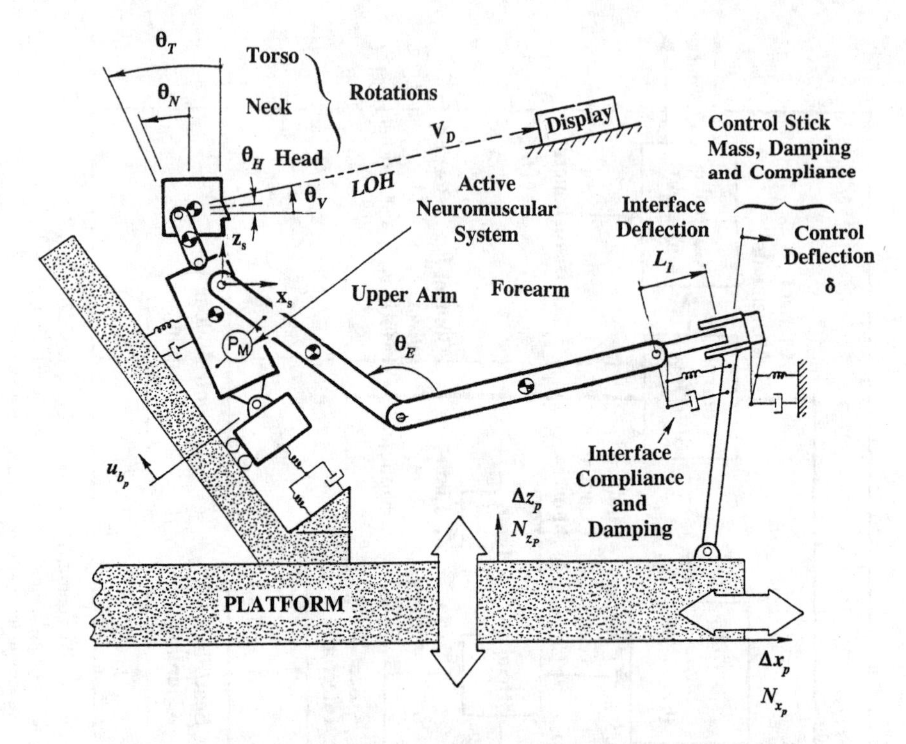

FIG. 8.20. A schematic representation of a biodynamic model of the pilot. Models such as this allow the computation of transfer functions between vertical and/or fore-aft vibration inputs and various biodynamic outputs such as motions of the torso, head, eyes, arms, or hands.

the model is created and then simulated as part of a pilot/vehicle system. In this simulation, a specified force input is applied to the control inceptor, for example, a step input. The resulting aircraft motion, accelerations, and so forth are fed back through the various biodynamic interfaces, and the desired force input is thus contaminated by this feedback. The resulting behavior of the pilot/vehicle system is then analyzed; for example, it is determined whether significant biodynamic feedback will occur and if such feedback will adversely affect the task performance. Such an approach was used by Hohne (1999) to analytically reproduce a biodynamic phenomenon known as "roll ratchet" that occurred in flight tests of a number of fighter aircraft.

An Algorithmic Model

Algorithmic models refer to those models of pilot control behavior that are created through the use of algorithmic synthesis techniques that have been successfully applied to the design of inanimate controllers. Although some isomorphism may be retained in the modeling procedure, algorithmic models basically attempt to replicate human input/output dynamics with little concern for the subsystem elements that may be creating these dynamics.

As was pointed out, the crossover model of the human pilot implies that the human adopts dynamics similar to those that would be selected by a control systems engineer attempting to design an inanimate controller to accomplish the task assigned to the human pilot. The *optimal control model* (OCM) of the human pilot extends this idea by using a time-domain feedback synthesis technique as opposed to a frequency-domain technique to obtain a model of the human pilot (Kleinman, Baron, & Levison, 1970). The limitations of the human are incorporated into the OCM synthesis procedure through (a) the introduction of *observation* and *motor noise* added to each observed variable and pilot inceptor input and (b) the incorporation of simplified neuromuscular dynamics through inclusion of inceptor input-rate terms in the *index of performance* that the synthesis technique minimizes.

Figure 8.21 shows the OCM in block diagram form. The directed line segments entering and leaving the blocks in Fig. 8.21 now represent a group of signals (often referred to as a vector) rather than one signal. The model itself represents the solution to a *linear, quadratic, Gaussian* (LQG) control problem. The name derives from the fact that the controller is *linear* and minimizes a *quadratic* index of performance (squares of time signals are involved), with the system disturbed by random inputs and injected noise signals, each of which possesses a Gaussian or normal amplitude distribution. In applications to modeling the

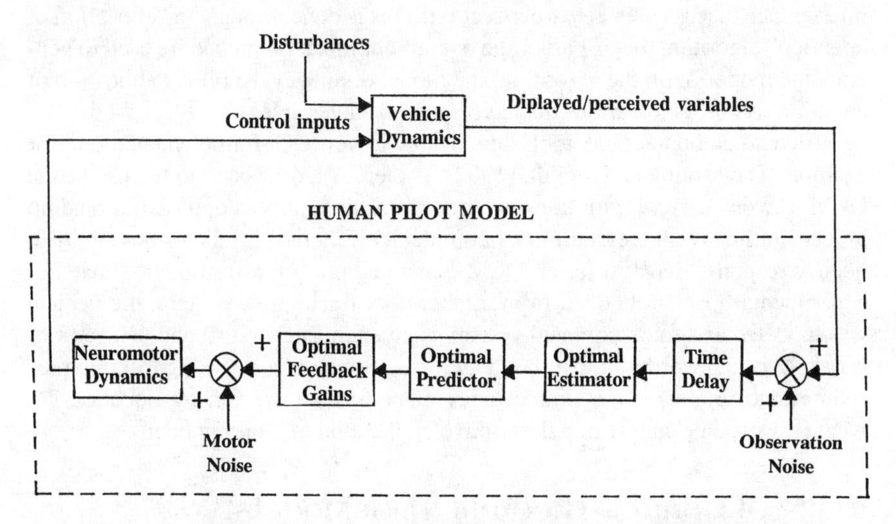

FIG. 8.21. A block diagram representation of the optimal control model (OCM) of the pilot. The directed line segments connecting the blocks in the figures represent a vector of system variables. Motor and observation noise inputs constitute the remnant effects in the OCM.

human pilot, the aforementioned index of performance is typically written as a weighted sum of the mean square values of one or more aircraft response variables and the rate of change of the inceptor input created by the pilot. As an example, consider the problem of aircraft pitch-attitude disturbance regulation while flying through a turbulence field. A possible index of performance to be minimized for this control problem might be given as

$$J = \lim_{T \to \infty} \int_0^T [\theta^2(t) + \rho\dot{\delta}^2(t)]dt \tag{20}$$

where ρ is a weighting factor chosen to provide a suitable neuromuscular system bandwidth (Kleinman, Baron, & Levison, 1970).

One basic assumption is employed in the OCM that differs from that of other models discussed. For each signal available to the pilot as a spatial displacement of elements in the visual field, the time derivative of that signal is also assumed to be perceived. Each of these displayed and perceived variables is assumed to be contaminated with additive white noise. The PSD of this noise is assumed to scale with the mean square value of the displayed or perceived variable to which it is added. The noise represents the remnant signals in the model. Although the OCM is a more complex model than those that have been discussed, it is also more powerful in that applications to multiloop problems are conceptually no more difficult than those for single-loop ones. Indeed, separate multiloop representations of the pilot such as that posed in Fig. 8.10a are no longer necessary in OCM applications. The OCM also possesses a natural means for handling the effects of task interference discussed in an earlier section. This is done through an "allocation of attention" algorithm that is part of the overall optimization procedure used to generate the model. With the algorithm, the model optimizes the pilot's allocation of attention resources to various displayed/perceived variables.

Efficient computational tools are available for OCM implementation, for example, Davidson and Schmidt, (1992). If there is a drawback to the use of the OCM, it would be use with transient, deterministic inputs as opposed to random inputs can involve increased model complexity. Figure 8.22 shows how the frequency response diagram for an OCM-generated pilot transfer function matches experimental results obtained from a laboratory tracking task. Here, the vehicle was modeled as a rate-command system as given in Equations 2 and 6. The fit of Fig. 8.22 is noteworthy in that the model was obtained from the algorithmic procedure that minimizes the simple index of performance given by Equation 20. Again, this is an example of a descriptive application of a pilot model.

Utility of Control–Theoretic Pilot Models

The pilot models just discussed are applicable to modeling the behavior of well-trained, well-motivated pilots in well-defined flight control tasks involving precise control of vehicle output variables. Such activity encompasses only a portion

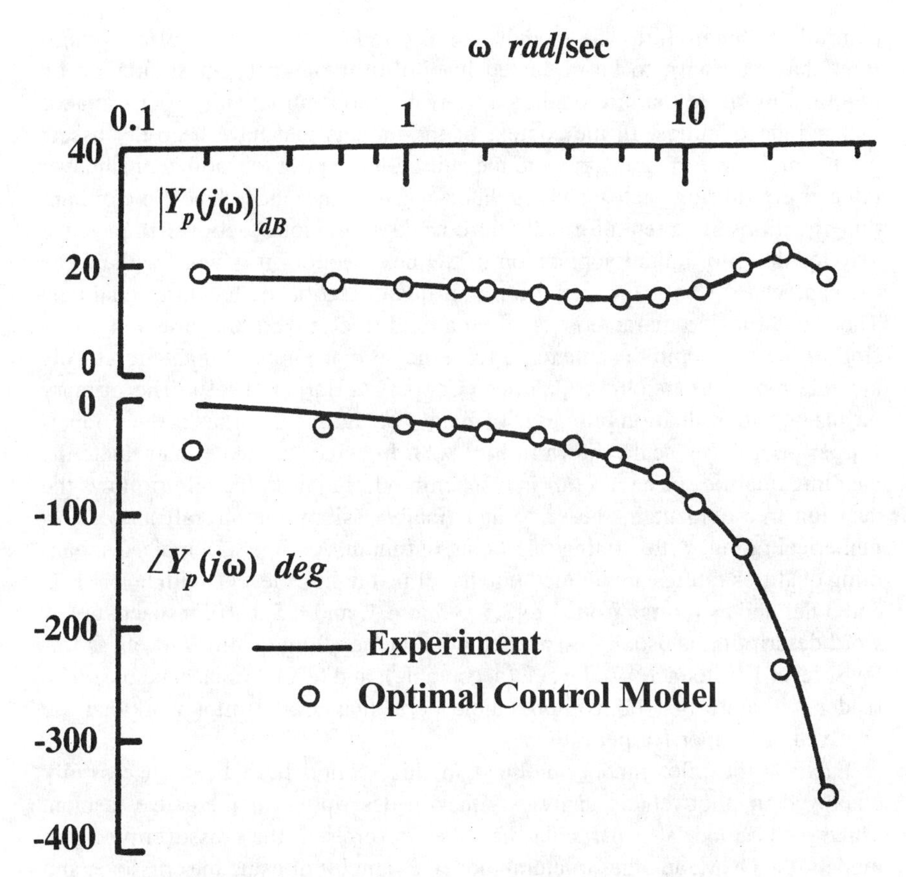

FIG. 8.22. A frequency response diagram indicating the ability of the Optimal Control Model of the pilot to match experimental results obtained from a laboratory tracking task.

of the flight control responsibilities of the pilot of a modern aircraft. Although the modern cockpit is becoming increasingly automated, the active, manual control of an aircraft will not disappear as a piloting requirement. This is obviously not only true in the case of military combat aircraft, but also true in the case of transport aircraft where manual backup systems are a necessity from the standpoint of safety. In addition, the maintenance of pilot proficiency would demand frequent flight in such manual modes.

From an engineering standpoint, the most useful application of the models just discussed involves prediction of pilot/vehicle performance and handling qualities while the aircraft is still in early stages of design or development. Here, pilot models can be used to indicate potential problem areas such as inadequacies in the flight control laws, or in the cockpit displays used by the pilot in controlling the aircraft. The models can also reduce the number of configurations that may have to be evaluated in pilot-in-the-loop flight simulation by highlighting those that are predicted to exhibit poor pilot/vehicle performance or handling qualities. In the context of the models that have been discussed, pilot/vehicle performance refers to the ability of the pilot to stabilize, maneuver (change equilibrium state), and regulate against disturbances. The topic of handling qualities has been informally introduced in previous sections. To pave the way for the pilot model application of the next section, it is worthwhile to be more precise in both the definition and quantification of handling qualities: Those qualities or characteristics of an aircraft that govern the ease and precision with which a pilot is able to perform the tasks in support of an aircraft role are referred to as handling qualities (Cooper & Harper, 1969). The primary means that an evaluation pilot uses to quantify handling qualities is the Cooper-Harper pilot rating scale shown in Fig. 8.23. In using this scale to evaluate the handling qualities of an aircraft in a specific piloting task, the pilot follows the decision tree evident in the scale and finally assigns the aircraft and task a numerical rating, with a rating of 1 being optimum. As Fig. 8.23 indicates, handling qualities ratings are divided into handling qualities levels, with levels 1, 2, and 3 defined as ratings from 1 to 3.5, 3.5 to 6.5, and 6.5 to 10, respectively. A word description is usually associated with the handling qualities levels as follows: level 1 (satisfactory), level 2 (acceptable), and level 3 (unacceptable). The reader will note that both pilot/vehicle performance and pilot workload are involved in Cooper-Harper ratings.

Each of the pilot models outlined in this section have been successfully employed in pilot/vehicle analyses either in descriptive or predictive fashion. Three of the models in particular have been exercised: the crossover/precision models, the OCM, and the structural model. Examples of using the crossover and precision models in performance and handling qualities investigations are summarized by McRuer and Krendel (1974). Similarly, an excellent overview of OCM application to performance and handling qualities prediction is provided by Innocenti (1988). Finally, the structural model has seen similar application, as described by Hess (1997b) and Zeyada and Hess (2000).

AN EXAMPLE PILOT/VEHICLE ANALYSIS

The Task and Pilot Model

At this juncture, an example of applying a control–theoretic model of the pilot to a realistic flight task will be demonstrated. To allow a more detailed discussion, only one of the models previously discussed will be exercised, the structural model. The flight task will be a helicopter "bob-up" maneuver. This task is of interest because it was also involved in a pilot-in-the-loop simulation study at NASA Ames Research Center (Schroeder, Chung, & Hess, 1999).

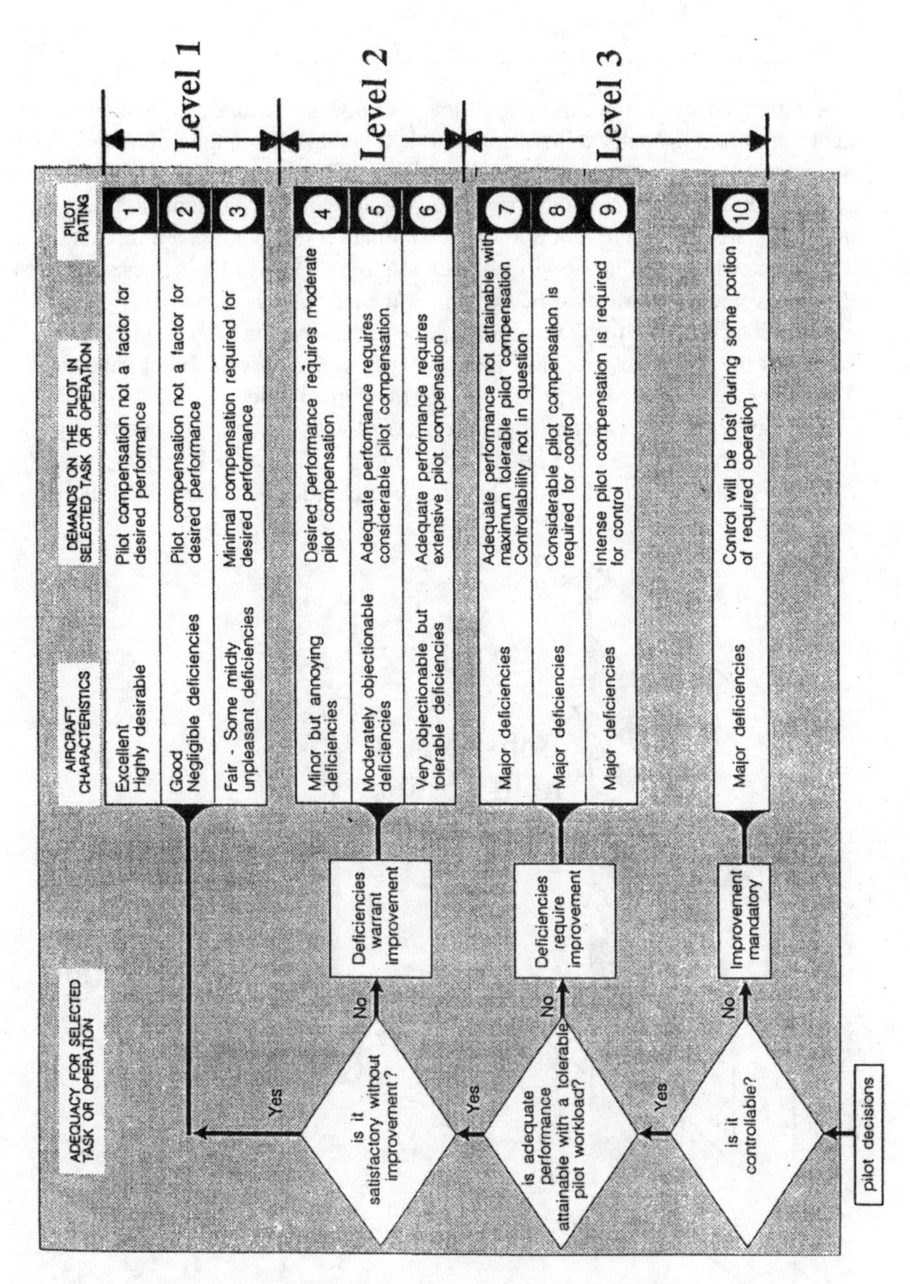

FIG. 8.23. The Cooper-Harper handling qualities rating scale. This scale is used by trained test pilots to evaluate the handling qualities of specific aircraft and flight tasks. Both pilot/vehicle performance and pilot workload are evaluated. Three handling qualities levels are indicated corresponding to handling qualities rated as "satisfactory" (level 1), "acceptable" (level 2), and "unacceptable" (level 3).

Figure 8.24 shows the geometry of the piloting task. It consisted of a rapid vertical translation from a stabilized hover to a target 32 ft from the initial helicopter position, followed by a second stabilized hover. This task represents a maneuver in which the equilibrium condition of the vehicle is being changed. Task performance standards that were required of the simulation pilots are given in Table 8.1. The manner in which these performance requirements will be incorporated into the pilot/vehicle analysis will be explained in what follows.

The block diagram for the complete pilot/vehicle system is shown in Fig. 8.25. Note that the structural model of the pilot only constitutes the block labeled $Y_{p\dot{h}}$. This inner loop is devoted to the control of vertical velocity, $\dot{h}(t)$, whereas the outer loop, with compensation Y_{p_h}, is concerned with the control of vertical position, or height, $h(t)$. Like the pilot representation in Fig. 8.10a, both $Y_{p\dot{h}}$ and Y_{p_h} constitute the pilot model, and outer-loop compensation is determined after the

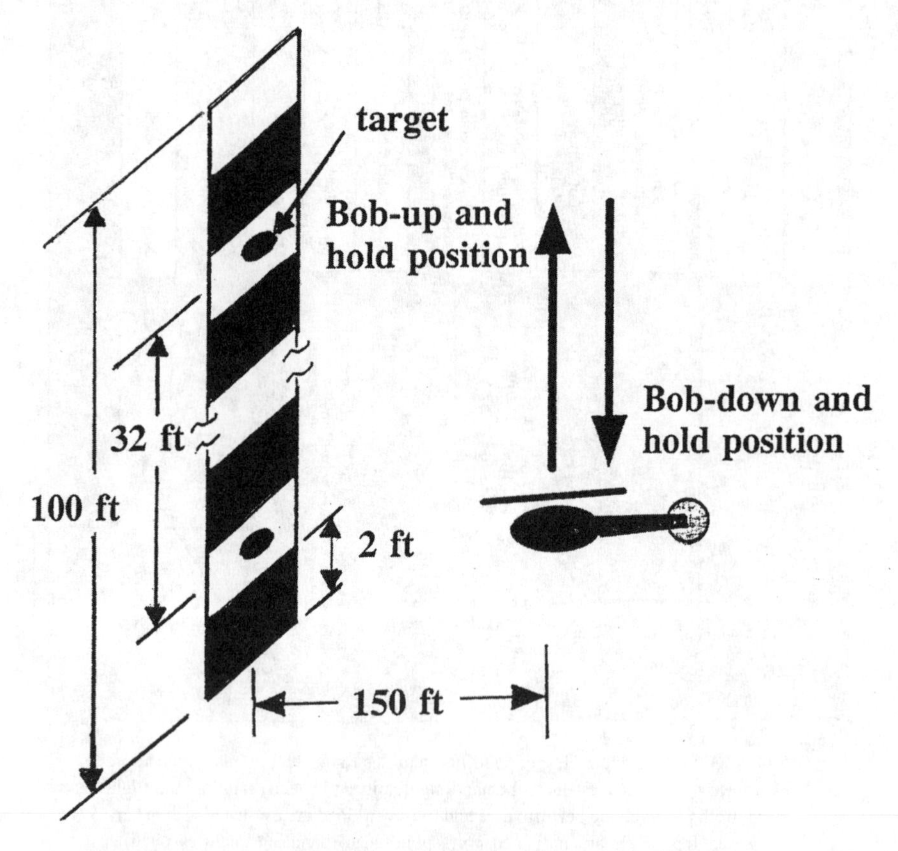

FIG. 8.24. The geometry of a helicopter vertical repositioning task (bob-up maneuver) as simulated at NASA Ames Research Center. Pilot/vehicle performance requirements used in the handling qualities evaluation are given in Table 8.1.

TABLE 8.1
Pilot-in-the-loop simulation bob-up performance requirements

Segment	*Satisfactory Performance*	*Adequate Performance*
Bob-up	Complete maneuver in less than 6 sec*	Complete maneuver in less than 10 sec*
Stabilized hover at target	No vertical excursions from target greater than ± 2 ft	No vertical excursions from target greater than ± 5 ft

*Maneuver completion defined as time to enter and stay within vertical excursion limit.

inner $\dot{h}$ loop has been "closed" through application of the structural model. Y_{p_h} is selected so that the desirable open-loop characteristics of Fig. 8.8 are obtained with the $\dot{h}$ loop closed and with the inner- and outer-loop crossover frequencies separated by a factor of three as suggested by Equation 18.

The mathematical model of the helicopter is quite simple, and only a single control inceptor is involved, the collective inceptor that changes the pitch of the helicopter's main rotor. The transfer function between vertical velocity $\dot{h}(t)$ in ft/sec and collective input $\delta_c(t)$ in inches of inceptor displacement in the cockpit for this vehicle is given by

$$\frac{\dot{h}}{\delta_c}(s) = \frac{(14.6s + 70.34)e^{-0.15s}}{s^2 + 13.022s + 1.5738} \text{ (ft/sec)/inch} \quad (21)$$

where the term $e^{-0.15s}$ represents a pure time delay of 0.15 sec in the vehicle dynamics. The parameters of Equation 21 are specific to the vehicle that was simulated. The dynamics of the collective inceptor are given by

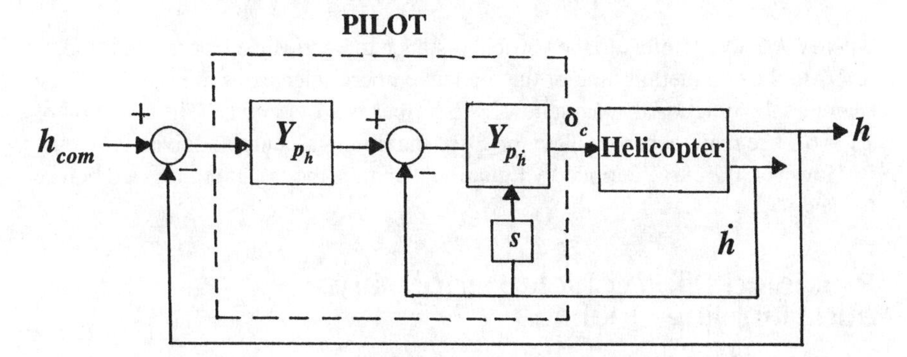

FIG. 8.25. A block diagram representation of the task of Fig. 8.24. The transfer function $Y_{p_{\dot{h}}}$ was implemented using the structural model of Fig. 8.14. The block labeled with the Laplace variable s represents time differentiation of the altitude rate $\dot{h}$ and indicates that vestibular feedback was employed in the structural model. The inner- and outer-feedback loops of the model imply a piloting strategy similar to that of Fig. 8.10.

$$\frac{\delta_c}{\delta_F}(s) = \frac{1}{0.1s + 1} \text{ inches/pound force} \tag{22}$$

where δ_F is the force applied by the pilot to the inceptor.

The structural model parameters were chosen via the computer-based procedure given by Zeyada and Hess (1998) and includes visual, proprioceptive, and vestibular feedback loops.

The Input to the Pilot/Vehicle System

In order to complete the pilot/vehicle model prior to exercising it in a computer simulation, one needs to specify an input command $h_{com}(t)$. The simplest choice would merely be a step command with magnitude of 32 ft. However, use of such abrupt commands leads to unrealistically large initial signal values in the model of the pilot/vehicle system. Problems such as these have also led researchers to consider a pilot model consisting of two modes of behavior: (a) an initial "target-acquisition" mode and (b) an "error reduction" mode. An alternative approach that eliminates these problems can be utilized that involves what can be termed "inverse dynamics." In this approach, we determine the input $h_{com}(t)$ that should be applied to the analytical pilot/vehicle system so that a "desired," smoothly varying height response meeting the ascent requirements in Table 8.1 results. This is accomplished using the closed-loop pilot vehicle transfer function h/h_c and the Laplace transform of the "desired" height response (Zeyada & Hess, 2000). The desired height response can be given in equation form as

$$h_{des}(t) = \left(\frac{DISTANCE}{16}\right) \cdot \left[\cos\left(\frac{3\pi t}{T_d}\right) - 9\cos\left(\frac{\pi t}{T_d}\right) + 8\right] ft \tag{23}$$

where *DISTANCE* refers to the vertical distance to be traversed, here 32 ft, and T_d refers to the completion time of the maneuver, here selected as 6.5 sec. It will be seen that the function that Equation 23 describes is an "s-curve." The selection of $T_d = 6.5$ sec may seem to violate the 6 sec maximum ascent time given previously. However, the curve defined by Equation 23 enters the $\pm$ 2 ft region well before 6.5 sec.

Predicted Pilot/Vehicle Performance and Handling Qualities

Figure 8.26 shows $h_{com}(t)$, $h_{des}(t)$ and the resulting $h(t)$ from a computer simulation of the pilot/vehicle system. Note how well $h(t)$ follows $h_{des}(t)$ and that $h(t)$ easily meets the 6 sec ascent and stabilization requirements. Had the pilot/vehicle system failed to meet the ascent requirement of Table 8.1, the T_d in Equation 23 would have to be decreased. These decrements would continue until either (a) the satisfactory ascent performance was met or (b) control inputs would be produced

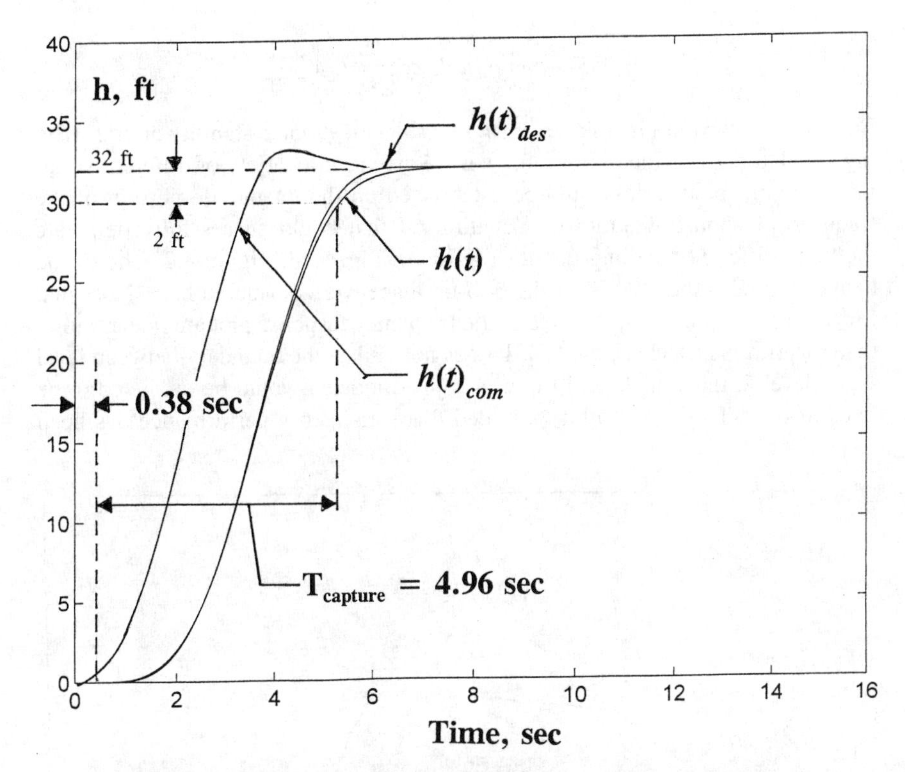

FIG. 8.26. Desired pilot/vehicle height response [$h(t)_{des}$], the commanded input to the pilot/vehicle system [$h(t)_{com}$], and the pilot/vehicle response [$h(t)$] in a computer simulation of the helicopter task of Fig. 8.24. The pilot/vehicle time-to-capture as defined in Table 8.1 is seen to be met by the pilot/vehicle system. A 0.38 sec interval was subtracted from the actual time-to-capture rather than advancing $h(t)_{des}$ by 0.38 sec, as required by the inverse dynamics analysis.

that exceeded the maximum allowable in the cockpit (or implied torque limits on the helicopter main rotor were exceeded). In case (b), the performance would likely fall only into the "acceptable" rather than the satisfactory category.

The explicit performance requirements called out in Table 8.1 mean that evaluation of RMS performance is not an issue here. In addition, the brief, transient nature of the task argued against the inclusion of remnant signals, although they could have been considered.

It has been demonstrated by Hess (1997a) that the structural model of the pilot can be used to predict handling qualities levels as defined in the rating scale of Fig. 8.23. This prediction is based on the magnitude portion of the frequency response diagram of the transfer function U_M/C in the structural model of Fig. 8.14. The magnitude of this transfer function has been called the *Handling Qualities Sensitivity Function* (HQSF),

$$HQSF = magnitude\ of \left[\frac{U_M}{C}(s)\Big|_{s=j\omega} \right] \tag{24}$$

By modeling the pilot in a series of flight tests with variable stability aircraft (flying simulators), in which pilot ratings were given for a variety of different simulated aircraft, boundaries could be established on the magnitude portion of the frequency response diagram of Equation 24 that could successfully delineate the three different handling qualities levels of Fig. 8.23. Figure 8.27 shows the bounds and also the HQSF for the bob-up maneuver considered here. Note that for ease of interpretation, the axes of the frequency response plot are linear, rather than logarithmic. Because the HQSF does not violate the boundary between level 1 and level 2, the vehicle and flight task just modeled would be predicted to be rated as level 1 by a test pilot, provided that satisfactory performance has been

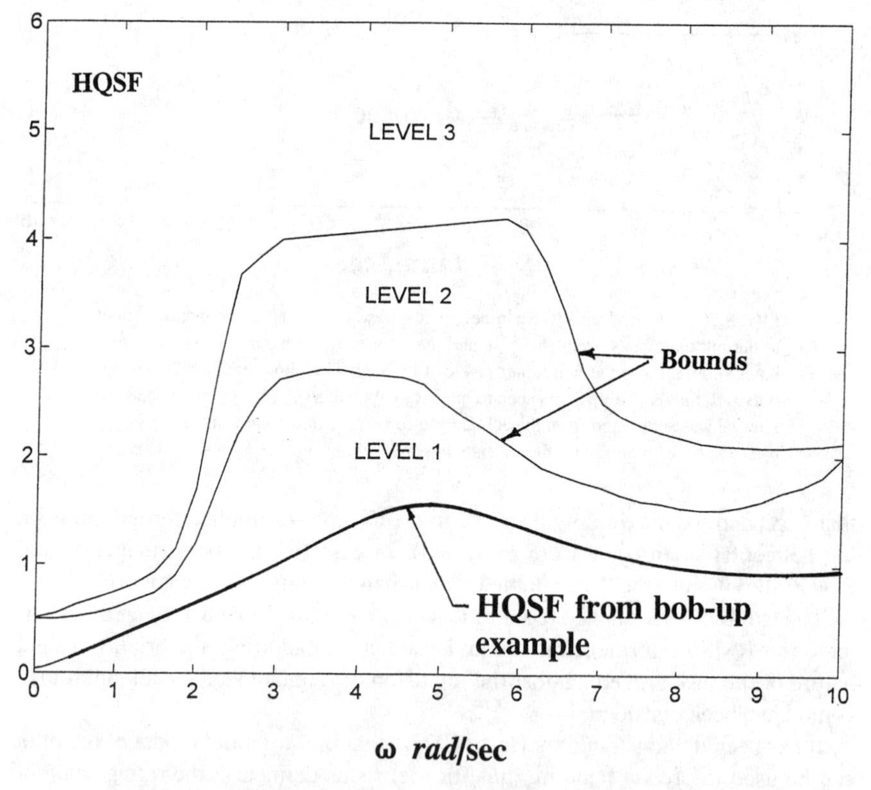

FIG. 8.27. The Handling Qualities Sensitivity Function (HQSF) of Equation 24 resulting from application of the structural model of the pilot to the task of Fig. 8.24. Because the HQSF does not violate either the level 2 or level 3 boundaries on the figure, the vehicle and task are predicted to receive level 1 handling qualities *provided that satisfactory performance as defined by Table 8.1* is also evident in a computer simulation of the feedback pilot/vehicle system of Fig. 8.25.

predicted, which it has. Had only acceptable performance been predicted, but level 1 handling qualities were generated by the model, the vehicle and task would be predicted to receive level 2 handling qualities.

AUTOMATION AND PILOT CONTROL

Motivated by the desire to make flight safer and more efficient, aircraft cockpits are becoming increasingly automated. Wickens (1992) lists the purposes for introducing automation into any system involving a human operator. These are repeated and briefly expanded upon:

1. To perform functions that the human operator cannot perform because of inherent limitations; in aircraft flight control, this would include automatic stabilization systems for aircraft that could not be controlled by the human alone. Some modern high-performance fighter aircraft fall into this category. Often, the aircraft has been aerodynamically configured to reduce radar reflectivity, leaving the vehicle very difficult or impossible to control by the pilot.
2. To perform functions that the human operator can do but performs poorly or at the cost of high workload. Here again, this could include aircraft artificial stabilization. Other examples include automatic terrain warning systems on aircraft and automatic advisories regarding fuel state on military fighters.
3. To augment or assist performance in areas in which humans show limitations. The limitations inherent in human pilot control of multiple tasks or feedback loops constitute an area in which cockpit automation has proven beneficial. For example, many modern transport aircraft possess "auto-throttles." These are systems that automatically adjust engine thrust to allow maneuvering flight, thus relieving the pilot of this task and allowing better human performance on the remaining control tasks.

Despite its benefits, automation is sometimes a mixed blessing, as it can bring problems of its own, as in, for example, Wiener (1988) and Wickens (1992). Among the problems is a requirement for increased monitoring of the performance and health of the automated system(s), a loss or atrophy of basic piloting skills, and the ability of automation to increase the severity of human error by causing it to go undetected or undiagnosed. These potential automation-induced problems have led to the concept of *appropriate automation* in human/machine systems. In the realm of piloted aircraft, appropriate automation in any task can be somewhat loosely defined as that (a) minimizes pilot workload and the possibility of human error, while (b) allowing sufficient pilot activity to maintain proficiency. The latter would imply optimizing handling qualities so that maintaining proficiency does not require extensive, continuous, manual operation. For a more complete discussion of automation issues, see chapter 9.

CONCLUDING REMARKS

The treatment of the subject of pilot control that has been presented has primarily focused on quantitative, mathematical representations of the pilot in a control–theoretic framework. This has been done because this modeling approach has been the most common and successful means of analyzing pilot behavior and designing systems that must act in concert with the pilot, for example, control laws for the flight control computer. Quantitative techniques for examining higher levels of pilot control behavior are far less common and advanced, where we interpret "control" in its most general sense. However, it is these latter techniques that will be increasingly called on to provide guidance for the design of future aircraft cockpits (see chap. 8). The automation of tasks formerly left to the pilot alone will obviously continue. An efficient means for determining the appropriate role of human and machine may be the most pressing need in the study of pilot control in the immediate future. In closing, it should be mentioned that in addition to the studies cited in the main body of this chapter, there exists a sizable literature on the subject of human control of dynamic systems such as aircraft. The following serve as excellent examples: Birmingham & Taylor (1954), Jagacinski, (1977), Wickens (1986), and Baron (1988).

REFERENCES

Baron, S. (1988). Pilot control. In E. L. Wiener & D. Nagel (Eds.), *Human factors in aviation* (pp. 347–385). San Diego, CA: Academic Press.

Birmingham, H. P., & Taylor, F. V. (1954). A design philosophy for man-machine control systems. *Proceedings of the Institute of Radio Engineers, 42,* 1748–1758.

Chan, S. Y., Cheng, P. Y., Myers, T. T., Klyde, D. H., Magdaleno, R. E., & McRuer, D. T. (1992). *Advanced aeroservoelastic stabilization techniques for hypersonic flight vehicles* (Tech. Rep. No. NASA CR 189702). Hampton, VA: NASA Langley Research Center.

Cooper, G. E., & Harper, R. P., Jr. (1969). *The use of pilot rating in the evaluation of aircraft handling qualities* (Tech. Rep. No. NASA TN D-5153). Mountain View, CA: NASA Ames Research Center.

Davidson, J. D., & Schmidt, D. K. (1992). *Modified optimal control pilot model for computer-aided design and analysis* (Tech. Rep. No. TM-4384). Hampton, VA: NASA Langley Research Center.

Glusman, S. I., Landis, K. H., & Dabundo, C. (1986). Handling qualities evaluation of the ADOCS primary flight control system. In *Proceedings of the 42nd Annual Forum of the American Helicopter Society* (pp. 727–737). Washington, DC: American Helicopter Society.

Hess, R. A. (1997a). Feedback control models-manual control and tracking. In G. Salvendy (Ed.), *Handbook of Human Factors and Ergonomics* (pp. 1276–1294). New York: Wiley.

Hess, R. A. (1997b). Unified theory for aircraft handling qualities and adverse aircraft-pilot coupling. *Journal of Guidance, Control, and Dynamics, 20*(6), 1141–1148.

Hess, R. A. (1998). Theory for roll-ratchet phenomenon in high-performance aircraft. *Journal of Guidance, Control, and Dynamics, 21*(1), 101–108.

Hohne, G. (1999). A biomechanical pilot model for prediction of roll ratcheting. AIAA paper no. 99-4092, *AIAA Atmospheric Flight Mechanics Conference* (pp. 187–196). Portland, OR.

Innocenti, M. (1988). The optimal control model and applications. *AGARD Lecture Series, 157,* 7-1–7-17.

Jagacinski, R. J. (1977). A qualitative look at feedback control theory as a style of describing behavior, *Human Factors, 19,* 331–347.

Jex., H. R. (1971). Problems in modeling man-machine control behavior in biodynamic environments. In *Proceedings of the Seventh Annual Conference on Manual Control* (Tech. Rep. No. NASA SP-281, pp. 3–13). Los Angeles: University of Southern California.

Johnston, D. E., & Aponso, B. L. (1988). *Design considerations of manipulator and feel system characteristics in roll tracking* (Tech. Rep. No. NASA CR-4111). Edwards, CA: Dryden Flight Research Facility, NASA.

Kleinman, D. L., Baron, S., & Levison, W. H. (1970). An optimal control model of human response, Part I. *Automatica, 6,* 357–369.

Levison, W. H., Kleinman, D. L., & Baron, S. (1969). A model for human controller remnant. *IEEE Transactions on Man-Machine Systems, MMS-10,* 101–108.

Maciejowski, J. M. (1989). *Multivariable feedback design.* Reading, MA: Addison-Wesley.

McRuer, D. T., Graham, D., Krendel, E., & Reisener, W., Jr. (1965). *Human pilot dynamics in compensatory systems* (Tech. Rep. No. AFFDL-65-15). Dayton, OH: U.S. Air Force Flight Dynamics Laboratory.

McRuer, D. T., & Krendel, E. (1974). Mathematical models of human pilot behavior. *AGARDograph No. 188.*

Mitchell, D. G., Aponso, B. L., & Klyde, D. H. (1992). *Effects of cockpit lateral stick characteristics on handling qualities and pilot dynamics* (Tech. Rep. No. NASA CR 4443). Edwards, CA: NASA Dryden Flight Research Facility.

Nise, N. (2000). *Control systems engineering* (3rd ed.). Menlo Park, CA: Addison-Wesley.

Phillips, C. A. (2000). *Human Factors Engineering.* New York: Wiley.

Proctor, R. W., & Proctor, J. D. (1977). Sensation and perception. In G. Salvendy (Ed.), *Handbook of Human Factors and Ergonomics* (pp. 43–88). New York: Wiley.

Reidel, S., Jex. H. R., & Magdaleno, R. (1980). Description/demonstration of "biodyn-80." In *Proceedings of the 16th Annual Conference on Manual Control* (pp. 317–339). Cambridge, MA: Massachusetts Institute of Technology.

Schroeder, J. A., Chung, W. W. Y., & Hess, R. A. (1999). Spatial frequency and platform motion effects on helicopter altitude control. AIAA paper no. 99-4113. *AIAA Modeling and Simulation Conference and Exhibit,* Portland, OR.

Tustin, A. (1947). The nature of the operator's response in manual control and its implication for controller design. *Journal of the Institute of Radio Engineers, 94,* Part IIA, No. 2.

van Paassen, R. (1994). *Biophysics in aircraft control: A model of the neuromuscular system of the pilot's arm.* Ph.D. dissertation, Delft University of Technology, Faculty of Aerospace Engineering, Delft, The Netherlands.

Wickens, C. D. (1986). The effects of control dynamics on performance. In K. Boff, L. Kaufman, & J. Thomas (Eds.), *Handbook of perception and performance* (Vol. II, chap. 39). New York: Wiley.

Wickens, C. D. (1992). *Engineering Psychology and Human Performance* (2nd ed.). New York: HarperCollins.

Wiener, E. L. (1988). Cockpit automation. In E. L. Wiener & D. C. Nagel (Eds.), *Human Factors in Aviation* (pp. 433–461). San Diego, CA: Academic Press.

Zeyada, Y., & Hess, R. A. (1998). Pilot/vehicle dynamics—an interactive computer program for modeling the human pilot in single-axis linear and nonlinear tracking tasks. Department of Mechanical and Aeronautical Engineering, University of California, Davis.

Zeyada, Y., & Hess, R. A. (2000). Modeling human pilot cue utilization with applications to simulator fidelity assessment. *Journal of Aircraft, 37*(4), 588–597.

9

Automation and Human Performance in Aviation

Raja Parasuraman
The Catholic University of America

Evan A. Byrne
National Transportation Safety Board

BACKGROUND

Technology has revolutionized commercial aviation in the 20th century. The major development was the introduction since the 1970s of advanced computers into the aircraft cockpit and, less extensively, in air traffic control (ATC). Rapid growth in the availability of cheap and powerful microprocessors, as well as in sensor and control technologies, led to the widespread introduction of automation in aviation. The trend toward greater implementation of advanced automation will continue well into the new millennium.

Automation has markedly improved efficiency and allowed for greater flexibility in aviation operations. The accident rate is also lower than for the previous generation of aircraft (e.g., Boeing Commercial Airplane Group, 1998). Nevertheless, over the years, numerous incidents and new types of accidents have raised safety concerns about cockpit automation (Aviation Week & Space Technology, 1995; Billings, 1997). A major additional concern is that although the accident rate has remained relatively stable for many years, the volume of air traffic is projected to increase by up to 100% over the next decade. Consequently, it is estimated that in the near future one major commercial airline accident may occur *every week* on average in some part of the world (Aviation Week & Space Technology, 1998;

Flight Safety Foundation, 1997; Huettner, 1996). This is undoubtedly one of the reasons behind the call for a fivefold improvement in safety by the Gore Commission on Aviation Safety and Security (1997). In response to this challenge, the Federal Aviation Administration (FAA) and the National Aeronautics and Space Administration (NASA) have outlined a joint plan to improve safety by a factor of five and capacity by a factor of three over the next decade (Huettner, 1996).

Given that the trend toward greater automation is likely to persist, safety will depend critically on further understanding of human interaction with automation. If the ambitious safety and capacity goals that have been set by the FAA and by NASA are to be achieved, this knowledge must be translated into effective design of automation. New and appropriate training procedures for users of automated systems will also have to be developed (see, for example, Air Transport Association, 1989; Wiener, Chute, & Moses, 1999). These needs apply not only to automated systems in the flight deck, but also to ATC and to other components of the National Airspace System (NAS).

The technical aspects of aircraft automation—automation functions and their implementation, the underlying algorithms, and the characteristics of the associated sensors, controls, and software—have been widely described. The *human* aspects of automation have been less well studied. In recent years, however, a small but growing body of research has examined the characteristics of human operator interaction with automation (Amalberti, 1999; Bainbridge, 1983; Billings, 1997; Chambers & Nagel, 1985; Lewis, 1998; Parasuraman, 2000; Parasuraman & Riley, 1997; Parasuraman, Sheridan, & Wickens, 2000; Rasmussen, 1986; Sarter, Woods, & Billings, 1997; Satchell, 1998; Sheridan, 1992; Wickens, Mavor, Parasuraman, & McGee, 1998; Wiener, 1988; Wiener & Curry, 1980).

Factors Contributing to Effective Human-Automation Interaction

In this chapter, we review the previous decade of research on human performance in automated systems in relation to applications in aviation. We also analyze a number of aviation incidents and accidents that have been associated with automation usage. We discuss several factors that have been found to be important for effective human interaction with automated systems.

What are the general principles of effective human-automation interaction? It has been suggested that automation must be *human centered* to be successful. There are many meanings of this term (see Wickens et al., 1998). The definition proposed by Billings (1997, p. 7), "automation designed to work cooperatively with human operators in pursuit of stated objectives," emphasizes that automation functionality should be designed to support human performance and human understanding of the system (see also Riley, 1997). Billings (1997) suggested a number of broad requirements for human-centered automation, beginning with the axiom that pilots must remain in command of their flights. From this axiom, he drew several corollary principles. Chief among them is that the pilot must be actively involved in the system and adequately informed about the status of the automation. Billings (1997) also proposed that humans and automation must understand each other's intent in complex systems (see also Woods, 1996).

Empirical studies of human performance in automated systems generally support these principles but also provide evidence for somewhat more specific instantiation of these principles. The empirical research literature has identified several factors that appear to contribute to effective human-automation interaction. We organize our review of the research and of aviation incidents around the following framework of six general issues:

1. Identifying types (stages) of information processing to which automation is applied
2. Choosing the appropriate level of automation
3. Making automation state indicators and behaviors salient and understandable
4. Calibrating operator trust in automation
5. Balancing operator mental workload
6. Considering dynamic function allocation, or adaptive automation

Overview of Chapter

The structure of this chapter is as follows. We first offer a definition of automation and provide an overview of aviation automation. We then describe a model for the types and levels of automation (Parasuraman, Sheridan, & Wickens, 2000). The model covers the first two factors that we have identified as contributing to effective human-automation interaction. The model also provides a framework for evaluating the human performance benefits and costs of automation. We then discuss automation-related incidents from a broad perspective that considers not only advanced automation but also simpler automated aids and warnings. As previously noted, such incidents have been a primary motivation for research into human interaction with automated systems. The next section describes this research and considers the human performance benefits and costs of automation. The topics covered include mode errors, situation awareness, decision biases, trust in automation, including overtrust (complacency) and undertrust (automation disuse), and mental workload. This section is followed by a brief overview of research on adaptive automation. We outline a scheme for adaptive automation based on matching adaptation to balanced operator workload. Finally, we examine a relatively recent research development, computational and formal modeling of human-automation interaction.

Our coverage emphasizes empirical studies that advance theoretical understanding of human performance in automated systems, with frequent reference to basic research and to work outside aviation proper whenever appropriate.

Automation has been implemented widely in other domains, and further reductions in the size and cost of computers will result in far-reaching applications to virtually all aspects of everyday life (Rawlins, 1996). Understanding the common features as well as the differences between aviation and these other domains will advance both basic and applied research on human-automation interaction (Parasuraman & Mouloua, 1996).

Despite the number of topics discussed, our coverage is not comprehensive and is limited as follows. We refer to specific automated systems but do not describe such systems in detail (e.g., see Billings, 1997, chap. 5; Spitzer, 1987). Also, we focus on aircraft automation, although automation is increasingly affecting ATC (Parasuraman, Duley, & Smoker, 1998; Smith et al., 1999; Wickens et al., 1998). Automation is also being implemented with the move toward Free Flight, in which pilots will be granted greater authority in routing and separating their aircraft from others in the airspace (RTCA, 1995). Free Flight has significant human factors implications both in the air (Wickens, 2000) and on the ground (Galster, Duley, Masalonis, & Parasuraman, 2001), but these are not discussed here because of space limitations (see Parasuraman, Hilburn, & Hoekstra, 2001). Finally, we do not consider the impact of automation on team performance in the cockpit (Bowers, Oser, Salas, & Cannon-Bowers, 1996) or on organizational behavior and structure (Satchell, 1998).

AVIATION AUTOMATION

Definition

Automation can be defined as the execution by a machine agent of a function previously carried out by a human (Parasuraman & Riley, 1997). The machine that takes over a human function will generally be a computer. Furthermore, given the rapid growth in the speed, capacity, and budding "intelligence" of computers, it is likely that most future automation will involve computer hardware and software of one kind or another.

This definition of automation can be linked to the classic human factors issue of *allocation of function* between human and machine (e.g., see Kantowitz & Sorkin, 1987). As traditionally defined and used, function allocation procedures have not significantly impacted on system design in practice, including automated system design. Numerous reasons for this failure have been discussed (e.g., Billings, 1997; Jordan, 1963; Sheridan, 1998), but a major factor must be that the traditional concept assumes a *static* allocation process that is conducted once prior to system development. It has been pointed out, however, that *dynamic* function allocation during system operations and not just during system development will provide for effective harmonization of human and machine capabilities (Hancock & Scallen, 1996; Parasuraman, 1990). Such a dynamic conceptualization is more closely associated with the definition of automation we propose than a static approach because it allows for the complementary sharing of functions between human and machine (see also Moray, Inagaki, & Itoh, 2000). We discuss dynamic function allocation in a later section when we describe research on *adaptive* automation.

Trends in Aviation Automation

Applying our definition of automation to the domain of aviation reveals the gradual and increasing replacement by machines and computers of functions once carried out by the flight crew. The evolution of aircraft automation can be roughly classified as involving two phases, automation of primary flight control and automation of flight management (Billings, 1997). Increasingly, capable autopilots were the most visible changes in the automation of primary flight control. Integrated navigation instruments were also introduced, the most prominent being the horizontal situation indicator and, later, enhanced map displays. The introduction of the flight management system (FMS) in the 1980s marked a significant turning point in aircraft automation, as it fundamentally changed the nature of piloting by allowing multistep, horizontal, vertical, and velocity navigation commands to be executed autonomously. In addition to increased "intelligence" provided by autoflight systems, the "flight envelope," or the airplane's speed and state limitations, also became subject to automated control and monitoring. Most recently, fly-by-wire systems have further changed the nature of the interface between the pilot and aircraft control mechanisms.

The result of the implementation of these systems has led to a steady but growing "distancing" of the pilot from the direct, inner-loop control of the aircraft. Figure 9.1, taken from Billings (1997), illustrates this trend in cockpit automation from the 1920s to the 1980s. The increased complexity, when coupled with the fact that automation states and behaviors are often not well understood by the flight crew (Parasuraman, 2000; Sarter & Woods, 1994; Sherry & Polson, 1999), has led to concerns about safety when unexpected or emergency conditions develop (Aviation Week & Space Technology, 1995).

The recent focus of much research on human interaction with automation in aviation has been on the FMS and its accompanying complexity (Sarter & Woods, 1995). Nevertheless, there remain many forms of less-advanced automation on the flight deck, such as automated aids and warnings, that also deserve consideration. Aircraft stall warnings represent a simple example. Other examples include the Ground Proximity Warning System (GPWS), a system for alerting pilots when the aircraft is dangerously close to terrain (when not landing), and the Traffic Alert and Collision Avoidance System (TCAS), which warns pilots of other aircraft in the immediate airspace.

Many systems currently under development will add to this list. Among the first new systems likely to be incorporated in the aircraft cockpit is the Cockpit

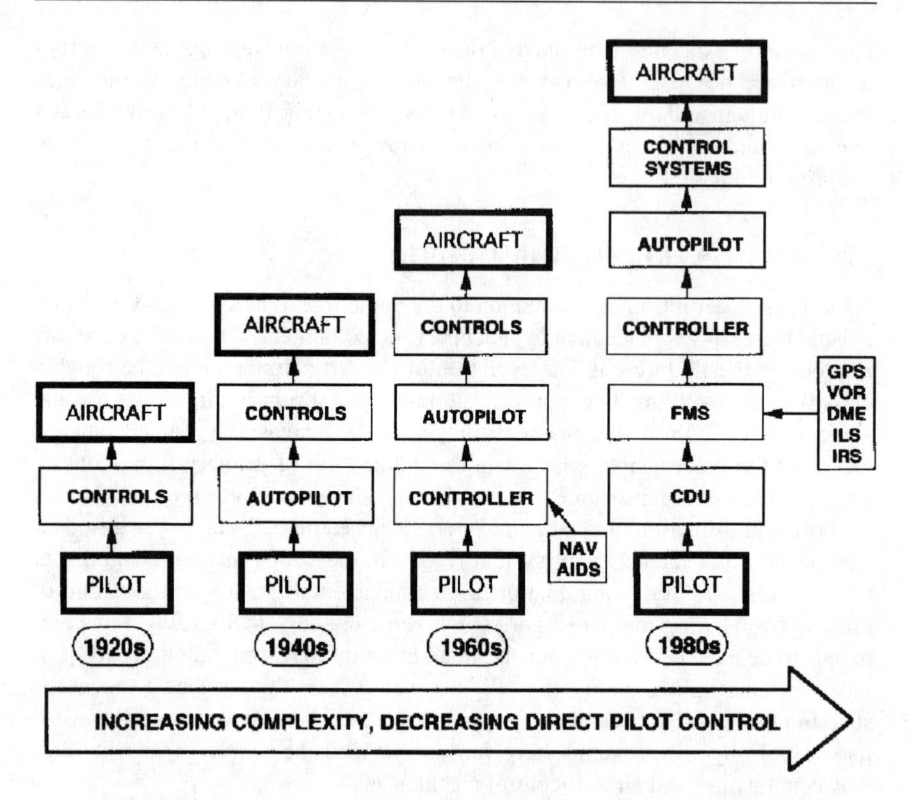

FIG. 9.1. Trends in cockpit automation showing greater complexity and increased "distancing" of the pilot from the aircraft control systems. *Note.* From *Aviation Automation: The Search for a Human-Centered Approach* (p. 36), by C. E. Billings, 1997. Mahwah, NJ: Lawrence Erlbaum Associates, Inc. Copyright 1997 by Lawrence Erlbaum Associates, Inc. Reprinted with permission.

Display of Traffic Information (CDTI), which represents a maturation of the current TCAS and will provide a representation of air traffic in the airspace surrounding the aircraft. CDTI is anticipated to be needed given the move toward Free Flight. Two-way electronic data link will also alter the nature of pilot–controller communications. Other technologies that support Free Flight include the Global Positioning System (already increasingly used in all facets of aviation) and Automatic Dependent Surveillance—Broadcast (ADS—B), which will allow aircraft positions to be determined accurately by other aircraft and by ground-based systems, independent of less-accurate radar. Finally, these airborne automated systems will be complemented by ground systems to assist air traffic controllers, such as the Center Tracon Automation System (CTAS) and the User Evaluation and Request Tool (URET).

A MODEL FOR TYPES AND LEVELS OF HUMAN INTERACTION WITH AUTOMATION

Our brief overview indicates that aviation automation has taken many forms, from simple to complex. At the same time, the discussion to date has implied that automation is all or none, that is, that a function is either automated or not. However, automation can also vary in level. For example, Sheridan (1992) proposed that several levels can be identified between the extremes of total manual control and full automation (see Table 9.1). From the perspective of our definition of automation, higher levels of automation correspond to increasing replacement of functions formerly carried out by the human by the machine. Under Sheridan's (1992) scheme, the different levels of automation also vary in their degree of authority. At the highest level, the automation has complete authority. At this level, the human does not have access to the operations of the automation, only its products. At moderate levels, on the other hand, the automation might suggest alternatives for decision, but the human retains authority for executing that alternative or choosing another one (see Table 9.1). Broadly similar schemes for levels of automation authority have been proposed by others (Billings, 1997; Endsley & Kaber, 1999; Riley, 1989).

It is also useful to dissect automation by functional dimensions (Parasuraman, Sheridan, & Wickens, 2000; Wickens et al., 1998). This leads to different *types* of automation, each of which can have several levels (see Fig. 9.2). The scale in Table 9.1 applies mainly to automation of *decision* and *action* selection, or *output* functions. However, automation may also be applied to *input* functions, that is, to information gathering, analysis, and integration. For example, a fault-management system might sense the values of several aircraft system variables, integrate them in some manner, and provide a predictive display of system state to the pilot. This type of automation qualifies as information automation because the display

TABLE 9.1
Levels of automation of decision and action selection.

HIGH	10. The computer decides everything, acts autonomously, ignoring the human.
	9. Informs the human only if it, the computer, decides to
	8. Informs the human only if asked, or
	7. Executes automatically, then necessarily informs the human, and
	6. Allows the human a restricted time to veto before automatic execution, or
	5. Executes that suggestion if the human approves, or
	4. Suggests one alternative
	3. Narrows the selection down to a few, or
	2. The computer offers a complete set of decision/action alternatives, or
LOW	1. The computer offers no assistance: human must take all decisions and actions.

Adapted from Sheridan (1992).

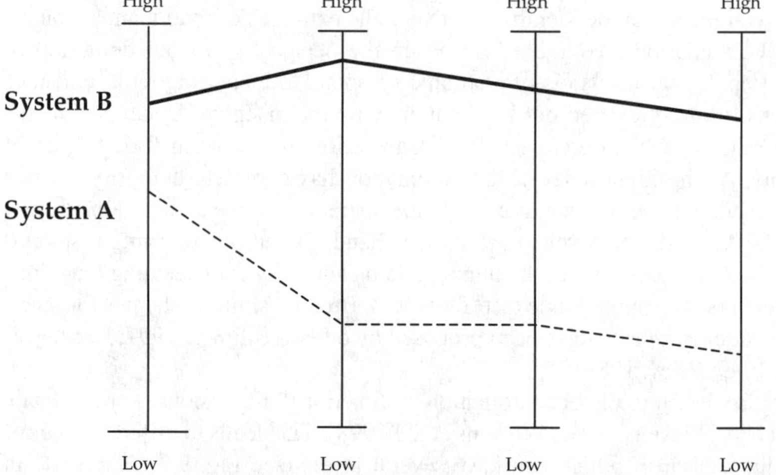

FIG. 9.2. Levels of automation for independent functions of information acquisition, information analysis, decision selection, and action implementation. Examples of systems with different levels of automation across functional dimensions are also shown. *Note.* From "A Model for Types and Levels of Human Interaction with Automation," by R. Parasuraman, T. B. Sheridan, and C. D. Wickens, 2000, *IEEE Transactions on Systems, Man, and Cybernetics, 30,* pp. 286–297. Copyright 2000 by the Institute of Electrical and Electronics Engineers, Inc. Reprinted with permission.

provides information relevant to some decision or action that might be taken, but does not directly provide a decision choice or carry out the action (although another automated system might do that). The distinction between these types of automation is similar to the difference between control automation and information automation proposed by Billings (1997), except that we distinguish four types of automation, and in our conceptualization the different types of automation vary in level:

1. Information acquisition
2. Information analysis
3. Decision and action selection
4. Action implementation

Each of these functions can be automated to differing degrees, or many levels. The 10-level classification shown in Table 9.1 applies only to decision automation. A more general classification applicable to all types of automation might involve, say, five levels, as follows: (a) none, (b) low, (c) medium, (d) high, and (e) full.

Automation of information acquisition applies to the sensing and registration of input data, operations that are equivalent to the first human information-processing stage, sensory processing. At the lowest level, such automation may simply consist of using and integrating sensors in order to obtain data for further observation and analysis. A moderate level of automation may involve *organization* of incoming information according to some criteria, for example, a priority list and *highlighting* of some part of the information. Organization and highlighting preserve the visibility of the original information ("raw" data). This is not necessarily the case with the higher level automation function of *filtering,* in which certain items of information are selected and brought to the operator's attention (Wickens, 2000). Filtering automation is exemplified by systems for decluttering displays according to the phase and type of flight (see chap. 5).

Automation of information analysis involves higher cognitive functions such as working memory and inference. At a low level of this type of automation, algorithms can be applied to incoming data to allow for their extrapolation over time, or *prediction,* for example, a cockpit display to show the projected future course of another aircraft in the neighboring airspace (Hart & Wempe, 1979). A higher level of automation involves *integration,* in which several input variables are combined into a single value, as in safety parameter displays in process control (Moray, 1997).

Automation of decision and action selection can vary in level according to the degree of autonomy given to the computer, as described previously and in Table 9.1. The various forms of autopilots represent examples of this type of automation. Low-level autopilots replace the pilot's need to compensate for turbulence by automatically adjusting the aircraft attitude. Higher level automation is represented by the Flight Management Computer (FMC), which replaces the pilot's need to decide on the flight path corrections required to achieve a particular heading or altitude, and the Flight Management System (FMS), which allows for automation of control inputs to achieve complex flight goals such as intercepting a navigational fix at a specified time, entering a holding pattern, or even landing the airplane.

Finally, the action implementation scale refers to the actual execution of the action choice (e.g., "pressing the button"). Often this scale will have only two levels, manual and automatic, as opposed to a large number of levels in the other scales.

The four functional dimensions correspond roughly to stages of human information processing. Automation in each dimension can vary across several levels, from low to high. Different systems will involve automation of one or more of these three dimensions at different levels. Thus, for example, a given system (A) could be designed to have moderate to high information automation, low decision automation, and low action automation. Another system (B), on the other hand, might have high levels of automation across all three dimensions (see Fig. 9.2).

This model of types and levels of automation can be used to examine the impact of different automation designs, including full automation, on system performance. The model also provides a framework to examine the human performance consequences of different forms of automation. Distinguishing between stages of information processing to which automation is applied represents the first of the six factors we have proposed. The second proposes that the choice of the level of automation within each of these stages, but particularly within the decision stage, is also critically important (Parasuraman, Sheridan, & Wickens, 2000). The implications of both these factors are discussed further in a later section when we review the human performance costs and benefits of automation. Before describing this research, however, we discuss a number of aviation incidents and accidents that have pointed to the critical role of automation in aviation safety.

AUTOMATION-RELATED INCIDENTS IN AVIATION

Accident Investigation

The system performance benefits of aviation automation have been proven, and the safety record of automated aircraft has been good. Nevertheless, several highly publicized incidents and accidents involving automated aircraft, coupled with the quest for even greater safety levels, has motivated greater scrutiny of automation by the aviation industry. In June 1996, the FAA released a report that examined automation with respect to design, flight crew training/qualifications, and operational issues for potential safety problems (Abbott et al., 1996). The report outlined several issues concerning potential vulnerabilities in flight crew management of automation. Several recommendations were made to reduce potential sources of error that may occur in automated flight decks. The report has contributed to the establishment and training of "automation policies" that address the appropriate use and management of automation for some air-carrier operations (see, for example, reports from the Air Transport Association Human Factors Automation Subcommittee [Aviation Week & Space Technology, 1999], and Wiener et al., 1999).

Although potential problems have been identified in the flight crew management of automation, a succinct definition of an automation-related incident or accident is not easily achieved. We have offered a definition of automation that is linked to system functions, and we have emphasized the human use of automation. Unfortunately, there is an inherent lack of specificity in the term *automation* as it is used in industry and in accident investigation. Because there are many variants of technology in the flight deck related to automation, it is difficult to readily generate statistics that adequately capture the frequency of accidents and incidents involving automation. This task is further complicated because most accidents and incidents, especially those involving human factors, are the result of multiple precipitating occurrences and conditions ultimately leading to the event (e.g., Reason, 1990; see also Wiegmann & Shappell, 1997).

The National Transportation Safety Board (NTSB) has an Aviation Coding Manual that is used to document findings and assign causal and contributing factors to the accidents in its Aircraft Accident Database (www.ntsb.gov/avdata/codman.pdf). This database contains hundreds of codes and subcodes that analysts select from in classifying accidents. These codes, which involve both specific aircraft systems as well as pilot behaviors and characteristics (e.g., autopilot, FMS, training, fatigue), are also used in the search and retrieval of information from this database. For example, accidents involving deficient pilot training are easily surveyed by using the index terms for training. However, the coding manual has no single category for "automation." As a result, a user must probe for the specific hardware component. (e.g., FMC, TCAS, GPWS) and cross these with pilot operator factors (i.e., workload, lack of situation awareness, complacency) to identify automation-related incidents.

Notwithstanding these difficulties, the NTSB database can be used to identify automation-related accidents. Another resource is the Aviation Safety Reporting System (ASRS), which can be examined with appropriate keywords to identify automated-related incidents (Funk et al., 1999). We focus on incidents and accidents related to (a) feedback about automation, (b) misuse of automation, and (c) misunderstanding of automation. Collectively, these illustrate the consequences of violation of the principle that automation state indicators and behaviors should be salient (Parasuraman & Riley, 1997). These three issues were also among the top five of over 100 automation-related issues identified in a survey of aviation experts by Funk et al. (1999) (the other two issues were poor display design and inadequate automation training).

"Fail Silent" Problems and Unforeseen Limits

In 1972 an L-1011 crashed in the Florida Everglades while the crew attempted to troubleshoot a problem with a landing gear indication light and did not recognize that the altitude hold function of the autopilot had been inadvertently disconnected (National Transportation Safety Board [NTSB], 1973). Although this accident had several contributing factors, one factor that has captured the attention of human factors psychologists was the poor feedback on automation state provided by the system (e.g., Norman, 1990). The disengagement of automation should be clearly signaled to the human operator so that it can be validated as an intended or unintended disengagement. In the L-1011 accident, the principle that automation states should be salient was clearly violated. Most current autopilots now provide an aural and/or a visual alert on disconnect. The alert remains active for a few seconds or requires a second disconnect command input by the pilot before it is silenced. Persistent warnings such as these, especially when they require additional input from the pilot, are intended to decrease the chance of an autopilot disconnect or failure going unnoticed.

In 1997, 25 years after the L-1011 accident, an A300 experienced an in-flight upset off the coast of Florida (NTSB, 1998c). During descent into the terminal area, flight recorder information indicated that the autothrottles were holding speed constant at the start of the descent. However, after level off at an intermediate altitude, the autothrottles were no longer controlling airspeed, and the airplane slowed gradually to almost 40 knots below the last airspeed set by the pilots and subsequently experienced an in-flight upset after the stall warning activated. No evidence was found that the autothrottles malfunctioned, and therefore it appears that the pilots believed that the automated system was controlling airspeed. After further examination, it was learned that the autothrottle system could be disengaged by a single press of the disconnect button. When the system was disengaged, the green mode annunciator in the primary flight display would change to amber, and the illuminated button on the glare shield used to engage the system would extinguish. The NTSB noted in a letter to the FAA that although the change in the annunciators could function as a warning, the format of the displays did not command attention because they were passive and persistent. In its letter to the FAA, the NTSB pointed to autothrottle disconnect warning systems in other transport category airplanes that required positive crew action to silence or turn off. These systems incorporated flashing displays and, in some cases, aural alerts that would serve to help capture the pilot's attention in the case of an inadvertent disconnect. These systems more rigorously adhere to the principle of automation state saliency.

Misuse of Automation

Alarms and Alerts. The modern commercial aircraft cockpit has, in addition to advanced flight management automation, numerous less-complex automated aids and warnings. Alarms and alerts are pervasive and, if not heeded appropriately, can lead to adverse situations. Such alarms have been installed in aircraft because humans are not very good at monitoring infrequently occurring events because of declines in vigilance, particularly during long-duration flight (Parasuraman, 1987). If the alarm is sufficiently salient (e.g., a loud sound or a flashing light), then pilots are more likely to notice and attend to the hazard that led to the alarm. Unfortunately, many automated alerting systems can elicit false alarms. Early versions of the GPWS and TCAS systems were plagued with this problem. False alerts can lead to two difficulties. First, believing the alarm to be true, the pilot can take an evasive action that results in an incident. Second, due to the frequency of false and nuisance alarms, the pilot may ignore or even turn off a warning system (Satchell, 1993). This "cry wolf" syndrome may not lead to any untoward incident except on the one occasion when the alarm is true but the pilot continues to ignore it. In a later section we discuss research that has addressed this issue and that has provided options for minimizing this problem (Parasuraman, Hancock, & Olofinboba, 1997). First, however, we describe some incidents related to inappropriate use of alarm systems. The stall warning serves as a good example of a relatively simple alerting system. We begin with an incident that involved a false alarm that the pilot responded to inappropriately.

In 1992 an L-1011 ran off the end of the runway at JFK Airport (NTSB, 1993). Immediately after takeoff, the stick-shaker activated, and the crew responded by landing on the remaining runway, but the airplane could not be stopped before running off the end. The investigation determined that the alert was the result of a malfunctioning angle-of-attack sensor and therefore represented a false alarm. However, the pilots perceived that the airplane was about to stall. Several years later, the flight crew of a Beach 1900 commuter airplane experienced a similar situation. In this accident (NTSB, 1997b), the crew received a false stall warning at takeoff, and the airplane was damaged in the resulting attempt to land on the remaining runway. In both situations, the pilots' actions suggest that the validity of the alarm was never confirmed using other sources of information. For example, in each situation, airspeed, pitch, and power indications would have been available to indicate to the pilot that the stall warning was a false alarm.

Automated alerting systems can make two types of error—false alarms and omissions. A more catastrophic accident occurred in 1996 that involved an omitted alert, in contrast to the false alarm situations just described. In this accident (NTSB, 1997a), a DC-8 was being evaluated after it had undergone modifications and maintenance. One of the requirements was to stall the airplane to record both the speed at which the stick-shaker was triggered and the actual speed of the stall. The airplane had a three-person crew, and the pilot flying had been trained how to execute and recover from this maneuver in a simulator but had never performed it in actual flight. When the pilot initiated the maneuver at altitude, aerodynamic indications of an approaching stall began, and the flight engineer, who had experienced several of these flights before in an airplane, called out "that's a stall right there . . . ain't no [stick] shaker." The pilot increased power to start recovery but never moved the control column forward to reduce the pitch attitude and break the stall. For several seconds, the airplane's pitch attitude remained constant as altitude and airspeed decreased. The airplane then entered a deep stall condition and an unarrested descent into the ground below. It is notable that the stick-shaker, for undetermined reasons, failed to alert in this situation because the recovery procedure that the pilot used in the simulator practice sessions (in which the stick-shaker was always functional) stated "At stick-shaker or initial buffet, apply and call out Set Max Power. Simultaneously reduce back pressure sufficiently to stop the stick-shaker" (NTSB, 1997a). Therefore, the pilot would have learned to rely on the stick-shaker not only as a warning device (for which it was designed) but also as an integrated aid or tactile/aural display that provided an indication of performance during the recovery. Even though the primary cues associated with deteriorating airplane performance (loss of altitude and airspeed) were available, they were not heeded, and the pilot never released the control column.

These accidents are examples of a decision bias known as automation misuse, or automation bias (Mosier, Skitka, Heers, & Burdick, 1998; Parasuraman & Riley, 1997; Skitka, Mosier, & Burdick, 1999). This refers to the tendency of human operators to substitute acceptance of the output of an automated system for vigilant checking of information sources. (We discuss the characteristics of this phenomenon and the related concept of automation complacency in more detail in later sections of the chapter.) The failure to monitor instruments and other information sources as a result of overreliance on automation has been associated with numerous aviation incidents and accidents. In an analysis of 200 ASRS incident reports involving monitoring failures, Sumwalt, Morrison, Watson, and Taube (1997) found that 30% of the errors occurred while pilots were engaged in programming the FMS.

The source of automation bias in the aforementioned accidents may have its place in the way pilots are trained. For example, in primary flight training (i.e., private pilot certificate), students are first exposed to the aerodynamic effects and precursors to a stall. Most light airplanes have some sort of stall warning device to alert the pilot to an impending stall. Because the warning alerts before the actual stalled condition of the wing, flight instructors often need to encourage the student to delay the initiation of their recovery until the actual stall occurs so that they are able to learn the aerodynamic cues associated with the onset of a stall (loss of control effectiveness, pitch change). Indeed, the standards by which the student is evaluated state ". . . [the student] recognizes and announces the first aerodynamic indications of the oncoming stall, i.e., buffeting or decay of control effectiveness" (Federal Aviation Administration [FAA], 1997).

But, as a pilot earns additional ratings and certificates and advances to more complex airplanes, the nature of stall training and stall recognition change. The standards by which the Airline Transport Pilot (ATP) candidate is evaluated states ". . . [the pilot] announces the first indication of an impending stall (such as buffeting, stick-shaker, decay of control effectiveness, and any other cues related to the specific airplane design characteristics) and initiates recovery" (FAA, 2001). In this case, stall recognition now becomes approach to stall, and a stall-warning device, the stick-shaker, is the basis of starting the recovery. In addition, the stick-shaker is often used during recovery procedures from other conditions, such as windshear avoidance, as a performance indicator to gauge the effectiveness of recovery maneuvers. Therefore, as compared to its use during initial flight training, the stall warning changes from an early warning device of impending condition to the actual condition itself and, in some cases, as a method to determine the effectiveness of a recovery maneuver.

Overreliance on Automation. Overreliance on automation was cited in the probable cause of a DC-10 landing accident in 1984 in New York (NTSB, 1984). The airplane touched down just over halfway down the runway and was unable to stop on the remaining runway. The NTSB determined that the probable cause of this accident was the flight crew's (a) disregard for prescribed procedures for monitoring and controlling of airspeed during the final stages of the approach, (b) decision to continue the landing rather than to execute a missed approach, and (c) overreliance on the autothrottle speed control system, which had a history of recent malfunctions. This example, like the A300 autothrottle example previously described, is consistent with pilot overreliance on control automation.

There are also several instances of accidents and incidents involving overreliance on navigational support automation. Perhaps most notable in the last few years is the 1995 crash of a Boeing 757 near Cali, Colombia. The airplane was under autoflight guidance, and it crashed into a mountain after the wrong navigational beacon was inadvertently selected from a list in the navigational database. The errant selection guided the airplane into hazardous terrain instead of toward the airport, and the pilots failed to detect this error. The government of Colombia's Aeronautica Civil determined that one of the probable causes of this accident was the "failure of the flightcrew to revert to basic radio navigation at the time when the FMS-assisted navigation became confusing and demanded an excessive workload in a critical phase of flight" (Aeronautica Civil, 1996). This incident has been cited as exemplifying the effect of automation on the pilot's situation awareness (Endsley & Strauch, 1997). We discuss research on this phenomenon in a later section.

Similar incidents have been documented in which pilots have failed to recognize an inadvertent selection when using information automation. For example, one pilot reported, "I departed on an IFR flight plan with an IFR-approved GPS [Global Positioning System]. I was cleared direct to ABC, at which time I dialed ABC into the VOR portion of the GPS and punched 'direct.' The heading was 040 degrees. After a few minutes, Approach inquired as to my routing, heading etc. I stated direct ABC, 040 degrees. They suggested turning to 340 degrees for ABC. I was dumbfounded. My GPS receiver had locked to ADC, 3,500 miles away [in Norway]! Closer inspection revealed that my estimated time en route was 21 hours" (ASRS, 1997).

A recent general aviation accident illustrates a similar dependency on navigational automation. In 1997 (NTSB, 1998b) a single-engine airplane was cruising at 10,500 feet over central Virginia when the engine failed. The airplane was destroyed during the subsequent off-airport landing, but fortunately the occupants were not seriously injured. After the engine failure, the pilot headed toward an airport that was 20 miles away because it appeared on the "Nearest Airport" feature of the LORAN equipment. However, the airplane was unable to glide that distance, and at 5,000 feet the pilot elected to make an off-airport landing. At the time that the engine failed, there was a private-use airport located about 4 miles away and other airports between the site of the failure and the pilot's intended destination that would have been noted on the aeronautical charts.

Another accident exemplifies the danger of overreliance on GPS. In 1997 a single-engine airplane with a noninstrument rated pilot at the controls took off under instrument meteorological conditions (NTSB, 1998a). The airplane impacted trees at the top of a ridge 46 miles away about 1 hour and 43 minutes later, after following a meandering course that included course reversals and turns of more than 360 degrees. No mechanical problems with the airplane's controls, engine, or flight instruments were identified. A person who spoke with the pilot before departure stated that the pilot ". . . was anxious to get going. He felt he could get above the clouds. His GPS was working and he said as long as he kept that instrument steady [attitude indicator] he'd be all right. He really felt he was going to get above the clouds." Many factors undoubtedly played a role in this accident. But the apparent reliance on GPS technology, perhaps to compensate for having no instrument rating, stands out as a compelling factor.

Misunderstanding or Lack of Understanding of System

Misunderstanding of automated systems can also result in adverse consequences. Much previous research has focused on misunderstandings of complex automated systems in the cockpit such as the FMS (Sarter & Woods, 1995). It has been suggested that misunderstandings arise because of a mismatch between the mental model of the pilot and the behaviors of the automated system as programmed by the designers (Sherry & Polson, 1999). Several examples of incidents and accidents resulting from these system misunderstandings have been detailed elsewhere (Billings, 1997; Funk et al., 1999; Sarter & Woods, 1995). Although some have had benign outcomes and become "lessons learned," others have involved serious loss of life. For example, in 1994, an A300 crashed in Nagoya, Japan, after the pilots inadvertently engaged the autopilot's go-around mode. The pilots countered the unexpected pitch-up by making manual inputs, which were ineffective.

Misunderstandings arise with simpler automated systems as well. For example, most large aircraft have some mechanical switch that is activated by wheel spin-up or weight on the wheels to signal to other systems whether the airplane is on the ground or in the air. This air-ground sensing system has been characterized as an abuse of automation because it arbitrarily takes control of certain systems from the pilot (e.g., Parasuraman & Riley, 1997). However, the scope of its integration into several airplane systems means serious consequences can occur if pilots attempt to circumvent and attempt to manually control this apparently simple switching device.

In 1996 (NTSB, 1996a), a DC-9 had difficulties after takeoff with pressurization, and the pilots received a takeoff configuration warning horn when the flaps and slats were retracted. These symptoms indicated that the ground-shift mechanism was still in ground mode, even though the airplane was in the air and enroute to Nashville. To remedy the situation, the pilots pulled the circuit breakers controlling the ground shift mechanism and continued along their route of flight. When approaching to land at their intended destination, the pilots armed the spoilers as was required by the checklist. However, as the airplane descended through about 100 feet the captain reset the ground shift mechanism circuit breakers. This action shifted the airplane systems from air mode to ground mode; as a result, the ground spoilers deployed, which led to an increased rate of descent, and the airplane touched down hard short of the runway. Although this accident was caused by the convergence of several factors, it is notable that the pilot's choice of pulling and then resetting the circuit breakers for the ground shift system was not based on any guidance material/checklists in their possession. This, coupled with their apparent lack of understanding of the intricacies and scope of the relatively simple automated device they were disabling, contributed to the outcome.

A recent MD-11 upset (NTSB, 1996b) occurred after one of the pilots adjusted the pitch thumb wheel on the autopilot control panel in an attempt to ensure that the airplane leveled off at an assigned altitude. The action did not have the anticipated effect and after repeated attempts, the pilot elected to move the control column to accomplish the level off, which subsequently resulted in the upset. What the pilot did not know was that the autopilot required a 2-second delay after movement of the pitch thumb wheel before it would resume altitude acquisition, and therefore the pilot actions that were an attempt to facilitate the autopilot's acquisition of the altitude actually prevented the automated system from achieving that goal.

To conclude, we have presented several incidents and accidents associated with aspects of human interaction with automated systems. Instead of extensive discussion of widely known accidents (e.g, Cali), we selected accidents and types of automation that have received less attention in the literature. Human-automation interface issues can affect flight safety along the full continuum of flight deck automation. Although much attention in the past has been given to high-end automation issues, particularly related to mode awareness and FMS complexity, automation is broad in scope, and many basic characteristics of automation have been associated with incidents and accidents.

HUMAN PERFORMANCE IN AUTOMATED SYSTEMS

The aviation incidents and accidents discussed resulted from a multitude of factors. As our discussion also indicates, it is rare for any single factor to be responsible. Thus, although we have called these incidents "automation related," automation is only one factor. Sometimes it may be possible to say that automation was the major factor. But even such an assertion can have little explanatory value. "Blaming" automation can be as unproductive as blaming the human operator. The specific design characteristics of the automation, associated procedural constraints, and training and experiential issues must all be examined. Examining the *interaction* between humans and automation is necessary to gain an appreciation of the contributory factors. A growing body of research has examined this issue. We describe some of the main findings from this work, using as a framework the six factors we have identified as contributing to effective human-automation interaction.

Situation Awareness

In our overview of the history of aviation automation, we noted that advanced automation has led to a greater distancing of the pilot from the direct control of the aircraft to higher level goals that lead ultimately to control, as exemplified by the pilot's use of the FMS (Sarter & Woods, 1995). One unanticipated outcome of the FMS and other high-level automated systems is the problem of new types of errors that did not exist with less-capable autoflight systems. Mode errors are a prime example of this new error form (Sarter & Woods, 1994). A mode refers to the setting of a system in which inputs to the system result in outputs specific to that mode but not to other modes. Mode errors can be relatively benign when the number of modes is small and transitions between modes do not occur without operator intervention. For example, in using a remote controller for a TV/VCR, it is commonplace to make mistakes such as pressing functions intended to change the TV display while the system is in the VCR mode or vice versa. When the number of modes is large, however, as in the case of the FMS, the consequences of error can be more significant.

Mode errors arise when the operator executes a function that is appropriate for one mode of the automated system but not the mode that the system is currently in (Sarter et al., 1997). Furthermore, some systems may be designed to be operated at a high level of automation (cf. Table 9.1). As a result, mode transitions can occur autonomously without being immediately commanded by the pilot, who may therefore be unaware of the change in mode. If the pilot then makes an input to the FMS that is inappropriate for the current mode, an error can result. Several aviation incidents and accidents have involved this type of error (Billings, 1997). Violation of the principle that automation states and behaviors should be salient and understandable will inevitably lead to this type of error. Where systems are in place that are vulnerable to mode errors, procedural constraints (such as limiting the number of modes routinely used and requiring that mode changes be announced and confirmed by both pilots) can become effective tools to mitigate the potential for mode errors to arise.

Mode awareness problems in highly automated systems raise the general issue of the impact of automation on situation awareness (Endsley, 1999). Situation awareness refers to the dynamic extraction of information about the environment and the use of this information in anticipating future events (see chap. 4). The concept arose from studies of military aviators. Because the term is not easily defined (Sarter & Woods, 1991), some researchers have wondered whether it is necessary (Flach, 1994). Despite these criticisms, the concept remains useful to refer to the extraction of environmental information relevant to an understanding of the behavior of automated systems (Durso & Gronlund, 1999).

The nature of the impact of automation on situation awareness depends on the first two factors in our framework; the type and level of automation. Automation of information acquisition, for example, data filtering or integration, can improve the pilot's situation awareness (see chap. 7). At the same time, a high level of automation, particularly of decision-making functions, may reduce the operator's awareness of certain system and environmental dynamics which can impair performance if the automation fails (Endsley & Kiris, 1995; Kaber, Omal, & Endsley, 1999). Humans tend to be less aware of changes in environmental or system states when those changes are under the control of another agent (whether that agent is automation or another human) than when they make the changes themselves (Wickens, 1994). One possibility, therefore, is that if a decision aid, expert system, or other type of decision automation consistently selects and executes decision choices in a dynamic environment, the human operator may also not be able to sustain a good "picture" of the information sources in the environment because he or she is not actively engaged in evaluating the information sources leading to a decision. The studies by Endsley and colleagues (see Endsley, 1999, for a review), suggest that a moderate level of decision automation (e.g., level 4 or 5 in Table 9.1), in which the automation provides decision support but the human remains in charge of the final choice of a decision option, is optimal for maintaining operator situation awareness.

However, this conclusion may not generalize to all systems. Lorenz, Di Nocera, Röttger, and Parasuraman (in press) showed that a high level of decision automation is not necessarily associated with poor operator performance when automation fails. This study investigated operator use of an intelligent fault management decision aid in interaction with an autonomous, space-related, atmospheric control system. Lorenz et al. found that a high level of automation (level 6 in Table 9.1) offloaded operators from workload-intensive recovery implementation when a fault occurred, therefore allowing operators to engage in fault assessment activities as evidenced by their information sampling behavior. Lorenz et al. concluded that the potential cost of high-level automation, in which decision advisories are automatically generated, need not inevitably be counteracted by choosing a low or medium level of automation. Instead, freeing operator cognitive resources by automatic implementation of recovery plans can promote better fault comprehension and situation awareness, so long as the automation interface is designed to support efficient information sampling.

Decision Biases

Human decision makers exhibit a variety of biases in reaching decisions under uncertainty. Many of these biases reflect decision *heuristics* that people use routinely as a strategy to reduce the cognitive effort involved in solving a problem (Wickens, 1992). Heuristics are generally helpful, but their use can cause errors when a particular event or symptom is highly representative of a particular condition and yet is extremely unlikely (Tversky & Kahneman, 1974). This can lead to another unintended consequence of high-level automation (see chap. 6). Systems that automate decision making may reinforce the human tendency to use heuristics and result in a susceptibility to "automation bias" (Mosier & Skitka, 1996). Although reliance on automation as a heuristic may be an effective strategy in many cases, overreliance can lead to errors, as in the case of any decision heuristic.

Automation bias may result in omission errors when the operator fails to notice a problem or take an action because the automation fails to inform the operator to that effect. Commission errors occur when operators follow an automated directive that is inappropriate, as in the 1992 L-1011 incident discussed earlier. Evidence for both types of errors was reported in studies by Mosier et al. (1998) and Skitka et al. (1999).

Finally, automation bias is likely to be related to the issue of trust in automation. As we discuss further, operator trust in automation has been shown to be an important contributor to the manner in which humans use automated systems, and evidence for both overtrust and undertrust has been reported (Lee & Moray, 1992; Parasuraman, Molloy, & Singh, 1993; see also Parasuraman & Riley, 1997 for a review). Automation bias may reflect inappropriate *calibration* of operator trust in a particular automated system.

Trust in Automation

Trust is an important aspect of human interaction with automation. As a research issue, trust has generally been examined by social psychologists in the context of human interaction with other humans (e.g., Rempel, Holmes, & Zanna, 1985). As automation becomes more "intelligent" and humanlike in its capabilities, however, it is quite natural to examine trust between humans and machines. The topic is important because operators may not use a well-designed, reliable automated system if they believe it to be untrustworthy. Conversely, they may continue to rely on automation even when it malfunctions and may not monitor it effectively (e.g., the 1972 L-1011 and 1997 A300 accidents discussed earlier).

What are the factors that influence an operator's trust in automation? An obvious first candidate is the reliability of an automated system. It appears intuitively clear that operator trust in automation should increase as automation reliability increases, and this has been found in studies in which subjective measures of trust were obtained (Lee & Moray, 1992, 1994). Performance measures that indirectly index operator trust also support this finding. For example, it has been argued that in a multitask environment, human monitoring and detection of failures in an automated system reflects trust in that system (Parasuraman et al., 1993). If so, then monitoring performance should improve as the reliability of the automation decreases (and operator trust declines). In confirmation of this prediction, an almost linear inverse relationship was found between detection performance and automation reliability in a study in which reliability was varied systematically over a number of levels (May, Molloy, & Parasuraman, 1993).

The influence of automation reliability is typically moderated by other factors. In a simulation study of process control, Lee and Moray (1992) found that use of automation and subjective trust in automation were correlated. However, they also found that subjects chose manual control if their confidence in their own ability to control the process exceeded their trust of the automation and chose automation otherwise. Thus, if trust in automation is greater than self-confidence, operators will use automation, but not otherwise.

In general, trust in automation can be conceptualized as an intervening variable between automation reliability and automation use. Several other factors may be important, in addition to operator self-confidence. Riley (1996) suggested that the interaction between trust and self-confidence could itself be moderated by the risk associated with the decision to use or not use automation. Cohen, Parasuraman, and Freeman (1997) proposed that operator trust in automation is likely to be context-dependent, that is, whether a given situation matches that under which the performance of the automation has been verified. They also proposed that operator trust and usage of an automated aid would depend on operator workload and the time criticality of the situation. For a recent review of the factors mediating the relationship between trust and automation use, see Masalonis and Parasuraman (1999).

Overtrust and Automation Complacency. As mentioned previously, variations in operator trust affect operator monitoring of automation. Several studies have shown that humans are not very good at monitoring automation states for occasional malfunctions if their attention is occupied with other manual tasks (Parasuraman et al., 1993; Singh, Molloy, & Parasuraman, 1997). Parasuraman et al. (1993) had subjects perform three tasks of the Multi-Task Attribute Battery (MAT; Comstock & Arnegard, 1992), a low-fidelity simulation of flight tasks: engine-systems monitoring, two-dimensional compensatory tracking, and fuel systems management. The tracking and fuel management tasks were controlled manually, whereas the engine-systems task was automated. However, the automation failed from time to time, and subjects were required to detect the failure and reset the system. Parasuraman et al. (1993) found that operator monitoring of the automated system was poor compared to a manual control condition when the reliability of the automation was relatively high and unchanging and the operator was engaged in other manual tasks. The influence of concurrent manual task load on this "complacency" effect (Billings, Lauber, Funkhouser, Lyman, & Huff, 1976; Wiener, 1981) has also been noted in aviation incident analyses. Sumwalt et al. (1997) found that 60% of ASRS reports involving monitoring failures occurred when pilots were carrying out other nonmonitoring tasks.

Parasuraman et al. (1993) also found that the automation complacency effect was eliminated when the reliability of the automation was variable, alternating between high and low. Figure 9.3 shows the results of a replication of the original study by Parasuraman et al. (1993) carried out by Singh, Molloy, Mouloua, Deaton, & Parasuraman (1998). This study differed from the original in two factors, both aimed at increasing the ecological validity of the automation complacency effect. First, instead of an artificial engine-monitoring task, a simulation of Boeing's Engine Indicator and Crew Alerting System (EICAS) display was used as the automated engine-systems task. Second, experienced pilots were tested. As Fig. 9.3 shows, detection performance of engine malfunctions on the EICAS display was greater under manual than automation control for the constant-reliability condition. For variable-reliability automation, performance was stable across 10-min blocks spent under automation control. The monitoring performance of pilots in these studies supports the view that reliable automation engenders trust. This leads to a reliance on automation that is associated with only occasional monitoring of its efficiency, suggesting that a critical factor in the development of this phenomenon might be the constant, unchanging reliability of the automation. Conversely, automation with inconsistent reliability does not induce overtrust and is therefore monitored more closely.

The first and second factors in our organizing framework for effective human-automation interaction propose that the types and levels of automation should be considered in evaluating automation. As discussed previously, a high level of automation of decision-making functions (e.g., higher than level 5 in Table 9.1) can have a negative effect on the operator's system awareness, at least in some cases. The automation complacency effect represents another adverse outcome. Furthermore, although the evidence is not as strong as for situation awareness and complacency, high-level decision automation, including full automation, might also result in a reduction in manual skill (Rose, 1989; Wiener, 1988).

Collectively, reduced situation awareness, complacency, and manual skill degradation have been referred to as reflecting "out-of-the-loop unfamiliarity" (Wickens, 1994) due to high-level decision automation. What about information automation? There is some preliminary evidence to indicate that although complacency can occur with both information automation and decision automation, its effect on performance may be greater with the latter. Crocoll and Coury (1990) found that both forms of automation benefited performance in a simulated decision-making task when the automation was perfectly reliable. When the automation was unreliable, however, performance suffered much more when unreliable recommendations were given by decision automation than when

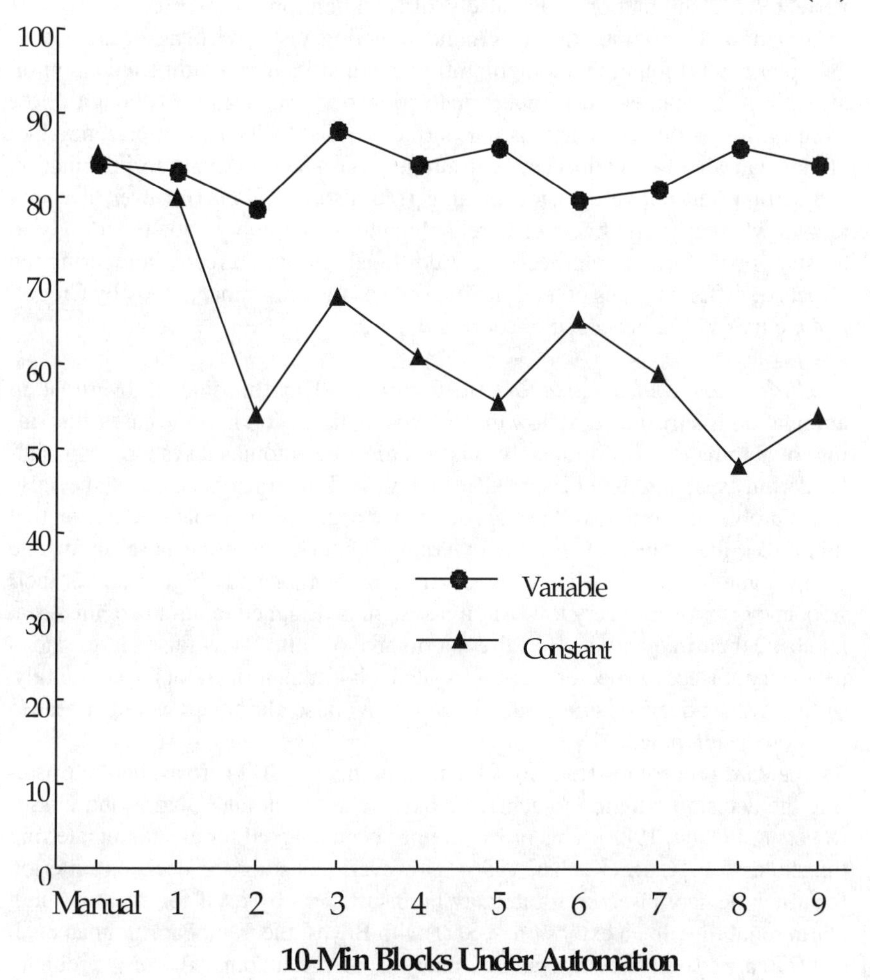

FIG. 9.3. Detection rates by pilots of malfunctions as displayed by a simulation of the Engine Indicator and Crew Alerting System (EICAS) during performance of the Multi-Attribute Task (MAT) battery under manual control and nine 10-min. blocks of automation control, for constant-reliability and variable-reliability automation. Based on data in Singh, Molloy, Mouloua, Deaton, and Parasuraman (1998).

incorrect status information was provided by information automation. However, this is not to say that unreliability in information automation is inconsequential. If the algorithms underlying information filtering, prediction, integration, or other operations characteristic of information automation are reliable but not perfectly so, automation may falsely direct attention to incorrect sources of information. In a simulated air-ground targeting task, Wickens, Conejo, and Gempler (1999) found that a highlighting cue that incorrectly directed attention away from the target led to poorer detection performance, even though pilots were informed that the cue was not perfectly reliable. Thus, complacency-like effects can also be obtained even if automation is applied only to information acquisition and not to decision making (see also chap. 7). However, it is not known whether such effects of unreliable automation apply equally strongly to all stages of information processing. Additional research directly comparing the effects of different types of automation, such as the preliminary study by Crocoll and Coury (1990), is needed to address this issue.

Undertrust and Disuse of Automation. The flip side of overtrust in automation is a trust level so low that it leads to disuse (or, in some cases, disarming) of automation. Unfortunately, mistrust of some automated systems, especially alerting systems, is widespread in many work settings because of the false alarm problem. Alerting systems are set with a decision threshold or criterion that minimizes the chance of a missed warning while keeping the false alarm rate below some low value. Missed signals have a phenomenally high cost, yet their frequency is typically very low. But if a system is designed to minimize misses at all costs, then frequent device false alarms may result. A low false alarm rate is necessary for acceptance of warning systems by human operators. Accordingly, setting a strict decision criterion to obtain a low false alarm rate would appear to be good design practice.

Standard procedures from signal detection theory (SDT) are available for setting the decision criterion to achieve a balance between false alarms and misses (Swets & Pickett, 1982). This procedure has been adapted for examining alerting thresholds for TCAS (Kuchar, 1996). However, adjusting the decision criterion for a low device false alarm rate may be insufficient by itself for ensuring high alarm reliability. In an extension of SDT with Bayes' theorem, Parasuraman et al. (1997) carried out a computational analysis of the automated alarm problem. They showed that despite the availability of the most advanced sensor technology and the development of very sensitive detection algorithms, the low a priori probability or base rate of most hazardous events may limit the effectiveness of many alerting systems. If the base rate is low, as it often is for many real events, then the *posterior* probability of a true alarm—the probability that given an alarm a hazardous condition exists—can be quite low even for very sensitive warning systems. As a result, operators may not trust or use the system or may attempt to disable it (Satchell, 1993; Sorkin, 1988). Even when operators do attend to the alarm, a low posterior probability may elicit a very slow response (Getty, Swets, Pickett, & Gounthier, 1995).

Designers of automated alerting systems must therefore take into account not only the decision threshold at which these systems are set but also the a priori probabilities of the condition to be detected. Parasuraman and Hancock (1999) outlined a set of equations that could be used by designers to determine the appropriate decision threshold that maximizes the posterior probability of a true alarm. For example, the system parameters could be set so that the posterior probability of a true alarm is at least 0.8. This means that of every 10 alarms that occur, 8 point to a true hazard. Applying this solution may work in a number of cases, so long as accurate information on the base rate is available. But in other cases, maximizing the posterior probability of a true alarm may result in an unacceptable increase in the miss rate. If so, the designer will be faced with a dilemma. When the hazardous event to be detected has a very low base rate of occurrence and the cost of a missed hazard is very high, difficult design trade-offs may need to be made. Other possibilities that could be pursued in this case include developing even more sensitive detection sensors and algorithms, as has been achieved with newer versions of TCAS, and using a "likelihood alarm display," in which the alarm output is partitioned into two or more levels of confidence or certainty (Sorkin, Kantowitz, & Kantowitz, 1988).

Finally, alerting systems like GPWS and TCAS provide not only information about a hazard but also indicate the appropriate course of action to avoid the hazard. An additional human performance concern is that pilots may not always follow the directive of the system when a hazard is detected. Pilots do not always follow the directives of GPWS and TCAS. Such nonconformance can be problematic because of the limited time available to the pilot to decide on and pursue alternative actions. Recent work has examined whether conformance can be increased by making explicit to pilots the underlying logic of the alerting system (Pritchett & Hansman, 1997).

Mental Workload

The final human performance issue we examine, mental workload, is in some sense the broadest and also represents one of the first areas of concern in research on human interaction with automation (Wiener & Curry, 1980). Moreover, mental workload is a fundamental issue because many automated systems when first introduced are touted as workload-saving moves, and the justification for automation is to reduce mental workload and hence human error. As we shall see in this section, this does not always occur.

Benefits and Costs of Automation. A number of studies have examined how automation has changed the mental workload of pilots (Kantowitz & Campbell, 1996; Wiener, 1988). This has been an area for scrutiny because of the relevance of operator workload to safety (Kantowitz, 1986; Wickens, 1992). The evidence indicates that automation can decrease, increase, or have no measurable impact on pilot workload. Increased use of automated systems in the modern cockpit can reduce some of the operational workload of the pilot. However, for a variety of reasons, automation may actually increase other aspects of workload.

When properly designed, automation can reduce the human operator's mental workload to a manageable level under peak loading conditions. A good example is the navigational map display in the new generation of automated aircraft and a common feature on newer GPS units used in general aviation. As discussed previously, this system has enhanced pilot situation awareness of their flight path in relation to their flight plan. The system has also considerably reduced the pilot's physical and cognitive workload in flight planning and route evaluation (Wiener, 1988). Simulator studies have also shown that automation of flight control, for example, through the use of an autopilot, reduces the overall mental workload associated with other flight tasks, as indexed by both performance (Kantowitz & Campbell, 1996) and physiological measures (Masalonis, Duley, & Parasuraman, 1999).

Although workload reduction during times of peak task loading can be beneficial, automation to reduce workload should not be pursued as an end in itself, but must be considered in relation to other possible consequences for human performance. As discussed previously, higher levels of automation applied to decision making (e.g., level 5 and higher in Table 9.1) can reduce the operator's workload but may lead to loss of situation awareness, complacency, and skill loss. These costs may offset any workload benefit. Wickens et al. (1998) point to this possibility in an advanced design option that could be considered for data link systems. It would be technically feasible for an uplinked clearance from a controller to be routed directly to the FMS in the cockpit, and the pilot would be required only to press a key to approve its execution. This would represent a relatively high level of automation that would reduce the pilot's workload, but would make the system vulnerable to a loss of situation awareness on the part of the pilot as to the type of clearance and its impact on the aircraft's future flight path.

Although workload reduction is obtained with some automated systems, instances of automation *increasing* workload have also been found (Parasuraman & Riley, 1997). First, if automation is implemented in a "clumsy" manner, for example, if executing an automated function requires extensive data entry or "reprogramming" by human operators at times when they are very busy, workload reduction may not occur (Wiener, 1988). Second, if engagement of automation requires considerable "cognitive overhead" (Kirlik, 1993), that is, extensive cognitive evaluation of the benefit of automation versus the cost of performing the task manually, then users may experience greater workload in using the automation. Alternatively, they may decide not to engage automation. Finally, if automation involves a safety-critical task, then pilots may continue to monitor the automation because of its potential unreliability (McDaniel, 1988). As Warm, Dember, and Hancock (1996) have shown, enforced monitoring can increase mental workload, even for very simple tasks. Thus, any workload benefit due to the allocation of the task to the automation may be offset by the need to monitor the automation.

Workload Management. The relationships between task load, mental workload, and automation usage are complex. It might seem plausible that an operator is more likely to choose to use automation when his or her workload is high than when it is low or moderate (see Huey & Wickens, 1993; see also chap. 4). Surprisingly, the evidence in favor of this assertion is mixed at best (Harris, Hancock, & Arthur, 1993; Riley, 1996). Nevertheless, human operators often cite excessive workload as a factor in their choice of automation. Riley, Lyall, and Wiener (1993) reported that workload was cited as one of the two most important factors (the other was the urgency of the situation) in pilots' choice of such automation as the autothrottle, autopilot, FMS, and flight director during simulated flight. However, these data also showed substantial individual differences. Riley et al. (1993) also asked the pilots in their study how often, in actual line performance, various factors influenced their automation use decisions. For most factors examined, many pilots indicated that a particular factor, say fatigue, influenced their decisions very infrequently, whereas an almost equal number said that the same factor influenced their decisions quite often; very few pilots gave an answer in the middle. Simulator studies of human use of automation also typically find large individual differences. These results suggest that different people employ different strategies when making automation use decisions and are influenced by different considerations.

Subjective perceptions and objective measurement of performance are sometimes dissociated (Yeh & Wickens, 1988). It would therefore seem worthwhile to run additional studies in which task load manipulations are used that are shown to have clear effects both on performance and on other indices of workload. Furthermore, workload in the cockpit can be fundamentally different from laboratory manipulations. For example, pilots are often faced with deciding whether to fly an ATC clearance using (either alone or in combination) the FMS, simple heading or altitude changes on the glare shield panel, or manual control. Casner (1994) found that pilots consider the predictability of the flight path and the demands of other tasks in reaching their decision. When the flight path is highly predictable, overall workload may be reduced by accepting the higher short-term workload associated with programming the FMS, whereas when the future flight path is more uncertain, such a workload investment may not be warranted. To fully explore the implications of workload on automation use, the workload attributes of particular systems should be better represented; in the flight deck, this would include a trade-off between short-term and long-term workload investments. Such trade-offs involve what has been termed *cockpit task management* (Funk, 1991), in which pilots use strategies (e.g., task prioritizing and planning) to maintain their workload within some middle range. Under conditions of time pressure, however, pilots may adopt maladaptive strategies, leading to possible incidents or accidents (Funk, 1991).

In conclusion, the research on automation and workload indicates that automation is not the same as simple *task shedding* and hence does not always translate into a reduction in workload. This is because automation sometimes creates new demands, poses new coordination challenges, may be difficult to manage, and may need to be closely monitored (Parasuraman & Riley, 1997; Sarter et al., 1997). As a result, workload can increase, thereby countering any workload reduction initially achieved by automating a function previously carried out manually. Thus, the consequences of automation of a function for pilot workload must involve a consideration not only of the particular function, but also of other functions that the pilot will carry out in the resulting system. The challenge is to find ways to reach the goal of balancing the pilot's workload between the extremes of overload and underload. Cockpit task management has been proposed as one method for achieving this balance, but this method should also include training to cope with time pressure for it to be fully successful. Complementing this training strategy is an alternative approach to automation design, namely adaptive automation. We discuss some of the characteristics of adaptive automation and provide evidence that it can provide for balanced operator workload.

ADAPTIVE AUTOMATION

Characteristics of Adaptive Systems

Systems in which automation is implemented in a fixed manner—so-called static automation—may be particularly prone to some of the human performance costs we have discussed in the preceding section, if, for example, the automation design promotes out-of-the-loop unfamiliarity on the part of the human operator. As a result, there has been considerable interest in developing alternative approaches to the implementation of automation. Adaptive automation represents one such alternative design approach. In adaptive systems, the division of labor between human and machine agents is not fixed during system design but dynamic. Computer aiding of the human operator and task allocation between the operator and computer systems are flexible and context dependent during system operation (Hancock, Chignell, & Lowenthal, 1985; Parasuraman, Bahri, Deaton, Morrison, & Barnes, 1992; Rouse, 1988; Scerbo, 1996; Wickens, 1992). In contrast, in static automation, provision of computer aiding is predetermined at the design stage, and task allocation is fixed once the system becomes operational.

Scerbo (2001) has also emphasized the distinction between *adaptable* and *adaptive* systems. In the former, the user initiates changes in the display or interface. In truly adaptive systems, *both* the user and the system can initiate changes among system states or modes. The distinction is one primarily one of authority. In an adaptable system, the human always maintains authority to invoke or change the automation, whereas this authority is shared in an adaptive system. Inagaki's (1999) design concept of "situation-adaptive autonomy" is related to this view of an adaptive system, but in his approach, control of a process is traded off between human and computer in real time based on time criticality and the expected costs of human and machine performance.

The adaptive automation concept is not a new one, having been proposed about 25 years ago (Rouse, 1976). However, the technologies needed for its effective implementation were not readily available until recently. The idea of adaptive automation also falls naturally out of the levels of automation concept. It is also related to the classic problem in human factors of allocation of function as discussed earlier. Because decisions about task allocation cannot always be made based on stereotypical characteristics of human and computer capabilities, the need for an adaptive computer aid that responds to task demands and operator performance becomes a logical design alternative for increasing system effectiveness. Although previous attempts to solve the allocation of function problem met with mixed results (Jordan, 1963; Sheridan, 1998), developments in computer technology, particularly in machine intelligence and parallel-distributed processing, have led to greater optimism for the success of adaptive systems (Hancock et al., 1985).

Techniques for Adaptive Automation

One of the key issues in adaptive automation concerns the method by which adaptation is implemented. In other words, how is the adaptation to be done? What properties (of the human operator, the task environment, or both) should the system adapt to? What tasks should in fact be automated, and when? A number of methods for adaptive automation have been proposed. Parasuraman et al. (1992) reviewed the major techniques and found that they fell into five main categories:

1. Critical events
2. Operator performance measurement
3. Operator physiological assessment
4. Operator modeling
5. Hybrid methods

In the critical-events method, the implementation of automation is tied to the occurrence of specific tactical events (Barnes & Grossman, 1985). For example, in an air defense system, the beginning of a "pop-up" weapon delivery sequence may lead to the automation of all aircraft defensive measures.

This method of automation is adaptive because if the critical events do not occur, the automation is not invoked. Such an adaptive automation method is inherently flexible because it can be tied to current tactics and doctrine during mission planning.

A disadvantage of the critical-event technique is its insensitivity to actual system and human operator performance. For example, this method will invoke automation irrespective of whether or not the pilot needs the automation at the time. The operator measurement technique attempts to overcome this limitation. In this method, various pilot mental states (e.g., mental workload, or more ambitiously, intentions) may be inferred on the basis of performance or other measures and then input to adaptive logic. For example, performance and physiological measurements may allow the inference that a pilot is dangerously fatigued or experiencing extremely high workload. An adaptive system could use these measurements to provide computer support or advice to the pilot that would mitigate the potential danger. Alternatively, human operator states and performance may be modeled theoretically, with the adaptive algorithm being driven by the model parameters. Intelligent systems that incorporate human intent inferencing models have been proposed (Geddes, 1985).

The measurement and modeling methodologies each have merits and disadvantages. Measurement has the advantage of being an "online" technique that can potentially respond to unpredictable changes in human operator cognitive states. Kaber and Riley (1999), for example, used a secondary-task measurement technique to assess operator workload in a target acquisition task. They found that adaptive computer aiding based on the secondary-task measure enhanced performance on the primary task. However, the measurement method is only as good as the sensitivity and diagnosticity of the measurement technology (see chap. 4). Modeling techniques have the advantage that they can be implemented off-line and easily incorporated into rule-based expert systems. However, this method requires a valid model, and many models may be required to deal with all aspects of human operator performance in complex task environments.

Operator physiological assessment offers another potential input for adaptive systems (Byrne & Parasuraman, 1996; Parasuraman et al., 1992). Physiological measures can provide additional information that can be tapped for control of adaptive systems. Technology is available to measure a number of physiological signals from the pilot, from autonomic measures such as heart rate variability to central nervous system measures such as the EEG and event-related potentials or ERPs, as well as measures such as eye scanning and fixations. There is now a substantial literature indicating that different psychophysiological measures can be used for real-time assessment of mental workload (Gevins et al., 1998; Jorna, 1999; Kramer, Trejo, & Humphrey, 1996; Parasuraman, 1990; Scerbo, Freeman, Mikulka, Di Nocera, Parasuraman, & Prinzel, 2001). Prinzel, Freeman, Scerbo, Mikulka, and Pope (2000) have also specifically demonstrated the feasibility of an adaptive system based on EEG measures.

Benefits of Adaptive Automation

Apart from procedures that might be used to implement adaptive automation, a basic question concerns the effects of this form of automation on pilot performance. Several authors have suggested that adaptive systems can regulate operator workload and enhance performance, while preserving the benefits of static automation (Hancock et al., 1985; Parasuraman et al., 1992; Rouse, 1988). The performance costs of certain forms of static automation described in the preceding section—reduced situation awareness, complacency, skill degradation—may also be mitigated. These suggestions have only recently been tested empirically (Hilburn, Jorna, Byrne, & Parasuraman, 1997; Kaber & Riley, 1999; Parasuraman, 1993; Parasuraman, Mouloua, & Molloy, 1996; Scallen, Hancock, & Duley, 1995; see also Scerbo, 1996, 2001, for reviews). The benefits of adaptive automation have been proposed with respect to automation both in the cockpit (Parasuraman, 1993) and in ATC (Duley & Parasuraman, 1999; Hilburn et al., 1997).

Empirical evaluations of adaptive automation have focused primarily on the performance and workload effects of either (a) adaptive aiding (AA) of the human operator, (b) adaptive task allocation (ATA) from the human to the machine (ATA-M), or (c) ATA from the machine to the human (ATA-H). Each of these forms of adaptive automation have been shown to enhance human-system performance (e.g., AA: Hilburn et al., 1997; ATA-M: Parasuraman, 1993; Scallen et al., 1995; ATA-H: Parasuraman et al., 1996). We briefly review these studies.

Hilburn et al. (1997) examined the effects of adaptive aiding (AA) on the performance of air traffic controllers. The controllers were provided with a decision aid for determining optimal descent trajectories of aircraft—the Descent Advisor (DA) of CTAS. Hilburn et al. (1997) found significant benefits for controller workload (as assessed using pupillometric and heart rate variability measures) when the DA was provided adaptively during high traffic loads, compared to when it was available throughout (static automation) or only at low traffic loads. Parasuraman (1993) examined the performance effects of dynamic allocation from the human to the machine (ATA-M) using the MAT multitask environment. Allocation of one or two tasks for a 10-min period during a 30-min session benefited performance of the remaining tasks. This performance benefit was obtained even though subjects were required to monitor the automated task(s), a condition that was imposed so as to ensure that the effects of dynamic automation were not simply due to task shedding (Tsang & Johnson, 1989). Finally, with respect to adaptive task allocation from the machine to the human (ATA-H), Parasuraman et al. (1996), again using the MAT, showed that temporary return of an automated engine-systems task to manual control benefited subsequent monitoring of the task when it was returned to automated control. This benefit was found for either of two adaptive algorithms, a model-based approach in which the return to manual control was initiated at a particular time specified by the model and a performance-measurement approach in which the adaptive change was triggered only when the operator's performance on the engine-systems task fell below a specified level. A subsequent study showed that the operator (and system) performance benefit could also be sustained for long periods of time, in principle indefinitely, by repetitive or multiple adaptive task allocation at periodic intervals (Mouloua, Parasuraman, & Molloy, 1993).

These findings point to the performance benefits of the different forms of adaptive automation. However, to date, these benefits have been demonstrated separately in different studies. For adaptive systems to be effective, many forms of adaptive automation need to be examined jointly in a single work domain. Furthermore, if adaptive systems are designed in a manner typical of "clumsy automation"—that is, providing aiding or task reallocation when they are least helpful (Wiener, 1988)—then performance may be degraded rather than enhanced (see also Billings & Woods, 1994).

As discussed previously, one of the drawbacks of some automated systems—for example, the FMS—is that they may require reprogramming and impose added workload during high task load phases of flight, such as final approach, while doing little to regulate workload during the low-workload phase of cruise. This suggests the need for linking the provision of aiding to the level of operator workload or performance. A model for effective combination of adaptive aiding and adaptive task allocation, based on matching adaptation to operator workload, is shown in Fig. 9.4 (Parasuraman, Mouloua, & Hilburn, 1999). Fig. 9.4 plots operator workload against work period or time. As workload fluctuates due to variations in task load or operator strategies, high-workload periods represent good times for AA (or ATA-M) and poor times for ATA-H; the converse is true for low-workload periods. According to this model, if AA and ATA-H are to be combined in an adaptive system, they should be provided at the times indicated in order to regulate operator workload and maximize performance (see also Parasuraman & Hancock, 2001).

Parasuraman et al. (1999) carried out a study to test this model by examining the effects of AA and ATA-H in the same participants. They had pilots perform a modified version of the MAT in which the engine-systems task used a realistic simulation of an EICAS display, as in the study by Singh et al. (1998) described previously. A 60-min session comprising three phases of high-low-high task load was used to simulate a profile of takeoff/climb, cruise, and approach/landing. Twenty-four pilots were assigned to three groups of 8 each, a workload-matched adaptive group, a "clumsy automation" adaptive group, and a control group. For the workload-matched group, AA in the form of automation of the vertical dimension of the two-dimensional tracking task of the MAT was provided during the high task load phases at the beginning and end of the session. Also, ATA-H, temporary return of the automated EICAS task to manual control, was used in the middle of the second, low task load phase. For the clumsy automation group,

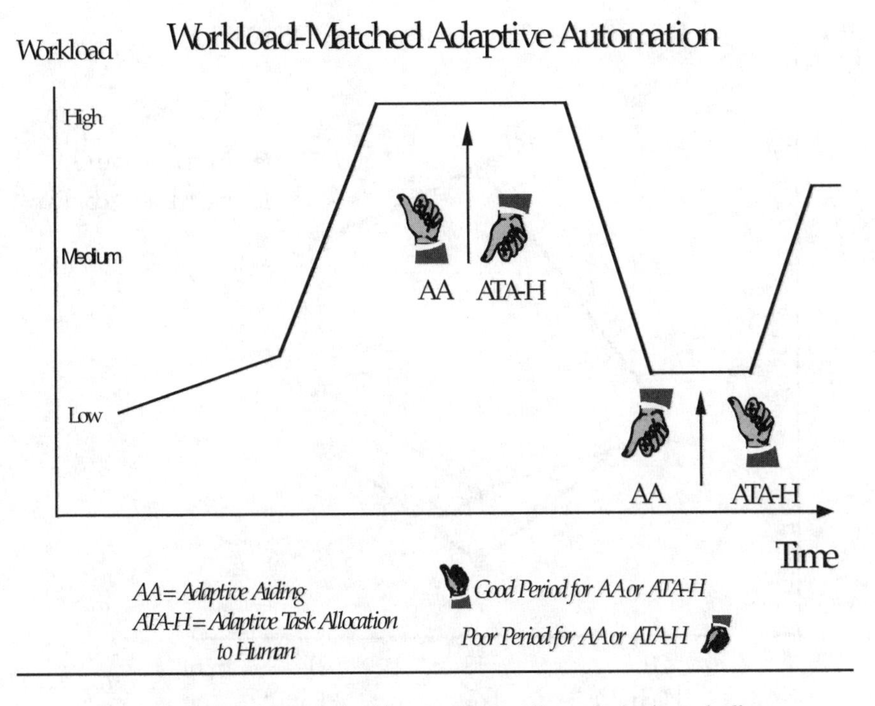

FIG. 9.4. A model for workload-based adaptive aiding and adaptive task allocation. *Note.* From *Automation Technology and Human Performance: Current Research and Trends* (pp. 119–123), by M. Scerbo and M. Mouloua (eds.), 1999, Mahwah, NJ: Lawrence Erlbaum Associates, Inc. Copyright 1999 by Lawrence Erlbaum Associates, Inc. Reprinted with permission.

these contingencies were reversed, so that they received aiding when task load was low and were allocated the automated EICAS task when task load was high. For the control group, neither AA nor ATA-H was provided. Overall performance on all three tasks of the MAT was higher in the workload-matched group than in the other two groups. Figure 9.5 shows the results for the EICAS task. The detection rate of automation failures was significantly higher for the workload-matched adaptive group in the second phase (when ATA-H was introduced) and was maintained even when the EICAS task was returned to automated control in the last phase. Thus, whereas the clumsy-automation and control groups showed evidence of automation complacency, the workload-matched adaptive group did not.

These results show that workload-based adaptation can balance pilot workload and reduce automation complacency. These findings validate an approach to adaptive automation based on adapting to pilot workload. However, the results of Parasuraman et al. (1999) also show that performance benefits can be eliminated if adaptive automation is implemented in a clumsy manner.

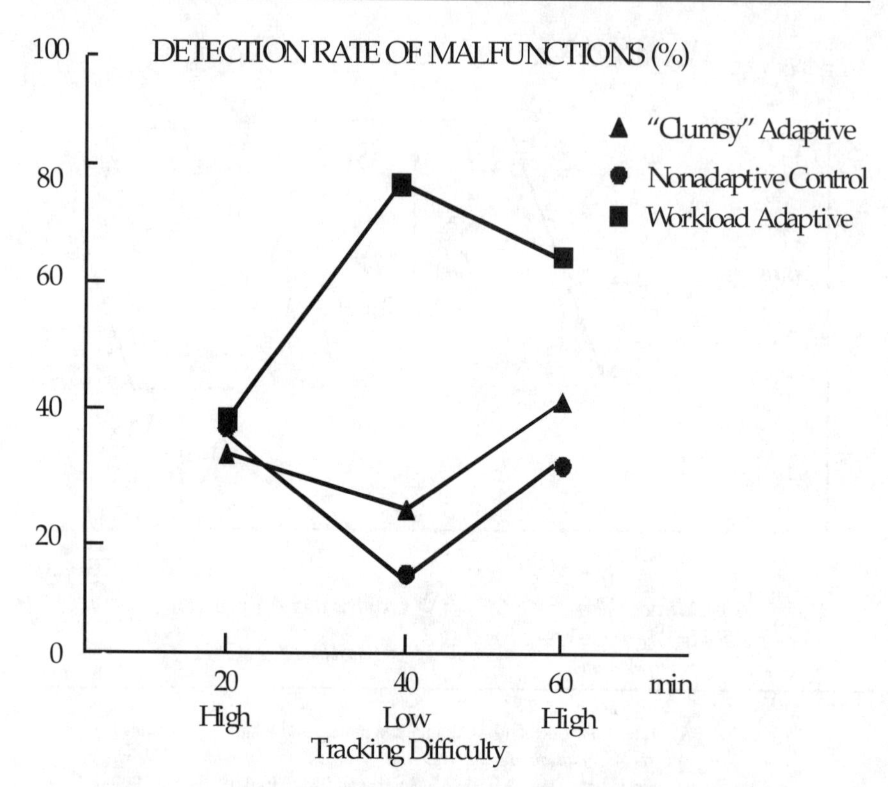

FIG. 9.5. Detection rates by pilots of malfunctions as displayed by a simulation of the Engine Indicator and Crew Alerting System (EICAS) during performance of the Multi-Attribute Task (MAT) battery. Performance values are shown for three groups, a workload-matched adaptive automation group, a "clumsy automation" group, and a nonadaptive control group. *Note.* From *Automation Technology and Human Performance: Current Research and Trends* (pp. 119–123), by M. Scerbo and M. Mouloua (eds.), 1999, Mahwah, NJ: Lawrence Erlbaum Associates, Inc. Copyright 1999 by Lawrence Erlbaum Associates, Inc. Reprinted with permission.

COMPUTATIONAL AND FORMAL METHODS FOR STUDYING HUMAN-AUTOMATION INTERACTION

The human performance findings discussed above provide a knowledge base on which to approach automation design and training. As discussed previously, proposals for implementing automation in a human-centered manner have also been put forward (e.g., Billings, 1997). Parasuraman, Sheridan, & Wickens, (2000) also proposed a model for automation design emphasizing selecting types and levels of automation that meet several primary human performance criteria and other secondary criteria. For the most part, these proposals represent conceptual frameworks that are qualitative in nature. It would be desirable to have quantitative models that could inform automation design for human-machine systems (Pew & Mavor, 1998). For example, such models could be used to address the first two factors in our framework, that is, to determine the appropriate type and level of automation in a particular system. Several computational models have been put forward very recently.

Standard and Fuzzy Signal Detection Theory

SDT models have been applied to the analysis of automated alerting and warning systems, as discussed previously. SDT can be used to determine the appropriate decision threshold β for this trade-off and to measure the sensitivity d' of the automated system (Green & Swets 1966; Swets & Pickett 1982), a procedure that has been adapted for examining alerting thresholds for the TCAS system (Kuchar, 1996). Also as discussed previously, Parasuraman et al. (1997) also showed that automated alerts must be designed for a high posterior probability to be effectively used by human operators. Parasuraman and Hancock (1999) showed how this could be achieved by combining SDT and Bayesian analysis. Recently, however, a variant of SDT has been developed, fuzzy signal detection theory (Parasuraman, Masalonis, & Hancock, 2000), that can also be applied to automation design issues.

The use of SDT to assess automated information acquisition systems rests on the assumption that "states of the world" can be cleanly divided into one of two nonoverlapping categories, signal or noise (e.g., hazardous and nonhazardous events). The definition of a signal in many real settings, however, varies with context and over time. Consequently, a signal can be defined as having a value that falls in a range in between unequivocal presence and unequivocal absence. Fuzzy logic represents an alternative to conventional set theory and allows the possibility that a particular event may belong to *both* a given category (e.g., signal) and its opposite (e.g., noise), where the degree of set membership varies from 0 to 1 (Zadeh, 1965). Accordingly, Parasuraman, Masalonis, & Hancock, (2000) combined the methods of fuzzy logic with SDT, yielding what they termed fuzzy SDT. They proposed that fuzzy SDT could be especially useful in analyzing automated detection systems in real, noisy environments.

The question "What is a signal?" does not usually arise in the laboratory, because the signal is whatever the experimenter defines it to be for the participant. In contrast, in most real settings, the definition of a signal to the human operator is often context dependent and varies with several factors. For example, a sonar operator trying to detect the electronic return of an approaching submarine on a visual display has to look for a line with a luminance, contrast, and spatial frequency that will depend on the submarine speed, ocean currents, and the presence of other nearby objects. The signal will also vary over time as the submarine approaches. Even when the formal or legal definition of a signal is clearly specified in terms of some measurable physical event, the human operator may treat the event variably in different contexts. For example, the legal definition of a "conflict" in the flight paths of two aircraft being monitored by ATC is fixed. FAA regulations mandate a "signal" in ATC when two aircraft come within 5 nautical miles (nm) horizontally and 1,000 ft. vertically of each other. However, the separation distances that the air traffic controller will consider a signal requiring action will often exceed these minimum values. The controller's definition of signal will vary depending on the complexity of the traffic, the nature of the sector being controlled, and other factors.

The mathematical details of the fuzzy SDT analysis developed by Parasuraman, Masalonis, & Hancock, (2000) cannot be reproduced here because of space limitations. However, an application is briefly discussed to illustrate the theory. Masalonis and Parasuraman (2001) applied fuzzy SDT to the analysis of a recently developed automated system, URET, which is a system for aiding air traffic controllers in determining conflict-free trajectories for aircraft. URET evaluates the paths of all the aircraft that are or shortly will be in the controller's sector and generates either a red or a yellow alert with respect to any pair of aircraft that will approach within certain distances. Masalonis and Parasuraman (2001) conducted both a conventional ("crisp") and a fuzzy SDT analysis of the detection performance of this automated system for recorded air traffic, based on data reported by Brudnicki and McFarland (1997). The fuzzy SDT analysis showed more clearly than crisp SDT which characteristics of the automated warning system lead to more effective performance. For example, the fuzzy analysis showed that there were a relatively high number of instances when the separation distance was less than 5 nm (below the legal limit) but the URET alert status was yellow, indicating that in these cases, controllers were not receiving the strongest possible alert for a situation that was in actuality worthy of attention. Although Brudnicki and McFarland (1997) reported that controllers found some utility for yellow alerts, it may be that in busy situations, controllers would attend mainly to the red alerts, or at least attend to the red alerts first. Therefore, the fuzzy SDT analysis provided clues as to where room for improvement might lie in the algorithms used by URET and other systems. More generally, fuzzy SDT can provide estimates of sensitivity and criterion that may better capture the temporal and contextual variability inherent in real-world signals that need to be detected by automated warning systems.

Expected-Value Models

Expected-value models have been developed by Sheridan and Parasuraman (2000) and by Inagaki (1999). Both considered the decision to automate in economic terms, that is, in terms of the anticipated gains of having a task performed by automation or by the human. Sheridan and Parasuraman (2000) applied standard expected-value analysis to a failure-response task involving two true but uncontrollable states of the world, namely *failure* and *normal,* and two responses, one appropriate to failure and one appropriate to normal operation. Expected values were computed for either human or automated control of the task in the standard way, combining three variables: the a priori probability or base rate of the actual state, the probability of correct or incorrect decision/response, and the benefit or cost of correct or incorrect decision/response. Their analysis showed that the decision to automate could be linked to human-machine differences in the probabilities of a false negative—not responding when the system state is failed—and of a false positive—responding when the system state is not failed. Specifically, one should automate if the difference in false negative probabilities between automation and human control is less than the difference in false positive probabilities between human and automation control, weighted by a factor *K*. The weighting factor *K* is dependent on the a priori probability of a failure state and the benefits and costs of decision outcomes.

The expected-value analysis proposed by Sheridan and Parasuraman (2000) considered only the *accuracy* of decision choices. In many systems, the *speed* of such responses may be important; that is, the timeliness of response by human or automation may be a critical factor in the decision to automate. Inagaki (1999) carried out a related expected-value analysis comparing human versus automation control in time-critical conditions. Time criticality was defined as the window of time following an abnormal event (e.g., an engine failure just prior to V1 speed during takeoff) within which corrective action must be applied. Inagaki's model showed that control of a task should be allocated to automation when the human cannot respond effectively within the time window, but to human operators otherwise.

Other Formal Models

Formal models have also been applied to determining the appropriate level and form of interaction between human and machine. Wei, Macwan, and Wieringa (1998) described a quantitative model of the degree of automation based on the number and complexity of tasks that exist in a system. Tasks were weighted differentially in terms of their effects on system performance, the demand they imposed on the operator, and their effect on perceived mental workload. Wei et al. (1998) defined an index of the degree of automation as the ratio of the sum of the weights of all the automated tasks to the sum of the weights of all tasks. In an experimental validation of their computational index, they showed that system performance did not increase beyond a certain high level of automation. The authors suggested that their model could be used to ascertain the appropriate level of automation for optimizing system performance while maintaining an appropriate role for the human in the system. Given its basis in task load measurement, this model could also be considered in an adaptive automation design to balance workload.

The model by Wei et al. (1998) considers tasks as a whole and examines their overall impact on operator mental workload and system performance. Other models have attempted more detailed examination of the cognitive processes underlying both individual task performance and coordination processes between tasks. One such model is the Man Machine Integrated Design and Analysis System (MIDAS) model (Corker, Pisanich, & Bunzo, 1997), which incorporates a human-information processing model (e.g., including such components as a working memory, a declarative memory, sensory and motor buffers) and a model of the task environment. The model has been applied with some success to the analysis of pilot-automation interaction and to air-ground integration issues involving multiple human (e.g., pilot, air-traffic controller) and machine (e.g., cockpit and ATC automation) agents to predict flight crew times to initiate evasive maneuvers to maintain separation under different conflict scenarios. For example, Corker et al. (1997) showed that MIDAS could be used to determine the appropriate look-ahead times for automated detection of aircraft-to-aircraft conflicts, taking into account the processing times of both the human and machine agents involved.

A functional analysis of human-automation interaction is also a feature of a model by Degani, Shafto, and Kirlik (1999). They used as a formalism, *Statecharts,* which is a type of state-transition network, in which the behavior of a system can be represented by states and actions that move the system from one state to another. They developed the model to attack the problem of designing effective displays and interfaces to minimize mode errors in human-automation interaction and more generally to support effective human performance. They illustrated the application of the model by discussing the FMS, although their model is more generally applicable to any system with automation that has the capability of transitioning between states without immediate operator input.

CONCLUSION

The past decade of research on pilot interaction with automated systems has provided rich results. This empirical progress has not been matched by the development of theories of human-automation interaction. Nevertheless, much is known of the factors that contribute to effective human performance in automated environments. In this chapter, we have outlined six such factors, each of which has a basis in local psychological theories and is supported by a body of empirical evidence. The factors can be summarized as requiring a coordinated analysis of two agents, that is, the types of automation from an information-processing view and the cognitive responses of the pilot to automation states and behaviors. The very recent development of computational models of this interactive relationship is a welcome sign of maturity of research on humans and automation. The factors we have outlined can form the basis for enhancing automation design and for developing training procedures that promote effective use of automation.

REFERENCES

Abbott, K., Slotte, S., Stimson, D., Amalberti, R. R., Bollin, G., Fabre, F., Hecht, S., Imrich, T., Lalley, R., Lydanne, G., Newman, T., & Thiel, G. (1996). *The interfaces between flightcrews and modern flight deck systems* (Report of the FAA Human Factors Team). Washington, DC: Federal Aviation Administration.

Aeronautica Civil of the Republic of Columbia. (1996). *Aircraft Accident Report. Controlled flight into terrain, American Airlines Flight 965, Boeing 757–223, Near Cali, Colombia, December, 20, 1995.* Bogota, Columbia: Author.

Air Transport Association. (1989). *National plan to enhance aviation safety through human factors improvements.* Washington, DC: Air Transport Association.

Amalberti, R. R. (1999). Automation in aviation: A human factors perspective. In D. J. Garland, J. A. Wise, & V. D. Hopkin (Eds.), *Handbook of aviation human factors* (pp. 173–192). Mahwah, NJ: Lawrence Erlbaum Associates.

ASRS. (1997). *Callback #217. July.* Aviation Safety and Reporting System.

Aviation Week & Space Technology. (1995, January 30). *Automated cockpits, special report part 1, 142*(5), 52–65.

Aviation Week & Space Technology. (1998, February 2). *Air traffic control outlook, 148*(5), 42–62.

Aviation Week & Space Technology. (1999, July 26). *ATA group guides training, 151*(4), 28–29.

Bainbridge, L. (1983). Ironies of automation. *Automatica, 19,* 775–779.

Barnes, M., & Grossman, J. (1985). *The intelligent assistant concept for electronic warfare systems* (Technical Report NWC-TP-5585). China Lake, CA: U.S. Naval Warfare Center.

Billings, C. E. (1997). *Aviation automation: The search for a human-centered approach.* Mahwah, NJ: Lawrence Erlbaum Associates.

Billings, C. E., Lauber, J. K., Funkhouser, H., Lyman, G., & Huff, E. M. (1976). *NASA Aviation Safety Reporting System* (Technical Report TM-X-3445). Moffett Field, CA: NASA Ames Research Center.

Billings, C. E., & Woods, D. D. (1994). Concerns about adaptive automation in aviation systems. In R. Parasuraman & M. Mouloua (Eds.), *Human performance in automated systems: Current research and trends* (pp. 264–269). Hillsdale, NJ: Lawrence Erlbaum Associates.

Boeing Commercial Airplane Group. (1998). *Statistical summary of commercial jet aircraft accidents: Worldwide operations 1959–1997* (Technical Report, Boeing Airplane Safety Engineering B-210B). Seattle, WA: Author.

Bowers, C. A., Oser, R. J., Salas, E., & Cannon-Bowers, J. A. (1996). Team performance in automated systems. In R. Parasuraman, & M. Mouloua (Eds.), *Automation and human performance: Theory and applications* (pp. 243–263). Mahwah, NJ: Lawrence Erlbaum Associates.

Brudnicki, D. J., & McFarland, A. L. (1997). *User Request Evaluation Tool (URET) conflict probe performance and benefits assessment* (Technical Report). McLean, VA: MITRE Corporation.

Byrne, E. A., & Parasuraman, R. (1996). Psychophysiology and adaptive automation. *Biological Psychology, 42,* 249–268.

Casner, S. (1994). Understanding the determinants of problem-solving behavior in a complex environment. *Human Factors, 36,* 580–596.

Chambers, N., & Nagel, D. C. (1985). Pilots of the future: Human or computer? *Communications of the Association for Computing Machinery, 28,* 1187–1199.

Cohen, M. S., Parasuraman, R., Freeman, J. (1997). *Trust in decision aids: A model and a training strategy* (Technical Report USAATCOM TR 97-D-4). Arlington, VA: Cognitive Technologies.

Commission on Aviation Safety and Security. (1997, February 12). *Final report to President Clinton. Vice President Al Gore, Chairman* Washington, DC: Author.

Comstock, J. R., & Arnegard, R. J. (1992). *The multi-attribute task battery for human operator workload and strategic behavior research* (Technical Memorandum 104174). Hampton, VA: NASA Langley Research Center.

Corker, K., Pisanich, G., & Bunzo, M. (1997). Empirical and analytic studies of human/automation dynamics in airspace management for Free Flight. In *Proceedings of the 10th International CEAS Conference on Free Flight* (pp. 176–182). Amsterdam: CEAS.

Crocoll, W. M., & Coury, B. G. (1990). Status or recommendation: Selecting the type of information for decision aiding. In *Proceedings of the Human Factors and Ergonomics Society 34th Annual Meeting* (pp. 1524–1528). Santa Monica, CA: Human Factors and Ergonomics Society.

Degani, A., Shafto, M., & Kirlik, A. (1999). Models in human-machine systems: Constructs, representation, and classification. *International Journal of Aviation Psychology, 9,* 125–138.

Duley, J. A., & Parasuraman, R. (1999). Adaptive information management in future air traffic control. In M. Scerbo & M. Mouloua (Eds.), *Automation technology and human performance: Current research and trends* (pp. 86–90). Mahwah, NJ: Lawrence Erlbaum Associates.

Durso, F. T., & Gronlund, S. D. (1999). Situation awareness. In F. T. Durso, R. S. Nickerson, R. W. Schvanaveldt, S. T. Dumais, & M. T. H. Chi (Eds.), *Handbook of applied cognition* (pp. 283–314). New York: Wiley.

Endsley, M. (1999). Situation awareness in aviation systems. In D. J. Garland, J. A. Wise, & V. D. Hopkin (Eds.), *Handbook of aviation human factors* (pp. 257–276). Mahwah, NJ: Lawrence Erlbaum Associates.

Endsley, M. R., & Kaber, D. B. (1999). Level of automation effects on performance, situation awareness and workload in a dynamic control task. *Ergonomics, 42,* 462–492.

Endsley, M., & Kiris, E. O. (1995). The out-of-the-loop performance problem and level of control in automation. *Human Factors, 37,* 381–394.

Endsley, M., & Strauch, B. (1997). Automation and situation awareness: The accident at Cali, Columbia. In *Proceedings of the Ninth International Symposium on Aviation Psychology* (pp. 877–881). Columbus: The Ohio State University.

Federal Aviation Administration. (1997, April). *Private Pilot Practical Test Standards for Airplanes* (FAA-S-8081-14). Washington, DC: Author.

Federal Aviation Administration. (2001, February). *Airline Transport Pilot and Aircraft Type Rating Practical Test Standards* (FAA-S-8081-5D). Washington, DC: Author.

Fitts, P. M. (Ed.). (1951). *Human engineering for an effective air navigation and traffic control system.* Washington, DC: National Research Council.

Flach, J. (1994). Situation awareness: The emperor's new clothes. In M. Mouloua & R. Parasuraman (Eds.), *Human performance in automated systems: Current research and trends* (pp. 241–248). Mahwah, NJ: Lawrence Erlbaum Associates.

Flight Safety Foundation. (1997). *Aviation statistics.* Washington, DC: Author.

Funk, K. (1991). Cockpit task management: Preliminary definitions, normative theory, error taxonomy, and design recommendations. *International Journal of Aviation Psychology, 1,* 271–286.

Funk, K., Lyall, B., Wilson, J., Vint, R., Niemczyk, M., Suroteguh, C., & Owen, G. (1999). Flight deck automation issues. *International Journal of Aviation Psychology, 9,* 125–138.

Galster, S., Duley, J. A., Masalonis, A., & Parasuraman, R. (2001). Air traffic controller performance and workload under mature Free Flight: Conflict detection and resolution of aircraft self-separation. *International Journal of Aviation Psychology, 11,* 71–93.

Geddes, N. (1985). Intent inferencing using scripts and plans. *Proceedings of the First Annual Aerospace Applications of Artificial Intelligence Conference* (pp. 36–41). Washington, DC.

Getty, D. J., Swets, J. A., Pickett, R. M., & Gounthier, D. (1995). System operator response to warnings of danger: A laboratory investigation of the effects of the predictive value of a warning on human response time. *Journal of Experimental Psychology: Applied, 1,* 19–33.

Gevins, A., Smith, M. E., Leong, H., McEvoy, L., Whitfield, S., Du, R., & Rush, G. (1998). Monitoring working memory during computer-based tasks with EEG pattern recognition. *Human Factors, 40,* 79–91.

Green, D. M., & Swets, J. A. (1966). *Signal detection theory and psychophysics.* New York: Wiley.

Hancock, P. A., Chignell, M. H., & Lowenthal, A. (1985). An adaptive human-machine system. *Proceedings of the IEEE Conference on Systems, Man and Cybernetics, 15,* 627–629.

Hancock, P. A., & Scallen, S. F. (1996). The future of function allocation. *Ergonomics in Design, 4*(4), 24–29.

Harris, W., Hancock, P. A., & Arthur, E. (1993). The effect of taskload projection on automation use, performance, and workload. In *Proceedings of the Seventh International Symposium on Aviation Psychology* (pp. 361–365). Columbus: The Ohio State University.

Hart S. G., & Wempe, T. E. (1979). *Cockpit display of traffic information: Airline pilots opinions about content, symbology and format* (NASA Technical Memorandum 78601). Moffett Field, CA: NASA Ames Research Center.

Hilburn, B., Jorna, P. G., Byrne, E. A., & Parasuraman, R. (1997). The effect of adaptive air traffic control (ATC) decision aiding on controller mental workload. In M. Mouloua & J. Koonce (Eds.), *Human-automation interaction* (pp. 84–91). Mahwah, NJ: Lawrence Erlbaum Associates.

Huettner, C. (1996). *Towards a safer 21st century* (NASA Technical Report). Washington, DC: NASA Headquarters.

Huey, B. M. & Wickens, C. D. (Eds.). (1993). *Workload transition: Implications for individual and team performance.* Washington, DC: National Academic Press.

Inagaki, T. (1999). Situation-adaptive autonomy for time-critical takeoff decisions. *International Journal of Modeling and Simulation, 19*(4), 241–247.

Jordan, N. (1963). Allocation of functions between man and machines in automated systems. *Journal of Applied Psychology, 47,* 161–165.

Jorna, P. G. A. H. (1999). Automation and Free Flight: Exploring the unexpected. In M. W. Scerbo, & M. Mouloua (Eds.), *Automation technology and human performance* (pp. 107–111). Mahwah, NJ: Lawrence Erlbaum Associates.

Kaber, D. B., Omal, E., & Endsley, M. R. (1999). Level of automation effects on telerobot performance and human operator situation awareness and subjective workload. In M. Scerbo & M. Mouloua (Eds.), *Automation technology and human performance* (pp. 165–170). Mahwah, NJ: Lawrence Erlbaum Associates.

Kaber, D. B., & Riley, J. M. (1999). Adaptive automation of a dynamic control task based on workload assessment through a secondary monitoring task. In M. Scerbo & M. Mouloua (Eds.), *Automation technology and human performance: Current research and trends* (pp. 129–133). Mahwah, NJ: Lawrence Erlbaum Associates.

Kantowitz, B. H. (1986). Mental workload. In P. A. Hancock (Ed.), *Human factors psychology* (pp. 81–122). New York: North-Holland.

Kantowitz, B. H., & Campbell, J. L. (1996). Pilot workload and flightdeck automation. In R. Parasuraman and M. Mouloua (Eds.), *Automation and human performance: Theory and applications* (pp. 117–136). Hillsdale, NJ: Lawrence Erlbaum Associates.

Kantowitz, B. H., & Sorkin, R. D. (1987). Allocation of function. In G. Salvendy (Ed.), *Handbook of human factors* (pp. 355–369). New York: Wiley.

Kirlik, A. (1993). Modeling strategic behavior in human-automation interaction: Why an "aid" can (and should) go unused. *Human Factors, 35,* 221–242.

Kramer, A., Trejo, L. J., & Humphrey, D. G. (1996). Psychophysiological measures of workload: Potential applications to adaptively automated systems. In R. Parasuraman & M. Mouloua (Eds.), *Automation and Human Performance* (pp. 137–162). Mahwah, NJ: Lawrence Erlbaum Associates.

Kuchar, J. (1996). Methodology for alerting-system performance evaluation. *Journal of Guidance, Control, and Dynamics, 19,* 438–444.

Lee, J. D., & Moray, N. (1992). Trust, control strategies, and allocation of function in human-machine systems. *Ergonomics, 35,* 1243–1270.

Lee, J. D., & Moray, N. (1994). Trust, self-confidence, and operators' adaptation to automation. *International Journal of Human-Computer Studies, 40,* 153–184.

Lewis, M. (1998, Summer). Designing for human-agent interaction. *Artificial Intelligence Magazine,* 67–78.

Lorenz, B. D., Nocera, F., Röttger, S., & Parasuraman, R. (in press). Automated fault management in a simulated spaceflight micro-world. *Aviation, Space, and Environmental Medicine.*

Masalonis, A. J., Duley, J. A., & Parasuraman, R. (1999). Effects of manual and autopilot control on mental workload and vigilance during general aviation simulated flight. *Transportation Human Factors, 1,* 187–200.

Masalonis, A. J., & Parasuraman, R. (1999). Trust as a construct for evaluation of automated aids: Past and future theory and research. In *Proceedings of the Human Factors and Ergonomics Society 43rd Annual Meeting* (pp. 184–188). Santa Monica, CA: Human Factors and Ergonomics Society.

Masalonis, A. J., & Parasuraman, R. (2001). Fuzzy signal detection theory: Analysis of human and machine performance in air traffic control, and analytical considerations. Manuscript submitted for publication.

May, P., Molloy, R., & Parasuraman, R. (1993, October). *Effects of automation reliability and failure rate on monitoring performance in a multitask environment.* Paper presented at the 37th Annual Meeting of the Human Factors and Ergonomics Society, Seattle, WA.

McDaniel, J. W. (1988). Rules for fighter cockpit automation. In *Proceedings of the IEEE National Aerospace and Electronics Conference* (pp. 831–838). New York: IEEE.

Moray, N. (1997). Human factors in process control. In G. Salvendy (Ed.), *Handbook of human factors and ergonomics* (pp. 1944–1971). New York: Plenum Press.

Moray, N., Inagaki, T., Itoh, M. (2000). Adaptive automation, trust, and self-confidence in fault management of time-critical tasks. *Journal of Experimental Psychology: Applied, 6,* 44–58.

Mosier, K. L., & Skitka, L. J. (1996). Human decision makers and automated decision aids: Made for each other? In R. Parasuraman & M. Mouloua (Eds.), *Automation and human performance: Theory and applications* (pp. 201–220). Mahwah, NJ: Lawrence Erlbaum Associates.

Mosier, K. L., Skitka, L. J., Heers, S., & Burdick, M. (1998). Automation bias: Decision making and performance in high-tech cockpits. *International Journal of Aviation Psychology, 8,* 47–63.

Mouloua, M., Parasuraman, R., & Molloy, R. (1993). Monitoring automation failures: Effects of single and multiadaptive function allocation. In *Proceedings of the Human Factors and Ergonomics Society 37th Annual Meeting* (pp. 1–5). Santa Monica, CA: Human Factors and Ergonomics Society.

National Transportation Safety Board. (1973). *Eastern Air Lines, Inc., L-1011, N310EA, Miami, Florida, December 29, 1972* (AAR-73–14). Washington, DC: Author.

National Transportation Safety Board. (1984). *Aircraft Accident Report. Scandinavian Airlines System Flight 901, McDonnell Douglas DC-10–30, John F. Kennedy Airport, Jamaica, New York, February 28, 1984* (AAR-84–15). Washington, DC: Author.

National Transportation Safety Board. (1993). *Aircraft Accident Report. Aborted Takeoff Shortly After Liftoff, Trans World Airlines Flight 843, Lockheed L-1011, N11002, John F. Kennedy International Airport Jamaica, New York July 30, 1992* (AAR-93-04). Washington, DC: Author.

National Transportation Safety Board. (1996a). *Aircraft Accident Report. Ground Spoiler Activation in Flight / Hard Landing, Valujet Airlines Flight 558, Douglas DC-9-32, N922VV, Nashville, Tennessee, January 7, 1996* (AAR-96-07). Washington, DC: Author.

National Transportation Safety Board. (1996b). *Brief of Accident NYC96LA148.* Washington, DC: Author.

National Transportation Safety Board. (1997a). *Aircraft Accident Report. Uncontrolled Flight Into Terrain. ABX Air (Airborne Express), Douglas DC8–63, N827AX Narrows, Virginia December 22, 1996* (AAR-97-05). Washington, DC: Author.

National Transportation Safety Board. (1997b). *Brief of Accident. NYC97FA045.* Washington, DC: Author.

National Transportation Safety Board. (1998a). *Brief of Accident NYC98FA020.* Washington, DC: Author.

National Transportation Safety Board. (1998b). *Brief of Accident NYC98LA056.* Washington, DC: Author.

National Transportation Safety Board. (1998c). *Safety Recommendation Letter A-98-3 through -5. January 21, 1998.* Washington, DC: Author.

Norman, D. A. (1990). The problem with automation: Inappropriate feedback and interaction, not "over-automation." *Philosophical Transactions of the Royal Society (London), B237,* 585–593.

Parasuraman, R. (1987). Human-computer monitoring. *Human Factors, 29,* 695–706.

Parasuraman, R. (1990). Event-related brain potentials and human factors research. In J. W. Rohrbaugh, R. Parasuraman, & R. Johnson (Eds.), *Event-related brain potentials* (pp.279–306). New York: Oxford University Press.

Parasuraman, R. (1993). Effects of adaptive function allocation on human performance. In D. J. Garland and J. A. Wise (Eds.), *Human factors and advanced aviation technologies* (pp. 147–157). Daytona Beach, FL: Embry-Riddle Aeronautical University Press.

Parasuraman, R. (2000). Designing automation for human use: Empirical studies and quantitative models. *Ergonomics, 43,* 931–951.

Parasuraman, R., Bahri, T., Deaton, J., Morrison, J., & Barnes, M. (1992). *Theory and design of adaptive automation in aviation systems* (Progress Report No. NAWCADWAR-92033-60). Warminster, PA: U.S. Naval Air Warfare Center.

Parasuraman, R., Duley, J. A., & Smoker, A. (1998). Automation tools for controllers in future air traffic control. *The Controller: Journal of Air Traffic Control, 37,* 7–13.

Parasuraman, R., & Hancock, P. A. (1999). Using signal detection theory and Bayesian analysis to design parameters for automated warning systems. In M. Scerbo & M. Mouloua (Eds.), *Automation technology and human performance: Current research and trends* (pp. 63–67). Mahwah, NJ: Lawrence Erlbaum Associates.

Parasuraman, R., & Hancock, P. A. (2001). Adaptive control of workload. In P. A. Hancock and P. Desmond (Eds.), *Stress, workload, and fatigue* (pp. 305–320). Mahwah, NJ: Lawrence Erlbaum Associates.

Parasuraman, R., Hancock, P. A., & Olofinboba, O. (1997). Alarm effectiveness in driver-centered collision-warning systems. *Ergonomics, 39,* 390–399.

Parasuraman, R., Hilburn, B., & Hoekstra, J. (2001). Free Flight. In W. Karwowski (Ed.), *International encyclopedia of ergonomics and human factors* (pp. 1005–1008). New York: Taylor & Francis.

Parasuraman, R., Masalonis, A. J., & Hancock, P. A. (2000). Fuzzy signal detection theory: Basic postulates and formulas for analyzing human and machine performance. *Human Factors 42,* 636–659.

Parasuraman, R., Molloy, R., & Singh, I. L. (1993). Performance consequences of automation-induced "complacency." *International Journal of Aviation Psychology, 3,* 1–23.

Parasuraman, R., Mouloua, M., & Hilburn, B. (1999). Adaptive aiding and adaptive task allocation enhance human-machine interaction. In M. Scerbo & M. Mouloua (Eds.), *Automation technology and human performance: Current research and trends* (pp. 119–123). Mahwah, NJ: Lawrence Erlbaum Associates.

Parasuraman, R., Mouloua, M., & Molloy, R. (1996). Effects of adaptive task allocation on monitoring of automated systems. *Human Factors, 38,* 665–679.

Parasuraman, R., & Riley, V. A. (1997). Humans and automation: Use, misuse, disuse, abuse. *Human Factors, 39,* 230–253.

Parasuraman, R., Sheridan, T. B., and Wickens, C. D. (2000). A model for types and levels of human interaction with automation. *IEEE Transactions on Systems, Man, and Cybernetics, 30,* 286–297.

Pew, R. W., & Mavor, A. S. (1998). *Modeling human and organizational behavior: Application to military simulations.* Washington, DC: National Academy Press.

Prinzel, L. J., Freeman, F. G., Scerbo, M. W., Mikulka, P. J., & Pope, A. T. (2000). A closed-loop system for examining psychophysiological measures for adaptive automation. *International Journal of Aviation Psychology, 10,* 393–410.

Pritchett, A. R., & Hansman, R. J. (1997). Pilot non-conformance to alerting system commands. In *Proceedings of the Ninth International Symposium on Aviation Psychology* (pp. 274–279). Columbus: The Ohio State University.

Rasmussen, J. (1986). *Information processing and human-machine interaction.* Amsterdam: North-Holland.

Rawlins, G. J. E. (1996). *Moths to the flame: The seductions of computer technology.* Cambridge, MA: MIT Press.

Reason, J. T. (1990). *Human error.* Cambridge, England: Cambridge University Press.
Rempel, J. K., Holmes, J. G., & Zanna, M. P. (1985). Trust in close relationships. *Journal of Personality and Social Psychology, 49,* 95–112.
Riley, V. (1989). A general model of mixed-initiative human-machine systems. In *Proceedings of the Human Factors Society 33rd Annual Meeting* (pp. 124–128). Santa Monica, CA: Human Factors Society.
Riley, V. (1996). Operator reliance on automation: Theory and data. In R. Parasuraman, & M. Mouloua (Eds.), *Automation and human performance: Theory and applications* (pp. 19–35). Mahwah, NJ: Lawrence Erlbaum Associates.
Riley, V. (1997). What avionics engineers should know about pilots and automation. In *Proceedings of the Digital Avionics Systems Conference* (pp. 53–57).
Riley, V., Lyall, B., and Wiener, E. (1993). *Analytic methods for flight-deck automation design and evaluation. Phase two report: Pilot use of automation* (Technical Report). Minneapolis, MN: Honeywell Technology Center.
Rose, A. M. (1989). Acquisition and retention of skills. In G. McMillan, D. Beevis, E. Salas, M. H. Strub, R. Sutton, & L. Van Breda (Eds.), *Applications of human performance models to system design* (pp. 419–426). New York: Plenum Press.
Rouse, W. B. (1976). Adaptive allocation of decision making responsibility between supervisor and computer. In T. B. Sheridan & G. Johannsen (Eds.), *Monitoring behavior and supervisory control* (pp. 271–283). New York: Plenum Press.
Rouse, W. B. (1988). Adaptive aiding for human/computer control. *Human Factors, 30,* 431–438.
RTCA. (1995). *Report of the RTCA Board of Director's Select Committee on Free Flight.* Washington, DC: Author.
Sarter, N., & Woods, D. D. (1991). Situation awareness: A critical but ill-defined phenomenon. *International Journal of Aviation Psychology, 1,* 45–57.
Sarter, N., & Woods, D. D. (1994). Pilot interaction with cockpit automation II: An experimental study of pilots' model and awareness of the flight management system. *International Journal of Aviation Psychology, 4,* 1–28.
Sarter, N., & Woods, D. D. (1995). How in the world did we ever get into that mode? Mode error and awareness in supervisory control. *Human Factors, 37,* 5–19.
Sarter, N., Woods, D. D., & Billings, C. E. (1997). Automation surprises. In G. Salvendy (Ed.), *Handbook of human factors and ergonomics* (2nd ed, pp. 1926–1943). New York: Wiley.
Satchell, P. (1993). *Cockpit monitoring and alerting systems.* Aldershot, England: Ashgate.
Satchell, P. (1998). *Innovation and automation.* Aldershot, England: Ashgate.
Scallen, S., Hancock, P. A., & Duley, J. A. (1995). Pilot performance and preference for short cycles of automation in adaptive function allocation. *Applied Ergonomics, 26,* 397–403.
Scerbo, M. W. (1996). Theoretical perspectives on adaptive automation. In R. Parasuraman & M. Mouloua (Eds.), *Automation and human performance: Theory and applications* (pp. 37–63). Mahwah, NJ: Lawrence Erlbaum Associates.
Scerbo, M. W. (2001). Adaptive automation. In W. Karwowski (Ed.), *International encyclopedia of ergonomics and human factors* (pp. 1007–1009). London: Taylor & Francis, Inc.
Scerbo, M., Freeman, F., Mikulka, P. J., Di Nocera, F., Parasuraman, R., & Prinzel, L. J. (2001). *The efficacy of physiological measures for implementing adaptive technology* (NASA Technical Paper 211018). Norfolk, VA: Old Dominion University.
Sheridan, T. B. (1992). *Telerobotics, automation, and supervisory control.* Cambridge, MA: MIT Press.
Sheridan, T. B. (1998). Allocating functions rationally between humans and machines. *Ergonomics in Design, 6*(3), 20–25.
Sheridan, T. B., & Parasuraman, R. (2000). Human vs. automation in responding to failures: An expected-value analysis. *Human Factors, 42,* 403–407.
Sherry, L., & Polson, P. G. (1999). Shared models of flight management system vertical guidance. *International Journal of Aviation Psychology, 9,* 139–153.

Singh, I. L., Molloy, R., Mouloua, M., Deaton, J. E., & Parasuraman, R. (1998). Cognitive ergonomics of cockpit automation. In I. L. Singh & R. Parasuraman (Eds.), *Human cognition: A multidisciplinary perspective* (pp. 242–254). New Delhi, India: Sage.

Singh, I. L., Molloy, R., & Parasuraman, R. (1997). Automation-related monitoring inefficiency: The role of display location. *International Journal of Human-Computer Studies, 46,* 17–30.

Skitka, L. J., Mosier, K. L., & Burdick, M. (1999). Does automation bias decision-making? *International Journal of Human-Computer Studies, 51,* 991–1006.

Smith, P. J., Woods, D. D., Billings, C. E., Denning, R., Dekker, S., McCoy, E., & Sarter, N. (1999). Conclusions from the application of a methodology to evaluate future air traffic management systems. In M. Scerbo & M. Mouloua (Eds.), *Automation technology and human performance: Current research and trends* (pp. 81–85). Mahwah, NJ: Lawrence Erlbaum Associates.

Sorkin, R. D. (1988). Why are people turning off our alarms? *Journal of the Acoustical Society of America, 84,* 1107–1108.

Sorkin, R. D., Kantowitz, B. H., & Kantowitz, S. C. (1988). Likelihood alarm displays. *Human Factors, 30,* 445–459.

Spitzer, C. R. (1987). *Digital avionics systems.* Englewood Cliffs, NJ: Prentice Hall.

Sumwalt, R. L., Morrison, R., Watson, A., & Taube, E. (1997). What ASRS data tell about inadequate flight crew monitoring. In *Proceedings of the Ninth International Symposium on Aviation Psychology* (pp. 977–982). Columbus: The Ohio State University.

Swets, J. A., & Pickett, R. M. (1982). *Evaluation of diagnostic systems: Methods from signal detection theory.* New York: Academic Press.

Tsang, P. S., & Johnson, W. W. (1989). Cognitive demands in automation. *Aviation, Space, and Environmental Medicine, 60,* 130–135.

Tversky, A., & Kahneman, D. (1974). Judgment under uncertainty: Heuristics and biases. *Science, 185,* 1124–1131.

Warm, J. S., Dember, W., & Hancock, P. A. (1996). Vigilance and workload in automated systems. In R. Parasuraman & M. Mouloua (Eds.), *Automation and human performance: Theory and applications* (pp. 183–200). Mahwah, NJ: Lawrence Erlbaum Associates.

Wei, Z., Macwan, A. P., & Wieringa, P. A. (1998), A quantitative measure for degree of automation and its relation to system performance. *Human Factors, 40,* 277–295.

Wickens, C. D. (1992). *Engineering psychology and human performance* (2nd ed.). New York: HarperCollins.

Wickens, C. D. (1994). Designing for situation awareness and trust in automation. In *Proceedings of the IFAC Conference on Integrated Systems Engineering* (pp. 171–176). Baden-Baden, Germany.

Wickens, C. D. (2000). *Imperfect and unreliable automation and its implications for attention allocation, information access, and situation awareness* (Technical Report ARL-00-10/NASA-00-2). Savoy: University of Illinois, Aviation Research Laboratory.

Wickens, C. D., Conejo, R., & Gempler, K. (1999). Unreliable automated attention cueing for air-ground targeting and traffic maneuvering. In *Proceedings of the Human Factors and Ergonomics Society 43rd Annual Meeting* (pp. 21–25). Santa Monica, CA: Human Factors and Ergonomics Society.

Wickens, C. D., Mavor, A., Parasuraman, R., & McGee, J. (1998). *The future of air traffic control: Human operators and automation.* Washington, DC: National Academy Press.

Wiegmann, D. A., & Shappell, S. A. (1997). Human factors analysis of post-accident data: Applying theoretical taxonomies of human error. *International Journal of Aviation Psychology, 7,* 67–81.

Wiener, E. L. (1981). Complacency: Is the term useful for air safety? In *Proceedings of the 26th Corporate Aviation Safety Seminar.* Denver, CO: Flight Safety Foundation.

Wiener, E. L. (1988). Cockpit automation. In E. L. Wiener & D. C. Nagel (Eds.), *Human factors in aviation* (pp. 433–461). San Diego, CA: Academic Press.

Wiener, E. L., Chute, R. D., & Moses, J. H. (1999). *Transition to glass: Pilot training for high-technology transport aircraft* (Technical Report No. NASA/CR-1999-208784), Moffett Field, CA: NASA Ames Research Center.

Wiener, E. L., & Curry, R. E. (1980). Flight-deck automation: Promises and problems. *Ergonomics, 23,* 995–1011.

Woods, D. D. (1996). Decomposing automation: Apparent simplicity, real complexity. In R. Parasuraman & M. Mouloua (Eds.), *Automation and human performance: Theory and applications* (pp. 1–17). Mahwah, NJ: Lawrence Erlbaum Associates.

Yeh, Y. Y., & Wickens, C. D. (1988). The dissociation of subjective measures of mental workload and performance. *Human Factors, 30,* 111–120.

Zadeh, L. A. 1965, Fuzzy sets. *Information and Control, 8,* 338–353.

ACKNOWLEDGMENTS

Dr. Parasuraman was supported by NASA Grants NAG 2–1096 (Ames Research Center, CA), NAG 3–2103 (Langley Research Center, VA), and NAG 5–8761 (Goddard Space Center, MD). The views expressed in this chapter are solely those of the authors and not necessarily those of NASA or the NTSB.

10

Pilot Selection Methods

Thomas R. Carretta
Air Force Research Laboratory

Malcolm James Ree
Our Lady of the Lake University

The quality of the box matters little. Success depends upon the man who sits in it.
—Baron Manfred von Richthofen, *The Red Baron.*

This chapter consists of seven parts. The first part describes pilot selection, why it is important, and the knowledge, skills, abilities, and other characteristics typically considered during selection. Part two introduces "validity" and the steps involved in a validation study. Part three reviews common methodological issues that make the interpretation of pilot selection studies more difficult and offers "best practices" advice for researchers and practitioners. Parts four and five review military and commercial pilot selection. Where available, information about the construct and predictive validity of the selection methods is provided. Part six examines future trends in the measurement of pilot aptitude. Finally, the conclusion provides recommendations for researchers and practitioners.

WHAT IS PILOT SELECTION AND WHY IS IT IMPORTANT?

Organizations need people to serve in various capacities and people need jobs. In military aviation, the goal is achieving and maintaining a high level of mission readiness. To do so, enough qualified pilots must be available to accomplish mission requirements. This is done by training new and experienced pilots and improving retention of experienced pilots to achieve mission readiness. Other organizational goals in military aviation include reducing training costs, avoiding loss of aircraft/loss of life, and achieving diversity in the workforce. In commercial aviation, organizational goals emphasize public safety, low training and operating costs, and customer satisfaction. Cascio (1982) provides several examples of the impact of personnel selection on training costs and organizational productivity. In the U.S. Air Force, estimates of the cost of each person who fails to complete undergraduate pilot training range from $50,000 (Hunter, 1989) to $80,000 (Siem, Carretta, & Mercatante, 1988). Obviously, even a small reduction in training attrition could result in large cost avoidance savings. To achieve this, the needs of the organization and the job applicants must be matched (Guion, 1976). Making the right selection decision reduces training costs, improves job performance, and enhances organizational effectiveness.

Overview of Pilot Selection Process

Since World War I, personnel specialists in both military and commercial aviation have spent a great deal of time, money, and effort attempting to identify the characteristics needed to be a good pilot and the means to accurately measure those characteristics. The military has gone even further, attempting to determine whether a pilot would be better suited to fly fighter or nonfighter aircraft (Carretta, 1989).

In military aviation, pilot applicants typically have little or no prior flying experience and may not have had prior exposure to the military. Commonly used selection factors include measures of ability (e.g., standardized test scores, college grade point average and major), medical qualification, indicators of "officership" (e.g., commander's ratings from an officer training program), and prior flying experience (e.g., number of hours flown, private pilot's license). Personality assessment is done in some military organizations (e.g., psychological interview), but is less common.

Some commercial airlines have ab initio (from the beginning) training programs, where carefully selected applicants with little or no flying experience are put through intensive pilot training courses. However, most commercial airlines prefer to hire experienced pilots to avoid the time and expense of training. When selecting applicants for ab initio training, indicators of ability (i.e., trainability) are emphasized. When selecting from experienced pilots, commercial carriers tend to emphasize indicators of prior experience (e.g., certificates and licenses, log book hours, military pilot experience, recommendations) and flying competence (e.g., check flight performance, simulator performance).

Selection into a military or civilian pilot training program typically is a multistage process. Multistage selection is the process in which decisions are made at several points. The first stage might be an evaluation of credentials such as flight hours and letters of recommendation. A second stage might be a written test, an interview, or a simulator flight evaluation. The third stage might be flying an aircraft and the last stage might be a final interview.

Are Effective Pilots "Selected" or "Trained"?

Both selection and training play important roles in producing effective pilots who will allow the organization to meet its goals. Effective selection procedures will produce cost-avoidance savings through reduced attrition and reduced training requirements and will lead to improved job performance and improved organizational effectiveness. Poor selection will result in increased training attrition, training requirements, and cost, and lead to poor job performance and poor organizational effectiveness. Effective training methods can help reduce training attrition and contribute to improving organizational effectiveness (chap. 11; Smallwood & Fraser, 1995; Walter, 1998).

VALIDITY AND VALIDATION STUDIES

What Is Validity?

Validity is the most fundamental testing and selection issue. As described by Jensen (1980), ". . . validity is the extent to which scientifically valuable or practically useful inferences can be drawn from the scores" (p. 297). However much effort is made to develop selection methods based on theories of the relations between personnel characteristics and performance, they will come to nothing without validity. Theory without proof is worthless.

Historically, we have acknowledged three types of validity: content, construct, and criterion (predictive). A test has *content validity* to the extent that its items are judged to represent some clearly specified area of knowledge, skill, ability, or characteristic. This judgment is often based on the consensus of subject-matter-experts (SMEs). For example, psychologists might be SMEs for making judgments about tests of cognitive processes and experienced pilots might be appropriate SMEs for measures of flying job knowledge or performance.

Whereas content validity is based on expert judgment, *construct validity* is concerned with the scientific attempt to determine what a test actually measures. Construct validity becomes an important issue when we have a theory about the nature of the trait that we measured. A theoretical foundation allows us to develop and test hypotheses about what will happen under specified conditions. A test is said to have construct validity if it predicts behavior in specific situations that would be inferred from our theory.

Criterion or predictive validity is the ability of test scores to predict performance in some activity (criterion) external to the test itself. Typically, in personnel selection the criterion consists of one or more measures of training or job performance. Though content and construct validity are very desirable for enhancing our understanding of the tests and criteria, neither is essential for criterion validity. All that is needed for criterion validity is that the test predict the criteria. An important related concept is *incremental validity.* A test has incremental validity if it improves prediction of the criteria beyond that provided by some baseline test. Although all three types of validity are important in personnel measurement and selection, criterion validity will be emphasized here, due to its greater use in pilot selection.

What Is a Validation Study?

The description is based on best professional practice and legal requirements. The legal requirements come from case law, especially *Griggs et al. v. Duke Power Co.* (1971) and from the federal *Uniform Guidelines on Employee Selection Procedures* (Equal Employment Opportunity Commission, 1978). The general standards for validity studies are described in §1607.5 of the *Uniform Guidelines on Employee Selection Procedures.*

Selection necessarily implies screening of job applicants and rejection of some. As noted by Jensen (1980), there are two justifications for selection. The first is when the pool of applicants is larger than the number of training or job positions. The second is when the predictive validity of the selection procedures can be demonstrated. Tests or other selection methods (e.g., biodata, interviews, recommendations) are said to have *predictive validity* to the extent that they would distinguish between the performance of selectees and rejectees if all of them had been selected.

In their pursuit of an ideal pilot selection system, aviation psychologists have examined a variety of personnel constructs and measurement methods (Hunter & Burke, 1995). A formal validation study is required to determine the utility of these constructs and methods for predicting training and job performance. Guion (1976) describes a four-step procedure for forming and testing hypotheses about personnel selection.

Perform a Job Analysis. The first step is to identify important job performance constructs, usually through job analysis (Cascio, 1991; Gael, 1988; McCormick, 1976, 1979). The goal of job analysis is the establishment of job, task, and cognitive requirements or Knowledge, Skills, Abilities, and Other

(KSAO) requirements. It can be accomplished many different ways. Cascio (1991) provides a good discussion of the methods. Results from the job analysis can lead to the development of a structural taxonomy and specification of predictor and criteria measures.

Develop Operational Definitions of Important Constructs. The second step is to develop operational definitions of these job performance constructs and ensure that they show acceptable construct validity. As previously discussed, construct validity is based on theory and is determined by testing hypotheses about the relations between the tests and performance criteria. Construct validity of a psychomotor test could be examined by administering it along with marker tests whose properties are well known and examining the relations between the psychomotor test and marker tests.

Identify a Set of Predictors and Criteria. The third step is to propose a set of predictor and criteria variables. The choice of predictors should be guided by theory and evidence. They should be developed using psychometric techniques to ensure appropriate content, difficulty, and precision of measurement. Pilot job performance criteria must be established using the same psychometric guidelines. The criterion is usually some measure of occupational performance such as training accomplishments, hands-on job ratings, work samples, job knowledge, or productivity. Examples include supervisory ratings, accident reports, or direct indicators of job performance such as percent of on-time arrivals or percent of enemy targets destroyed. Performance ratings are the most frequently used criterion measure (Pulakos, 1997). In practice, most criterion variables are positively correlated. This means that most criterion variables measure aspects of the same underlying construct, and the main feature of criterion development is specifying a sufficient number of measures, avoiding criterion contamination from extraneous features, and covering the breadth of the criterion construct.

Examine Predictive Validity. In the fourth step, select predictor and criterion measures and examine predictive validity. When the predictor and criterion measures have been deemed suitable, they can be administered to an appropriate sample in either a predictive or a concurrent validation design. In a predictive design, the appropriate sample is a large group of applicants. The predictor measures are administered during application, and the criteria are collected after those selected have completed training or been on the job for some period such as 3, 6, or 12 months. In a concurrent design, a large sample of job incumbents is administered the predictor and the criterion measures simultaneously (concurrently). In both validity designs, the criteria data are available for only a sample of those selected for training or employment. This leads to a selected sample and the artifact of range restriction as described later.

During validation, the data are analyzed and inferences are drawn about the relationship of the predictor and criterion. Typically, one or more predictors are correlated with one or more criteria. The index used to assess predictive validity is usually the correlation coefficient (r) or the multiple correlation (R) if there is more than one predictor. The *Griggs et al. v. Duke Power Co.* (1971) decision established the commonly used $p < .05$ significance level. The final part of the analysis is the reporting of the validity study results. Whetzel and Oppler (1997) provide a good overview of this process.

Other Considerations. In addition to the selection system's utility for identifying those likely to be successful, there are other important considerations in personnel selection. These include whether or not the selection methods predict training and job performance equally well for members of different sex and ethnic/racial groups (i.e., predictive bias) and whether or not a test differentially qualifies members of different subgroups (i.e., adverse impact).

COMMON METHODOLOGICAL ISSUES IN PILOT SELECTION

Courses in research methods and statistics are common for personnel specialists, and there are established guidelines for conducting studies of personnel measurement and selection (American Psychological Association, American Educational Research Association, & National Council on Measurement in Education, 1985; Society for Industrial-Organizational Psychology [SIOP], 1987). Despite this, many studies of personnel measurement and selection embody methodological issues that cloud their interpretability. We have identified four major methodological issues that can influence conclusions about construct and criterion-related (predictive) validity. There are more, but these four are common and lead to incorrect conclusions and decisions about the effectiveness of pilot selection systems. The four issues are misunderstanding constructs, lack of statistical power, failure to estimate cross-validation effects, and misinterpretation of correlations and regression. Our purposes in discussing them in this chapter is to raise the readers' awareness so they will be able to read the published literature with a critical eye and to offer "best practices" solutions for researchers and practitioners. Carretta and Ree (2001) provide a more detailed discussion.

Misunderstanding Constructs

Abstractions such as "airmanship," "intelligence," "situational awareness," or "workload" are called constructs. They cannot be observed directly and must be inferred from some test, measurement scale, or questionnaire that operationalizes the components of the construct. There is no scientific value to a construct that cannot be measured.

It is important to remember that a construct can be measured by many means. Hunter and Hunter (1984) have demonstrated this, and Spearman (1927) noted it in his idea of "the indifference of the indicator" as Jensen (1993) has pointed out. Walters, Miller, and Ree (1993) have labeled the specious reasoning that because two tests look different they must measure different constructs, the "topographical fallacy." The appearance of the items or tasks in a test is not a reliable indicator of what is being measured. Only results of a construct validation provide a good indicator of what is being measured.

Consider the following example. Several NATO countries use interviews as part of their military pilot selection process (Hansen, 1999). On the basis of the topographical fallacy (i.e., differing appearances), the U.S. Air Force (USAF) considered using a structured interview in their pilot selection process (Walters et al., 1993). Structured interviews have predetermined rules for obtaining, observing, and evaluating responses. In a recent review of the literature on employment interviews, Whetzel and McDaniel (1997) concluded that structure tends to make the interview more reliable and valid. In their meta-analytic review of the utility of personnel selection methods, Schmidt and Hunter (1998) concluded that structured interviews were incrementally valid to measures of general cognitive ability for predicting training and job performance. Walters et al. (1993) reported that the highly structured USAF pilot selection interview measured educational background, motivation to fly, self-confidence and leadership, and flying job knowledge. Additionally, three ratings of probable success in pilot training, bomber-fighter flying, and tanker-transport flying were made. The sample was 223 USAF pilot trainees who were administered the structured interview, Air Force Officer Qualifying Test (AFOQT; Skinner & Ree, 1987), and computer-based cognitive "information processing" (verbal classification, mental rotation, and short-term memory) and personality (self-confidence and attitudes toward risk) tests. Passing-failing pilot training was the criterion. The seven structured interview scores had an average validity of .21. When the seven scores were added to regression equations containing the AFOQT scores (subtest average validity of .28) and the computer-based test scores (average validity of .18), no incremental validity was found. Despite the difference in appearance between the paper-and-pencil AFOQT and the interview, the interview was not able to account for any unique prediction of pilot performance.

It is suggested that researchers and practitioners in pilot selection administer their selection instruments along with a battery of tests of known constructs to have a better understanding of what is being measured by their selection instruments.

Lack of Statistical Power

Statistical power is the probability of detecting a significant effect, such as a difference between means or a nonzero correlation, when present. Specifically, it is the probability of rejecting a false null hypothesis (Cohen, 1987). Although the topic is covered in most introductory statistics classes, relatively few published studies report power for the test statistics (such as *F, t,* or *z,*) used. See, for example, Ree and Earles (1991, 1993) and Walters et al. (1993). Sedlmeier and Gigerenzer (1989) reported two surveys of a prestigious applied psychology journal and showed that the average statistical power for published studies was only .46 and fell to .37 two decades later. This means that the researchers could only detect an existing effect 46% (or 37%) of the time. On the other hand, 54% of the time (or 63% of the time) they would fail to find the existing effect! No researcher should conduct a study with less than a high chance of detecting a significant result. Low statistical power makes it more likely to draw incorrect conclusions from studies. Too many pilot selection studies have been performed with small samples that inevitably yield low statistical power.

Cohen (1987), noted that statistical power is a joint function of the degree to which the sample values reflect their true values in the population (i.e., the reliability of the sample values), effect size, Type I error rate (significance level), and sample size. Before conducting a study, power tables (Cohen, 1987) should be consulted. An informative discussion of sample size requirements is given by Schmidt and Hunter (1978, see especially p. 222). Schmidt, Hunter, and Urry (1976) provide tables showing sample sizes necessary for sufficient statistical power in validation studies given varying selection ratios, reliabilities, and effect sizes. A single rule of thumb cannot be given. In general, the higher the selection ratio and the less reliable the variables are, the larger the sample must be.

Failure to Cross-Validate

Cross-validation refers to the use of a regression equation computed in one sample being applied in another sample. This is done to determine the extent to which the initial regression result can be expected to generalize. The stability and generalizability of the regression solution are important because the results of the regression equation will be used to make selection decisions about people who were not in the original validation sample. In general, the correlation of the several predictors with the criterion will go down in this second or cross-validation sample (Wherry, 1975). This reduction of the correlation is called shrinkage from overfitting.

The classic paradigm for cross-validation was provided by Mosier (1951) in which a single sample is drawn from a population and then divided into separate validation and cross-validation samples. Murphy (1983) has pointed out that the correlation in the cross-validation sample is still a consequence of overfitting, as there was only one sampling. Moreover, even if there were two samplings from the population, the validation and cross-validation multiple correlations would be only two values out of a virtually infinitely large set of values. That is why we recommend the use of Stein's operator (Stein, 1960). Kennedy (1988) demonstrated the accuracy of Stein's formula for estimating the mean of the distribution of all possible cross-validated correlations from the population from which the sample was selected. Stein's operator has the advantage of allowing estimates on the largest available sample while offering an estimate of cross-validity.

Misinterpretation of Correlations and Regression

Selection studies use correlation and regression as a general model and analytic technique. Predictors of success in pilot job performance are correlated with measures of success in pilot job performance. There is an extensive literature on correlation and regression (Cronbach, 1971; Messick, 1989). In the next sections, we discuss several issues that can lead to the misinterpretation of correlations and regression, and provide solutions.

Holding Job Experience Constant. Ability research is generally correlational, and the interpretation of correlations can be fraught with hazards. As an example, consider the correlation of an ability test and ratings of pilot job performance. "Artificially" low correlations that could lead to inappropriately abandoning the ability test can occur for a variety of reasons, including the effects of range restriction, unreliability, and the influence of moderating variables. Range restriction and unreliability will be discussed later in this section. It is noted in the *Principles for the Validation and Use of Personnel Selection Procedures* (Society for Industrial-Organizational Psychology, 1987) that the relationship between ability (or any other measure) and occupational criteria is best represented with the effect of job experience removed. This can be done easily by using partial correlation and "partialing-out" experience from the relationship between ability and the occupational criteria. Carretta, Perry, and Ree (1996) provided an example. The observed correlation between ability test scores and ratings of situational awareness (SA) for 171 F-15 pilots was .10. After F-15 flying experience was partialed out, the correlation was .17, an increase of 70% in predictive efficiency. It would have been incorrect to report the correlation of ability and SA as .10.

The idea of partial correlation can be subsumed under "mediation," which means that one variable acts through another variable to exert its influence on a third variable. For instance, "A → B → C" indicates that variable A acts through variable B to exert its influence on variable C. In this instance, there is no *direct* influence of A on C and we do not specify "A → C." This does not mean that variable A has no influence on variable C, but rather that A works through B to influence C. An informative model of mediation in the area of job performance has been provided by Hunter (1986). He demonstrated that job knowledge mediated the relationship between ability and job performance. Ree, Carretta, and Teachout (1995) illustrated this mediation for pilot trainees (see Fig. 10.1). Ree et al. examined the influence of general cognitive ability (g) and prior job knowledge (JK_p)

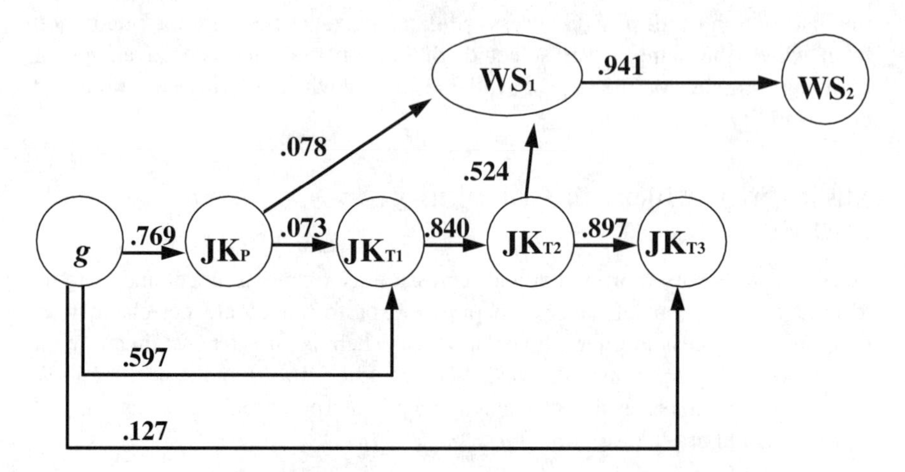

FIG. 10.1. Ree, Carretta, and Teachout (1995) model of the influence of general cognitive ability (g) and prior job knowledge (JK_P) on the acquisition of additional job knowledge (JK_{T1} to JK_{T3}) and sequential training performance (WS_1 and WS_2).

on the acquisition of job knowledge acquired during early, middle, and late pilot training (JK_{T1} to JK_{T3}) and early (T-37) and late (T-38) hands-on flying performance (WS_1 and WS_2). In their study, general cognitive ability (g) had both direct and indirect influences on the acquisition of aviation job knowledge and hands-on flying performance during pilot training. It is necessary to partial out the effect of job experience to know the true relationship of a predictor to job performance.

Range Restriction. Studies of training and job performance frequently use censored samples. When the variance of one or more variables has been reduced due to prior selection, censoring occurs. Range restriction is the name given to this reduction in variance. For example, military organizations do not admit all those who apply, and commercial airlines do not hire all pilot applicants. In military pilot selection, applicants may have been screened on the basis of aptitude test scores, completion of a college degree, completion of an officer commissioning program, medical fitness, prior flying experience, selection interview, and vocational interest. Censored range-restricted samples have been shown to cause artifacts that may lead to erroneous conclusions (Morrison & Morrison, 1995) and to inappropriately abandoning predictive measures.

Range-restricted samples can produce estimates of correlations that are much less than they would be in an uncensored sample (Martinussen, 1997; Thorndike, 1949). In some instances, correlations based on censored samples can even change signs from their population value (Ree, Carretta, Earles, & Albert, 1994; Thorndike, 1949).

Damos (1996) argued against the use of corrections for range restriction, noting that organizations do not administer selection tests to a completely unscreened population. She contends that the uncorrected correlation provides the most accurate estimate of the strength of the relationship between two variables. The following example disagreed.

Thorndike (1949, pp. 170–171) provided a dramatic illustration of the detrimental effects of range restriction. An experimental group of 1,036 U.S. Army Air Corps aircraft pilot applicants was admitted to training in 1944 without regard to their scores on five aptitude tests. Correlations were computed with the training criterion for all participants (n = 1,036) and for those pilot candidates (n = 136 out of 1,036) that would have been selected had the strict standards in effect been used. The Pilot Stanine composite derived from the five tests had a correlation of .64 with training outcome in the unrestricted sample. In the range-restricted sample it dropped to .18. The most dramatic change from the unrestricted to the range-restricted sample occurred for a psychomotor test where the correlation changed sign from + .40 to − .03. In the range-restricted sample, the average decrease in the five validity coefficients was .29. It is clear that the validity estimates were adversely affected by range restriction. Wrong decisions would have been made had only the range-restricted correlations been reported.

Range-restricted samples are common in pilot selection research. Goldberg (1991) observed that ". . . one can always filter out or at least greatly reduce the importance of a causal variable, no matter how strong that variable, by selecting a group that selects its members on the basis of that variable" (p. 132). He noted that the more restricted the variance of a variable, the less its apparent validity.

Statistical corrections for range restriction are available and should be applied to provide better statistical estimates. "Univariate" corrections described by Thorndike (1949) are appropriate if censoring has occurred on only one variable. However, the multivariate correction (Lawley, 1943; Ree et al., 1994) is more appropriate if censoring has occurred on more than one variable. These corrections provide better statistical estimates and tend to be conservative (Linn, Harnish, & Dunbar, 1981). The corrections still tend to *underestimate* the population values. Johnson and Ree (1994) offered free windows-based software to perform either univariate or multivariate corrections.

Unreliability of Measures. Reliability refers to the stability of measurement of a test, interview, or other selection device. Reliability can be estimated by correlation between test forms (test-retest) and under certain conditions from the single administration of a test (internal consistency) (McDonald, 1999). The method used depends on several factors including the content of the test or scale and the question being asked. Internal consistency estimates are appropriate if the items are independent of one another and the question of interest is whether or not the items measure the same construct. Test-retest estimates are appropriate if the test is speeded, the items are not independent of one another (e.g., most psychomotor tests), or the question of interest involves stability of performance over time or across alternate forms.

The use of *unreliable* measures will lead to incorrect conclusions (Spearman, 1904). The magnitude of the correlation between variables is limited by their reliabilities. Correcting for the unreliability of variables informs us about the true relationship between predictors and criteria. Correlations between predictors and criteria that change from low to moderate or high after correction suggest that the predictor could help increment validity if it (or the criteria) were more reliable. If validities remain low to moderate after correction for unreliability it is likely that the criterion has other sources of variance that are not being predicted.

Carretta and Ree (1995) provided an example when they examined the validity of the 16 Air Force Officer Qualifying Test (AFOQT) tests against five USAF undergraduate pilot training criteria. The validities of the AFOQT tests were examined as observed, corrected for range restriction, and corrected for both range restriction and unreliability (of the predictor and criterion). The average magnitude of the correlations of the 16 tests with the five criteria varied from .03 to .13 for the observed correlations, from .10 to .25 for the range-restriction-corrected correlations, and from .25 to .58 for the fully corrected correlations. The use of appropriate range restriction and unreliability corrections removes artifacts that are inevitable in all studies (Hunter & Schmidt, 1990).

Unreliability also has an effect on regression coefficients. Fuller (1987) provided a mathematical demonstration that unreliability biases the estimate of β, the population regression parameter. The estimate is reduced because the less-than-perfect reliability causes the value to be lower than if a more reliable measure had been used.

One solution to increase reliability is to add test questions or job performance ratings to unreliable measures. Other solutions would be to improve instructions and to remove ambiguity from existing items and scoring. If these remedies are ineffective, it may be necessary to discard the measure. Nonetheless, reliability should be estimated for all predictors and training or job performance criteria.

Dichotomization of Criteria. In training studies of pilots or other aviation occupations, it is not unusual to have criteria that have been artificially divided into two categories. Did the student pass or fail the pilot course? Did the mechanic pass or fail the airframe certification test? Dichotomization of the criteria causes correlations to appear lower than they should and places an upper limit on the magnitude of the correlation that depends on the proportion in each of the pass and fail categories. With proportions of 50–50 there is no biasing effect on the correlation, but when the proportions deviate from 50–50, there is a downward bias on the correlation. For example, if a correlation between two variables is .50 before dichotomization and the dichotomized criterion has proportions of 50–50, the correlation in the study will be .50. However, if the proportions are 90–10, 80–20, 70–30, or 60–40, the correlations will be .29, .35, .38, and .39, respectively. This has long been recognized as a problem, and a statistical correction for the dichotomization (Cohen, 1983) provides an estimate of the correlation had the variable *not been* dichotomized.

Weighting of Variables. Typically, aviation job applicants are given a series of tests, or multiple interviews, or a simulation task with many scores, or a combination of these. Then a decision is made by the selecting agency, frequently combining the scores and other applicant information by addition to form a composite. Frequently, the various parts of the composite will be given greater importance by weighting them more. The score on the composite, rather than its parts, will be used to make a decision. Variable weighting to create composites has been the subject of both analytic and empirical study. Two common weighting methods include unit weighting and criterion-based regression weighting. Unit weighting assigns each score the same weight and simply adds the scores together to create a composite. Criterion-based regression weighting uses the best-fitting weights when a criterion is regressed on a set of predictors. Criterion-based regression weighting is used frequently in pilot selection (Walters et al., 1993), even though several studies argue for unit or simple weighting. Three decades ago, Aiken (1966) thought the controversy over the use of simple weights was settled and was surprised to find colleagues arguing for regression-based weights on an *intuitive basis.* Two decades ago, Wainer (1976, 1978) showed only small losses in predictive efficiency from equal weights when compared with regression weights. He noted that selection usually involved ranking and top-down selection rather than predictive efficiency, making weighting schemes of little importance. Wilks (1938) proved a mathematical theorem showing that under very common circumstances, almost all weighted composites of a set of variables are very strongly correlated. In other words, if two different sets of weights were applied to a set of variables to produce two composites, the correlation between the two composites will be very high.

Ree, Carretta, and Earles (1998) demonstrated the consequences of Wilks' (1938) theorem through multiple examples. They also provided numerous examples from published studies showing near identical rankings for composites based on various weighting schemes (e.g., unit weights, regression weights, factor weights, and policy-capturing weights).

When considering weighting variables, it is sufficient to know which are important and then use simple or unit weights. Considering other than simple or unit weights, Wainer's (1976) said it pithily, "it don't make no nevermind."

Recommendations for Researchers and Practitioners

Ability research is fraught with pitfalls that can lead to incorrect inferences. We offered recommendations for each of the methodological issues raised earlier. It is worth noting that many or all of them can occur in any study, and the effects will be compounded. Ree (1995) provided an example study with multiple problems as caused by multiple issues. This example study is reminiscent of many others and shows how incorrect conclusions can occur.

MILITARY PILOT SELECTION

Historical Overview

Dockeray and Isaacs (1921) reported that Italy, prior to World War I, was the first country with a pilot selection research program. The Italians used measures of reaction time, emotional reaction, equilibrium, perception of muscular effort, and attention. At the same time, the French were investigating reaction time and emotional stability.

In the World War I era, Yerkes (1919) showed that measures of intelligence were valid predictors of pilot training success. Between the wars, Flanagan (1942) noted that the American aviation selection exam was a general mental battery testing comprehension and reasoning.

Most of the World War I research reflected Spearman's (1904) two-factor theory that demonstrated the existence of a general cognitive factor and test-unique factors. The work of Thurstone (1938) changed the emphasis from Spearman's two-factor theory to the theory of multiple aptitudes (i.e., innate or acquired capabilities; talents). Thurstone's theory was eventually reified in multiple-aptitude batteries such as the Differential Aptitude Test (Bennett, Seashore, & Wesman, 1982), General Aptitude Test Battery (Dvorak, 1947; Hunter, 1980), Armed Services Vocational Aptitude Battery (ASVAB; Earles & Ree, 1992), and the Air Force Officer Qualifying Test (AFOQT; Carretta & Ree, 1995).

World War II brought a renewed interest in pilot selection. Influenced by Thurstone's multiple-aptitude theory, the U.S. Army (Melton, 1947) and U.S. Navy (Fiske, 1947; Viteles, 1945) used several ability measures for pilot selection. These included intelligence, psychomotor skill, mechanical comprehension, and spatial measures. The British (Parry, 1947) and the Canadians (Signori, 1949) employed tests similar to those of the Americans. Hilton and Dolgin (1991) recorded that the Germans used ability measures similar to those used by the Allies. Geldard and Harris (1946) investigated the pilot selection system used by the Japanese during World War II and found that they were using tests based on the American Army Alpha, a paper-and-pencil derivative of the Stanford-Binet test.

The quarter century following World War II saw little change in pilot selection methods. Most countries were limited to producing new forms of paper-and-pencil multiple-aptitude tests with some exceptions (e.g., Gopher & Kahneman, 1971). The field of personality measurement saw most of the research innovation (see Dolgin & Gibb, 1989). Since 1970, studies of multiple aptitudes and psychomotor abilities (Carretta, 1990; Imhoff & Levine, 1981) have been prevalent. For a more complete review, see Hilton and Dolgin (1991) and Hunter (1989).

Recent Validation Studies

Pilot selection procedures used in NATO-member countries vary in content, focus, and method of administration. However, all NATO-member countries employ some form of psychometric testing as part of military pilot selection (Burke, 1993). Psychometric testing involves the measurement of mental traits, abilities, and processes. Examples of psychometric testing approaches include aptitude, simulation-based, and personality tests. Hilton and Dolgin (1991) and Li (1993) provide detailed reviews.

We have previously defined *aptitude* as "innate or acquired capabilities." Common examples of aptitude tests include traditional paper-and-pencil tests (Bartolo Ribeiro, 1992; Carretta & Ree, 1995; Martinussen & Torjussen, 1998; Orgaz & Loro, 1993; Prieto et al., 1996), psychomotor tests (Bartolo Ribeiro, 1992; Bartolo Ribeiro, Martins, Vicoso, Carpinteiro, & Estrela, 1992; Burke, Hobson, & Linsky, 1997; Carretta & Ree, 1993, 1994; Gibb & Dolgin, 1989), and computer-based tests (Bailey & Woodhead, 1996; Boer, 1992; Carretta & Ree, 1993).

Methods such as performance tests (Boer, Harsveld, & Hermans, 1997; Delaney, 1992) and simulation-based tests of pilot work samples (Gress & Willkomm, 1996; Spinner, 1991) are less common. Finally, the emphasis placed on personality assessment varies widely in military pilot selection (Burke, 1993; Dolgin & Gibb, 1989).

Aptitude Tests. In a recent review of Royal Air Force (RAF) aircrew selection methods, Bailey and Woodhead (1996) stated that historically the RAF has relied heavily on ability for job specialties such as pilot and on measures of personality/character and biographical information for overall officer suitability. The RAF takes a "domain-centered" approach to test battery construction. The emphasis is on first identifying the appropriate ability domains for a particular occupation (e.g., pilot, navigator, weapons director, air traffic controller), then choosing one or more tests to represent the critical domains. As the result of task analyses, the RAF identified five aircrew-ability domains: attentional capacity, mental speed, psychomotor, reasoning, and spatial. The current RAF Pilot Aptitude composite samples all five domains.[1] Bailey and Woodhead report the predictive validity of the RAF Pilot Aptitude composite against Basic Flying Training[2] outcome as $r = .52$ after correction for statistical artifacts. This value is considered good and is consistent with general personnel selection results

[1]As the result of factor-analytic studies of the RAF test battery, the RAF has modified its ability domains to include Attentional Capacity, Psychomotor, Reasoning (Numerical), Reasoning (Verbal), Spatial, and Work Rate (M. Bailey, personal communication, February 21, 2001).

[2]At the time of this study, RAF Basic Flying Training (BFT) consisted of Elementary Flying Training (up to 65 hours) in the Bulldog aircraft followed by BFT (up to 120 hours) in the Tucano aircraft.

(Schmidt & Hunter, 1998). The RAF computer-based test system is commercially available and has been purchased by several civilian airlines and military services. As of 1997, it was being used for all military pilot selection in the United Kingdom (Burke et al., 1997).

Burke et al. (1997) reported a meta-analysis of the validity of three tests from the RAF pilot selection battery. Meta-analysis techniques allow researchers to combine validity estimates from multiple studies and correct for the effects of statistical and measurement artifacts (Hunter & Schmidt, 1990). Meta-analytic studies are valuable because they provide more accurate estimates of validity than do individual studies. Burke et al. examined the validity of the RAF Control of Velocity, Instrument Comprehension, and Sensory Motor Apparatus tests and a summed composite of the three tests. The sample consisted of 1,760 pilot trainees (RAF fixed wing, n = 849; Turkish Air Force fixed wing, n = 570; and British rotary wing, n = 341). The criterion was a dichotomous pass/fail training score. Observed validities were corrected for range restriction (Thorndike, 1949) and dichotomization (Cohen, 1983), as suggested by Hunter and Schmidt (1990). Validities for the tests ranged from .15 to .16 as observed and from .28 to .29 after correction for range restriction and dichotomization of the criterion. The observed validity of the composite was .24 and increased to .40 after correction. These are good values considering the lack of perfect reliability of the predictor and criterion variables.

Support for the utility of various aptitude tests has been provided in three recent meta-analyses (Hunter & Burke, 1994; Martinussen, 1996; Martinussen & Torjussen, 1998). Hunter and Burke (1994) conducted a "bare-bones" analysis of validities for 68 pilot selection studies published between 1940 and 1990. A bare-bones analysis corrects for sampling error, but usually does not correct for other artifacts such as reliability and range restriction. In general, bare-bones analyses are less informative than studies fully corrected for artifacts (Hunter & Schmidt, 1990). Mean validities were estimated for 16 predictor categories: general ability, verbal, quantitative, spatial, mechanical, general information, aviation knowledge, gross dexterity, fine dexterity, perceptual speed, reaction time, biodata inventory, age, education, job sample, and personality. All categories except age and personality are examples of aptitude. Biodata inventories may include indicators of aptitude (e.g., education or job-related experience), attitudes, personality, and life experiences. Cumulative sample sizes varied by category and ranged from 2,792 (fine dexterity) to 52,153 (spatial). The predictor categories with the highest observed mean validities were job sample (.34), gross dexterity (.32), mechanical (.29), and reaction time (.28). The predictor groups with the lowest observed mean validities were education (.06), age (-.10), fine dexterity (.10), and personality (.10). However, because these results represent a bare-bones meta-analysis, the validities represent conservative estimates. For instance, measures of general ability, which had a mean observed validity of .13 in the Hunter and Burke meta-analysis, have shown much greater validity when corrected for statistical artifacts such as range restriction and reliability (Ree & Carretta, 1996, 1997).

Martinussen (1996) conducted a meta-analysis of 50 studies published between 1919 and 1993. Most of the studies were published after World War II, with 1973 as the median year of publication. The studies were from 11 different countries with about half from the United States and about 74% from English-speaking countries. Martinussen grouped the predictors into nine categories. Aptitude categories included cognitive, intelligence, psychomotor/information processing, aviation information, combined index (a combination of several tests, usually cognitive and psychomotor), academics, and flying training experience. Nonaptitude categories included personality and biographical. Cumulative sample sizes for the predictor categories ranged from 3,736 (aviation information) to 17,900 (cognitive). Although Martinussen corrected the validities for dichotomization of the criterion, as with Hunter and Burke (1994), due to a lack of data, correlations were not corrected for range restriction or for reliability of the predictors or criterion. Martinussen's validities were similar to those reported by Hunter and Burke. The highest validities were for the combined index (.37) and flying training experience (.30). The next highest validities were for cognitive (.24), psychomotor (.24), aviation information (.24), and biographical (.23). The categories with the lowest validities were intelligence (.16), academics (.15), and personality (.14). Again, these validities should be considered conservative estimates because they were not corrected for either range restriction or unreliability. The effects of range restriction on the validity of measures of intelligence and conscientiousness (personality) are notable, as these constructs are pervasive in personnel selection contexts (Schmidt & Hunter, 1998). Pilot training applicants typically have completed a college degree and have been directly selected for training based on academic achievement (e.g., college grades, major) and indicators of conscientiousness (e.g., life experiences, college graduation, direct observation).

What conclusions can be drawn from these meta-analyses? Clear interpretation is difficult because the validities usually were not corrected for factors that would reduce their magnitudes (e.g., range restriction, unreliability). Despite this limitation, measures of technical/job knowledge, pilot work samples, and flying experience showed predictive validity against pilot training performance. Measures of psychomotor ability also showed predictive validity. Measures of cognitive ability and personality were less valid. However, as previously noted, this is not surprising, as measures of cognitive ability (Carretta & Ree, 1996b; Hunter & Burke, 1995, Ree & Carretta, 1996) and conscientiousness (Weeks & Zelenski, 1998) are mainstays in military pilot selection procedures, thus leading to restriction of range on these constructs.

Simulation-Based Tests. Although the use of multiple aptitude tests is common in pilot selection, others have proposed using simulator-based tests to improve selection procedures (Gress & Willkomm, 1996; Long & Varney, 1975; Spinner, 1991). Simulator-based tests have an "intuitive appeal" as they look like some part of the job (e.g., instrument flight) for which applicants are being selected.

The U.S. Air Force (Long & Varney, 1975) examined the utility of a computer-based apparatus approach for pilot selection known as the Automated Pilot Aptitude Measurement System (APAMS). The APAMS system was very crude by today's standards (see Ree & Carretta, 1998 for a description). It could accommodate only one participant, and test presentation/response materials and was limited in motion to pitch, roll, and yaw. The APAMS syllabus consisted of a 5-hour sample of flying tasks based on the USAF syllabus for the T-41 light aircraft (single-engine high-wing monoplane) training program. The APAMS tasks were intended to reflect individual differences in psychomotor abilities, learning rates, multitask integration, and performance under overload. They were not intended to train pilot skills. Participants in the validation study were 178 student pilots. Criteria from light aircraft (T-41) training were four dichotomous outcome variables based on flying grades (e.g., students rated as "waivered" or "excellent" versus students rated as "good," "fair," or "deficient"). Pilot training criteria for primary (T-37) jet training (twin turbo fan two-seat jet) were two dichotomous variables (graduates versus all eliminees and graduates versus only those eliminated for "flight training deficiencies"). The average multiple correlations for APAMS scores with training performance were .49 for the T-41 criteria and .30 for the T-37 criteria. Despite good predictive validity for training performance, the USAF decided not to pursue full-scale development and operational implementation of the APAMS system due to its cost and poor utility for decentralized testing. However, it should be noted that the Canadian Air Force, which uses centralized testing for pilot selection, has developed and operationally implemented a system known as the Canadian Automated Pilot Selection System (CAPSS; Okros, Spinner, & James, 1991; Spinner, 1991), which is largely based on APAMS.

CAPSS is a moving-based simulator of a single-engine light aircraft (Okros et al., 1991; Spinner, 1991). The test system records up to 250,000 instrument readings per candidate. CAPSS testing includes five, 1-hour "flights" that are performed over a 2.5-day period. In addition to test administration time, participants must spend time preparing for each test session (i.e., reviewing instructions and a flight plan). Over the five test sessions, participants are instructed on, practice, receive feedback, and perform eight basic flight maneuvers. Spinner (1991) examined the validity of CAPSS scores for predicting completion (pass/fail) of preliminary flying training (PFT) for 172 participants. PFT consists of classroom instruction and 27 hours on a CT-134 Musketeer. Spinner reported a multiple correlation of .47. Using discriminant analysis,[3] Okros et al. (1991) subsequently examined

[3]Discriminant analysis is useful for situations where it is desirable to build a predictive model of group membership based on observed characteristics of each case. The procedure generates a discriminant function based on linear combinations of the predictor variables that provide the best discrimination between the groups. If there are more than two groups, a set of discriminant functions is required. The functions are generated from a sample of cases for which group membership is known. The functions can then be applied to new cases with data for the predictor variables, but unknown group membership.

the utility of CAPSS scores for identifying graduates and failures in both PFT (using the Spinner, 1991 sample) and basic flying training (BFT). BFT is the initial jet-training course. CAPSS scores correctly classified 75% (129 of 172) of the pilot trainees attending PFT and 80% (154 of 192) of those attending BFT.

Another example of a simulator-based pilot aptitude test is the FPS 80 (Gress & Willkomm, 1996). The FPS 80 is used as part of a sequential selection strategy (Hansen & Wolf, 2000). Phase 1, Military Aptitude and Academic Fitness, includes computerized cognitive testing, written composition, oral presentation, group discussion, physical fitness, and an interview with two officers and a psychologist. Phase 2, Psychological and Medical Selection, includes computerized testing (cognitive, psychomotor, personality), assessment center activities (behavior rated by two psychologists and an officer), another interview, and a flight physical. Phase 3, FPS 80 Testing, consists of lectures on aerodynamics, navigation, and the FPS 80 system, and FPS 80 testing.

The FPS 80 is a low-fidelity simulator of a single-engine propeller-driven aircraft that consists of a control center and two cockpits. The flight model is based on the Piaggio 149D (single-engine low-wing monoplane). Pilot candidates complete four "missions" on the FPS 80 over a 2-week period. Prior to performing the missions, pilot candidates are given a training guide that includes detailed descriptions of the missions to be flown. They also must complete two lessons on basic aerodynamic principles. Prior to the first mission, candidates must pass a written test on mission-relevant material. FPS 80 performance is graded in two ways: a computer-generated score based on data from the check ride and an observation-based rating by an aviation psychologist. The psychologist rates each candidate on several factors (e.g., aggressiveness, concentration, coordination, stress tolerance, training progress). A single composite score is generated across all four missions combining the computer-generated and observation-based scores. Gress and Willkomm (1996) evaluated the validity of the FPS 80 and the basic psychological selection tests for student pilots attending flight screening. The criteria consisted of academic grades (n = 310) and a final flying score (n = 267) during flight screening. Results indicated that the basic psychological tests had uncorrected validities of .24 and .30 against academic grades and final flying score, respectively. Using the FPS 80 grade along with the basic psychological tests increased the validities to .42 and .54, values that are consistent with meta-analytic findings (Schmidt & Hunter, 1998). Although Gress and Willkomm were encouraged by the results of the validation study, they identified several obstacles to the use of simulator-based tests for pilot selection including cost of the test system and test administration (e.g., centralized testing, amount of time needed).

To summarize, the validity of simulation-based approaches for pilot selection appears comparable to that for general cognitive ability (g). Further, simulation-based tests may significantly increment the validity of cognitive tests when the two are used together (Gress & Willkomm, 1996). These results are consistent with a large-scale meta-analysis of 19 commonly used personnel selection methods across many occupations (Schmidt & Hunter, 1998). Schmidt and Hunter reported meta-analytically derived validities of .51 for *g* and .54 for work sample tests for predicting job performance. When used together, the multiple correlation was .63.

Despite their apparent validity and incremental validity, simulator-based tests have drawbacks involving the costs associated with test development and administration (centralized testing, single-administration, preparation and administration time). These drawbacks make simulation-based tests impractical for evaluating large numbers of applicants or for pilot selection programs that rely solely on decentralized testing. Simulation-based testing probably has its greatest value in multiple-stage selection situations where applicants first could be screened on inexpensive group-administered paper-and-pencil cognitive tests. Those who "passed" this screen could be brought to a centralized location for simulator-based testing. We are not aware of any studies to determine the cost-benefit trade-offs of simulator-based tests for pilot selection. To be useful, the cost of test development and administration would need to be made up by a reduction in training costs (e.g., reduced attrition, reduced training requirements).

Personality. The relation between personality and military pilot performance has been the subject of many studies (see Dolgin & Gibb, 1989 for a descriptive review). NATO-member countries vary substantially in the emphasis placed on personality assessment in pilot selection, as well as on assessment methods (Burke, 1993). Some countries, such as the United Kingdom and the United States, do not directly measure personality during selection. In these cases, personality assessment may find its way into the selection process indirectly through its influence on training commander's ratings of cadets on "officership" and "military bearing" or through interviews and observer ratings. The range of explicit personality measures that have been evaluated and are in use within NATO includes a variety of paper-and-pencil questionnaires, projective tests, clinical interviews, and computer-based measures that appear to combine ability and personality assessment. Some examples include the Eysenck Personality Inventory (Jessup & Jessup, 1966), Jackson's Personality Research Form (Retzlaff & Gilbertini, 1987), and the Defense Mechanisms Test (Harsveld, 1991; Martinussen & Torjussen, 1993). Often, personality tests are used during an interview with a psychologist (e.g., Evdokimov, 1988).

Despite the large number of personality characteristics examined using a variety of instruments and methods, empirical support regarding the role of personality in pilot performance is lacking (Dolgin & Gibb, 1989). It is not unusual to find studies that have used the same personality instrument to predict pilot performance yielding contradictory results. An illustrative example is provided by the Defense Mechanism Test (DMT), a projective test in which the participant is shown a picture for a short exposure using a tachistoscope. The participant is asked to describe or draw their impression of the images shown. The responses are scored according to Freudian defense mechanisms. Although validations reported for Scandinavian researchers yield impressive results (Torjussen & Vaernes, 1991), a British study of RNAF pilots found zero validity against flying training attrition (Harsveld, 1991). Burke (1993) speculated that cultural factors might have caused these contradictory results for the DMT.

Developments in personality theory in the late 1980s indicated that past reviews of the personality-performance literature suffered from a lack of a conceptual framework for evaluating results from different studies. A consensus model of personality emerged, based on the observation that five global factors describe individual differences in personality traits (Digman, 1990; Tupes & Christal, 1961). These factors, which are known as the "Big Five," are Agreeableness, Conscientiousness, Extraversion, Neuroticism, and Openness. Each of these broad factors includes several facets. For example, Conscientiousness includes the facets of Achievement Striving, Competence, Dutifulness, Deliberation, Order, and Self-Discipline. The utility of the Big Five framework has been demonstrated in several meta-analytic studies of the relations between personality and job performance (Barrick & Mount, 1991; Tett, Jackson, & Rothstein, 1991). In another meta-analytic study, Schmidt and Hunter (1998) demonstrated that measures of Conscientiousness are incrementally valid for predicting training and job performance when paired with measures of general cognitive ability.

The study of personality in military pilots generally follows two lines. In one, personality profiles of pilot applicants, trainees, or pilots are compared to the general population (e.g., Callister, King, Retzlaff, & Marsh, 1999). In the other, personality scores are validated against some indicator of training or job performance (Dolgin & Gibb, 1989). Callister et al. (1999) used the Revised NEO-PI (Costa & McCrae, 1992), a measure of the Big Five, to develop personality profiles for male and female USAF student pilots. Participants were 1,198 male and 103 female student pilots tested during a flight screening program. Compared with the general population, student pilots scored high on Extraversion (83rd percentile), Openness (60th percentile), and Conscientiousness (58th percentile), and low on Neuroticism (42nd percentile) and Agreeableness (20th percentile).

Siem and Murray (1994) used the Big Five framework to investigate personality factors affecting pilot combat performance. Participants were 100 USAF pilots. Most (90%) were Captains with a minimum of 6 years service. Several (43%) had combat experience in Operation Desert Storm. Participants rated the importance of 60 personality traits for each of six flying performance dimensions. The 60 traits were selected from unipolar markers of the Big Five developed by Goldberg (1992). The six flying performance dimensions were (a) flying skills and knowledge, (b) compliance, (c) crew management and emotional support, (d) leadership, (e) situational awareness, and (f) planning. Conscientiousness was rated as the most important factor for five of the six performance criteria. Openness was rated slightly higher than Conscientiousness for the planning dimension. No predictive validation study was done.

As previously noted, the utility of cognitive and psychomotor abilities for predicting military pilot training performance has been illuminated from the application of meta-analytic techniques (e.g., Hunter & Burke, 1994; Martinussen, 1996; Martinussen & Torjussen, 1998). We believe that the role of personality factors in pilot performance would benefit from the joint application of the Big Five framework and meta-analysis.

Current Research

The previous section described several approaches to pilot selection. This section concentrates on examining what underlying constructs are measured by pilot selection tests and what about them is predictive of training and job performance. We chose to focus on a leader in this area, the USAF.

Our discussion is guided by the seminal work of Schmidt and Hunter (1998) who examined the validity of general mental ability (*g*) and 18 other selection procedures for predicting training and job performance across many occupations. On the basis of meta-analytic findings, Schmidt and Hunter concluded that the three combinations of predictors with the highest validity and utility for job performance were *g* plus a work sample test, *g* plus an integrity test (or a conscientiousness test), and *g* plus a structured interview. They also concluded that the latter two methods were appropriate for both entry-level selection and selection of experienced employees. It should be noted that Schmidt and Hunter did not consider measures of job knowledge or work sample performance for entry-level jobs, as these methods were not commonly used for that purpose. However, as already discussed, in pilot selection job knowledge and work sample tests are fairly common for entry into ab initio flying training programs.

The construct of general cognitive ability, *g* was developed in the early 20th century by Charles Spearman. Every test or measure of ability measures *g* and specific ability or knowledge, *s*. A long history of research findings has shown *g* to be the most valid predictor of academic and job performance, and for numerous other human characteristics (Brand, 1987; Jensen, 1998; Schmidt & Hunter, 1998). The predictive validity of *s* is mostly due to specific knowledge, not specific ability. General cognitive ability is usually defined as the common source of variability among a set of cognitive measures. For practical purposes, it can be thought of as the main factor of intelligence. Jensen (1998) provides the most complete presentation and discussion of *g*.

Despite these consistent results, the subject of *g* remains contentious (McClelland, 1993; Ree & Earles, 1992, 1993; Schmidt & Hunter, 1993; Sternberg & Wagner, 1993). Some have proposed that to understand human characteristics and job performance, it is necessary to measure noncognitive traits, specific abilities, and knowledge, but have offered no empirical proof. For example, McClelland (1993) suggested that under some circumstances noncognitive traits such as motivation might be better predictors of job performance than cognitive abilities.

Sternberg and Wagner (1993) proposed using measures of tacit knowledge and practical intelligence instead of measures of "academic intelligence." They define tacit knowledge as "the practical know how one needs for success on the job" (p. 2). Practical intelligence is defined as a more general form of tacit knowledge. Schmidt and Hunter (1993) noted that Sternberg and Wagner's concepts of tacit knowledge and practical intelligence are redundant with the well-established construct of job knowledge.

Currently, the Air Force Officer Qualifying Test (AFOQT; Carretta & Ree, 1996a; Skinner & Ree, 1987) is an important USAF pilot selection component. As shown in Table 10.1, it consists of 16 cognitive and pilot job knowledge tests. The tests are Verbal Analogies (VA), Arithmetic Reasoning (AR), Reading Comprehension (RC), Data Interpretation (DI), Word Knowledge (WK), Math Knowledge (MK), Mechanical Comprehension (MC), Electrical Maze (EM), Scale Reading (SR), Instrument Comprehension (IC), Block Counting (BC), Table Reading (TR), Aviation Information (AI), Rotated Blocks (RB), General Science (GS), and Hidden Figures (HF). The tests are combined into three academic composites used primarily to assess "officership" (Verbal [V], Quantitative [Q], and Academic Aptitude [AA]) and two aviation-related composites (Pilot [P] and Navigator-Technical [N-T]). The Pilot composite is unique in that it is the only one that includes tests of job knowledge (i.e., IC and AI tests).

Confirmatory factor analysis (CFA; Jöreskog & Sörbom, 1996; Kim & Mueller, 1988) is a statistical technique that allows investigators to specify and test hypotheses about the relations among a set of variables. Recent CFAs (Carretta & Ree, 1996a) have found that the AFOQT has a hierarchical structure similar to other multiple aptitude tests (Jensen, 1994; Ree & Carretta, 1994b; Vernon, 1969). See Figure 10.2. The higher order factor was identified as general cognitive ability (*g*). All 16 tests contributed to the measurement of *g* The proportion of common variance due to *g* was 67%. The remaining common variance (33%) in the residualized (Schmid & Leiman, 1957) lower order factors was 11%[4] for verbal, 9% for aviation interest/aptitude, 4% for perceptual speed, 4% for spatial, and 4% for math. These proportions are similar to that found in other multiple-aptitude batteries (Jensen, 1980). Most of the predictive utility of the AFOQT against pilot training performance can be attributed to its measurement of *g* and aviation interest/aptitude (Olea & Ree, 1994; Ree et al., 1995).

Another important component in USAF pilot selection is the computer-administered Basic Attributes Test (BAT; Carretta, 1992a). The BAT system is fairly representative of computer-based pilot aptitude tests. The test apparatus consists of a computer and monitor built into a testing carrel. The carrel has side, back, and top panels designed to minimize glare and distractions. Participants respond by manipulating individually or in combination, a dual-axis right-hand control stick, a single-axis left-hand control stick, and a specialized response keypad. The

[4] Adds to less than 33% due to rounding.

TABLE 10.1.
Composition of AFOQT Aptitude Composites

Test	*V*	*Q*	*AA*	*P*	*N-T*	*Description*
			Composite			
Verbal Analogies	X		X	X		Ability to reason & recognize word relationships
Arithmetic Reasoning		X	X		X	Understanding of arithmetic relationships expressed as word problems
Reading Comprehension	X		X			Reading skill
Data Interpretation		X	X		X	Ability to extract data from graphs & charts
Word Knowledge	X		X			Understanding of written language through the use of synonyms
Math Knowledge		X	X		X	Use of mathematical terms, formulas, & relationships
Mechanical Comp.				X	X	Understanding of mechanical functions
Electrical Maze				X	X	Spatial ability based on choice of a path through a maze
Scale Reading				X	X	Ability to read dials & scales
Instrument Comp.				X		Ability to determine aircraft attitude from illustrations of flight instruments
Block Counting				X	X	Spatial ability through analysis of three-dimensiona representations of blocks
Table Reading				X	X	Ability to quickly & accurately extract information from tables
Aviation Information				X		Knowledge of general aviation technology & concepts
Rotated Blocks					X	Spatial aptitude through mental manipulation & rotation of objects
General Science					X	Knowledge of scientific terms, concepts, & principles
Hidden Figures					X	Spatial ability to find simple figures embedded in complex drawings

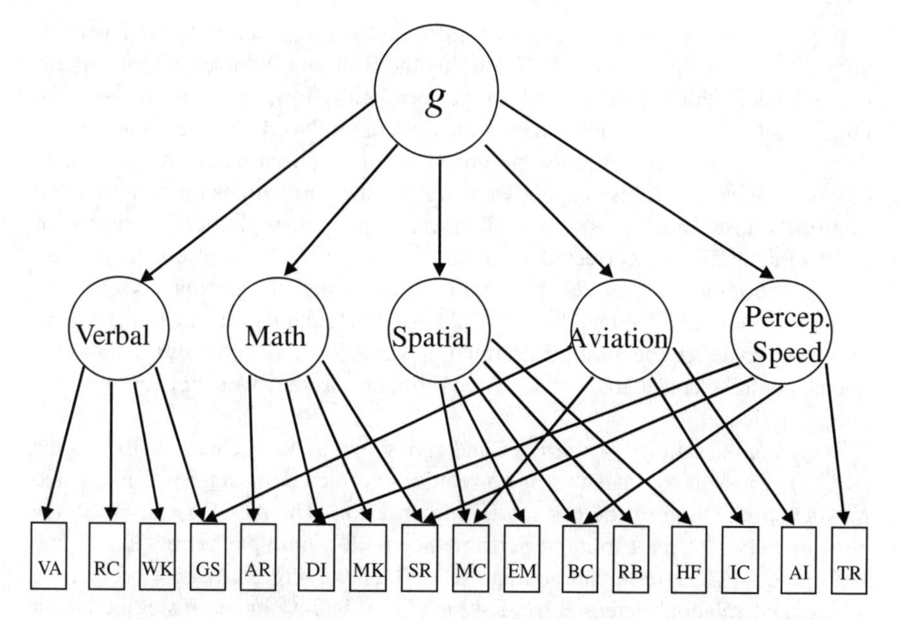

FIG. 10.2. Hierarchical factor structure of the AFOQT with *g* as the higher order factor and five lower-order factors of Verbal, Math, Spatial, Aviation Knowledge, and Perceptual Speed.

BAT battery provides psychomotor test scores (multilimb coordination, pursuit tracking, rate control, response time), cognitive scores (short-term memory), and a personality measure (attitudes toward risk).

The U.S. Air Force combines the AFOQT Pilot composite; BAT psychomotor, cognitive, and personality measures; and a self-report of flying experience to create a measure of pilot aptitude known as the Pilot Candidate Selection Method (PCSM; Carretta, 1992a). The BAT psychomotor scores are used in a unit-weighted composite as are the tests in the AFOQT Pilot composite. These psychomotor and AFOQT Pilot composites are regression weighted along with the other PCSM components of cognitive, personality, and flying experience to predict pilot training criteria. Several studies have demonstrated the validity of PCSM scores for a variety of pilot training criteria. High PCSM scores are associated with greater probability of completing jet training (Carretta, 1992a, 1992b, 2000), fewer flying hours needed to complete training (Duke & Ree, 1996), higher class ranking (Carretta, 1992b), and greater likelihood of being fighter-qualified (Weeks, Zelenski, & Carretta, 1996).

Carretta and Ree (1994) evaluated the validity and incremental validity of the components of a preoperational form of PCSM on a sample of 678 pilot trainees. Analyses showed the following (uncorrected for range restriction or dichotomization) correlations with passing-failing pilot training: AFOQT Pilot composite .17,

BAT psychomotor .15, BAT cognitive (information processing) .06, BAT personality .10, and flying experience .17. Adding the BAT and flying experience scores to the AFOQT Pilot composite raised the correlation from .17 to .30. It should be noted that the correlation magnitudes were reduced due to the severe dichotomization of the criterion produced by the high pass rate in this sample (85.7% passing). For this chapter, we corrected the correlations for dichotomization of the criterion. As expected, all correlations increased after correction for dichotomization. The corrected correlations were AFOQT Pilot composite .26, BAT psychomotor scores .23, information processing .09, personality .16, and flying experience .26. When the AFOQT Pilot, BAT, and flying experience scores were used together, the multiple correlation was .46. These correlations, although good, should be considered conservative estimates, as they were not corrected for range restriction.

Incremental validity of the BAT and flying experience scores relative to the AFOQT tests was estimated using correlations corrected for range restriction and for dichotomization of the passing/failing criterion. The AFOQT tests were the best predictors of pilot training performance with a multiple correlation of .42. Using the AFOQT tests and adding the BAT and flying experience scores, the multiple correlation increased from .42 to .52. This 24% increase and increment of .10 above the AFOQT tests represents potentially large cost-avoidance savings. Cost estimates of each USAF undergraduate pilot training failure range from $50,000 (Hunter, 1989) to $80,000 (Siem et al., 1988). These should be considered conservative estimates, as they are over a decade old. Obviously, even a small reduction in training attrition due to improved selection could produce large training cost-avoidance savings.

Studies were done to determine the causes of the level of the incremental validity of the various predictors. The BAT psychomotor scores were examined in the presence of a highly *g*-loaded battery of verbal and math tests (Ree & Carretta, 1994a). Confirmatory factor analyses on a sample of 354 enlisted personnel showed the BAT psychomotor scores to have three lower order factors reflecting its three psychomotor tests. The cognitive battery had two lower order factors of verbal and math. A higher order psychomotor factor influencing all psychomotor scores was found and unexpectedly, a higher order *g* was found to influence all scores, both cognitive and psychomotor. Subsequently, Wheeler and Ree (1997) examined the validity of measures of general and specific psychomotor abilities extracted from the BAT psychomotor tests for predicting USAF pilot training performance. Results indicated that the validity of the BAT psychomotor tests comes from their measurement of a general psychomotor factor and *g*.

The choice of the personality trait of attitude toward risk in the PCSM equation was made before the current prevalence of the Big Five (Digman, 1990) model of personality. Our finding of small incremental validity for personality variables is similar to the findings of McHenry, Hough, Toquam, Hanson, and Ashworth (1990).

That information-processing speed was not predictive was surprising (Carretta & Ree, 1994), as it has been found to be predictive of pilot training achievement in other studies (Carretta, 1992a). We suspect that the lack of validity in this study may have been a consequence of sampling error.

Consider another example. Olea and Ree (1994) compared the validity of general cognitive ability, *g*, and specific abilities (including pilot job knowledge), $s_1 \ldots s_n$ for predicting several pilot criteria in samples ranging from 1,867 to 3,942. General cognitive ability, specific abilities, and pilot job knowledge were estimated from the AFOQT. The criteria included academic performance; work samples of landings, loops, and rolls; and an overall performance composite. The overall criterion was a sum of the individual criteria providing a global measure of training performance. The work sample criteria were more like job performance measures of core technical task proficiency (i.e., core content specific to the job) and general task proficiency (i.e., general or common tasks not job specific), such as used by McHenry et al. (1990). Multiple correlations were compared to estimate the predictive efficiency of *g* and *s* for each criteria. Notwithstanding the apparent differences among the criteria, *g* was the best predictor whereas *s* contributed little. The validity for *g* ranged from .21 to .43 across all criteria with a mean of .31. The incremental validity for the specific abilities beyond *g* ranged from .07 to .14 with a mean of .10. Little incremental validity was found for the composite performance criteria (.09) or for the work sample criteria. For pilots, three predictors entered each equation: *g*, s_1, and s_3. Although the exact psychological nature of s_1 and s_3 cannot be assessed with certainty, the weights associated with the components emphasized special knowledge of aviation information and instrument comprehension. Results suggested that the incremental validity of specific measures for pilots was due to specific knowledge about aviation principles, aviation instruments, and aircraft controls, rather than specific abilities such as spatial or perceptual ability.

Research results point to *g* as the most important underlying construct in the prediction of pilot training success. Clearly, three others have been shown to be important but to a smaller degree: flying job knowledge, personality, and general psychomotor ability.

COMMERCIAL PILOT SELECTION

Almost all of the published literature on pilot selection methods concerns military pilot selection. However, there have been a few recent studies involving commercial aviation. Many commercial airlines rely on high-fidelity simulators because they are hiring trained pilots. This is different from the military's reliance on paper-and-pencil tests and computerized tests for a simple reason. Commercial pilot selection traditionally has relied on the military for trained pilots, a trend that appears to be ending, as the military currently is training fewer pilots. Commercial selection procedures will have to become like the military procedures, as the commercial airlines have to select untrained pilot applicants and provide initial training. The following section provides mostly descriptive information collected by survey. Such survey data should be seen as describing the current activities of commercial selection and not as scientific data from which causal explanations should be drawn as to proper practice.

In the United States, the federal government (Equal Employment Opportunity Commission, 1978) has issued a set of technical standards for validity studies that call for a job analysis to gather information about the job. The executive branch of the federal government (i.e., Department of Defense) has exempted itself from these standards. Commercial air carriers are not exempt, and they should conduct informative job analyses. McCormick (1976, 1979) and Gael (1988) provide detailed guidelines for conducting job analyses.

The Federal Aviation Administration (FAA) conducted surveys to identify trends in pilot hiring and selection for U.S. air carriers in 1994 (Suarez, Barborek, Nikore, & Hunter, 1994) and again in 1997 (D. R. Hunter, personal communication, August 27, 1998). The surveys focused only on hiring and selection procedures and did not assess the validity of the selection methods.

Suarez et al. (1994) sampled corporate operators, regional/major airlines, specialized air services, and commuters/air taxis. The most common sources cited by the regional or major carriers were other major carriers (59%), government/corporate (53%), and commuter airlines (53%). Relatively few regional or major airlines reported hiring pilots from either air taxi services (24%) or flight schools (6%).

Among the regional and major carriers, the most common selection methods were reference checks (100%), background checks (100%), and interviews (94%). Simulators (47%), aptitude tests (35%), clinical psychological assessment (25%), and flight checks (24%) were less common. The skills needed to fly an airplane can be checked by a work sample in the form of either a check flight or simulator. Suarez et al. (1994) noted that because skill levels can be tested, minimum qualification in combination with prior flying experience are the most widely used set of hiring variables. Smaller operators tend to assess pilot skills using flight checks, whereas larger operators use simulators. Overall, aptitude tests and psychological assessment were not commonly used selection methods for U.S. air carriers, perhaps because they are satisfied with indicators of pilot skills (e.g., log books, flight check, simulator) and biographical data (e.g., background and reference checks). To some extent, aptitude and personality factors are assessed in the background and reference checks.

Suarez et al. (1994) speculated that several factors could lead to an eventual shortage of experienced pilots. These include expected growth in the airline industry, fewer pilots entering the labor pool from the military, reduction in the number of pilots working their way up to the major/regional airlines from air taxi and commuter carriers, and mandatory age-based retirement. A 1997 survey of 29 regional and 10 major U.S. air carriers (D. R. Hunter, personal communication, August 27, 1998) reported similar results.

Hansen and Oster (1997) identified five pathways for civilian pilots. The first pathway, military pilot training, traditionally has accounted for about 75% of the new hires for major U.S. civilian air carriers. On-the-job training, collegiate training, ab initio training, and foreign hires account for the rest. Hansen and Oster note that U.S. air carriers have shown reluctance to use either ab initio programs or foreign hires. Although ab initio training is popular with foreign air carriers, U.S. carriers have been unwilling to pay for high-cost ab initio programs. This attitude will probably continue as long as applicants from other sources are plentiful. As the supply of former military pilots dwindles, U.S. carriers are likely to turn to on-the-job training and collegiate-based programs to pick up the slack.

Several recent studies have described pilot selection procedures for non-U.S. commercial air carriers (Bartram & Baxter, 1996; Doat, 1995; Hörmann & Luo, 1999; Manzey, Hörmann, Osnabrügge, & Goeters, 1990; Novis Soto, 1998; Stahlberg & Hörmann, 1993; Stead, 1991, 1995). However, little information was provided about their validity (Bartram & Baxter, 1996; Hörmann & Luo, 1999; Stahlberg & Hörmann, 1993; Stead, 1991). Due to page limitations, we discuss these studies elsewhere (Carretta & Ree, 2000).

Psychological Evaluations for Commercial Pilots. The use of psychological evaluations for commercial pilots is more formalized among European than among U.S. air carriers (Goeters, 1995). In 1991, the Flight Crew Licensing Medical Group of the European Civil Aviation Conference (ECAC) adopted psychological requirements for commercial pilots. The requirements state:

> The applicant for the holder of a Class I (Commercial) or Class II (Private) medical certificate shall have no established psychological deficiencies, particularly in operational aptitudes or any relevant personality factor which is likely to interfere with the safe exercise of the privileges of the applicable license(s). (Goeters, 1995, p. 149)

Goeters (1995), citing the ECAC draft guidelines, stated that a complete psychological evaluation includes an assessment of biographical data (e.g., general life history, family, work history, health), aptitude (e.g., cognitive, psychomotor), personality (e.g., decision making, motivation, stress coping), and a psychological interview. When the psychological evaluation was implemented, it was not a part of the routine medical examination, but was initiated when the Aeromedical Board received information that led to concerns about the aptitude or personality of the pilot. Although this psychological evaluation was not included in the initial Class I evaluation, some (Goeters, 1995) have suggested that it be included. It should be noted, however, that support for this type of testing is not universal (Johnston, 1996; Murphy, 1995). A major concern of commercial pilots (Murphy, 1995) seems to be that under the current regulations, the decision as to whether a pilot is "operationally fit" is made by aviation psychologists or physicians, not other pilots. Johnston noted that although psychological testing might provide some economic and training benefits (e.g., reduction in training attrition), such testing has many potential problems and risks. These include a possible shortage of qualified aviation psychologists to perform the assessments, a lack of an accepted test battery or standards of "acceptable performance," concerns with the psychometric properties of pilot assessment tests, cultural differences affecting test performance and interpretation of scores, and others.

The role of psychological assessment for licensing of commercial pilots is controversial. The debate over its implementation and use will continue for some time.

THE FUTURE OF PILOT SELECTION METHODS

General Cognitive Ability

Because general cognitive ability has been shown to be a versatile predictor in pilot selection, we discuss three emerging measurement methods that might be useful. These are chronometric measures, neural conductive velocity, and cognitive components.

Chronometric measures are typified by reaction time and choice reaction time. Jensen (1980, 1998) has shown that simple reaction time is correlated with measures of intelligence (i.e., *g*). This simple reaction-time task in which a finger is placed on a "home" button and moved to a "target" button when a light comes on, shows a low but positive correlation with measured intelligence. Choice reaction time, which requires pressing the one lighted button among many (for example eight), shows moderate correlation with measured intelligence.

The speed at which a neuron transmits an impulse is called neural conductive velocity. It requires no physically invasive procedure and is typically measured in the optic nerve. Electrodes are attached to the participant's head and a light is flashed which the participant sees. The verbal instructions are as simple as "look at the light," and no overt response is required by the participant. The head-mounted electrodes are connected to a computer with special software that measures the speed of the nerve impulse. Reed and Jensen (1992) have shown that neural conductive velocity in the optic nerve is correlated a moderate .37 with measured intelligence.

Cognitive components such as information processing speed and working memory capacity have been shown to mostly measure *g* (Kranzler & Jensen, 1991; Kyllonen & Christal, 1990; Miller & Vernon, 1992). The measurement of cognitive components is frequently done with computers. For example, participants may be shown a series of letters and sequentially told a set of rules governing the order of the letters. Following the rules, participants must state the proper order of the letters. Arthur, Barrett, and Doverspike (1990) have shown validity for these types of tests in an occupational setting. However, Stauffer, Ree, and Carretta (1996) have demonstrated that cognitive components tests measure mostly *g* so that no major improvements in validity can be expected. Not withstanding, Seamster, Redding, and Kaempf (1997) have speculated that cognitive task analysis could lead to identification of specific cognitive components that would be predictive of pilot performance. This is inconsistent with past results, and their speculations are highly unlikely to be fruitful, given the high *g*-loading of cognitive components measures (Stauffer et al., 1996). Further, Jones and Ree (1998) and Schmidt, Hunter, and Pearlman (1981) have shown that job task differences do not change the effectiveness of *g* as a predictor. Empirical results to support the speculations would be helpful.

Chronometric measures, neural conductive velocity, and cognitive components may be fruitful only because they measure mostly *g*. They offer the advantage of being content free, thereby potentially reducing mean test score differences that were the consequences of differential educational choices. With careful psychometric development yielding increased reliability, these measures could find a place in pilot selection.

Flying Knowledge and Skills

Another potential trend in pilot selection could emerge in the measurement of flying knowledge and skills. As previously discussed, although simulation-based tests of flying skills (Gress & Willkomm, 1996; Okros et al., 1991) have shown validity for pilot selection, their use is relatively rare due to development and operating costs and the need for centralized administration. Advances in computer hardware and software have resulted in progressively faster, smaller, and less-expensive computers, that in turn have made computer-based tests more common (Ree & Carretta, 1998). In recent applications, the U.S. Air Force has developed several experimental "work sample" tests for possible use in the selection of pilots and other aircrew specialties. These tests resemble some aspect of job performance (e.g., instrument flight, cross-check) and require learning and applying complex rules. Although these computerized tests exhibit face validity, their utility for pilot selection has yet to be determined.

Incrementing the Validity of *g*

Finally, the seminal work of Schmidt and Hunter (1998) that showed the job related validity of *g* and incremental validity of 18 other predictors should serve as a guide to future research and practice. They collected and meta-analyzed hundreds of validation studies involving training and job performance and noted the predictors used. Predictors included *g*, selection interviews, work samples, and personality. The three combinations of predictors with the highest validity and utility for job performance were *g* plus a work sample test, *g* plus an integrity test (or a conscientiousness test), and *g* plus a structured interview. They noted that the latter two methods were appropriate for both entry-level selection and selection of experienced employees.

CONCLUSION

Hunter and Burke (1995) offer a practical summary of pilot selection. They have created a how-to manual that embodies the current science and art of selection. Although we cannot agree with them on all details (e.g., their failure to correct studies for statistical artifacts), their book contains many practical ideas and should be studied by all concerned with pilot selection.

The military has a long history of research in the selection of pilots and other aircrew. Cumulative results suggest that general cognitive ability (*g*) has been a mainstay of military testing and will likely remain so. Measures of pilot job knowledge and psychomotor ability have shown incremental validity when used with measures of *g*.

American law requires job analyses for the development of job selection tests. The results of the analyses should be converted to good practice guided by cumulative knowledge. There is no single ideal pilot selection system, because not all pilots are hired the same way. Some are hired directly from the military with many flying hours, some from other airlines, and some directly from training. Although different, all the selection systems should have three common measurement elements: cognitive ability, conscientiousness (or perhaps "integrity"), and job knowledge (Schmidt & Hunter, 1998).

There is a dearth of studies reported by American commercial airlines. Most likely, this is a consequence of legal liability and competitive edge. Results from two recent FAA surveys suggest that U.S. commercial airlines rely heavily on recruiting applicants with prior pilot experience. Aptitude and personality testing have received relatively little emphasis. In the instances where airlines employ ab initio selection (e.g., Bartram & Baxter, 1996), test batteries similar to those commonly found in military pilot selection are used. In the few instances where validity studies are reported for commercial aviation, it is difficult to interpret the results due to a variety of methodological problems and a failure to report results in sufficient detail. Further, most of the reported commercial aviation studies provide conservative estimates of the value of the selection system. Publication of future studies involving commercial pilot selection is strongly encouraged for two reasons. First, to share information on the predictiveness of commercial pilot selection methods and improve them. Second, to provide sufficient detail such as means, variances, and correlation coefficients of all variables to allow secondary analyses (e.g., meta-analysis). The ability to pool validity data across multiple studies, correct for study artifacts such as range restriction, sampling error, and unreliability allows researchers to obtain better estimates of the effectiveness of selection methods.

Personality assessment is little used in the United States and the United Kingdom, but is more prevalent in continental Europe. Cumulative research suggests that incremental validity could be achieved by using measures of personality, particularly conscientiousness (Barrick & Mount, 1991) or integrity (Schmidt & Hunter, 1998).

The role of psychological evaluation in the licensing of airline pilots has been raised and debated in Europe. Proponents see it as a means of identifying psychological deficits of pilots and reducing potential risks to aviation safety. Opponents express fears of abuse and concerns with the use of tests in circumstances for which they were not designed. This topic is controversial and will receive attention for some time.

Most important in conducting pilot selection research is scientific rigor. Without it, results may be worse than meaningless leading to counterproductive practice. Before setting out to develop a pilot selection system, it is imperative to have a firm foundation in the published literature of human abilities, reliability, validity, job performance measurement, and meta-analysis. Cumulative research results should guide practice.

REFERENCES

Aiken, L. R., Jr. (1966). Another look at weighting test items. *Journal of Educational Measurement, 3,* 183–185.

American Psychological Association, American Educational Research Association, & National Council on Measurement in Education (Joint Committee). (1985). *Standards for educational and psychological testing.* Washington, DC: American Psychological Association.

Arthur, W., Jr., Barrett, G. V., & Doverspike, D. (1990). The validation of an information-processing-based test battery for the prediction of handling accidents among petroleum product transport drivers. *Journal of Applied Psychology, 75,* 621–628.

Bailey, M., & Woodhead, R. (1996). Current status and future developments of RAF aircrew selection. *Selection and Training Advances in Aviation: AGARD Conference Proceedings 588* (pp. 8-1-8-9). Prague, Czech Republic: Advisory Group for Aerospace Research and Development.

Barrick, M. R., & Mount, M. K. (1991). The Big Five personality dimensions and job performance. *Personnel Psychology, 44,* 1–26.

Bartolo Ribeiro, R. (1992). Predicao da performance em psicologia aeronautica: Validacao de uma bateria de seleccao. *Analise Psicologia, Serie X, 3,* 353–365.

Bartolo Ribeiro, R., Martins, A., Vicoso, A., Carpinteiro, M. J., & Estrela, R. (1992). Validacaocapacidade preditiva dos testes utilizados na seleccao de pilotos. *Revista depsicologiamilitar, Numero Especial,* 271–280.

Bartram, D., & Baxter, P. (1996). Validation of the Cathay Pacific Airways pilot selection program. *International Journal of Aviation Psychology, 6,* 149–169.

Bennett, G. J., Seashore, H. G., & Wesman, A. G. (1982). *Differential Aptitude Tests (Forms V and W): Administrators Handbook.* New York: Psychological Corporation.

Boer, L. C. (1992). *TASKOMAT: Meaning of test scores.* Soesterberg, Netherlands: TNO Institute for Perception.

Boer, L. C., Harsveld, M., & Hermans, P. H. (1997). The selective-listening task as a test for pilots and air traffic controllers. *Military Psychology, 9,* 137–149.

Brand, C. (1987). The importance of general intelligence. In S. Modgil & C. Modgil (Eds.), *Arthur Jensen: Consensus and controversy.* New York: Falmer Press.

Burke, E. (1993). Pilot selection in NATO: An overview. In R. S. Jensen & D. Neumeister (Eds.), *Proceedings of the Seventh International Symposium on Aviation Psychology* (pp. 373–378). Columbus: The Ohio State University.

Burke, E., Hobson, C., & Linsky, C. (1997). Large sample validations of three general predictors of pilot training success. *International Journal of Aviation Psychology, 7,* 225–234.

Callister, J. D., King, R. E., Retzlaff, P. D., & Marsh, R. W. (1999). Revised NEO Personality Inventory profiles for male and female U.S. Air Force pilots. *Military Medicine, 164,* 885–890.

Carretta, T. R. (1989). USAF pilot selection and classification systems. *Aviation, Space, and Environmental Medicine, 60,* 46–49.

Carretta, T. R. (1990). Cross-validation of experimental USAF pilot training performance models. *Military Psychology, 2,* 257–264.

Carretta, T. R. (1992a). Recent developments in U.S. Air Force pilot candidate selection and classification. *Aviation, Space, and Environmental Medicine, 63,* 1112-1114.

Carretta, T. R. (1992b). Understanding the relations between selection factors and pilot training performance: Does the criterion make a difference? *International Journal of Aviation Psychology, 2,* 95–105.

Carretta, T. R. (2000). U.S. Air Force pilot selection and training methods. *Aviation, Space, and Environmental Medicine, 71,* 950–956.

Carretta, T. R., Perry, D. C., Jr., & Ree, M. J. (1996). Prediction of situational awareness in F-15 pilots. *International Journal of Aviation Psychology, 6,* 21–41.

Carretta, T. R., & Ree, M. J. (1993). Basic Attributes Test (BAT): Psychometric equating of a computer-based test. *International Journal of Aviation Psychology, 3,* 189–201.

Carretta, T. R., & Ree, M. J. (1994). Pilot candidate selection method (PCSM): Sources of validity. *International Journal of Aviation Psychology, 4,* 103–117.

Carretta, T. R., & Ree, M. J. (1995). Air Force Officer Qualifying Test validity for predicting pilot training performance. *Journal of Business and Psychology, 9,* 379-388.

Carretta, T. R., & Ree, M. J. (1996a). Factor structure of the Air Force Officer Qualifying Test: Analysis and comparison. *Military Psychology, 8,* 29–42.

Carretta, T. R., & Ree, M. J. (1996b). U.S. Air Force pilot selection tests: What is measured and what is predictive? *Aviation, Space, and Environmental Medicine, 67,* 275–283.

Carretta, T. R., & Ree, M. J. (2000). *Pilot selection methods.* (Tech. Rep. No. AFRL/HE-WP-TR-2000-0116). Wright-Patterson Air Force Base, OH: Air Force Research Laboratory, Human Effectiveness Directorate, Crew Systems Interface Division.

Carretta, T. R., & Ree, M. J. (2001). Pitfalls of ability research. *International Journal of Selection and Assessment, 9,* 325–335.

Cascio, W. F. (1982). *Costing human resources: The financial impact of behavior in organizations.* New York: Van Nostrand Reinhold.

Cascio, W. F. (1991). *Applied psychology in personnel management* (4th ed.). Englewood Cliffs, NJ: Prentice Hall.

Cohen, J. (1983). The cost of dichotomization. *Applied Psychological Measurement, 7,* 249–253.

Cohen, J. (1987). *Statistical power analysis for the behavioral sciences* (rev. ed.). Hillsdale, NJ: Lawrence Erlbaum Associates.

Costa, P. T., Jr., & McCrae, R. R. (1992). *Revised NEO Personality Inventory (NEO-PI-R) and NEO Five-Factor Inventory (NEO-FFI) professional manual.* Odessa, FL: Psychological Assessment Resources.

Cronbach, L. J. (1971). Test validation. In R. L. Thorndike (Ed.), *Educational Measurement* (2nd ed., pp. 443–507). Washington, DC: American Council on Education.

Damos, D. L. (1996). Pilot selection batteries: Shortcomings and perspectives. *International Journal of Aviation Psychology, 6,* 199–209.

Delaney, H. D. (1992). Dichotic listening and psychomotor task performance as predictors of naval primary flight-training criteria. *International Journal of Aviation Psychology, 2,* 107–120.

Devore, J., & Peck, R. (1993). *Statistics: The exploration and analysis of data* (2nd ed.). Belmont, CA: Duxbury.

Digman, J. M. (1990). Personality structure: Emergence of the five-factor model. *Annual Review of Psychology, 41,* 417–440.

Doat, B. (1995). Need of new development in Air France selection. In N. Johnston, R. Fuller, & N. McDonald (Eds.), *Aviation psychology: Training and selection. Proceedings of the 21st Conference of the European Association for Aviation Psychology (EAAP)* (Vol. 2, pp. 182–187). Aldershot, England: Avebury Aviation.

Dockeray, F. C., & Isaacs, S. (1921). Psychological research in aviation in Italy, France, England, and the American Expeditionary Forces. *Journal of Comparative Psychology, 1,* 115–148.

Dolgin, D. L., & Gibb, G. D. (1989). Personality assessment in aviator selection. In R. S. Jensen (Ed.), *Aviation psychology* (pp. 288–320). London: Gower.

Duke, A. P., & Ree, M. J. (1996). Better candidates fly fewer training hours: Another time testing pays off. *International Journal of Selection and Assessment, 4,* 115–121.

Dvorak, B. J. (1947). The new U.S. E. S. General Aptitude Test Battery. *Journal of Applied Psychology, 31,* 372–376.

Earles, J. A., & Ree, M. J. (1992). The predictive validity of ASVAB for training grades. *Educational and Psychological Measurement, 52,* 721–725.

Equal Employment Opportunity Commission. (1978). *Uniform guidelines on employee selection procedures.* Title 29—Labor, Part 1607. *National Archives and Records Administration code of federal regulations.* Washington, DC: U.S. Government Printing Office.

Evdokimov, V. L. (1988). Socio-psychological selection in the system of professional training of pilots. *Psikocheskii Zhurnal, 9,* 71–74.

Fiske, D. W. (1947). Validation of naval aviation cadet selection tests against training criteria. *Journal of Applied Psychology, 31,* 601–614.

Flanagan, J. C. (1942). The selection and classification program for aviation cadets (aircrew-bombardiers, pilots, and navigators). *Journal of Consulting Psychology, 5,* 229–238.

Fuller, W. A. (1987). *Measurement error models.* New York: Wiley.

Gael, S. (1988). *The job analysis handbook for business, industry, and government.* Vols. 1 and 2. New York: Wiley.

Geldard, F. A., & Harris, C. W. (1946). Selection and classification of aircrew by the Japanese. *American Psychologist, 1,* 205–217.

Gibb, G. D., & Dolgin, D. L. (1989). Predicting military flight training success by a compensatory tracking task. *Military Psychology, 1,* 235–240.

Goeters, K. M. (1995). Psychological evaluation of pilots: The present regulations and arguments for their application. *Aviation psychology: Training and selection. Proceedings of the 21st Conference of the European Association for Aviation Psychology (EAAP)* (Vol. 2, pp. 149–156). Aldershot, England: Avebury Aviation.

Goldberg, L. R. (1992). The development of markers of the Big Five factor structure. *Psychological Assessment, 4,* 26–42.

Goldberg, S. (1991). *When wish replaces thought.* Buffalo, NY: Prometheus.

Gopher, D., & Kahneman, D. (1971). Individual differences in attention and the prediction of flight criteria. *Perceptual and Motor Skills, 33,* 1334–1342.

Gress, W., & Willkomm, B. (1996). Simulator-based test systems as a measure to improve the prognostic value of aircrew selection. *Selection and Training Advances in Aviation: AGARD Conference Proceedings 588* (pp. 15-1-15-4). Prague, Czech Republic: Advisory Group for Aerospace Research and Development.

Griggs *et al.* v. Duke Power Co., 401 U.S. 424.

Guion, R. M. (1976). Recruiting, selection, and job placement. In M. D. Dunnett (Ed.) *Handbook of industrial and organizational psychology* (pp. 777–828). Chicago: Rand McNally.

Hansen, D., & Wolf, G. (2000). Aircrew selection in the German Air Force. *Modern Simulation and Training, 1,* 26–30.

Hansen, I. (1999). *Use of interviews by Euro-NATO air forces in pilot selection* (Tech. Rep. No. AHFWG-3). Euro-NATO Aircrew Human Factors Working Group.

Hansen, J. S., & Oster, C. V., Jr. (Eds.). (1997). *Taking flight: Education and training in aviation careers.* Washington, DC: National Academy Press.

Harsveld, M. (1991). The Defense Mechanism Test and success in flying training. In E. Palmer (Ed.), *Human resource management in aviation* (pp. 51–65). Aldershot, England: Avebury Technical.

Hilton, T. F., & Dolgin, D. L. (1991). Pilot selection in the military of the free world. In G. Gal & A. D. Mangelsdorff (Eds.), *Handbook of military psychology* (pp. 81–101). New York: Wiley.

Hörmann, H. J., & Luo, X. L. (1999). Development and validation of selection methods for Chinese student pilots. In R. S. Jensen, B. Cox, J. D. Callister, & R. Lavis (Eds.), *Proceedings of the Tenth International Symposium on Aviation Psychology* (pp. 571–576). Columbus, OH.

Hunter, D. R. (1989). Aviator selection. In M. S. Wiskoff & G. M. Rampton (Eds.), *Military personnel measurement: Testing, assignment, evaluation* (pp. 129–167). New York: Praeger.

Hunter, D. R., & Burke, E. F. (1994). Predicting aircraft pilot training success: A meta-analysis of published research. *International Journal of Aviation Psychology, 4,* 297–313.

Hunter, D. R., & Burke, E. F. (1995). *Handbook of pilot selection.* Brookfield, VT: Avebury Aviation.

Hunter, J. E. (1980). *The dimensionality of the General Aptitude Test Battery (GATB) and the dominance of general factors over specific factors in the prediction of job performance.* Washington, DC: U.S. Employment Service, U.S. Department of Labor.

Hunter, J. E. (1986). Cognitive ability, cognitive aptitudes, job knowledge, and job performance. *Journal of Vocational Behavior, 29,* 340–362.

Hunter, J. E., & Hunter, R. F. (1984). Validity and utility of alternative predictors across studies. *Psychological Bulletin, 96,* 72–98.

Hunter, J. E., & Schmidt, F. L. (1990). *Methods of meta-analysis.* Newbury Park, CA: Sage.

Imhoff, D. F., & Levine, J. M. (1981). *Perceptual-motor and cognitive performance task battery for pilot selection* (Tech. Rep. No. AFHRL-TR-80-27). Brooks Air Force Base, TX: Air Force Human Resources Laboratory, Manpower and Personnel Division.

Jensen, A. R. (1980). *Bias in mental testing.* New York: Free Press.

Jensen, A. R. (1993). Commentary: Vehicles of *g. Psychological Science, 3,* 275–278.

Jensen, A. R. (1994). What is a good *g*? *Intelligence, 18,* 231–258.

Jensen, A. R. (1998). *The g factor: The science of mental ability.* Westport, CT: Praeger.

Jessup, G., & Jessup, H. (1966). Validity of the Eysenck Personality Inventory for pilot selection. *Journal of Occupational Psychology, 45,* 111–123.

Johnson, J. T., & Ree, M. J. (1994). RANGEJ: A Pascal program to compute the multivariate correction for range restriction. *Educational and Psychological Measurement, 54,* 693–695.

Johnston, N. (1996). Psychological testing and pilot licensing. *International Journal of Aviation Psychology, 6,* 179–197.

Jones, G. E., & Ree, M. J. (1998). Aptitude test validity: No moderating effects due to job ability requirements. *Educational and Psychological Measurement, 58,* 282–292.

Jöreskog, K., & Sörbom, D. (1996). *LISREL 8: User's reference guide.* Chicago: Scientific Software International.

Kennedy, E. (1988). Estimation of the squared cross-validity coefficients in the context of best subtest regression. *Applied Psychological Measurement, 12,* 231–237.

Kim, J. O., & Mueller, C. W. (1988). *Factor analysis: Statistical methods and practical issues.* Beverly Hills, CA: Sage.

Kranzler, J. H., & Jensen, A. R. (1991). The nature of psychometric *g*: Unitary process or a number of independent processes? *Intelligence, 15,* 397–422.

Kyllonen, P. C., & Christal, R. E. (1990). Reasoning ability is (little more than) working memory capacity?! *Intelligence, 14,* 389–433.

Lawley, D. N. (1943). A note on Karl Pearson's selection formulae. *Proceedings of the Royal Society of Edinburgh, Section A, 62, Part 1,* 28–30.

Li, L. (1993). On military pilot selection. *Psychological Science (China), 16,* 299–304.

Linn, R. L., Harnish, D. L., & Dunbar, S. (1981). Corrections for range restriction: An empirical investigation of conditions resulting in conservative corrections. *Journal of Applied Psychology, 66,* 655–663.

Long, G. E., & Varney, N. C. (1975). *Automated pilot aptitude measurement system* (Tech. Rep. No. AFHRL-TR-75-58). Lackland Air Force Base, TX: Air Force Human Resources Laboratory, Personnel Research Division.

Manzey, D., Hörmann, H. J., Osnabrügge, G., & Goeters, K. M. (1990). *International application of the DLR test-system: First year of cooperation with IBERIA in pilot selection* (DLR-FB-90-05). Hamburg, Germany: DLR Institut für Flugmrdizin, Abteilung Luft- und Raumfahrtpsychologie.

Martinussen, M. (1996). Psychological measures as predictors of pilot performance: A meta-analysis. *International Journal of Aviation Psychology, 6,* 1–20.

Martinussen, M. (1997). Pilot selection and range restriction: A red herring or a real problem? *Proceedings of the Ninth International Symposium on Aviation Psychology* (pp. 1314–1318). Columbus, OH.

Martinussen, M., & Torjussen, T. (1993). Does DMT (Defense Mechanism Test) predict pilot performance only in Scandinavia? *Proceedings of the Seventh International Symposium on Aviation Psychology* (pp. 398–403). Columbus: The Ohio State University.

Martinussen, M., & Torjussen, T. (1998). Pilot selection in the Norwegian Air Force: A validation and meta-analysis of the test battery. *International Journal of Aviation Psychology, 8,* 33–45.

McClelland, D. C. (1993). Intelligence is not the best predictor of job performance. *Current Directions in Psychological Science, 2,* 5–6.

McCormick, E. J. (1976). Job and task analysis. In M. D. Dunnett (Ed.), *Handbook of industrial and organizational psychology* (pp. 651–696). Chicago: Rand McNally.

McCormick, E. J. (1979). *Job analysis: Methods and applications.* New York: AMACOM.

McDonald, R. P. (1999). *Test theory: A unified treatment.* Mahwah, NJ: Lawrence Erlbaum Associates.

McHenry, J. J., Hough, L. M., Toquam, J. L., Hanson, M. A., & Ashworth, S. (1990). Project A validity results: The relationship between predictor and criterion domains. *Personnel Psychology, 43,* 335–354.

Melton, A. W. (Ed.). (1947). *Army Air Forces aviation psychology research reports: Apparatus tests* (Report No. 4). Washington, DC: GPO.

Messick, S. (1989). Validity. In R. L. Linn (Ed.), *Educational Measurement* (3rd ed., pp. 13–103). New York: Macmillan.

Miller, L. T., & Vernon, P. A. (1992). The general factor in short-term memory, intelligence, and reaction time. *Intelligence, 16,* 5–29.

Morrison, T., & Morrison, M. (1995). A meta-analytic assessment of the predictive validity of the quantitative and verbal composites of the Graduate Record Examination with graduate grade point average representing the criterion of graduate success. *Educational and Psychological Measurement, 55,* 309–316.

Mosier, C. I. (1951). Problems and designs of cross-validation. *Educational and Psychological Measurement, 11,* 5–11.

Murphy, K. R. (1983). Fooling yourself with cross-validation: Single-sample designs. *Personnel Psychology, 36,* 111–118.

Murphy, T. (1995). JAA psychological testing of pilots: Objections and alarms. *Aviation psychology: Training and selection. Proceedings of the 21st Conference of the European Association for Aviation Psychology (EAAP)* (Vol. 2, (pp. 157–163). Aldershot, England: Avebury Aviation.

Novis Soto, M. L. (1998). Los cuestionarios de personalidad en la selection de los pilotos de línea aérea. *Revista de Psicologia del Trabajo y de las Organizaciones, 14,* 113–128.

Okros, A. C., Spinner, B., & James, J. A. (1991). *The Canadian Automated Pilot Selection System* (Research Report 91–1). Willowdale, Ontario: Canadian Forces Personnel Applied Research Unit.

Olea, M. M., & Ree, M. J. (1994). Predicting pilot and navigator criteria: Not much more than *g*. *Journal of Applied Psychology, 79,* 845–851.

Orgaz, B., & Loro, P. (1993). La psicologia aeronautico-militar en Espana: breve apunte historico. *Revista de Historia de la Psicologia, 14,* 271–283.

Parry, J. B. (1947). The selection and classification of RAF aircrew. *Occupational Psychology, 21,* 158–167.

Prieto, G., Carro, J., Palenzuela, D. L., Pulido, R. F., Orgaz, B., Delgado, A. R., & Loro, P. (1996). Diferencias individuales y práctica profesional en el ámbito militar: Seleccion de pilotos áereos. In M. DeJuan-Espinosa, R. Colom, & M. A. Quiroga (Eds.), *La práctica de la psicologia diferencial en industria y organizaciones* (pp. 171–194). Madrid, Spain: Piramide.

Pulakos, E. (1997). Ratings of job performance. In D. L. Whetzel & G. R. Wheaton (Eds.), *Applied measurement methods in industrial psychology* (pp. 291–318). Palo Alto, CA: Davies-Black.

Ree, M. J. (1995). Nine rules for doing ability research wrong. *The Industrial-Organizational Psychologist, 32,* 64–68.

Ree, M. J., & Carretta, T. R. (1994a). The correlation of general cognitive ability and psychomotor tracking tests. *International Journal of Selection and Assessment, 2,* 209–216.

Ree, M. J., & Carretta, T. R. (1994b). Factor analysis of ASVAB: Confirming a Vernon-like structure. *Educational and Psychological Measurement, 54,* 457–461.

Ree, M. J., & Carretta, T. R. (1996). Central role of *g* in military pilot selection. *International Journal of Aviation Psychology, 6,* 111–123.

Ree, M. J., & Carretta, T. R. (1997). What makes an aptitude test valid? In R. F. Dillon (Ed.), *Handbook on testing* (pp. 65–81). Westport, CT: Greenwood Press.

Ree, M. J., & Carretta, T. R. (1998). Computerized testing in the United States Air Force. *International Journal of Selection and Assessment, 6,* 82–89.

Ree, M. J., Carretta, T. R., & Earles, J. A. (1998). In top-down decisions, weighting variables does not matter: A consequence of Wilks' theorem. *Organizational Research Methods, 1,* 407–420.

Ree, M. J., Carretta, T. R., Earles, J. A., & Albert, W. (1994). Sign changes when correcting for range restriction: A note on Pearson's and Lawley's selection formulas. *Journal of Applied Psychology, 79,* 298–301.

Ree, M. J., Carretta, T. R., & Teachout, M. S. (1995). Role of ability and prior job knowledge in complex training performance. *Journal of Applied Psychology, 80,* 721-730.

Ree, M. J., & Earles, J. A. (1991). Predicting training success: Not much more than *g*. *Personnel Psychology, 44,* 327–332.

Ree, M. J., & Earles, J. A. (1992). Intelligence is the best predictor of job performance. *Current Directions in Psychological Science, 1,* 86–89.

Ree, M. J., & Earles, J. A. (1993). *g* is to psychology what carbon is to chemistry: A reply to Sternberg and Wagner, McClelland, and Calfee. *Current Directions in Psychological Science, 2,* 11–12.

Reed, T. E., & Jensen, A. R. (1992). Conduction velocity in a brain nerve pathway of normal adults correlates with intelligence level. *Intelligence, 16,* 259–272.

Retzlaff, P. D., & Gilbertini, M. (1987). Air force pilot personality: Hard data on the "right stuff." *Multivariate Behavioral Research, 22,* 383–399.

Schmid, J., & Leiman, J. M. (1957). The development of hierarchical factor solutions. *Psychometrika, 22,* 53–61.

Schmidt, F. L., & Hunter, J. E. (1978). Moderator research and the law of small numbers. *Personnel Psychology, 31,* 215–232.

Schmidt, F. L., & Hunter, J. E. (1993). Tacit knowledge, practical intelligence, general mental ability, and job knowledge. *Current Directions in Psychological Science, 2,* 8–9.

Schmidt, F. L., & Hunter, J. E. (1998). The validity and utility of selection methods in personnel psychology: Practical and theoretical implications of 85 years of research findings. *Psychological Bulletin, 124,* 262–274.

Schmidt, F. L., Hunter, J. E., & Pearlman, K. (1981). Task differences and validity of aptitude tests in selection: A red herring. *Journal of Applied Psychology, 66,* 166–185.

Schmidt, F. L., Hunter, J. E., & Urry, V. W. (1976). Statistical power in criterion-related validation studies. *Journal of Applied Psychology, 61,* 473–485.

Seamster, T. L., Redding, R. E., & Kaempf, G. L. (1997). *Applied cognitive task analysis in aviation.* Brookfield, VT: Ashgate.

Sedlmeier, P., & Gigerenzer, G. (1989). Do studies of statistical power have an effect on the power of studies? *Psychological Bulletin, 105,* 309–316.

Siem, F. M., Carretta, T. R., & Mercatante, T. A. (1988). *Personality, attitudes, and pilot training performance: Preliminary analysis* (Tech. Rep. No. AFHRL-TP-87-62). Brooks Air Force Base, TX: Air Force Human Resources Laboratory, Manpower and Personnel Division.

Siem, F. M., & Murray, M. W. (1994). Personality factors affecting pilot combat performance: A preliminary investigation. *Aviation, Space, and Environmental Medicine, 65,* A45-A48.

Signori, E. I. (1949). The Arnprior experiment: A study of World War II pilot selection procedures in the RCAF and RAF. *Canadian Journal of Psychology, 3,* 136–150.

Skinner, J., & Ree, M. J. (1987). *Air Force Officer Qualifying Test (AFOQT): Item and factor analysis of Form O* (Tech. Rep. No. AFHRL-TR-86-68). Brooks Air Force Base, TX: Air Force Human Resources Laboratory, Manpower and Personnel Division.

Smallwood, T., & Fraser, M. (1995). *The airline training pilot.* Brookfield, VT: Ashgate.

Society for Industrial-Organizational Psychology. (1987). *Principles for the validation and use of personnel selection procedures* (3rd ed.). College Park, MD: Author.

Spearman, C. (1904). "General Intelligence," objectively determined and measured. *American Journal of Psychology, 15,* 201–293.

Spearman, C. (1927). *The abilities of man: Their nature and measurement.* New York: Macmillan.

Spinner, B. (1991). Predicting success in primary flying school from the Canadian Automated Pilot Selection System: Development and cross-validation. *International Journal of Aviation Psychology, 1,* 163–180.

Stahlberg, G., & Hörmann, H. J. (1993). *International application of the DLR test-system: Validation of the pilot selection for IBERIA* (DLR-FB-93-42). Hamburg, Germany: DLR Institut für Flugmrdizin, Abteilung Luft- und Raumfahrtpsychologie.

Stauffer, J. M., Ree, M. J., & Carretta, T. R. (1996). Cognitive-components tests are not much more than *g*: An extension of Kyllonen's analyses. *Journal of General Psychology, 123,* 193–205.

Stead, G. (1991). A validation study of the Quantas pilot selection process. In E. Farmer (Ed.), *Human resource management in aviation* (pp. 3–18). Aldershot, England: Avebury Technical.

Stead, G. (1995). Quantas pilot selection procedures: Past to present. In N. Johnston, R. Fuller, & N. McDonald (Eds.), *Aviation psychology: Training and selection. Proceedings of the 21st Conference of the European Association for Aviation Psychology (EAAP)* (Vol. 2, pp. 176–181). Aldershot, England: Avebury Aviation.

Stein, C. (1960). Multiple regression. In I. Olkin, J. Hoeffding, S. Ghurye, W. Madow, & H. Mann (Eds.), *Contributions to probability and statistics* (pp. 424–443). Stanford, CA: Stanford University Press.

Sternberg, R. J., & Wagner, R. K. (1993). The *g*-ocentric view of intelligence and job performance is wrong. *Current Directions in Psychological Science, 2,* 1–5.

Suarez, J., Barborek, S., Nikore, V., & Hunter, D. R. (1994). *Current trends in pilot hiring and selection* (Memorandum No. AAM-240–94–1). Washington, DC: Federal Aviation Administration.

Tett, R. P., Jackson, D. N., & Rothstein, M. (1991) Personality measures as predictors of job performance: A meta-analytic review. *Personnel Psychology, 44,* 703–742.

Thorndike, R. L. (1949). *Personnel selection.* New York: Wiley.

Thurstone, L. L. (1938). *Primary mental abilities.* Psychometric Monographs No 1.

Torjussen, T., & Vaernes, R. (1991). The use of the Defense Mechanism Test (DMT) in Norway for selection and stress research. In M. Olss, G. Dodaert, & H. Ursin (Eds.), *Quantification of human defense mechanisms* (pp. 172–206). Berlin: Springer-Verlag.

Tupes, E. C., & Christal, R. E. (1961). *Recurrent personality factors based on trait rankings* (Tech. Rep. No. ASD-TR-61–97). Lackland Air Force Base, TX: Personnel Laboratory, Aeronautical Systems Division.

Vernon, P. E. (1969). *Intelligence and cultural environment.* London: Methuen.

Viteles, M. S. (1945). The aircraft pilot: 5 years of research. A summary of outcomes. *Psychological Bulletin, 42,* 489–526.

Wainer, H. (1976). Estimating coefficients in linear models: It don't make no nevermind. *Psychological Bulletin, 83,* 213–217.

Wainer, H. (1978). On the sensitivity of regression and regressors. *Psychological Bulletin, 85,* 267–273.

Walter, D. C. (1998). *Air warriors: The inside story of the making of a navy pilot.* New York: Simon & Schuster.

Walters, L. C., Miller, M., & Ree, M. J. (1993). Structured interviews for pilot selection: No incremental validity. *International Journal of Aviation Psychology, 3,* 25–38.

Weeks, J. L., & Zelenski, W. E. (1998). *Entry to USAF undergraduate flying training* (Tech. Rep. No. AFRL-HE-AZ-TR-1998-0077). Brooks Air Force Base, TX: Training Effectiveness Branch, Warfighter Training Research Division.

Weeks, J. L., Zelenski, W. E., & Carretta, T. R. (1996). Advances in USAF pilot selection. *Selection and Training Advances in Aviation (AGARD-CP-588)* (pp. 1–1–1–11). Prague, Czech Republic: Advisory Group for Aerospace Research and Development.

Wheeler, J. L., & Ree, M. J. (1997). The role of general and specific psychomotor tracking ability in validity. *International Journal of Selection and Assessment, 5,* 128–136.

Wherry, R. J. (1975). Underprediction from overfitting: 45 years of shrinkage. *Personnel Psychology, 29,* 1–18.

Whetzel, D. L., & McDaniel, M. A. (1997). Employment interviews. In D. L. Whetzel & G. R. Wheaton (Eds.), *Applied measurement methods in industrial psychology* (pp. 185–205). Palo Alto, CA: Davies-Black.

Whetzel, D. L., & Oppler, S. H. (1997). Validation of selection instruments. In D. L. Whetzel & G. R. Wheaton (Eds.), *Applied measurement methods in industrial psychology* (pp. 355–384). Palo Alto, CA: Davies-Black.

Wilks, S. S. (1938). Weighting systems for linear functions of correlated variables when there is no dependent variable. *Psychometrika, 3,* 23–40.

Yerkes, R. M. (1919). Report of the psychology committee of the National Research Council. *Psychological Review, 26,* 83–149.

ACKNOWLEDGMENTS

The authors offer special thanks to Dr. David R. Hunter for his help in locating materials regarding commercial pilot selection practices. We also thank Dr. Monica Martinussen, Eugene Burke, and the book editors for their comments on a previous draft. Finally we thank Mark Bowler, Rui Bartolo Ribeiro, Hans-Jürgen Hörmann, Jose Puente, Paul Rioux, James Steindl, and Dr. Joseph Weeks for their help in this effort.

11

Training

John Patrick
Cardiff University, UK

Although training is a ubiquitous activity, enabling individuals and teams to perform tasks and jobs throughout society, it is particularly important when the costs of poor training are unacceptable. This is the case in the aviation industry where accidents can involve loss of life as well as huge financial costs. Because of this, considerable research and practice have been directed at developing and improving training in aviation. Studies have addressed issues that are critical to the training of pilots. For example: How do previous flying skills transfer when a pilot has to make the transition to another type of aircraft? To what extent do pilots have to experience the full range of veridical cues during training in a flight simulator in order to optimize the acquisition and transfer of skills to the actual aircraft? These sorts of questions can be answered not only from general principles in the training literature but also from training studies in aviation that provide specific solutions to more detailed and contextualized questions. Reference will be made to both these sources of evidence in this chapter.

Training development can be broken down into a series of tasks or stages and how this can be achieved is discussed. The three main stages of training development involve analysis, design, and evaluation, and these distinctions provide the remaining structure and objectives of this chapter. The first stage of training development involves identifying the training needs and analyzing the skills, knowledge, and attitudes required in order to establish the content of a training program. To achieve this, it is necessary to be competent at not only specifying training objectives, about which much has been written, but also using job and task analysis methods. Perhaps surprisingly, this first stage depends less on psychological factors than the second stage of training development, which concerns how the training content should be structured, sequenced, and organized into an effective and efficient learning program. This second stage of training development, which we will refer to as the "design of training," should not be confused with the superordinate term *training development* that embraces all stages. There are various psychological theories and principles that can guide the design of training, discussed later. These include theories about the qualitative changes from novice to expert performance and how they can be facilitated through the design of, for example, appropriate practice conditions. Last, and certainly not least, training development involves evaluation because training programs will never be perfect first time and usually require continual revision to their content, design, or indeed both on the basis of an evaluation of trainees' attitudes to and learning from the program. For an extensive discussion of general training principles, the students should consult Goldstein (1993) and Patrick (1992).

Before embarking on a discussion of each of these three stages of training development, it is important to consider not only the variation in the nature of the tasks that pilots have to master but also the effect of recent trends toward automation and greater crew coordination.

TRENDS AND CONTEXTS OF PILOT TRAINING

The tasks involved in flying an aircraft differ immensely between types of aircraft and between commercial and military aviation communities. Prince and Salas (1993) point to the greater variability in the nature of the tasks confronting a military pilot or air crew who not only have to engage in safe flying but may also have to, for example, outmaneuver and attack an enemy plane, gather intelligence information, or engage in a low-level bombing run. These tasks have to be accomplished with precision under severe time stress, possibly using additional equipment, such as night vision goggles. Also landing on and taking off from an aircraft carrier imposes finer performance requirements than would be experienced in normal commercial flights.

Not only is there therefore great variability in the tasks confronting pilots but information relevant to performance of the same task may be represented within the cockpit in a different manner. Single-variable instruments have been replaced by integrated computer displays, which can be selected by the pilot in the so-called glass cockpit. How such displays can best be designed for different tasks has been the subject of much research. Displays should be designed to support the pilot's information-processing requirements so that integrated displays will be particularly effective when information has to be integrated from diffuse sources (Haskell & Wickens, 1993). Recently Breda and Veltman (1998) confirmed that when fighter pilots use 3-D perspective displays in contrast to conventional plan-view radar displays, target acquisition is much faster. However, it is not always straightforward to predict what display is optimal for a particular task, and even when it can, compromises often have to be made such that generic displays are used that are reasonably compatible for a range of tasks.

Highlighting these variations in tasks and displayed information is important from a training perspective because they will result in concomitant variations in both the content, and, possibly, the design of the training programs. The most important rule of training is that any training program has to be focused on any significant mental and physical activities required of the task performer during execution of the actual tasks. Training will be ineffective to the extent that this linkage between the operational and training environments is degraded in terms of the psychological demands imposed on the task performer. Therefore, as has been mentioned, techniques of job and task analysis have to be employed to ensure that training content not only captures important task requirements but is comprehensive in its coverage. Armchair theorizing about what should form the basis of a training program will inevitably result in poor training. Similarly, the use of generic training courses will be problematic unless they are designed to target the specific activities that require training.

Out of the many changes in how aircraft are flown, two will be highlighted that have particular significance for training. These are, firstly, the inexorable development of automation and secondly, and not unconnected, the reemerging importance of effective teamwork amongst the flight crew.

Increasing Automation

The human factors literature concerning aviation is rife with reports of the difficulties caused by the introduction of increased automation in the cockpit (e.g., Amalberti, 1989; Mouloua & Koonce, 1997; Sarter, 1996; Wiener, 1993; Woods et al., 1994; see also chap. 9). This is somewhat ironic, as the goal of automation is to free the pilot from routine tasks and to provide easy to use and reliable control systems. However, pilot workload may under some circumstances be increased by automation as greater monitoring of systems is required together with more inferential reasoning concerning system state when human intervention is needed, particularly in critical situations (Kantowitz & Campbell, 1996). In addition, as Woods (1996) remarks, increased automation and computerization on the flight deck has increased what he terms the "coupling" between parts of the system, and this increased interaction between system components means that faults and disturbances can now have not only a wider range of effects but also some surprising ones. All of these developments in automation have important training ramifications which it seems from the literature are often overlooked. Rogalski (1996) examined how six experienced air crew performed during the last period of training using an automated cockpit for the A320 airbus when confronted with an engine fire at takeoff during a full-flight simulation. An analysis of the cooperative processes between members of the crew during this incident suggested that status information was not always shared enough in contrast to information concerning actions. The implications were that air crews needed to be trained to not only manage and review their awareness of the situation but also to improve their knowledge of the functionality of the automated systems. Similar findings come from a study by Sarter and Woods (1994) of pilots' interactions with the Flight Management System (FMS) that facilitates the tasks of flight planning, navigation, and monitoring of flight progress. They studied the performance of 20 airline pilots during a 1-hour scenario on a B-737-300 simulator. Their conclusions were as follows:

1. There are gaps in pilots' understanding of the functional structure of the automation.
2. The opaque interface between pilots and automation makes it difficult for pilots to track the state and activity of the automation.
3. Pilots may not be aware of the gaps in their knowledge about FMS function.
4. Pilots can escape from the CDU [Control Display Unit] to the MCP [Mode Control Panel] whenever a situation gets too complicated or time pressure is too high.
5. The flight situations in which these problems produce unmistakable performance difficulties may occur infrequently in line observations.

(Sarter & Woods, 1994, p. 27)

Each of these points can be associated with a training need. One of the classic negative effects of automation has been labeled "mode error" (Woods et al., 1994), where pilots can become unaware of the status and behavior of the automated system. This is particularly problematic in unusual situations when time may be critical. Sarter (1996) suggests that when mode error is associated with a pilot failing to take a particular action when the automated system has changed its status (omission error), this will be more difficult to detect and recover from than when the wrong action is taken (commission error). Therefore, decisions concerning the introduction of automated systems need to analyze and fully take account of projected pilot interactions, a point emphasized also in chapter 9.

It is also important that pilot opinions be assessed and accommodated. A recent survey of 132 pilots of Boeing 747–400, Douglas MD-11, and Airbus A-320 aircraft by Tenney, Rogers, and Pew (1998) examined opinions concerning existing and future automation on the flight deck. Pilots appreciated the benefits of automation, for example, in the cruise portion of the flight. Preference was given to automated systems that were simple and reliable rather than adaptive. There was a desire for more automation in preflight and taxi situations and other high mental workload situations. The authors noted that this is somewhat ironic as these more demanding situations will not only be more difficult to automate but will also result in automated systems that are not as simple and transparent as pilots would like. Training has to meet these new requirements that increased automation brings, even when they are not ideal from an information-processing perspective.

Crew Coordination

It is important to conceptualize the task of flying as a cooperative exercise between humans and automated systems rather than a pilot flying a piece of hardware. Hutchins (1995) emphasizes that analysis of flight operations should be at the level of a distributed sociotechnical system rather than an individual person. This joint human-technical system has evolved with changing job roles for pilots and other flight crew, all of which is supported by changes in regulations and procedures. The number of flight crew has reduced from three to two, although communication within and outside the cockpit has become increasingly important. Degani and Wiener (1997) emphasize the difficulties in developing safe and comprehensive procedures for managing such complex human-technical systems. They note that such difficulties are exacerbated by variations between airlines in their organizational philosophies concerning automation.

Training in CRM (Crew Resource Management) is therefore important and has been extensively discussed in the research literature (e.g., Wiener, Kanki, & Helmreich, 1993; see also chap. 13). Both Hackman (1993) and Wiener (1993) emphasize that concentrating training provision on individual technical competencies is insufficient and ignores aspects of team coordination, leadership, and communication that are vital to efficient and safe operation. Helmreich and Foushee (1993) provide some chilling examples where failures in team coordination resulted in accidents:

> A crew, distracted by the failure of a landing gear indicator light, failing to notice that the automatic pilot was disengaged and the aircraft descending into a swamp.
>
> A co-pilot concerned that take-off thrust was not properly set during a departure in a snowstorm, failing to get the attention of the captain with the aircraft stalling and crashing into the Potomac River.
>
> A breakdown in communications between a captain, co-pilot, and Air Traffic Control regarding fuel state and a crash following complete fuel exhaustion. (p. 6)

Helmreich and Foushee (1993) proceed to propose a model of flight crew performance (Fig. 11.1) that differentiates input, group process, and outcome factors. Input factors include individual aptitudes, emotional states, group composition, organizational culture, regulations, and aircraft and general environmental operating conditions. These inputs affect the nature of the group processes which in turn determine how well an air crew performs its various functions. For example, Orasanu (1993) studied the ingredients that contribute to effective crew decision

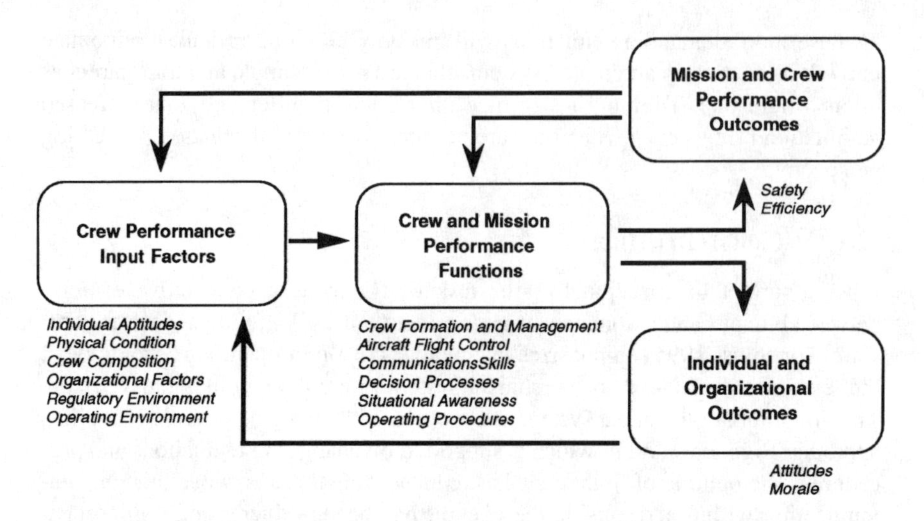

FIG. 11.1. A model of flightcrew performance. *Note.* From *Cockpit Resource Management* (p. 8), by E. L. Wiener, B. G. Kanki, and R. L. Helmreich, 1993, Orlando, FL: Academic Press. Copyright 1993 by Academic Press. Reprinted with permission.

making and highlighted the importance of situation awareness, planning, shared mental models, and resource management, all of which have training implications. One means of improving team coordination is through the use of "cross training" so that individuals experience and appreciate the role and task-related activities of other team members. The outcome factors in Helmreich and Foushee's model cover both the safety and efficiency of flight operations and also individual satisfaction, motivation, and so forth. These outcomes feed back to affect both input factors and group processes in a dynamic, iterative manner.

STAGES OF TRAINING DEVELOPMENT

There is a danger in large-scale organizations that the activities involved in the development of training programs become diffuse and fragmented (Hays, 1992). Those responsible for training development may not be experts and may themselves require training in how to perform these activities. It is for these reasons that models of training development, known as Instructional System Development (ISD) models, were developed a couple of decades ago. These models adopt what is known as a systems perspective where training development is viewed as a system that attempts to achieve a particular goal. This system can be broken down into its constituent subsystems, the interrelationships between them, and the function of each. Andrews and Goodson (1980) describe over 60 ISD models, and Patrick (1992) discusses four of these in detail. All of them cover the various tasks or activities that confront the training developer beginning with identification of training needs through to evaluation of the effectiveness of training.

There is variation between ISD models in how many tasks are involved in training development and their associated labels, although these differences are mostly superficial in nature. One of the most well known ISD models is the IPISD (Interservice Procedures for Instructional Systems Development) model developed by Branson et al. (1975) for training in the U.S. military (see Fig. 11.2). Summaries of the IPISD model can also be found in Logan (1979). The aim was to provide a generalizable, context-free framework for the development of good training programs. The model breaks the development of training into five phases: analyze, design, develop, implement, and control. (In some organizations these correspond to what is know as the "training development cycle.") These five phases are further divided into a total of 19 tasks. These tasks are in a logical sequence beginning with phase 1 that concerns identifying precisely the tasks that require training and analyzing them so that the content of training can be specified. (These training needs may be a consequence of new or changed jobs, including the introduction of new technology.) Information from phase 1 is then used in phase 2 to specify training objectives and to begin to map out the overall structure and sequence of a training programme, including trainees' entry and exit performance requirements. The third phase concerns not only trying to develop the training content into effective learning material by using as many principles and theories of training design as possible, but also considers the practical aspects of delivering the instruction, including the training of instructors, and pilot testing and revising the program as necessary. After this the training program is implemented (phase 4) and subsequently evaluated (phase 5). In order to improve the training program, there are feedback loops from the evaluation activities that enable revisions to be made to any of the five phases of training development (Fig. 11.2). It might be necessary to make changes to the training needs, the training content, or design. In this way the system of training development is capable of modifying and improving itself.

The main advantage of the ISD approach is that the tasks and their interrelationships confronting the training developer are made explicit in a context-free manner. This is not only helpful to those unfamiliar with training development but also provides a framework that large organizations can use to coordinate the inputs and outputs of the many personnel that are often involved. A further advantage of the ISD approach, less-often recognized, is that such models can be used to evaluate whether training is being developed systematically within an organization. Whereas evaluation typically focuses on the products of training (i.e., trainees' learning, attitudes), it is often revealing to examine the process of training development by ascertaining whether development tasks specified by an ISD model have been carried out, and if so, how. This might lead to evaluative questions such as:

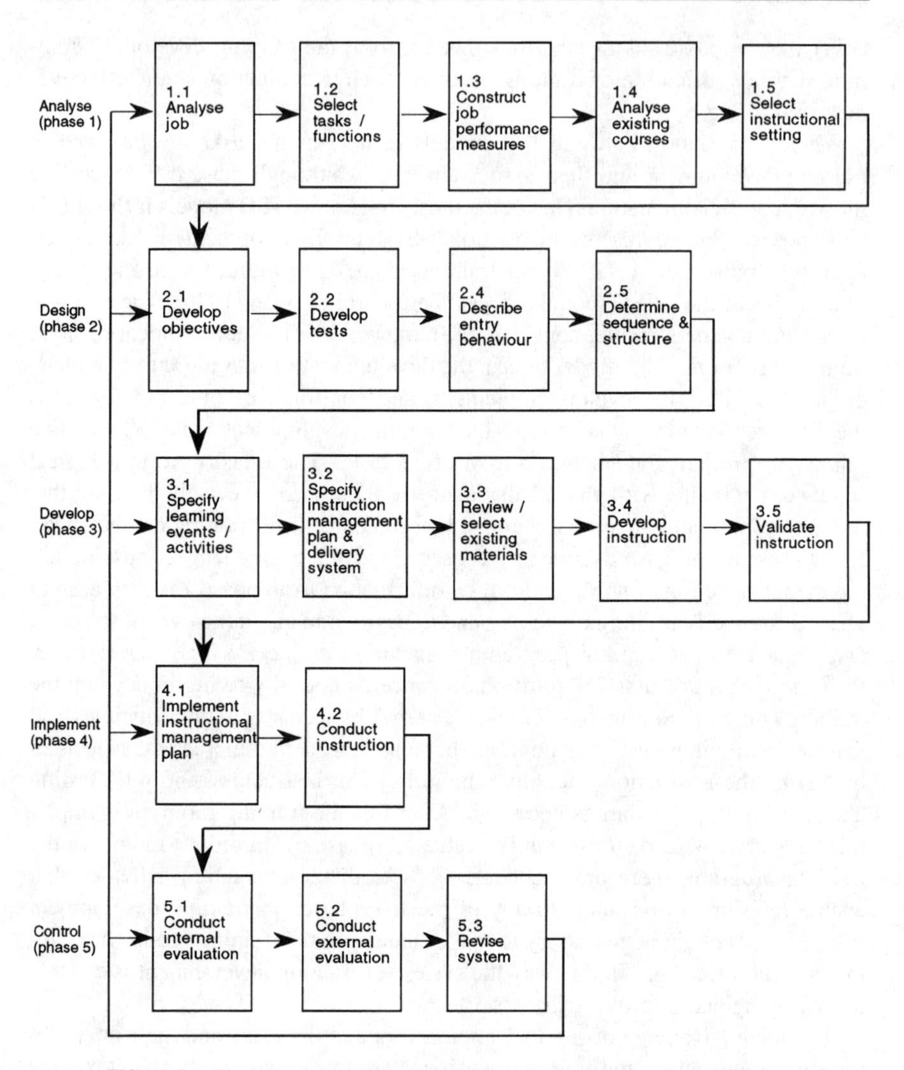

FIG. 11.2. The Interservices Procedures for Instructional Systems Development (IPISD). *Note.* From *Interservice Procedures for Instructional System Development: Executive Summary and Model,* by R. K. Branson, G. T. Rayner, L. Cox, J. P. Furman, F. J. King, and W. Hannum, 1975, Tallahassee, FL: Center for Educational Technology. Copyright 1975 by Florida State University. Reprinted with permission.

How were the tasks requiring training identified?

How was the job/task analysed?

At what level were training objectives specified and what format was used?

What learning principles were used to design the training materials?

How was the training program evaluated?

There are substantial criticisms of the ISD models. First and foremost is that although they specify the nature of the development tasks, how these tasks are to be accomplished is not addressed. Some attempts to remedy this for the IPISD model were made by Logan (1978, 1979) and O'Neil (1979a, 1979b) who surveyed the various techniques and procedures that could be used to fulfil each task in the IPISD model. Although these efforts were useful, it has to be noted that it is difficult, if not impossible, to specify precisely how each training development task should be carried out irrespective of context. This is not surprising, as otherwise training development could be fully proceduralized and therefore automated. Perhaps much to the relief of those involved in such activities this is not possible, not least because a complete psychology of training does not yet exist.

Given that on balance, an ISD approach has some merit, it is surprising how little it has permeated the culture of some organizations. Some argue that ISD models prescribe a sort of rationality that does not and cannot exist in training development, which is more opportunistic and idiosyncratic (Bunderson, 1977). Dipboye (1997) suggests that there are various organizational barriers to such an approach. These include the fact that unfortunately too much training practice is influenced by the superficial fads and fashions of consultants and many envisage that an ISD approach would be too costly and impersonal to implement. Also, the development of training and retraining programs are usually low on an organization's priorities due to their financial cost and perceived low value (Druckman & Bjork, 1994).

The remainder of this chapter discusses the main phases of training development suggested by the ISD approach, namely: analyzing pilots' skills and training needs, using psychological principles and theories to design the training, and evaluating the training program.

ANALYZING TASKS AND IDENTIFYING TRAINING NEEDS

The analysis phase is the cornerstone of training development, as suggested by the ISD approach, discussed earlier. Without good analysis it is not possible to identify precisely the training needs and the nature of the training content required to remedy them. Although there is unanimous agreement about this, there is very little concerning what techniques or methods should be used. Indeed, any student of training could be forgiven for being bewildered at the plethora of approaches and acronyms that abound in the literature. Introductions to job and task analysis techniques are given by Fleishman and Quaintance (1984), Gael (1988), Patrick (1991), and Kirwan and Ainsworth (1992). More recently, in 1998, a special issue of *Ergonomics* was devoted to various forms of hierarchical task analysis, including their development and application. In the same year, a practically oriented review of cognitive task analysis techniques by Seamster, Redding, and Kaempf (1998) discussed their application to aviation.

Before considering the different types of analysis that may be required from a training perspective, it is important to consider briefly some nomenclature and definitions in this area. First, the terms *job analysis* or *task analysis* are often used interchangeably, although generally a job is conceived as comprising a series of tasks and subtasks. In the IPISD model a task is defined as "the lowest level of behaviour in a job that describes the performance of a meaningful function in the job under consideration" (p. 171). A task can be represented as a goal that a person is trying to achieve through the performance of various physical and/or mental activities. Tasks can vary in their level of description and therefore in the range of activities that they embrace. So a wide-ranging task is to "fly an aircraft" whereas a lower level task/subtask is to "read the altitude." Task analysis is the term, particularly used in training development, for specifying the knowledge, skills, and attitudes required by the task performer in order to execute the task efficiently. In the last decade, the term *task analysis* has been prefaced by the word *cognitive* to further emphasise that it is important to identify the cognitions of the performer, including cognitive strategies, knowledge, information processing demands, and so forth. A task should therefore not be viewed as an entity that exists independent of the person who has to perform it. There is a complex interplay between the activities involved in the task, how a person attempts to perform these, and the information processing demands that they make. The same task can therefore be difficult for a novice but easy for a well-trained person who brings to bear appropriate knowledge and information-processing activities.

An important distinction that can be drawn between two types of analysis, both of which help to identify training needs and training content. The first type of analysis involves what might be described as a conventional form of job/task/cognitive analysis whereby an area of work or a job is broken down into an array of tasks, subtasks, and so forth. This type of analysis adopts a systems perspective and describes the tasks that have to be accomplished in a logical fashion. Very often these tasks are described in terms of goals. Some examples of this type of analysis are described next. The second type of analysis has as its starting point not an array of tasks but rather errors in task performance. These errors can be used by the trainer to identify where training can be profitably directed. This can be done at an intraindividual level and also at an interindividual level whereby errors can be aggregated so that frequent errors or misconceptions can be identified. Hence a feedback loop is established between evaluations of performance and the development or modification of training to remediate weaknesses. This iterative process of diagnosis of performance difficulties and provision of corresponding training is fundamental to instruction, and good instructors and teachers will have developed various strategies for achieving this (e.g., Collins, 1985). This second approach to analysis can take place in various ways. Accidents, minor incidents, and near misses can be analyzed, and they provide a rich source of information for subsequent training. Experimental data from performance in simulated situations is also revealing, and this may be supplemented by qualitative analyses of concurrent or retrospective verbal reports from the performer. The latter "process tracing" approach is useful when decision making or problem solving has to be analyzed in detail. Examples of this type of analysis are discussed below.

Traditional Task Analysis

Marti (1998) describes a form of hierarchical task analysis that is used to identify tasks of the tower controller in air traffic control that are generic between four airports (Fig. 11.3). This analysis provides an overview of the main functions of the tower controller.

The overall task of "guiding/controlling landing and takeoffs" is broken down into six subordinate tasks. The first subtask involves both visual monitoring of the airport plus, more importantly, developing a mental representation of the actual traffic situation. The second subtask of "planning traffic management" is broken down into an array of subtasks. This sort of analysis is very similar to Hierarchical Task Analysis (HTA) developed by Annett and Duncan (1967) and represents a traditional approach to task analysis whereby tasks are decomposed hierarchically, and their sequence of execution at any one level of description is specified where possible. From this task/subtask framework, it is possible to elaborate training content by specifying what has to be learned in order to carry out not only each task/subtask successfully but also all of the tasks in a planned and coherent sequence.

Another similar form of task analysis is outlined by Seamster, Redding, and Kaempf (1998) and labeled a "consistent component method." They use this method to analyze "performing a takeoff with engine failure at or after critical engine failure velocity." The task analysis technique involves the following six steps:

1. Verify subtask decomposition
2. Identify high-level skills for each subtask
3. Determine decision points for experts
4. Determine decision points for novices
5. Determine task performance difficulties for novices under high workload
6. Identify consistent categories of information. (p. 128)

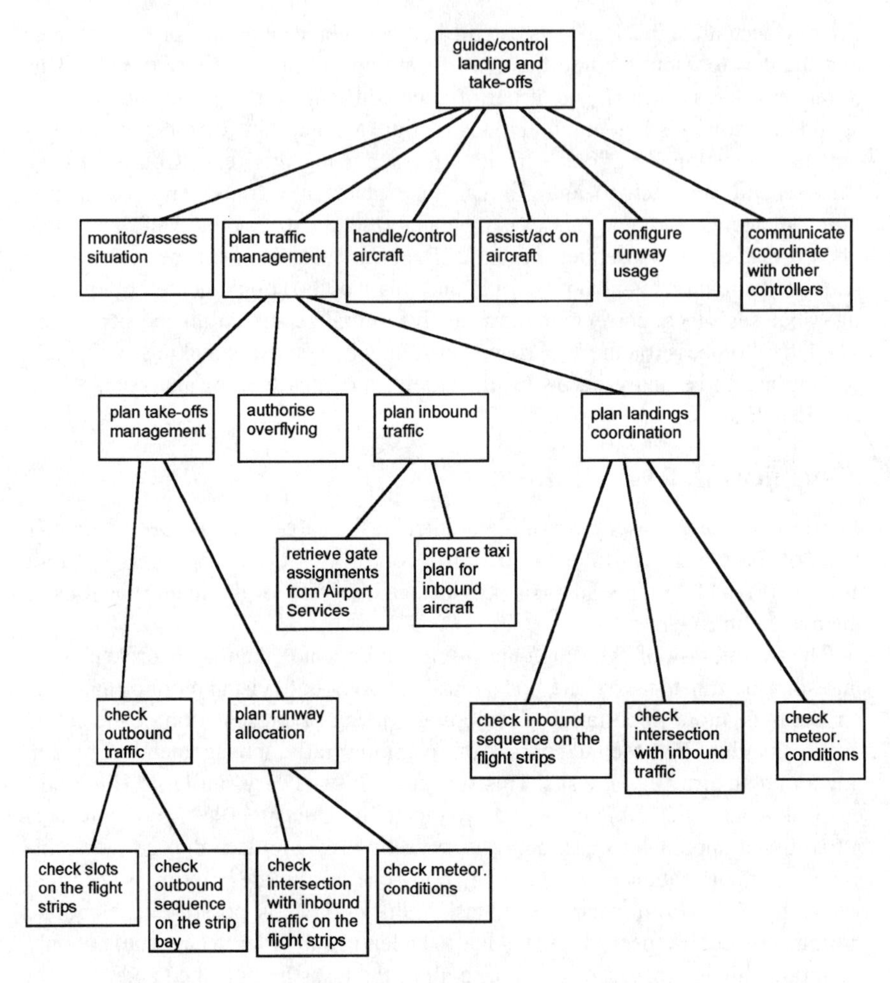

FIG. 11.3. Task analysis of a tower controller's job. *Note.* From "Structured Task Analysis in Complex Domains," by P. Marti, 1998, *Ergonomics, 41,* pp. 1664–1677. Copyright 1998 by Taylor & Francis Ltd. (http://www.tandf.co.uk/journals). Reprinted with permission.

The task was then broken down into seven subtasks, and the analyst and subject-matter expert then specified the skills involved in each subtask (see Table 11.1). Subsequently, each of the seven subtasks was then examined to identify expert and novice decision points. Finally, the two most difficult subtasks were further analyzed to identify their requirements. For example, "call gear up" had three consistent requirements for pilots, namely: a positive rate of climb, a stable airspeed and the allowance for a 5-second airspeed indicator lag.

TABLE 11.1
Skill components of performing a takeoff with engine failure.

Element	*Skills*
Recognize engine failure	Sensing a yaw. Hearing a change in sound or possible alarm. Determining which engine has failed.
Maintain directional control	Sensing change in heading. Determining correct rudder input. Implementing correct rudder input.
Achieve positive rate of climb	Determining appropriate attitude. Establishing appropriate attitude. Verifying that you have a positive rate.
Call "gear up"	Determining that pitch attitude will maintain positive rate. Verifying that you can maintain airspeed. Determining that the gear need to be raised.
Establish appropriate airspeed	Determining appropriate airspeed. Determining required changes to pitch attitude. Manipulating controls to new pitch attitude. Tuning until appropriate airspeed established.
Verify appropriate heading	Determining if heading change is required. Manipulating aileron and rudder to achieve heading change. Estimating displacement from desired track.
Determine if immediate shutdown is required	Determining if there is decreasing revolutions per minute (RPM). Monitoring for torque decrease. Monitoring for indications of fire.

From *Applied Cognitive Task Analysis in Aviation* (p. 132), by T. L. Seamster, R. E. Redding, and G. F. Kaempf, 1998, Aldershot, England: Ashgate Publishing Limited. Copyright 1998 by Ashgate Publishing Limited. Reprinted with permission.

These examples emphasize how important some form of task analysis is at the beginning of any training development. There is some variability between methods, and some do not provide a complete prescription of how the analyst should proceed, in which case some careful innovation is required. Nevertheless, most methods, as with the two illustrated above, begin with an identification of the tasks that is followed by their further subdivision into subtasks, and then those at the lower levels are elaborated in terms of their knowledge, skill, and attitude requirements.

Analysis of Incidents and Errors

Besides identifying tasks through traditional task analysis in order to specify training content, it is useful to analyze incidents, near misses, and the like to identify performance errors. These performance errors, particularly when they have been categorized, aggregated, and rated in terms of their importance, can then become the focus of remedial training. Analyses of performance errors vary in terms of the range of people sampled, the scope of the activities of interest, the methodologies employed, and whether performance is in a simulator or an actual aircraft. For example, O'Hare and Chalmers (1999) reported two surveys of pilots in New Zealand, one of which asked respondents the frequency with which they had experienced various hazardous events. Aggregation of such quantitative data and its decomposition by type of pilot, aircraft, and so forth begins to paint a high-level picture of general training priorities.

In contrast, a detailed qualitative analysis of the 1991 Gottröra aircraft accident in Sweden is reported by Martensson (1995). This examined how three pilots (captain, assisting captain, and copilot) coped with losing two engines immediately after takeoff and succeeded in crash landing the plane with 129 persons on board without fatalities. Martensson interviewed the pilots and used the findings from the Board of Accident Investigation together with a transcript from the cockpit voice recorder. This enabled a rich, qualitative description of the accident to be compiled together with identification of the various active and latent errors, after Reason (1990). In such qualitative analyses, it is common to utilize an error taxonomy to categorize and aggregate causal factors involved in an accident. Shappell and Wiegmann (1997) report the use of the Taxonomy of Unsafe Operations, which attempts to differentiate between the effects of supervisory practices, conditions of the human operator, and actions in their contribution to an accident. This is applied to AIA Flight 808, which crashed at Guantanamo Bay in August 1993. Although there were various factors contributing to the loss of this flight, inadequate training played a significant role. At the general level there was insufficient crew resource management training, which could have improved coordination between the flight crew. At a more specific level, the crew needed further training and guidance about the operation, and potential hazards, of landing at Guantanamo Bay. If this had been available, then they may have attempted an easier landing on an alternative runway.

Retrospective or concurrent verbal protocols from those involved in accidents can be used together with their actions in order to piece together individual and group activities and where errors occurred. This "process tracing" approach has been used a great deal by those trying to analyze cognitive processes involved in decision making and problem solving in applied settings. The process tracing can be used to identify not only the strategies of experts and experienced personnel but also how a pilot's actual decision making deviates from some "ideal" or "expert" reasoning path. One example of this type of approach to analysis is the Critical Decision Method developed by Klein, Calderwood, & MacGregor (1989) from Flanagan's (1954) Critical Incident Technique. The method uses a retrospective interview with a set of probes in order to identify the nature of the decision making in an accident or nonroutine situation. O'Hare, Wiggins, Williams, & Wong (1998) describe the use and modification of the Critical Decision Method to analyze the critical cues used by general aviation pilots to make decisions in poor weather conditions. The revised probes include the person's goals at various decision points, cues that are critical to the decision, available information at different times, alternative decisions and outcomes, and reliability and importance of different pieces of information (see also chap. 6).

Human performance data, relevant to training provision, can be gleaned not only from analysis of actual accidents and near misses but also from experimental studies of performance in simulated situations. Some such studies are undertaken after analysis of accident data in order to scrutinize and better understand a type of error so that appropriate training can be devised. For example, Chou, Madhavan, & Funk (1996) found that cockpit task management errors were involved in 23% and 49% of aircraft accident and incident reports that were reviewed, and a subsequent simulator study was set up to examine performance at some aspects of cockpit task management. The obvious advantage of experiments involving simulators is that a high level of experimental control can be achieved and comparisons can be made between various participants who experience an identical situation. This approach has been adopted to further understand some of the human performance problems associated with an increasingly automated flight system, as discussed earlier in this chapter. One negative effect of automation has been labeled "automation bias," which is a "term describing errors made when human operators use automated cues as a heuristic replacement for vigilant information seeking and processing" (Mosier, Skitka, Heer, & Burdick, 1998, pp. 50–51). Pilots become accustomed to depending on automated systems, which are reliable most of the time, for the annunciation of a problem, and as a consequence their usual information-gathering strategies are not used. Therefore, when an automated system fails to alert a pilot, an omission error may occur with appropriate action not being taken. Mosier et al. (1998) investigated this phenomenon using 25 commercial glass-cockpit pilots who flew using a part-task flight simulator. Four automation failures were injected into these flights, and the overall omission error rate was surprisingly high at about 55%. The altitude and heading failures, both of which were critical to operational safety, were undetected by nearly a half of the sample of pilots. Interestingly those pilots who reported a high sense of accountability for their actions, verified the functioning of the automated systems more frequently and therefore committed fewer errors than others.

A similar simulator study was carried out by Beringer and Harris (1999), who examined pilots' responses to various autopilot responses. The reaction times to detect some subtle automation failures were longer than the "reasonable period of time" expected by the Federal Aviation Regulations, and some of these failures were misdiagnosed with consequent jeopardy to "altitude loss, overstress of the airframe, disorientation of the pilot, or destruction of the aircraft" (Beringer & Harris, 1999, p. 155). Both this and the Mosier et al. study diagnose pilot performance weaknesses that should be the focus of subsequent training. Both simulator studies were designed to focus on human performance problems that were implicated in various aircraft accidents, and they provide a powerful platform for subjecting these problems to experimental scrutiny. The focus of such simulator studies can vary in the scope of activities of concern. At one extreme, both the flight crew interactions and individual responses during a complete flight may be of interest, whereas at the other extreme it may be an individual pilot's visual scanning strategies. An example of the latter type of study is reported by Bellenkes, Wickens, & Kramer (1997), who compared experts and novices in the number of times different instruments were scanned and for how long. The differences between novices and experts suggest that it may be possible to improve the development of scanning strategies through the introduction of training that focuses directly on the use of expert strategies.

Training Objectives

The preceding discussion has covered how the focus of a training program can be identified from either a task analysis or an analysis of performance deficiencies in real or simulated situations. The focus of a training program has to be both specific and exact so that the training achieves what is intended. To accomplish this, training objectives have to be established. Any person who has attended a course on training development will inevitably be familiar with this exhortation. Training objectives are very important, as can be seen from our previous discussion concerning ISD models. These objectives should provide a coherence and consistency between various instructional development activities. This perspective is highlighted by the Instructional Quality Profile developed by Merrill, Reigeluth, & Faust (1979). Training objectives should be consistent not only with the training needs but also with the tests administered after training that evaluate whether trainees have acquired the necessary skills (Fig. 11.4). Also, each part of the training program should be capable of being directly linked with a training objective, otherwise it is unnecessary. From this perspective, it is evident that training objectives have a key role in training development. Unfortunately, some training courses are devised without clear objectives, and consequently the training content is inappropriate for the needs being tackled.

The next question therefore concerns how training objectives should be specified. Although many have written on this topic, the seminal view remains that espoused by Mager (1962). In his classic account, he argued that the main goal of training objectives is to communicate in an unambiguous fashion. For example, what does "to know" or "to understand" imply in terms of human performance? Although these terms are used frequently, they suffer from ambiguity. Such ambiguity can be resolved, argued Mager, by specifying what evidence is necessary to demonstrate that a trainee has indeed mastered an objective. To collect evidence, it is necessary to make explicit three aspects of an objective: the type of activity or behavior in which the trainee should engage, the conditions in which performance is required, and the standards or levels of performance (e.g., levels of speed or accuracy). Slight changes to any of these elements will affect the nature of the skill being taught and therefore the training programme required. It is important to specify exactly what the trainee should be able to do after training. If a task analysis has been carried out, then the task statements will begin to clarify the

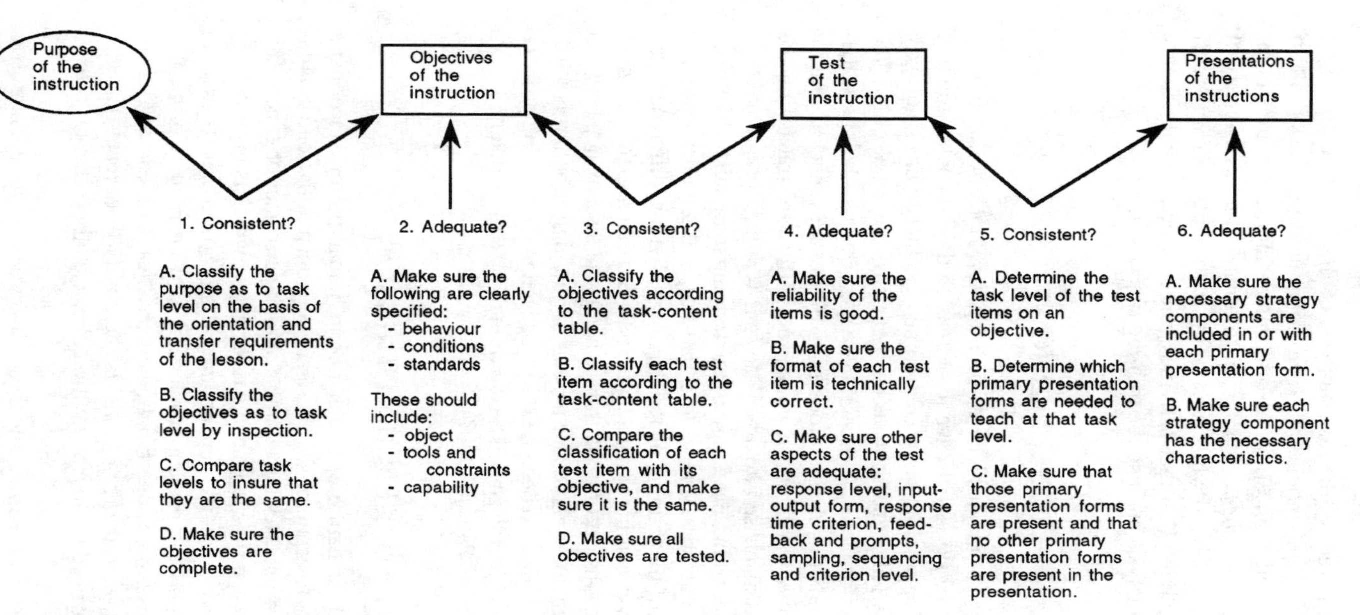

FIG. 11.4. A summary of aspects of instructional design covered by the Instructional Quality Profile. *Note.* From *Procedures for Instructional Systems Development* (p. 186), by M. D. Merrill, C. M. Reigeluth, and G. W. Faust, 1979, Orlando, FL: Academic Press. Copyright 1979 by Academic Press. Reprinted with permission.

nature of the behavior required. In the task analysis of the air traffic controller (Fig. 11.3), there are planning, monitoring, and communication tasks, all of which require further detailed explication. Also in this example there are a number of tasks that use the verb *to check,* which is unfortunate, as it has a variety of meanings. In some cases, it is worthwhile developing a vocabulary of verbs that have defined technical activities, so that activities can be specified consistently within not only the task analysis but also the training objectives.

TRAINING DESIGN

Design of training is the next major phase of training development that follows analysis of the tasks and the skills that have to be mastered. The goal of training design is to engineer an optimal learning environment so that trainees not only learn the appropriate knowledge and skills as easily as possible but also translate these into effective job performance. Although there are many psychological theories and principles to guide this design process, it still involves much ingenuity and creativity by the training designer. It is not straightforward translating the output from the task analysis phase into a well-designed training program. Decisions have to be made concerning the nature of the training materials, which may range from a text-based induction course to a part-task skills trainer involving simulation; the sequence of the training materials; how individual differences can best be accommodated by providing some adaptivity in the training; the design of adequate practice facilities that inform the trainee about the discrepancy between actual and desired performance; and ensuring that the training program is sufficiently motivating for the trainee to become fully engaged in the learning process. These training design issues are discussed next although for a more detailed treatment, the reader is referred to Patrick (1992).

Development of Expertise and Transfer

It is useful to understand the nature of the change in human performance that a training program attempts to accomplish. There are both quantitative and qualitative changes as a novice progresses to become a skilled performer or expert. Initial performance at any new task is ragged and error prone, which is in contrast to the coordinated and largely error-free performance of an expert. Experts excel in a particular domain and tend to represent problems at a "deeper" more principled level than novices, who use more superficial characteristics (Chi, Glaser, & Farr, 1988). Associated with a high degree of skill is the ability to execute that skill automatically without the need for much attention, thus enabling other tasks to be performed in parallel.

Psychological theories that describe how these changes occur have themselves changed over the past 50 years. The theories of skill development proposed by

Fitts (1962), developed from observations of pilots, and Anderson (1982) both suggest that there are three stages of learning. However Fitts' ideas are couched in the behavioral elements of stimuli and responses whereas Anderson is more concerned with the cognitive basis of skill. Anderson (1982) proposed that as skill develops, "declarative knowledge" (i.e., facts) is gradually transformed into "procedural knowledge" (i.e., knowing *how* to do something) through the process of knowledge compilation. In the first stage of skill development, the trainee receives information about the nature of the new task and its requirements. Performance initially is crude because general rather than task-specific procedures have to be used to apply this information to perform the new task. However, gradually, through practice, cognitive procedures develop, or in Fitts' terms, associations take place between task elements, which are gradually refined and become more task and context specific. Anderson describes how performance becomes smoother and more coordinated throughout what he terms the final "tuning" stage, which is similar to Fitts' "autonomous" stage.

From a training perspective Anderson's theory suggests that training will be more effective to the extent that the skill or procedures required to perform the task can be faithfully reproduced during the training program. This is why analysis of the task/skill, discussed earlier, is so important. Therefore, general training programs are less effective than specifically targeted ones because every task, although sharing some features with other tasks, inevitably has many unique elements that are typically easily underestimated. Another consequence of both of these theories is that skill or expertise can only be developed by *doing*. Training must enable the trainee to practice in order for specific procedures for performing the task to be compiled. In later publications, Anderson (e.g., 1993) stresses the importance of providing trainees with appropriate analogies and examples that facilitate the development of these specific procedures.

During the initial stage of learning, trainees are very dependent on the trainer's instructions and feedback concerning not only how to perform but also how performance deviates from some ideal. Naturally, there are high demands on working memory at this stage, and typically trainers make the mistake of bombarding trainees with too much information, which is not consolidated in a progressive manner. Also, errors of performance at this time are "mistakes," as discussed by Reason (e.g., 1987), where the trainee performs an action that does not have the intended effect. The fault therefore lies in an incorrect plan of action. In contrast, when expertise is developed, the task makes fewer demands on the trainee's working memory and behavior becomes "automatic." The person is able to discriminate important elements of the situation, formulate some action, and execute it without much need for conscious awareness. Because the cognitive procedures supporting this performance are comprehensive and well-practiced, expert performance appears smooth and unhurried. Occasionally, however, the very fact that the skill has become routinized may result in the sort of error that Reason labels a "slip." In this situation, the performer executes some action that was not intended, possibly due to a failure to pay sufficient attention to its execution. Slips can be characterized as "skilled" errors that are associated with "automatic" behavior and they are usually quickly detected by the performer. Therefore, as skill develops, the multitude of mistakes associated with initial performance gradually disappear and speed and accuracy increase although, when high levels of skill are attained, occasional slips occur.

Transfer Principles

Another topic in psychology that is not only relevant to training in any context, but particularly to aviation, is that of transfer of training. This can be defined as:

> whenever the existence of a previously established habit of skill has an influence upon the acquisition, performance or relearning of another habit or skill. (Department of Employment, 1971, p. 32)

Existing skills can either facilitate (i.e., positive transfer) or hinder (i.e., negative transfer) the development and performance of new ones. Transfer is therefore critically important to many aspects of a training program. At the input side of a new training program, a trainee's existing skills may either help or hinder the mastery of new ones. During training, various part-task training exercises may be used, involving text, simulation, and computer-based training, and these need to be designed in such a way that positive transfer from one to the other and to full task performance is maximized. At the end of training, the trainee has to be able to transfer all of the new skills developed under instructional support during training to unsupported performance by the end of the training program. It is important that all of the aids and techniques used to facilitate learning do not become themselves crutches to performance without which the trainee cannot perform. Finally, and most important, transfer is important after training in hopefully mediating the successful translation of new skills from a training environment to performance within a job in its natural context.

For all these reasons, transfer is important in any training situation. However, it is particularly so in the context of pilot training because of the unacceptable costs in terms of both money and human life if novice pilots were to begin their training in actual planes. Instead, flight simulators are developed, which may operate at approximately a tenth of the cost of aircraft. However, simulation is only a cost-effective training solution if there is a certain amount of positive transfer of training from performance using a flight simulator to piloting an aircraft. The manner in which this is assessed is usually to develop what is known as a "savings" measure of transfer. This measure refers to the percentage of training time in the real situation that is "saved" by being initially trained in a simulator. Chatelier, Harvey, and Orlansky (1982) estimated that as much as 50% of the time can be saved. However, this saving is only cost effective if the "cost" of the simulator training is a lot less than actual training.

Some time ago Povenmire and Roscoe (1973) addressed this issue. They were interested in the transfer effectiveness of the GAT-1 Link trainer with respect to training to fly the Piper Cherokee airplane. They organized four groups of trainees that received 0, 3, 7, and 11 hours of training in the GAT flight simulator. The number of training hours required in the airplane decreased as the number of hours in the flight simulator increased. Positive transfer, therefore, took place, and some training hours in the airplane were "saved." However, they calculated that the savings of the 7- and 11-hour groups were not cost effective given the relative costs of the two training regimes and performance savings. There will be a cutoff point at which transfer from simulator training will no longer be cost effective, although this will depend on the costs and savings in each particular training situation. Roscoe (1971) hypothesized that there will be diminishing returns in transfer as the amount of flight simulator training increases, resulting in the sort of theoretical relationship proposed in Fig. 11.5. Thus, successive hours of simulator training will result in less positive transfer than preceding ones.

Given the potential importance of simulator training in aviation, the next question to be addressed concerns how simulators should be designed to maximize positive transfer. In order to answer this, it is necessary to understand the mechanisms that determine the nature of transfer.

Basic theoretical notions of transfer have remained largely unchanged since Thorndike and Woodworth (1901), who proposed that transfer was determined by the extent to which two skills shared what they termed "identical elements." As

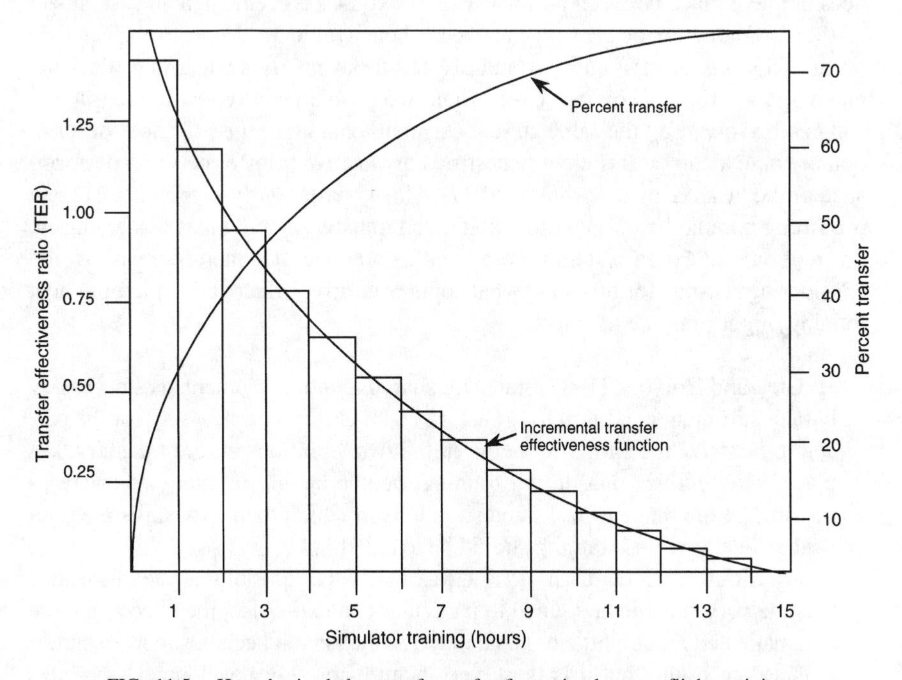

FIG. 11.5. Hypothesized degree of transfer from simulator to flight training. *Note.* From "Incremental Transfer Effectiveness," by S. N. Roscoe, 1971, *Human Factors, 13,* p. 563. Copyright 1979 by the Human Factors and Ergonomics Society. Reprinted with permission.

Thorndike and Woodworth were in the behaviorist tradition, they conceived of these elements in terms of stimuli and responses. Therefore, they proposed that the degree of commonality between any two tasks or skills determines the amount of transfer. However, as Osgood (1949) pointed out, greater commonality between tasks does not always result in *positive* transfer. For example, if a person has to produce different responses to the same stimuli, as for example when moving a joystick to the left rather than right, then interference or negative transfer will occur. Also, as the cognitive revolution took place in psychology, from the 1970s onward, so the "identical elements" that determined transfer became interpreted more in terms of covert cognitive elements. The classic account of this has been provided by Anderson (1987, 1993), who defined these cognitive elements as the rules that underpin procedural knowledge (i.e., knowing *how* to do something whether it is a motor or cognitive skill). These rules, called productions, are in the form of if-then rules that can be assembled into programs and can be shown to be capable of mimicking the performance of humans at many simple cognitive tasks. Therefore, the notion that it is the number of "identical elements" between skills that determines transfer has almost unanimous support in psychology, even though the manner in which these elements are defined has changed.

There is a surprising amount of research demonstrating that positive transfer does not take place between performance of two tasks, even when an assessment of the commonality of elements between them would lead one to expect it to occur. This has been found particularly for tasks involving learning, decision making, and problem solving. Even when two problems have a similar structure and can be solved by the same strategy, a small change in the "surface" description or context can be sufficient to destroy any positive transfer between performance of the tasks (Hayes & Simon, 1977). This is particularly surprising, as there is a strong popular belief in society that much transfer of skill takes place. Indeed, much of our education system is based on this premise. It is therefore worthwhile considering reasons for this somewhat counterintuitive effect and explaining how training might mitigate its effect.

1. Gick and Holyoak (1987) state that shared identical elements is a necessary but not sufficient condition for transfer to take place. It is necessary for the person to *perceive* the similarity between the two situations before transfer takes place. This implies that during training, people should be made as aware as possible of the range of tasks and situations in which their new skills are relevant. So-called blind training should be avoided.
2. From our previous discussion, it was noted that as skills are learned, they become very specific and tuned to particular contexts. Also, they become more automatic and require little or no attention as the person becomes more expert. It is therefore ironic that it is these very features of skill development that also make positive transfer less likely due to skill specificity and contextualization. Arguably, these features are an adaptive means by which humans avoid an unbearable workload in everyday situations. There is no training solution to this.

3. Some transfer does not take place because training has not embraced the full range of transfer situations and skill requirements that the person will subsequently encounter. The remedy for this is straightforward and involves carrying out a better task analysis before training takes place so that a comprehensive set of training objectives are specified that incorporate all of the necessary transfer requirements. If the transfer situations require a range of different but overlapping skills, then these need to be practiced during training.
4. Finally, transfer of skill from training to the on-the-job situation may break down because the skill is too fragile and has been insufficiently learned or overlearned to withstand the distractions and stress involved in job performance. This can be remedied by more training and also the development of a training regime that toward the end of training attempts to mimic the actual job situation with all of the elements that may reduce performance levels. One element that is notoriously difficult to recreate satisfactorily is stress involved in the job as, for example, associated with flying an aircraft or engaging in combat. One attempt to counter this problem in some military training involves the use of live ammunition, sometimes with fatal consequences.

Design of Simulations

Central to pilot training is the use of simulations for the reason, mentioned earlier, that this represents a substantial reduction in cost, both financial and in potential injury or loss of human life. Sometimes the use of simulations represents the only training solution because the operational situation is not available, as when astronauts have to practice a lunar landing.

As simulations involve training off-the-job, there are other advantages to using them for training purposes. First, it is possible to break down the complex task of flying into parts, each of which may require different skills, enabling part-task simulators to be developed to train each one. Second, the instructor or trainer is able to intervene and provide more advice than may be possible in the operational situation. As we will discuss below, the provision of opportunities for practice that are associated with feedback to the trainee are vital for the promotion of learning. Finally, using a simulation may enable the trainer to focus on particular aspects of the task that are problematic. For example, emergency procedures can be practiced and unusual, potentially dangerous, flight situations can be programmed to occur in simulations more frequently than would occur in actual flight.

Before discussing the role of transfer in the design of simulations, it is necessary to define what is meant by simulation. The term *simulation* has been used deliberately in preference to simulator in order to encompass situations that do not rely on equipment but involve, for example, role play. Gagné (1962) specified that a simulation has three features. It attempts to *represent* the real situation in which actions are performed; it provides the user with some *control* over the situation; and, by definition, some aspects of the real situation are *omitted.* All of these features have important ramifications for how simulations should be designed for training purposes. The sixty-four-thousand-dollar question is: What aspects of the real situation can be omitted while still preserving the integrity of the simulation such that positive transfer will flow from the training?

In order to begin to answer this question, it is necessary to introduce the concept of the "fidelity" of the simulation. The degree to which a simulation represents the real situation is termed its fidelity (see also chap. 12). However, interpretation of the term *fidelity* varies in the literature. Originally Miller (1954) thought of fidelity in engineering terms. Hence, the more the simulation duplicates the appearance and operation of the real equipment, the greater its fidelity and the greater the amount of positive transfer of training that can be expected. This view was challenged when it became evident that high physical fidelity was unnecessary for the training of procedures and cognitively oriented tasks involving decision making where cheap paper-and-pencil mock-ups were sufficient to represent the important elements of the situation. For example, Prophet and Boyd (1970) found that pilots trained on the start-up and shutdown procedures for a Mohawk aircraft using a wood and photograph mock-up were just as good as those trained in the actual aircraft. Therefore, rather than striving for a simulation with high physical fidelity, the argument was made that the critical factor determining its transfer value was its psychological fidelity, a view reiterated by Lintern, Sheppard, Parker, Yates, and Nolan (1989). The simulation has to represent the important psychological demands that the task makes of the pilot in terms of perception, attention, decision making, memory, and action. For some tasks, high physical fidelity may still be necessary, but for others this may not be the case. In order to specify the requirements for a simulation, the designer has to rely on the information gleaned from a full task analysis that identifies the psychologically salient aspects of the task. In this way, simulations with high psychological fidelity can be designed that have high transfer of training.

Unfortunately although this argument is theoretically sound, it is not as easy to implement as it might appear. The reason is that psychological knowledge is not as specific and clear-cut as one would like, and task analysis is itself not an exact science. Consequently, for some tasks the only means of determining with certainty what has to be simulated in order to maximize transfer effectiveness is to carry out empirical research studies that compare transfer outcomes from training under different simulated conditions. Collyer and Chambers (1979) embarked on the development of a full-scale comprehensive flight simulator with the aim of determining the importance of various parameters in the visual system to transfer of training. By carrying out empirical studies using the full-scale simulator, it was hoped to identify what parameters of the visual scene did not have to be realistically simulated so that smaller, less-realistic, and lower cost simulators could be designed and used without jeopardizing positive transfer.

This issue is still central to much recent research in aviation. A continuing debate over the last few decades is whether flight simulators should have motion cues (e.g., Burki-Cohen, Soja, & Longridge, 1998; see also chap. 12). This is important because of the considerable cost involved in comparison with a fixed-base simulator.

Transfer of training studies have found generally little adverse effect on transfer when motion cues were eliminated (e.g., Pohlman & Reed, 1978). However Caro (1979) points out that it rather depends on the nature of both the task and the motion cues available. He makes a distinction between *maneuver* and *disturbance* motion cues. Although maneuver motion, measured in most transfer of training studies, may provide redundant information given instrument readings, disturbance motion can alert pilots for the need to intervene, for example, in engine failure. Therefore, it is often dangerous to attempt to make generalizations in the transfer literature as the nature of the task, and how it is assessed may lead to different findings.

Another factor which, if it has to be simulated with high physical fidelity, has severe cost implications is the nature of the visual scene. Fidelity requirements for the Kiowa Warrior Crew Trainer were investigated to determine the minimum fidelity required in terms of visual display resolution and field of view to train successfully both gunnery and nongunnery tasks (Stewart, Cross, & Wright, 1998). Lintern, Roscoe, and Sivier (1990) found better transfer of training in terms of landing skills when pictorial rather than symbolic displays were used. This finding was confirmed by Lintern and Garrison (1992), who also found that the level of scene detail was unimportant. Similarly, Williams, Hutchinson, and Wickens (1996) found scene detail was not critical in navigation training.

These empirical research studies into transfer of training are important in informing trainers of the factors that need to be designed into simulations for different tasks. Unfortunately, this evidence typically lags behind the introduction of new aviation technologies such as weapon systems. Bell and Waag (1998) review the literature concerning the effectiveness of flight simulators for training combat skills. They bemoan the paucity of transfer of training studies in both air-to-surface and air-to-air combat tasks and the quality of some of the evaluation data, particularly when opinion surveys are the primary source of evidence. Bell and Waag also note that studies investigating air-to-surface weapons delivery involve manual delivery, and it is not clear how computer-aided delivery may change these findings.

Inevitably, computers are required to drive the complex simulations that have been discussed. With the development of powerful personal computers, the question arises as to how much they can be used to reduce training cost and provide greater flexibility of access. Koonce and Bramble (1998) consider the transfer of skills from personal computer-based aviation training devices (PCATDs). The same principles will determine transfer effectiveness, namely the degree of *psychological fidelity* provided by the PCATD for important dimensions of the task. However, Koonce and Bramble (1998) argue that the Federal Aviation Administration requires a PC hardware configuration that goes beyond that needed from a psychological perspective for some tasks. There is evidence that PCATDs can provide effective training and transfer for a range of tasks. Taylor et al. (1997) investigated the transfer savings from a PCATD for instrument flight tasks. They found a saving of between 15 to 40% of aircraft time that would have been needed without the use of the PCATD. This was cost effective given that the ratio in cost of aircraft time to that on the PCATD was approximately 14:1. This supports the finding of Pfeiffer, Horey, and Butrimas (1991). More recently, Jentsch and Bowers (1998) have argued that PCATDs can be used for training aircrew coordination. By now the reader will realize that it is not possible to judge the validity or effectiveness of training technologies per se, as this depends very much on *how* they are used rather than *what* they are.

Design of Practice

Simulations enable trainees to practice a task, and this is accompanied by appropriate instructional support from the trainer or computer. Practice can target both individual and team skills and is one of the most powerful training variables, as long as the context in which it takes place is well designed. From a human performance perspective, there are two aspects of practice that are important but difficult to disentangle, namely, cognitive and motivational. Cognitive issues include getting the trainee to understand what is required and then encouraging development of the necessary procedural knowledge by sequencing and structuring the practice sessions in the optimal manner. Typically, trainers assume that if practice is well designed from a cognitive perspective, then it follows that the trainee will be motivated. This assumption may not be correct, and motivation during a training program deserves consideration in its own right. Until recently, the area of motivation has been the neglected area of training research. Research has found that there are various individual and organizational factors that interact to determine a person's motivation to being trained or retrained, and these need to be considered in the design of practice (e.g., Tannenbaum & Yukl, 1992).

Cognitive Issues. Three main questions will be discussed. The first concerns what information the trainee should be given prior to any training or practice. The second question addresses the nature of the instructional support that should accompany practice of the task. Finally, what should be practiced? Given that pilot training involves a complex set of tasks, should these be separated or simplified during practice sessions and if so, what are the principles guiding their design?

Prior to training, it is important to provide the trainee with an overview of the scope and nature of the training that is to come. Some time ago, Ausubel advocated that a trainee should have a relevant conceptual framework available that will facilitate assimilation of subsequent information (Ausubel, 1960). He proposed that this framework be provided to the trainee by what he termed an advance organizer, which he defined as being at a "higher level of abstraction, generality and inclusiveness than the learning task." It has been found that advance organizers can be beneficial, particularly when unfamiliar conceptual material has to be mastered (Mayer, 1979), although testing their effectiveness is an area rife with methodological difficulties. The notion of an advance organizer was extended by Reigeluth and Stein (1983), who proposed, in their Elaboration Theory of training design, that training should proceed from the general to the specific, beginning with an "epitome" (or advance organizer), which is elaborated progressively into various types and levels of detail. They provide the analogy of a "zoom-lens" camera where:

> the student starts with a wide-angle view of the subject matter and proceeds to zoom in for more detail on each part of that wide-angle view, zooming back out for context and synthesis. (p. 217)

This theoretical notion emphasises that it is not only useful to provide an overview at the beginning of practice, but provision of types and levels of overview throughout practice can also have cognitive benefits for the trainee.

The sine qua non of learning is the provision of extra feedback to the trainee concerning the adequacy of performance during practice. As Bartlett famously said: "It is not practice but practice the results of which are known that makes perfect." This is important for the development of not only perceptual-motor skills (e.g., stick control) but also cognitive and crew coordination skills. Formally, this extra information is known as "extrinsic feedback" and is an addition to the "intrinsic feedback" that is available from normal performance of the task. Extrinsic feedback can be provided via any of the trainee's sense organs and can take many forms. It might, for example, involve verbal advice from the instructor during on-the-job or simulator training, debriefing from a video recording of task performance, or an additional on-screen visual cue indicating cumulative accuracy in a gunnery task. It is important that such feedback is available to the trainee *before* the next attempt at the task, as it should be used by the trainee to determine the discrepancy between current and desired performance and thus how future action should be changed. The important features and dangers of extrinsic feedback are well summarized in the following quotation from Wheaton, Rose, Fingerman, Karotkin, and Holding (1976):

> Effectiveness is greatest when the information is clearly and simply related to the action performed. Any distortion or equivocation in the information fed back to the trainee will reduce its effectiveness.
>
> Unduly full or complex information may be partly ignored or may confuse the trainee.
>
> The information given should indicate the *discrepancy* between what is required and what has been achieved, rather than merely give a reminder of requirements or some broad measure of achievement.
>
> The trainee must have *some* cues to the results of his actions if he is to perform accurately at all and training procedures will be effective insofar as they help him to observe and use such cues as are inherent in the task for which he is being trained. They will fail insofar as they provide him with extra cues on which he comes to rely but which are not available when he changes from training to the actual job. (p. 78. The word *trainee* has been substituted for *S* in this quotation.)

It is straightforward to provide such feedback that complies with these recommendations for simple tasks, and the evidence is that learning will be substantial. However, for complex, particularly cognitive tasks, in which the trainee engages in reasoning or problem solving, there are a variety of strategies that may guide task performance. In this case, it is much more of a challenge to both identify these strategies and provide the trainee with appropriate feedback in a timely fashion. Without some form of task analysis carried out beforehand, provision of such feedback is problematic. (An example of how a complete task can be analyzed and feedback provided with respect to its constituent subskills is given in a study by Frederiksen & White (1989), discussed below). In the last paragraph of the preceding quotation, Wheaton et al. pinpoint some of the dilemmas of providing such feedback. It is important that although this extra information during training supports trainee performance, it must not become a crutch, such that without it, performance returns to pretraining level. One solution to this is to gradually reduce its availability (e.g., by providing it on fewer practice sessions) so that its sudden withdrawal at the end of training is avoided (Winstein & Schmidt, 1990).

Finally, let us consider two issues that concern *what* is practiced during training. First, training should ideally be adaptive to differences between trainees. Rather than practicing the same task throughout training, it is sensible if more time is spent mastering those parts that the trainee finds difficult. Thus, in adaptive training, the nature of the task being practiced will vary according to the accuracy and speed of learning. Adaptive training has been developed successfully when the task can be broken down into a series of simple discrete elements such as in learning keyboard skills or simple perceptual discriminations (e.g., learning the names of symbols). However, adaptive training has had a more checkered history in the context of learning perceptual-motor control where a continuously adjustive response is required to a changing target. In an insightful review, Lintern and Gopher (1978) concluded that there was little compelling evidence of its positive effect. The basic problem is determining on what basis the task is adapted so that positive transfer occurs, because even slight changes in tracking tasks can result in negative transfer.

The second issue, not unconnected with the preceding discussion, concerns whether to practice the whole task or parts of the task and, in the latter instance, how these parts are defined. Inevitably, the task of flying an aircraft involves part-task training on procedures, navigation, and stick control, each of which is practiced separately before being assembled into whole-task training. However, the question can be reiterated, at a lower level of description, as to whether there is any advantage in further breaking down these main tasks of flying into parts and, if so, how. A task can be divided into parts according to the principle of partitioning, in which the nature of the parts is preserved, or simplification, in which some aspect is transformed in order to make it easier for the trainee to master. The difficulty with either approach is being able to predict whether positive transfer will ensue when the parts are reassembled. The main, but somewhat old, principles that can guide the training developer are those espoused by Naylor and Briggs (1963). Essentially, they warned against using part-task training when the task had high organization (i.e., many interrelated elements). When the task has low organization, it is effective to practice the parts independently, and negative transfer will not occur when they are reassembled. This seems intuitively sensible. One interpretation is that if it is possible to identify the different skills that the task requires, then these can be practiced separately. This is the rationale underlying the use of concept trainers, principle trainers, and procedure trainers that are used in military training. Each trains a different skill that requires different learning conditions.

Evidence from a study by Frederiksen and White (1989) suggests that even when the task has high organization, in Naylor and Briggs' terms, part-task training can be effective if the parts are psychologically meaningful. Frederiksen and White used a computer-controlled space fortress game developed by Mané and Donchin (1989), which is a typical combat strategy game in which various items have to be destroyed while avoiding one's ship being hit. Despite the highly integrated nature of the task, various subskills were identified, including the perceptual motor skills of firing, aiming, and ship control and strategies for playing the game successfully. Practice sessions were devised to teach each of these different subskills. In two studies it was found that part-task training devised in this manner was superior to whole-task training. Part-task training may be effective in performance terms, but in the harsh reality of the training department, it is also necessary to calculate that it is cost effective given that extra time and resources may be required to devise and implement it.

Motivation Issues. To ensure training effectiveness, a training program not only has to provide appropriate practice and instructional support but also has to ensure that trainees are suitably motivated to learn. There is an interplay between cognitive and motivational factors that relate to training, and it is not possible to fully disentangle them. For example, provision of an advance organizer or overview prior to training will have both cognitive and motivational benefits. Similarly, Hesketh (1997) points out that giving trainees goals prior to training will also have cognitive and motivational effects. Knowing the goal enables attention to be directed more appropriately. Also, if it is made clear how the training course fits into the trainee's future job and career aspirations, then this will be motivating. Smith-Jentsch, Jentsch, Payne, & Salas (1996) carried out an interesting study that investigated the effects of pilots' having experienced negative events related to the goal of a training program, involving improving assertiveness in the cockpit. The negative events included life-threatening incidents caused by poor crew coordination and pressure to fly when uncomfortable about weather or mechanical problems. The results indicated that the more negative events experienced, the higher the assertiveness performance in a simulated exercise 1 week after training. This was interpreted as evidence that prior negative experiences related to the goal of training enhanced motivation to learn. The attitudes and perceptions of trainees concerning both the training program and its wider organizational context are also important determinants of motivation and, in turn, training success (Tannenbaum & Yukl, 1992).

Just as training objectives are derived and specified for any training program, Keller and Kopp (1987) argue that a similar effort should be made to develop a set of motivational objectives. One should begin with an analysis of motivational needs, from which motivational objectives can be specified. Subsequently, strategies have to be developed for meeting these objectives and maintaining or improving the motivation of trainees. The next stage is evaluation to ascertain whether these motivational objectives have been met, and if not, how the training program could be modified to meet them. These stages mirror those, discussed previously, that are used by ISD models to develop training programs. The only difference is that the focus is not on the content and design of training per se but rather on its motivational ramifications.

Two attempts have been made by Keller (1983) and Noe (1986) to synthesize from the literature principles concerning factors that determine motivation in the context of training. Noe's model is represented in Fig. 11.6. It suggests that locus of control determines whether trainees are likely to benefit from skill assessment during training. The more that trainees ascribe performance as being under their control (i.e., internal attribution), the more they are likely to accept and benefit from extrinsic feedback concerning their performance during training. Locus of control is also hypothesized to affect career planning and job involvement, which in turn affect motivation to learn. However Matthieu, Tannenbaum, and Salas (1992) found no effect of career planning or job involvement on motivation, although Noe and

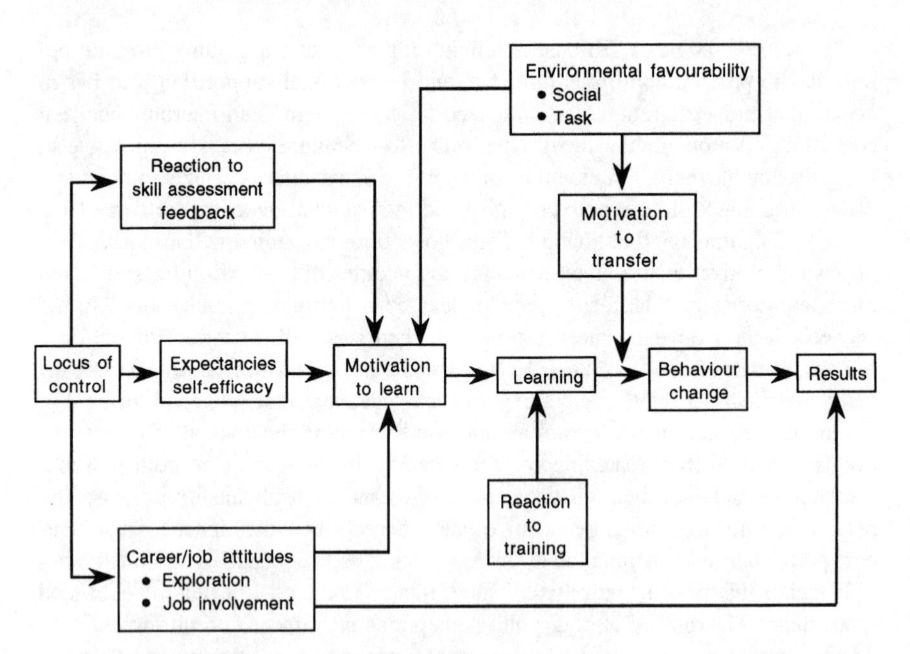

FIG. 11.6. Motivational factors affecting training effectiveness. *Note.* From "Trainees' Attributes and Attitudes: Neglected Influences on Training Effectiveness," by. R. A. Noe, 1986, *Academy of Management Review, 11,* pp. 736–749. Copyright 1998 by the Copyright Clearance Center, Inc. Reprinted with permission.

Schmitt (1986) found that those who were involved in their jobs and planning their careers benefited more from training. In Noe's model, self-efficacy also has a prominent role. A person's opinion concerning their ability to perform before, during, and after training will affect their motivation and commitment to learn. Self-efficacy will not only determine the effectiveness of training but will also be enhanced by training (Tannenbaum, Matthieu, Salas, & Cannon-Bowers, 1991). Also, if trainees believe that desirable outcomes, such as upgraded pay and job prospects, are associated with training, then there will be an increase in motivation to learn.

Keller and Kopp (1987) developed the ARCS (*A*ttention, *R*elevance, *C*onfidence and *S*atisfaction) theory of motivational design (Table 11.2). It was developed from work by Keller (1983) that identified 17 strategies for improving motivation during training derived from various theories in industrial and organizational psychology. In the ARCS model, the number of motivational strategies has been reduced to 12, and they are subsumed under four categories (Table 11.2). It is important to get and maintain the trainee's attention, establish the relevance of the training material to the trainee's goals, give the trainee

TABLE 11.2
The ARCS model of motivational design

A. Attention	*C. Confidence*
1. *Perceptual arousal.* Gain and maintain student attention by the use of novel, surprising, or uncertain events in instruction 2. *Inquiry arousal.* Stimulate information-seeking behavior by posing or having the learner generate questions or a problem to solve 3. *Variability.* Maintain student interest by varying the elements of instruction	7. *Expectancy for success.* Make learners aware of performance requirements and evaluate criteria 8. *Challenge setting.* Provide multiple achievement levels that allow learners to set personal standards of accomplishment and opportunities that allow them to experience performance success 9. *Attribution molding.* Provide feedback that supports student ability and effort as determinants of success
B. Relevance	*D. Satisfaction*
4. *Familiarity.* Use concrete language and use examples and concepts that are related to the learner's experience and values 5. *Goal orientation.* Provide statements or examples that present the objectives and utility of instruction, and either present goals for accomplishment or have the learner define them 6. *Motive matching.* Use teaching strategies that match the motive profiles of the students	10. *Natural consequences.* Provide opportunities to use newly acquired knowledge or skill in a real or simulated setting 11. *Positive consequences.* Provide feedback and reinforcements that will sustain the desired behavior 12. *Equity.* Maintain consistent standards and consequences for task accomplishment

From *Instructional Design Theories in Action: Lessons Illustrating Selected Theories* (pp. 289–320), by C. M. Reigeluth (Ed.), 1987, Mahwah, NJ: Lawrence Erlbaum Associates, Inc. Copyright 1987 by Lawrence Erlbaum Associates, Inc. Reprinted with permission.

sufficient confidence to participate in the training, and increase the trainee's satisfaction by rewarding new skills and appropriate performance. All of these suggestions are useful for the design of training programs and emphasize the importance of a trainee's motivation in skill development.

EVALUATION OF TRAINING

No chapter on training is complete without a brief discussion of the role of training evaluation. Unfortunately, there is agreement in the literature that evaluation of training is frequently neglected or poorly carried out. There are many reasons. Those carrying out training in organizations may lack both the political will or technical competence to discover the extent to which a training program they have designed meets, or rather doesn't meet, its objectives. Arguably, there are many techniques or methods in the literature that may also lead to some confusion. In addition, those with an academic interest in the effectiveness of particular training variables face a daunting task of disentangling the effect of one variable that may be dependent on factors concerning how both training content and design are specified.

Patrick (1992) proposed that there are four main approaches to the evaluation of training that differ in terms of their aims, criteria, and methods. The first, traditional, approach to evaluation has as its goal to identify whether training meets its objectives, and if not, how the program should be revised. The role of evaluation is therefore to control and regulate training, and this is integral to the ISD approach, discussed earlier in this chapter. Feedback from evaluation may result in revisions to the objectives, content, or design of training. The criteria used for this type of evaluation concern both the processes of training development and, more usually, the products of training as manifested in performance, at both individual and organizational levels. There should be a direct mapping between identification of needs, specification of objectives, and development of training content. Evaluation of the adequacy and consistency of these *processes* of training development will provide valuable insights. Kirkpatrick (1959), in his now-famous evaluation model, has specified four evaluation criteria that are used for judging the *products* of training (i.e., individual and organizational change). These concern the trainees' reactions, learning, job behavior, and organizational indices such as production quality and quantity and absenteeism. Trainees' reactions can be broken down into enjoyment, perceived usefulness, and difficulty of training (Warr & Bunce, 1995) and can be assessed through standard questionnaire and interview techniques. Although reaction data are easy to collect and trainees should have positive reactions to the training, this criterion is less important than the others. Assessment that the intended learning has taken place through a posttraining test is more important and is a prerequisite for an appropriate change in job behavior. However, identification of changes in job behavior due to training is notoriously difficult because of not only many potential confounding factors but also the need for the trainee to have the opportunity to put the newly acquired skills and knowledge to good effect.

A second approach to evaluation adopts an administrative perspective. It involves estimating the cost effectiveness or cost benefit of training. Costs include not only those involved in running and evaluating training but also those concerned with planning and development. Administrators adopt a production model for evaluating training programs and attempt to calculate whether a certain number of trainees on a course can be justified in terms of the resources required and the degree of improvement in specified knowledge and skills. Despite the various formulas that can be used for such a purpose, the difficulties lie in not only calculating the value of training to an organization but also, ultimately, in making a value judgment about whether the cost justifies the benefit.

A research-oriented approach to evaluation concerns investigating the effectiveness of training variables by using the scientific method. A hypothesis is developed, and a study is designed in as rigorous a manner as possible in order to collect and analyze data to determine whether there is support for the hypothesis. The research investigation should be internally valid, so that inferences drawn from the results are justified, and externally valid, so that the training effects observed can be generalized to relevant work contexts.

Finally evaluation may take place on an individual rather than a group basis. The purpose is to debrief a trainee about his or her strengths and weaknesses and possibly advise them about future training, career paths, and so forth. In this situation, rather than statistical analysis of quantitative data, evaluation involves a more qualitative interpretation of individual data.

CONCLUSION

This chapter has addressed three stages of training development as suggested by some of the ISD models that have been discussed. The first critical stage involves identifying training needs and analyzing tasks that need to be trained. This can be achieved by the use of traditional task analysis methods together with qualitative analyses of incidents and errors that provide a rich source of evidence for the focus of a training program. By the end of this analysis stage, it is important that training objectives are specified because they not only define relevant training content but also specify the measures to be used in order to evaluate whether the necessary skills and knowledge have been acquired through training. The second stage of training development involves designing the training program by using as many psychological theories and principles as possible to promote motivation, learning, and transfer. Central to much aviation training is the development and use of simulations in which part-skills can be practiced safely and with considerable cost savings. Theories and research findings have been reviewed that inform the designers of training how such simulations can be optimized to maximize transfer of training.

Undoubtedly, the most powerful training design variable is the opportunity to practice the necessary skills coupled with appropriate provision for advice from the trainer concerning any discrepancy between actual and desired performance. The third and final stage of training development involves evaluation, which assesses whether a training program achieves its objectives, and where it does not, necessary adjustments can be made to either the content or design of the training.

REFERENCES

Amalberti, R. (1989). Safety in flight operations. In B. Wilpert & T. Quale (Eds.), *Reliability and safety in hazardous work systems* (pp. 171–194). Hillsdale, NJ: Lawrence Erlbaum Associates.

Anderson, J. R. (1982). Acquisition of cognitive skill. *Psychological Review, 4,* 369–406.

Anderson, J. R. (1987). Skill acquisition: Compilation of weak-method problem solutions. *Psychological Review, 94,* 192–210.

Anderson, J. R. (1993). *Rules of the mind.* Hillsdale, NJ: Lawrence Erlbaum Associates.

Andrews, D. H., & Goodson, L. A. (1980). A comparative analysis of models of instructional design. *Journal of Instructional Development, 3,* 2–16.

Annett, J., & Duncan, K. D. (1967). Task analysis and training design. *Occupational Psychology, 41,* 211–221.

Ausubel, D. P. (1960). The use of advance organisers in the learning and retention of meaningful verbal material. *Journal of Educational Psychology, 51,* 267–272.

Bell, H. H., & Waag, W. L. (1998). Evaluating the effectiveness of flight simulations for training combat skills: A review. *International Journal of Aviation Psychology, 8,* 223–242.

Bellenkes, A. H., Wickens, C. D., & Kramer, A. F. (1997). Visual scanning and pilot expertise: The role of attentional flexibility and mental model development. *Aviation, Space, and Environmental Medicine, 68,* 569–579.

Beringer, D. B., & Harris, H. C., Jr. (1999). Automation in general aviation: Two studies of pilot responses to autopilot malfunctions. *International Journal of Aviation Psychology, 9,* 155–174.

Branson, R. K., Rayner, G. T., Cox, L., Furman, J. P., King, F. J., & Hannum, W. H. (1975). *Interservice procedures for instructional systems development: Executive summary and model.* Tallahassee: Center for Educational Technology, Florida State University. Distributed by Defense Technical Information Center, Alexandria, VA.

Breda, L. Van, & Veltman, H. A. (1998). Perspective information in the cockpit as a target acquisition aid. *Journal of Experimental Psychology: Applied, 4,* 55–68.

Bunderson, C. V. (1977). *Analysis of needs and goals for author training and production management systems* (Technical Report 1, MDA-903-76-C-0216). San Diego, CA: Courseware Inc.

Burki-Cohen, J., Soja, N. N., & Longridge, T. (1998). Simulator platform motion: The need revisited. *International Journal of Aviation Psychology, 8,* 293–317.

Caro, P. W. (1979). The relationship between flight simulator motion and training requirements. *Human Factors, 32,* 493–501.

Chatelier, R. R., Harvey, J., & Orlansky, J. (1982). *The cost-effectiveness of military training devices.* Working Group for Simulators and Training Devices. TTCP-UTP-2.

Chi, R., Glaser, M. T. H., & Farr, M. J. (1988). *The nature of expertise.* Hillsdale, NJ: Lawrence Erlbaum Associates.

Chou, C., Madhavan, D., & Funk, K. (1996). Studies of cockpit task management errors. *International Journal of Aviation Psychology, 6,* 307–320.

Collins, A. (1985). Teaching reasoning skills. In S. F. Chipman, A. W. Segal, & R. Glaser (Eds.), *Thinking and learning skills, Volume 2. Research and open questions* (pp. 579–586). Hillsdale, NJ: Lawrence Erlbaum Associates.

Collyer, S. C., & Chambers, W. S. (1979). *AWAVS: A research facility for defining flight trainer visual system requirements.* Orlando, FL: U.S. Naval Training Equipment Center.

Degani, A., & Wiener, E. L. (1997). Procedures in complex systems: The airline cockpit. *IEEE Transactions on Systems, Man and Cybernetics. Part A: Systems and Humans, 27,* 302–312.

Department of Employment. (1971). *Glossary of training terms* (2nd ed.). London: HMSO.

Dipboye, R. L. (1997). Organisational barriers to implementing a rational model of training. In M. A. Quiñones & A. Ehrenstein (Eds.), *Training for a rapidly changing workplace: Applications of psychological research* (pp. 31–60). Washington, DC: American Psychological Association.

Druckman, D., & Bjork, R. A. (Eds.). (1994). *Learning, remembering, believing: Enhancing human performance.* Washington, DC: National Academy Press.

Fitts, P. M. (1962). Factors in complex skill training. In R. Glaser (Ed.), *Training, research and education.* Pittsburgh: University of Pittsburgh.

Flanagan, J. C. (1954). The Critical Incident Technique. *Psychological Bulletin, 51,* 327–358.

Fleishman, E. A., & Quaintance, M. F. (1984). *Taxonomies of human performance.* Orlando, FL: Academic Press.

Fredericksen, T. R., & White, B. Y. (1989). An approach to training based upon principled task decomposition. *Acta Psychologica, 71,* 89–146.

Gael, S. (Ed.). (1988). *The job analysis handbook for business, industry and government.* New York: Wiley.

Gagné, R. M. (1962). Simulators. In R. Glaser (Ed.), *Training, research and education.* Pittsburgh: University of Pittsburgh Press.

Gick, M. L., & Holyoak, K. J. (1987). The cognitive basis of knowledge transfer. In S. M. Cormier & H. D. Hagman (Eds.), *Transfer of learning: Contemporary research and applications.* New York: Academic Press.

Goldstein, I. L. (1993). *Training in organisations: Needs assessment, development and evaluation* (3rd ed.). Monterey, CA: Brooks/Cole.

Hackman, J. R. (1993). Teams, leaders, and organisations: New directions for crew-oriented flight training. In E. L. Wiener, B. G. Kanki, & R. L. Helmreich (Eds.), *Cockpit resource management* (pp. 47–69). San Diego, CA: Academic Press.

Haskell, I. D., & Wickens, C. D. (1993). Two- and three-dimensional displays for aviation: A theoretical and empirical comparison. *International Journal of Aviation Psychology, 3,* 87–109.

Hayes, J. R., & Simon, H. A. (1977). Psychological differences among problem isomorphs. In N. Castellan Jr., D. Pisoni, & G. Potts (Eds.), *Cognitive theory* (Vol. 2, pp. 21–41). Hillsdale, NJ: Lawrence Erlbaum Associates.

Hays, R. T. (1992). Systems concepts for training systems development. *IEEE Transactions on Systems, Man and Cybernetics, 22*(2), 258–266.

Helmreich, R. L., & Foushee, H. C. (1993). Why crew resource management? Empirical and theoretical bases of human factors training in aviation. In E. L. Wiener, B. G. Kanki, & R. L. Helmreich (Eds.), *Cockpit resource management* (pp. 3–41). San Diego, CA: Academic Press.

Hesketh, B. (1997). Dilemmas in training for transfer and retention. *Applied Psychology: An International Review, 46*(4), 317–386.

Hutchins, E. (1995). How a cockpit remembers its speeds. *Cognitive Science, 19,* 265–288.

Jentsch, G., & Bowers, C. A. (1998). Evidence for the validity of PC-based simulations in studying aircrew coordination. *International Journal of Aviation Psychology, 8,* 243–260.

Kantowitz, B. H., & Campbell, J. L. (1996). Pilot workload and flightdeck automation. In R. Parasuraman & M. Mouloua (Eds.), *Automation and human performance: Theory and applications* (pp. 117–136). Mahwah, NJ: Lawrence Erlbaum Associates.

Keller, J. M. (1983). Motivational design of instruction. In C. M. Reigeluth (Ed.), *Instructional design theories and models. An overview of their current status* (pp. 386–429). Hillsdale, NJ: Lawrence Erlbaum Associates.

Keller, J. M., & Kopp, T. W. (1987). An application of the ARCS model of motivational design. In C. M. Reigeluth (Ed.), *Instructional design theories in action: Lessons illustrating selected theories* (pp. 289–320). Hillsdale, NJ: Lawrence Erlbaum Associates.

Kirkpatrick, D. L. (1959, November). Techniques for evaluating training programmes. *Journal of the American Society of Training Directors,* 3–9.

Kirwan, B., & Ainsworth, L. K. (1992). *A guide to task analysis.* London: Taylor & Francis.

Klein, G. A., Calderwood, R., & MacGregor, D. (1989). Critical decision making for eliciting knowledge. *IEEE Transactions of Systems, Man, and Cybernetics, 19,* 462–472.

Koonce, J. M., & Bramble, W. J., Jr. (1998). Personal computer-based flight training devices. *International Journal of Aviation Psychology, 8,* 277–292.

Lintern, G., & Garrison, W. V. (1992). Transfer effects of scene content and crosswind in landing instruction. *International Journal of Aviation Psychology, 2,* 225–244.

Lintern, G., & Gopher, D. (1978). Adaptive training of perceptual-motor skills: Issues, results and future directions. *International Journal of Man-Machine Studies, 10,* 521–551.

Lintern, G., Roscoe, S. N., & Sivier, J. E. (1990). Display principles, control dynamics, and environmental factors in pilot training and transfer. *Human Factors, 32,* 299–317

Lintern, G., Sheppard, D. J., Parker, D. L., Yates, K. E., & Nolan, M. D. (1989). Simulator design and instructional features for air-to-ground attack: A transfer study. *Human Factors, 3,* 87–99.

Logan, R. S. (1978). An instructional systems development approach for learning strategies. In H. F. O'Neil, Jr. (Ed.), *Learning strategies* (pp. 141–163). New York: Academic Press.

Logan, R. S. (1979). A state of the art assessment of instructional systems development. In H. F. O'Neil, Jr. (Ed.), *Issues in instructional systems development* (pp. 1–16). New York: Academic Press.

Mager, R. F. (1962). *Preparing instructional objectives.* Palo Alto, CA: Fearon Publishers.

Mané, A. M., & Donchin, E. (1989). The space fortress game. *Acta Psychologica, 71,* 17–22.

Martensson, I. (1995). The aircraft crash at Gottröra: Experiences of the cockpit crew. *International Journal of Aviation Psychology, 5,* 305–326.

Marti, P. (1998). Structured task analysis in complex domains. *Ergonomics, 41,* 1664–1677.

Matthieu, J. E., Tannenbaum, S. I., & Salas, E. (1992). Influences of individual and situational characteristics on measures of training effectiveness. *Academy of Management Journal, 35*(4), 828–847.

Mayer, R. E. (1979). Twenty years of research on advance organisers: Assimilation theory is still the best predictor of results. *Instructional Science, 8,* 133–167.

Merrill, M. D., Reigeluth, C. M., & Faust, G. W. (1979). The instructional quality profile: A curriculum evaluation and design tool. In H. F. O'Neil (Ed.), *Procedures for instructional systems development* (pp. 165–204). New York: Academic Press.

Miller, R. B. (1954). *Psychological considerations in the design of training equipment.* (Report No 54–563). Wright-Patterson Air Force Base, OH: Wright Air Development Center.

Mosier, K. L., Skitka, L. J., Heers, S., & Burdick, M. (1998). Automation bias: Decision making and performance in high-tech cockpits. *International Journal of Aviation Psychology, 8,* 47–63.

Mouloua, M., & Koonce, J. M. (1997). *Human automation interaction: Research and practice.* Mahwah, NJ: Lawrence Erlbaum Associates.

Naylor, J. C., & Briggs, G. E. (1963). Effects of task complexity and task organisation on the relative efficiency of part and whole training methods. *Journal of Experimental Psychology, 65,* 217–224.

Noe, R. A. (1986). Trainees' attributes and attitudes: Neglected influences on training effectiveness. *Academy of Management Review, 11*(4), 736–749.

Noe, R. A., & Schmitt, N. (1986). The influence of trainee attitudes on training effectiveness: Test of a model. *Personnel Psychology, 39,* 497–523.

O'Hare, D., & Chalmers, D. (1999). The incidence of incidents: A nationwide study of flight experience and exposure to accidents and incidents. *International Journal of Aviation Psychology, 9,* 1–18.

O'Hare, D., Wiggins, M., Williams, A., & Wong, W. (1998). Cognitive task analysis for decision centred design and training. *Ergonomics, 41,* 1698–1718.

O'Neil, H. F., Jr. (Ed.). (1979a). *Issues in instructional systems development.* New York: Academic Press.

O'Neil, H. F., Jr. (Ed.). (1979b). *Procedures for instructional systems development.* New York: Academic Press.

Orasanu, J. M. (1993). Decision-making in the cockpit. In E. L. Wiener, B. G. Kanki, & R. L. Helmreich (Eds.), *Cockpit resource management* (pp. 137–168). San Diego, CA: Academic Press.

Osgood, C. E. (1949). The similarity paradox in human learning: A resolution. *Psychological Review, 56,* 132–143.

Patrick, J. (1991). Types of analysis for training. In J. E. Morrison (Ed.), *Training for performance: Principles of applied human learning* (pp. 127–166). Chichester, England: Wiley.

Patrick, J. (1992). *Training: Research and practice.* London: Academic Press.

Pfeiffer, M. G., Horey, J. D., & Butrimas, S. K. (1991). Transfer of simulated instrument training to instrument and contact flight. *International Journal of Aviation Psychology, 1,* 219–229.

Pohlman, L. D., & Reed, J. C. (1978). *Air-to-air combat skills: Contribution of platform motion to initial training.* (Technical Report AFHRL-TR-78-53). Williams Air Force Base, AZ: Air Force Human Resources Laboratory.

Povenmire, H. K., & Roscoe, S. N. (1973). Incremental transfer effectiveness of a ground-based general aviation trainer. *Human Factors, 15*(6), 534–542.

Prince, C., & Salas, E. (1993). Training and research for teamwork in the military aircrew. In E. L. Wiener, B. G. Kanki, & R. L. Helmreich (Eds.), *Cockpit resource management* (pp. 337–364). San Diego, CA: Academic Press.

Prophet, W. W., & Boyd, H. A. (1970). *Device-task fidelity and transfer of training: Aircraft cockpit procedures training.* (Technical Report 70–10). Alexandria, VA: Human Resources Research Organization.

Reason, J. T. (1987). A preliminary classification of mistakes. In J. Rasmussen, K. D. Duncan, & J. Leplat (Eds.), *New technology and human error* (pp. 15–22). Chichester, England: Wiley.

Reason, J. T. (1990). *Human error.* Cambridge, England: Cambridge University Press.

Reigeluth, C. M., & Stein, F. S. (1983). The elaboration theory of instruction. In C. M. Reigeluth (Ed.), *Instructional design theories and models: An overview of their current status* (pp. 335–381). Hillsdale, NJ: Lawrence Erlbaum Associates.

Rogalski, J. (1996). Co-operation processes in dynamic environment management: Evolution through training experienced pilots in flying a highly automated aircraft. *Acta Psychologica, 91,* 273–295.

Roscoe, S. N. (1971). Incremental transfer effectiveness. *Human Factors, 13*(6), 561–567.

Sarter, N. B. (1996). Cockpit automation: From quantity to quality, from individual to multiple agents. In R. Parasuraman, & M. Mouloua (Eds.), *Automation and human performance: Theory and applications* (pp. 267–280). Mahwah, NJ: Lawrence Erlbaum Associates.

Sarter, N. B., & Woods, D. D. (1994). Pilot interaction with cockpit automation II: An experimental study of pilots' model and awareness of the flight management system. *International Journal of Aviation Psychology, 4,* 1–28.

Seamster, T. L., Redding, R. E., & Kaempf, G. L. (1998). *Applied cognitive task analysis in aviation.* Aldershot, England: Avebury/Ashgate.

Shappell, S. A., & Wiegmann, D. A. (1997). A human error approach to accident investigation: The taxonomy of unsafe operations. *International Journal of Aviation Psychology, 7,* 269–291.

Smith-Jentsch, K. A., Jentsch, F. G., Payne, S. C., & Salas, E. (1996). Can pretraining experiences explain individual differences in learning? *Journal of Applied Psychology, 81,* 110–116.

Stewart, J. E., II, Cross, K. D., & Wright R. H. (1998). *Fidelity analysis for the OH-58D Kiowa Warrior Crew Trainer.* Technical Report 1083. Alexandria, VA: U.S. Army Research Institute for the Behavioral and Social Sciences.

Tannenbaum, S. I., Matthieu, J. E., Salas, E., & Cannon-Bowers, J. A. (1991). Meeting trainees' expectations: The influence of training fulfilment on the development of commitment, self-efficacy, and motivation. *Journal of Applied Psychology, 76,* 759–769.

Tannenbaum, S. I., & Yukl, G. (1992). Training and development in work organisations. *Annual Review of Psychology,* 43, 399–441.

Taylor, H. L., Lintern, G., Hulin, C. L., Talleur, D., Emanuel, T., & Phillips, S. (1997). *Transfer of training effectiveness of personal computer-based aviation training devices.* (Technical Report DOT/FAA/AM-97/11). Washington, DC: Office of Aviation Medicine.

Tenney, J. J., Rogers, W. H., & Pew, R. W. (1998). Pilot opinions on cockpit automation issues. *International Journal of Aviation Psychology, 8,* 103–120.

Thorndike, E. L., & Woodworth, R. S. (1901). The influence of improvement in one mental function upon the efficiency of the other functions. *Psychological Review, 8,* 247–261; 384–395; 553–564.

Warr, P., & Bunce, D. (1995). Trainee characteristics and the outcomes of open learning. *Personnel Psychology, 48,* 347–375.

Wheaton, G., Rose, A. M., Fingerman, P. W., Karotkin, A. L., & Holding, D. H. (1976). Evaluation of the effectiveness of training devices: Literature review and preliminary model. *Research Memo,* 76–6. Washington, DC: U.S. Army Research Institute for the Behavioral and Social Sciences.

Wiener, E. L. (1993). Crew coordination and training in the advanced-technology cockpit. In E. L. Wiener, B. G. Kanki, & R. L. Helmreich (Eds.), *Cockpit resource management* (pp. 199–226). San Diego, CA: Academic Press.

Wiener, E. L., Kanki, B. G., & Helmreich, R. L. (Eds.). (1993). *Cockpit resource management.* San Diego, CA: Academic Press.

Williams, H. P., Hutchinson, S., & Wickens, C.D. (1996). A comparison of methods for promoting geographic knowledge in simulated aircraft navigation. *Human Factors, 38,* 50–64.

Winstein, C. J., & Schmidt, R. A. (1990). Reduced frequency of knowledge of results enhances motor skill learning. *Journal of Experimental Psychology: Learning, Memory and Cognition, 16,* 677–691.

Woods, D. D. (1996). Decomposing automation: Apparent simplicity, real complexity. In R. Parasuraman, & M. Mouloua (Eds.), *Automation and human performance: Theory and applications.* Mahwah, NJ: Lawrence Erlbaum Associates.

Woods, D. D., Johannesen, L. J., Cook, R. I., & Sarter, N. B. (1994). *Behind human error: Cognitive systems, computers, and hindsight.* Dayton, OH: Crew Systems Ergonomic Information and Analysis Center (CSERIAC).

12

Flights of Fancy: The Art and Science of Flight Simulation

Mary K. Kaiser
Jeffery A. Schroeder
NASA Ames Research Center

A few months after joining the Aerospace Human Factors Research Division at NASA Ames, a young perceptual psychologist was given her first ride in an FAA-certified, Boeing-727 simulator. (Only the most clever of readers will deduce who this psychologist might be.) The year was 1985, an auspicious time for full-mission transport simulation at Ames. The Man-Vehicle Systems Research Facility (MVSRF) had opened only the year before. The state-of-the-art simulators housed there boasted the best new computer-generated imagery (CGI) visual systems available; only a few years earlier, the simulators at Ames had utilized cameras and terrain boards to generate the pilot's out-the-window views. The rooms surrounding the MVSRF's main bay were filled with computers hosting the two flight simulators (the Boeing-727 and an "Advanced Concepts" simulator), as well as a simulated air-traffic control (ATC) room. Remarkable efforts had been taken to enhance the verisimilitude of the simulation: voice-altering software was integrated into the ATC communications systems so that pilots would hear "different" controllers as they transitioned from one airspace sector to another and a pilots' lounge, similar to those found at major airports, was available for briefings and breaks between legs of the "flight."

Entering the flight deck of the B-727 simulator, the psychologist was amazed by the realism of the cockpit environment. Every gauge and toggle switch was in the correct location and appeared to be fully functional. The yoke controls, rudder pedals, and center throttles looked and moved exactly like those of the actual aircraft. The seats for the pilot, copilot, and flight engineer even had proper sheepskin covers, as favored by most flight crews.

There were differences from an actual flight deck, of course. Aft of the jump seat, there was an experimenter's control station. The cockpit windows also appeared unusual. Although the physical apertures were located in the proper geometric positions, nothing could be seen through them while standing in the flight deck. The psychologist concluded the graphics computers driving the displays must not be running. But upon eagerly accepting the pilot's invitation to sit in the copilot's seat, she was able to see a nighttime view of the San Francisco airport through her forward and right-side windscreens. Hoping to experience the full panorama, she looked towards the pilot's windows. The pilot's forward window provided a wildly distorted view; the image out the left-side window was less warped, but still clearly noncanonical. The psychologist decided to focus on her own two windows, front and side.

The pilot, demonstrating amazing intuition, asked if she would like to go for a short flight. Following a few brief communications with the operator in the control room, the engines suddenly rumbled and the instrument gauges jumped to life. No one was manning the ATC simulation, so the pilot announced his plan to take off from Runway 28R, make a slow loop of the area, and land on the same runway. As he advanced the throttles, the engines powered up to speed, sending a satisfying vibration through the cockpit. The plane rolled down the runway, building up speed. About two-thirds of the way down, the pilot pulled back on the yoke, and the plane lifted into the air. The psychologist noticed her linked control inceptors (the yoke and pedals) move in perfect synchrony with those of the pilot. The runway receded below, and the lights of San Francisco grew closer.

The pilot initiated a gentle right bank; the ridge lines defining the horizon tilted in the window, and the psychologist was certain she felt the gentle G-forces associated with such a turn. She commented that the simulator's motion platform was quite impressive. The pilot replied that indeed it was . . . when it was operational. Before the psychologist could generate another insightful comment, the pilot was turning the plane onto final approach for landing. The runway lights were clearly visible in the distance; the psychologist studied the scene, hypothesizing which cues the pilot used to verify proper glide slope. His voice interrupted her thoughts. "Harry, it's awfully clear up here. Want to give me some fog?" A disembodied voice responded in the affirmative. Suddenly, the view out the windscreen was obscured by a diffuse fog. The runway lights were barely discernable. "Now that's San Francisco!" the pilot pronounced with obvious approval. Even under these degraded conditions, the pilot brought the aircraft in for what felt (or at least looked) like a perfect landing.

As they taxied towards the gate, the pilot invited the psychologist to take control of the aircraft. "Just steer with your feet," he instructed her. The pilot expertly started slowing the aircraft as it approached the terminal, but then had a change of heart. "Heck, let's go on through," he suggested. "Huh?" the psychologist articulately inquired. "Just go on through the building. It's a kick," he assured her. She made every effort not to flinch as the terminal building loomed in the forward windscreen. It came closer, and closer . . . and then disappeared, revealing the tarmac on the other side of the terminal.

The pilot smiled over at her. "Fun, wasn't it? Harry, give us a reset on 28R, please. We'll do one more takeoff. Oh, and remove the fog; give us dusk lighting." There was a blackout of the windows, then the plane was back in its original position, but at an earlier time of the evening. The engines were back to idle position. As before, the pilot advanced the throttle, accelerated the aircraft to take-off speed, and executed a flawless rotation. The B-727 rose gracefully into the evening sky, the lights of the city once again glowing in the distance.

The aircraft passed through 1,000 feet, then 1,200 feet. Suddenly, the view out the windows went black. "We just lost visuals," announced the disembodied voice from the control room. "Yeah, I sorta noticed that," the pilot replied in his best Chuck Yeager voice. Oh no, the psychologist thought, how would the pilot ever be able to safely land the plane without the visuals? Fortunately, she didn't voice her frantic concern. "Okay," the pilot continued, "let's shut it down, Harry." Several seconds later, the cockpit became eerily quiet; the instruments stilled at their neutral positions. The flight deck was flooded with light when the door to the stairwell opened and the psychologist's boss entered. "So," he asked, "what did you think?"

We think this anecdote is far too long, but it serves to introduce a number of important issues about flight simulators. What are their capabilities? What are their limitations? How did they mature from the time they were first introduced, shortly after the Wright brothers' first flight, until the time they were heavily used for pilot training in World War II? How was flight simulation technology advanced by the introduction of digital computer systems (and, subsequently, CGI systems)? What recent refinements have occurred? What further advances are needed?

THE PURPOSE OF FLIGHT SIMULATION

Broadly speaking, flight simulators have been designed and constructed to meet user requirements in three areas: (a) entertainment; (b) personnel training, selection, and testing; and (c) aircraft systems research and development. Entertainment applications are outside the scope of this chapter (taxiing through airport terminals notwithstanding). Nonetheless, the design of simulators for entertainment, from the Billing's trainer in 1910 to the FighterTown™ facilities mounted in several cities today, often reflects a fair degree of technical sophistication (leveraged off the expertise invested in more "serious" simulation efforts). In fact, many of the flight-simulation software packages available for the home computer utilize aircraft and flight models initially developed for government, military, and commercial aviation facilities. Nonetheless, criteria critical for effective entertainment are often quite different from those required for veridical simulation (Schell, 1998); our focus is the latter.

Training, Selection, and Testing

Historically, the most common use of flight simulators has been for training. It is common to divide the training of pilot expertise into the domains of initial and recurrent training, that is, the original acquisition of expertise versus the maintenance of this expertise. However, this parsing likely holds more validity for the determination of training standards and certification than for the development of effective flight simulation. Recently, other characterizations of training have been proposed (Hennessy, 1999) that instead focus on determining what degree of simulation fidelity is likely required to teach the targeted pilot capabilities. For example, if a simulator is to support the development of specific sensorimotor pilotage skills, it is likely that high levels of physical and visual fidelity are required. However, if the simulator is being used to hone the pilot's judgmental expertise, a lower fidelity system that nonetheless conveys the important scenario parameters may prove fully sufficient.

Further, it is important to consider what aspects of the flight environment need to be realistically simulated to support the acquisition (or maintenance) of specific skill sets. For example, if the primary training goal is to teach a pilot proper flare technique during landing, a simulator that excises most of the landing sequence in order to provide repeated presentations of the final minute prior to touchdown may be desired. If, however, the training goal is to ensure the pilot can properly manage and orchestrate all required landing activities (i.e., establish contact and communicate with the tower control, enter the pattern, complete prelanding checklist, execute landing and runway exit, contact and communicate with ground control, and execute taxi), then such temporal editing is not desirable and may, in fact, prove disruptive.

Personnel selection and testing can be seen as complementary activities to training. If it's possible to use simulators to develop and enhance flight-critical skills, it's reasonable to propose that such systems also be used to screen potential flight-training candidates or test for maintenance of skills (or transfer of skills to a different class of aircraft). Still, the goal of developing effective screening tasks has been a challenging one. Because pilot candidates are not yet trained in flying tasks, it is necessary to identify a battery of general tasks that tap into the skills and capabilities necessary to be a proficient pilot. The search for such a battery has proven elusive (see Sanders, 1991, and the references cited therein); further, it is not clear that a flight simulator is the best venue in which to implement such a test battery. (See chap. 10 for a more in-depth treatment of pilot selection.)

The use of simulators to test pilots' proficiency levels is more straightforward. In this case, the simulator is used to assess whether a fully trained pilot is performing within acceptable limits. Assuming such limits have been defined (which is necessary for the pilot to be objectively evaluated) and the simulation provides an adequate level of fidelity (a topic we consider more thoroughly in the next section), then it is reasonable to assume that meaningful proficiency checks can be performed in simulators.

Research and Development

It's highly desirable to evaluate aircraft systems in simulation rather than on actual aircraft in flight. By systems, we include everything from a new avionics display to novel crew-communications protocols. Ideally, one would like to test the impact a new system has in the safe and controlled environment of the simulator and feel confident that the test findings generalize to actual flight environments.

Progressively, more aviation design and evaluation is being done in simulation. Outside the flight deck, this can mean heavy reliance on theoretical modeling (computational fluid dynamics, structural finite element methods) and experimental findings (wind tunnel tests, engine test stand results) for the construction of a mathematical model of aircraft performance. For flight deck design, crew procedure development, and airspace system management, this means that high-fidelity flight simulators are used to evaluate crew performance.

As with training and selection, the effective use of simulators for research and development is dependent on the ability to build aircraft simulators that adequately recreate the critical aspects of the actual flight environment. This goal has been pursued for the better part of a century, yet the answer remains elusive. Before examining where simulation technology has been, is now, and is likely to go in the future, we consider the critical issues of defining simulator fidelity and determining simulators' effectiveness for their critical aviation missions.

FIDELITY OF FLIGHT SIMULATORS

Today, flight simulators range from simple desktop systems to moving-base high-fidelity simulators and in-flight simulators. Each level of technological sophistication affords a different balance of fidelity, flexibility, and cost. How does one determine an optimal (or at least acceptable) balance for one's simulation requirements? (See Hopkins, 1975, for an early debate.)

First, it is useful to distinguish different aspects of fidelity. One of the most obvious is *physical fidelity.* That is, to what extent does the simulator's displays, controls, and other physical components look and feel like the actual aircraft? A second issue is the simulator's *visual fidelity:* when the pilot looks through the simulator's window, to what extent does the visual scene resemble that seen through the cockpit window? Similarly, one can assess the simulator's *motion fidelity:* the extent to which the motion-induced forces experienced in the simulator reflect those of the actual flight environment. Finally, *cognitive fidelity* refers to the extent to which the simulation environment engages the pilot in the same sort of cognitive activities (e.g., juggling multiple tasks, supervising automated subsystems, and maintaining situational awareness and an accurate mental model of aircraft dynamics) as the actual modern flight deck.

The Boeing-727 simulator described at the beginning of this chapter possessed extraordinary physical fidelity. During full-mission simulations, every effort is made to optimize the simulator's cognitive fidelity; during demonstrations, cognitive fidelity is blithely compromised (when, for example, the simulation is reset from midflight to initial takeoff configuration in order to allow the next visitor a turn). When its motion base was activated, the simulator's hexapod platform afforded motion cueing in all six degrees of freedom; when in fixed-base mode, it had essentially no motion fidelity. And, although state of the art for its time, the visual fidelity of the system was rather limited. The challenge, then, is to determine the appropriate level of each type of fidelity given the specific purpose for the simulator's use.

Validating Simulator Fidelity: How Good Is Good Enough?

The abstract ideal for simulated flight is clear: the simulator should create a virtual flight environment in which training and/or observed behaviors generalize fully to the actual flight environment. The Federal Aviation Administration (FAA) has established qualification guidelines for both airplane (FAA, 1993) and helicopter (FAA, 1994) simulators. These delineate specific levels of capabilities the simulator should possess in order to legally function in lieu of the actual aircraft for training and/or proficiency checks. Table 12.1 lists some of the visual, motion, and sound requirements to meet the FAA's four levels of simulation fidelity. Many of these fidelity guidelines depend on functional definitions; it is typically incumbent on the simulator designer to demonstrate that their system meets the required level of training/performance transfer.

How, then, do we measure the extent to which a simulator achieves these goals? Two main methodologies have been employed: transfer-of-training exper-

TABLE 12.1
Visual, Motion, and Sound Requirements for Simulators by FAA Classification Level

Simulator Level	*Visual Scenes*	*Visual Field of View*[1]	*Motion*	*Sound*
A	Night	45 × 30	3 Axis	
B	Night	45 × 30	3 Axis	
C	Night & Dusk	75 × 30	6 Axis	Cockpit Noise
D	Night, Dusk, & Day	75 × 30	6 Axis	Realistic Cockpit Noise

[1]Per pilot simultaneously.

iments and rating methods. Adams (1979) and others (e.g., Blaiwes, Puig, & Regan, 1973; Mudd, 1968; Sanders, 1991) have suggested that both these methods of simulator evaluation possess inherent limitations.

Transfer-of-training studies are expensive. The experimenter must have access to the actual aircraft as well as the simulator and be able to test pilots in both environments. Further, it is difficult (and potentially unsafe) to impose proper experimental protocol, especially if the simulator is to be used to train or test pilot performance under dangerous conditions. A unique advantage of simulators is that they can be used to train pilots to deal with such hazardous situations, but safety concerns impose ethical and practical constraints on validating the transfer of training to actual aircraft. Further discussion of empirical studies with the transfer-of-training paradigm can be found in chapter 11.

Rating methods can take two forms. The first is an engineering evaluation of the simulator, which determines (in a fairly objective manner) the extent to which the physical properties of the simulator meet the design specifications. Unfortunately, this evaluation has little bearing on whether the simulation environment effectively recreates the critical aspects of the flight environment.

The second form of rating consists of subjective evaluations performed by pilots familiar with the aircraft being simulated. The ratings generated by these pilots reflect their impression of the similarity between the experience of flying the simulator and the actual aircraft. Such measures suffer from the limitations usually associated with subjective evaluation techniques, although (as with handling-quality scales) pilots are carefully instructed concerning the assignment of ratings (Cooper & Harper, 1969).

Sanders (1991) suggests some new methods of simulator validation that merit consideration, notably back-to-back experimentation (Gopher & Sanders, 1984). This methodology involves the systematic extension of fundamental performance models (developed in constrained laboratory settings) to more complex, real-world settings. Thus, for example, course-keeping performance should be jointly examined in instrumented vehicles, moving-base simulators, fixed-base simulators, and traditional compensatory tracking studies in the laboratory. Such a convergent-measures approach balances the experimental control (and lower costs) of laboratory experimentation with critical validation in operational settings.

Ultimately, the key lies in understanding what perceptual and cognitive experiences the simulator must provide (and at what level of fidelity) in order to support performance that generalizes to the flight environment. As Adams (1979) argued, simulator design should incorporate sound psychological theory. The field of human factors optimizes its contribution when it supports the principled design of systems rather than post hoc evaluation and validation of each new point design. To that end, after briefly reviewing the evolution of simulator technology, we will focus on simulator technologies that support two perceptual systems deemed critical for pilot-vehicle control: vision and motion. Fundamental psychological research not only demonstrates the critical role visual and vestibular cues play in vehicle control; it can also help guide the development of intelligent design requirements.

FLIGHT SIMULATORS: A BRIEF HISTORY

Even before the era of powered flight, aviators like the Wright brothers appreciated the value of a safe, controlled environment in which pilots could acquire the fundamental skills to control airborne vehicles. They and others utilized tethered gliders as training devices. Once it became clear that powered aircraft would depend on the pilot's control inputs to maintain equilibrium (Hooven, 1978), the race was on to design devices that would permit the acquisition of at least some of the required skills while still safely linked to the ground.

Early Efforts

Two of the earliest flight simulators were the Sanders Teacher and the Antoinette trainer (a.k.a. the "apprenticeship barrel"), both circa 1910. These two devices took very different design approaches. The Sanders Teacher was effectively a modified aircraft mounted atop a ground-based universal joint; by pointing the machine into the wind, the pilot could experience a sense of how the controls functioned. The Antoinette apprenticeship barrel had no effective control surfaces. Instead, the instructor(s) would physically move the device to introduce disturbances, which the student then compensated for using controls connected through wires and a pulley to the base. This can be seen as the first attempt to give the instructor some degree of control over the pilotage task presented to the student.

Despite such early efforts, simulators played little part in the selection and training of aviators during World War I. Both the British and French training systems were largely dependent on flying actual aircraft (although the French did utilize a modified monoplane, called the Penguin, whose sawn-off wings resulted in a craft that "hopped" along the training field at about 40 mph).

It was, in fact, the high attrition (and accident) rate of World War I flight training that engendered the field of aviation psychology. In addition to spurring the development of part-task screening devices to assess candidates' skills in presumed flight-critical domains (e.g., reaction time, coordination, vestibular-based orientation), the postwar era marked the genesis of a systematic, engineering-based approach to flight simulation.

A Systematic Approach

Simply stated, an effective flight simulator requires the following three elements (Rolfe & Staples, 1986): (a) a model (i.e., equations) of the response of the aircraft to inputs, both from the pilot (i.e., control inputs) and the environment (i.e., disturbances); (b) a means of solving these equations in real time in order to animate the model; and (c) a means of presenting the resulting aircraft response in a manner similar to that experienced by a pilot in the actual aircraft (i.e., changes in the visual, tactile, auditory, and/or vestibular "scene"). As previously mentioned, the degree of fidelity required in these elements will vary as a function of the simulator's use. Initially, the "model" used by early trainers such as the Antoinette apprenticeship barrel was simply the instructor's flight-honed intuitions.

By the 1920s, sufficient advances occurred in mathematical modeling of flight (Bairstow, 1920) and controls and motion platforms (Lender & Heidelberg, 1917, 1918) to set the stage for the development of a realistic trainer. The first simulator to achieve the feel of an actual aircraft is generally considered to be the Link Trainer. Edwin Link built on the engineering expertise of his family's company, the Link Piano and Organ Company. By his own admission, Link's first fixed-base trainer was "part piano, part organ, and a little bit of airplane" (Parrish, 1969). In 1930, the patent for his first aeronautical trainer was filed. Pneumatic bellows, used in other Link products to coax music from pipe organs, now served as actuators to create pitch, yaw, and roll in response to the student pilot's stick-and-rudder inputs.

Functional flight instruments were soon introduced into the Link General Aviation Trainer (GAT). The trainer's utility was further increased by the integration of a course plotter, which enabled the instructor to monitor the student's simulated flight. Thousands of the Link instrument flight trainers, affectionately (and not so affectionately) called the "Blue Box" for its brightly painted cab housings, were used to train U.S. and British aviators during World War II (see Fig. 12.1).

The Modern Era

The postwar years saw electronic technologies replace mechanical and pneumatic systems. Analog computers supported true computational solutions of flight equations to supplant the empirical "cheats" many simulators had employed. But likely the most significant advance in simulator technology was the introduction of digital computing. Not only did digital computers tremendously improve the fidelity of flight-dynamics modeling, they eventually enabled a new era of visual displays for flight simulation.

VISUAL SYSTEMS

Although the critical role visual information plays in flight control has long been recognized, only recently have display technologies supported the simulation of compelling out-the-window scenes. Early attempts at such "contact" displays entailed surrounding the simulator with static backdrops, which provided the pilot little more than attitude information. Other early attempts employed point-light projection systems. One of the most impressive of such systems was developed by Link for the specific purpose of teaching celestial navigation to flight

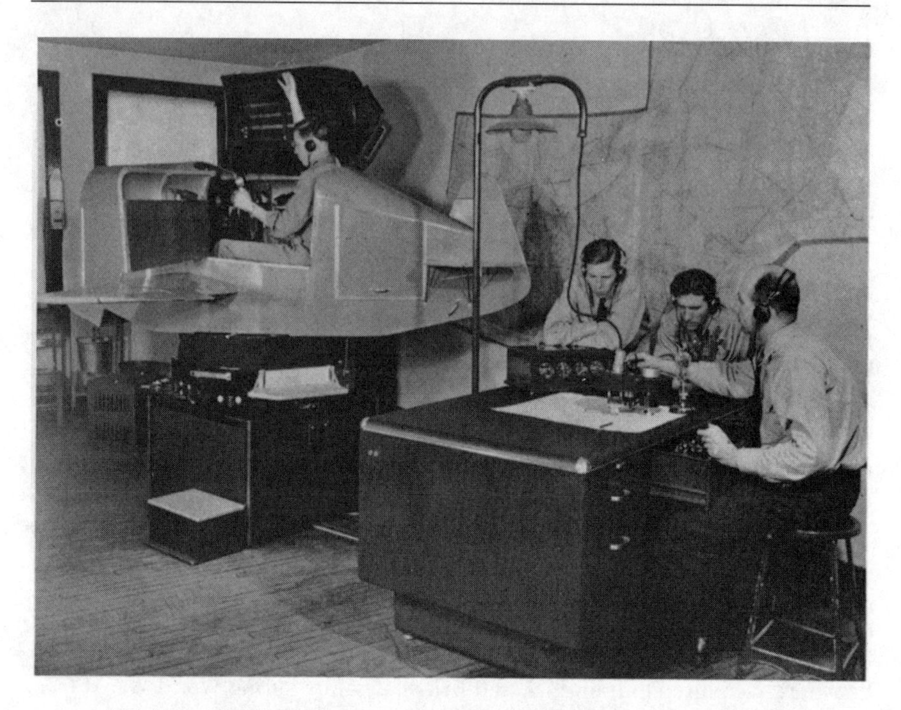

FIG. 12.1. The World War II-era Link "Blue Box" flight simulator (at left), connected to the instructor's course-plotting table (at right). Courtesy of Raytheon Systems Company.

crews (see Fig. 12.2). But like the backdrop displays, the Celestial Navigation Trainer provided none of the rich visual scene available to a pilot during normal, daylight flight. To better understand why it is so difficult to create compelling out-the-window simulations, it is helpful to consider the rich information humans glean from the visual world, particularly in the flight environment.

Characterizing the Visual Capabilities of the Pilot

What visual information does a pilot use to fly a plane? The question seems straightforward, yet finding a satisfactory answer has proven elusive. The human visual system is remarkably complex. In fact, it may be more accurate to think in terms of humans possessing a suite of visual systems with which we perform myriad functions on the information contained in the light that reaches our eyes. An in-depth discussion of human visual processing is beyond the scope of this chapter. Many excellent sourcebooks are available; those whose focus is especially appropriate to the problem of visual information for flight control include Bruce, Green, and Georgeson (1996); Milner and Goodale (1996); and the vol-

FIG. 12.2. The Link Celestial Navigation Trainer, designed to teach aircrews to "steer by the stars" for ferrying and bombing missions during WWII. Courtesy of Raytheon Systems Company.

ume edited by Epstein and Rogers (1995; see especially the chapters by Cutting and Warren).

Some visual activities of the pilot are obvious. Instruments are read; these readings are cognitively integrated to extract desired information concerning the aircraft's state. Of course, far more than instruments are included in the pilot's visual scan. Much time is spent looking out the windscreen at the surrounding world. Here, too, the pilot reads alphanumeric data and symbols (e.g., runway numbers, taxiway signs, displaced threshold markings). But in addition to such artificially structured information, the visual world provides the pilot with a wealth of information concerning the aircraft's heading, altitude, attitude, speed, glide slope, and other flight-relevant parameters (Flach & Warren, 1995; Kleiss, 1995).

When it comes to creating a realistic simulation of the necessary and sufficient visual information used by pilots, designers are faced with a daunting challenge. Clearly, it is not yet possible to render the full tapestry of the visual world. How then, does one select the truly critical threads?

A number of perceptual psychologists have suggested that it is useful to identify two classes of function for the visual system. This two-system parsing has been given different names by different researchers, reflecting their theoretical biases and emphases (e.g., Bridgeman, Lewis, Heit, & Nagle, 1979; Creem & Proffitt, 1998; Hubel & Wiesel, 1968; Liebowitz, 1982; McCarthy, 1993; Milner & Goodale, 1996). Nonetheless, common themes emerge from this body of work. First, one domain of visual processing (alternately called focal vision, the "what" system, or the parvo system) is primarily concerned with the analysis of form and objects in our central (foveal and parafoveal) region. Second, a fairly independent domain of visual analysis is devoted to processing motion and low spatial-frequency information in our parafoveal and peripheral regions. This system (referred to as ambient vision, the "where" system, or the magno system) provides our sense of orientation, self-motion (a.k.a. ego-motion), and spatial awareness. These two systems may not fully share information and may differ in the extent to which they inform our motor actions and verbal judgments (Milner & Goodale, 1996). Alternately, it may be necessary for the visual system to integrate information from both systems to form a rich, spatial representation of the environment.

Clearly, the successful control of an aircraft engages both systems, and does so to a challenging degree. Thus, it is critical that the flight simulator adequately recreate the visual stimuli each system requires for the normal extraction of information. As one might imagine, meeting the two systems' requirements creates goals that are sometimes in conflict with each other. For example, tuning a simulator display to provide sufficient visual detail to support the "what" system may compromise the display's update rate such that it no longer provides the motion flow information needed by the "where" system. Simulator designers must be innovative in their approach in order to satisfy both.

Modeling the Visual Information of the Flight Environment

Designers have employed two interesting cheats in simulating the visual flight environment. The first is to use actual components of the environment, most notably the cockpit interior. By so doing, the pilot experiences the physical stimuli (visual and tactile) associated with the controls and instruments; proper readings fed from the mathematical model to the instruments, and force feedback to the controls, complete the emulation. This creates an important distinction between high-fidelity flight simulators and general virtual reality [VR] systems. The flight simulator utilizes a combination of actual and virtual components to create the simulation; in VR systems, the entire simulated environment is computer rendered (Nemire, Jacoby, & Ellis, 1994).

The second cheat simulators have used in creating visual flight environments is, quite simply, to not even try. Many training simulators of the 1940s through 60s were designated "instrument trainers." As such, pilots were expected to fly based solely on instrument readouts, with no reference to the external visual scene. The Link GAT is one example of such devices. Similarly, even simulators with the capability to generate external visual scenes acknowledge that their systems are inadequate to generate sufficient visual quality for certain phases of flight (e.g., taxi) or visibility conditions. For example, the Boeing-727 simulator described at the beginning of this chapter was capable of only night and dusk visual scene simulation—not more visually complex daylight conditions.

Generally speaking, most modern flight simulators use computer generated imagery (CGI) to simulate the out-the-window scenes. Such was not always the case, of course. As mentioned earlier, a common technique employed in the 1960s and 70s was to use the pilot's control inputs to drive a camera over a model terrain board (Fig. 12.3; see Haber, 1986 for a detailed description). These terrain

FIG. 12.3. The track-mounted camera and model board components of the Link Mark IV simulator visual system, circa 1980.

boards were often remarkably realistic, and the camera tracking systems could emulate most of the aircraft dynamics. However, the limitations of such a system are readily apparent. The size of the boards was physically limited; usually, only operations near the airport were simulated. Further, any desired change in the terrain or structures required time-consuming, physical modifications of the model board. Then, too, unless the pilot's inputs had limits placed on them, the camera could "crash" into the board, causing damage (and necessitating painstaking repairs). There were also clear restrictions on the "weather" and lighting conditions that could be simulated. Finally, the model boards presented only static environments; no air traffic (or other dynamic phenomena) could be simulated (save, perhaps, flashing approach lights).

One alternative to using the model board/camera technique for external scene generation was to use prefilmed flight sequences. The most sophisticated example of this approach was Link's Variable Anamorphic Motion Picture (VAMP) system. Full-color, 70-mm films were taken of nominal approach, landing, and taxi operations. The VAMP could simulate limited deviations from the recorded trajectories by utilizing two methods of image processing: (a) subsampling the area of the film frame to correspond to the aircraft's pitch and heading and (b) optically warping the image with anamorphic lenses to simulate vehicle roll.

The advantages of such a visual system are apparent. The level of visual fidelity was unprecedented, both in scene resolution and dynamic realism. Still, the limitations are equally striking. Due to the logistical difficulties associated with collecting footage, only a limited number of flight sequences were available. Pilots became familiar with them and often found extraneous cues to aid them in their pilotage (e.g., using highway traffic events to "time" their approach trajectory). Plus, because optical distortions occurred if one deviated too far from the recorded flight path, pilots sometimes employed a strategy of nulling optical error to correct path errors. Clearly, such strategies do not generalize to actual flight environments, and hence are not desirable ones to engender in simulation-based training.

CGI seems, then, to offer the best possible solution for scene generation. It affords a degree of flexibility and configurability unavailable with earlier systems. Plus, with the advent of high-quality real-time rendering engines, it would seem that CGI could rival filmed sequences in terms of realism. But there are still limits to CGI rendering, particularly given the requirement of real-time generation. What are these limits? We will focus on three that seem to hold particular relevance for visual flight control: resolution, motion depiction, and visual field.

Scene Detail. This dimension deals with how much detail is presented in the visual scene. The ability to depict adequate detail with CGI is dependent on two major factors. First, the objects in the graphical database must contain sufficient detail. Second, the graphic objects must be imaged on a display media with enough picture elements (i.e., pixels) to represent the detail. As an example, consider simulating air traffic (i.e., other aircraft). Each aircraft needs to be rendered in adequate detail such that pilots can perceive the same critical information about that plane (at the same depicted distances) as they do in actual flight. Thus, during taxi, a plane must be sufficiently detailed that its type and affiliation is clear (because the tower might instruct the pilot to "follow TWA MD-80"). Such detail may require that hundreds of polygons (i.e., graphical building blocks) be specified in the database for rendering that aircraft. However, it is inefficient to always render each object at its full resolution; to do so can quickly overwhelm a graphical system's capability. Thus, it is a common technique to have several different versions of an object, each rendered with a different level-of-detail (LOD). Objects that are far from the pilot's eye point are rendered at a low LOD, whereas those in near range are rendered with high LOD. One challenge for visual simulation is to transition among the various LODs in a manner that is not noticeable to the observer, that is, in a manner that emulates the natural emergence of detail as objects grow closer (Montegut, 1995).

Once it is determined how many graphical building blocks are to be used to render the object, there is still the question of how fine a mesh will be used to "draw" them on the display. Ideally, the mesh should be so fine that the individual elements are not discernable (i.e., the size of the elements is below the resolution limit of the human eye, or approximately 120 pixels per visual degree in the foveal region). Unfortunately, it is still generally cost prohibitive to provide such high pixel counts on large-area displays. One solution that has been implemented is to exploit the fact that the spatial resolutions of the parafoveal and peripheral regions are significantly lower. Thus, if you design a display that provides a high-resolution image to the fovea, the lower resolution throughout the rest of the visual field is not noticeable. Both CAE, Inc., and Boeing (formerly McDonnell-Douglas) have created such display systems. The CAE system is head mounted and head tracked; the center of each eye's display has a high-resolution inset. The Boeing system utilizes a projection system. Superimposed onto the pilot's current area of interest is a higher resolution projection; a head-tracker placed on the pilot's helmet serves to drive the inset projector. Although both systems offer elegant (albeit expensive) solutions, they do require that pilots keep their eyes fixated on the center of the visual field in order to benefit from the higher-resolution imagery. This may cause pilots to subtly alter their visual scan strategies, moving their heads rather than just their eyes to capture detailed information.

Given the current state of technology, it's impossible to render imagery across the pilot's entire visual field at the high resolution the human eye can resolve. This leaves several options: (a) restrict simulation scenarios such that pilots need not act on highly detailed information; (b) alter visual scene properties to compensate for limits in display resolution (e.g., enlarge airport signs and markings in the simulation so that pilots can read them at "nominal" distances); or (c) utilize specialized display insets to provide higher resolution imagery to the pilots' foveal vision. All these choices compromise the completeness and fidelity of the simulation. However, used in concert (and in conjunction with other creative solutions), they can greatly alleviate visual resolution limitations.

Motion Depiction. CGI creates the impression of motion by presenting a sequence of static images in rapid succession. As psychologists have known for over a century (and artists for some time longer), the displacement of images from one frame to the next creates a compelling "illusion" of motion. We use the term *illusion* parenthetically because sampled motion sequences, such as those used in movies, television, and CGI, possess many of the same stimulus features as a true motion sequence (see chap. 8 of Bruce et al., 1996, for a useful overview of human motion perception). However, if such sampled sequences are used to simulate real motion, it is important to focus on the ways in which sampled motion differs from true motion.

CGI consists of images rendered on a display screen made up of discrete pixels. Each pixel is redrawn many times per second, limited by the refresh rate (or redrawing rate) of the display medium. Most computer driven displays have refresh rates of 60 or 72 Hz. These rates are generally well above people's ability to discern screen flicker. However, the other critical image display rate is determined by how quickly the computer can recompute what image should be drawn. The number of times per second a new image is drawn is referred to as frame rate or update rate. Despite the breathtaking advances in graphic chip design, even high-end CGI engines stress their limits to render realistic-looking, high-resolution images at a frame rate equal to the display's refresh rate. Thus, it is common for each frame to be shown multiple times (i.e., for several display refresh cycles).

The time course for double-buffered computer animation is as follows. The first generated frame of the sequence (Frame 1, stored in Graphics Buffer 1) is projected on the Display Medium. At the same time, the Graphics Engine is calculating the image for Frame 2 and storing it in Graphics Buffer 2. (The introduction of dual-buffer graphics systems in the 1980s greatly improved the quality of real-time animation.) When it is time to refresh the display, the computer checks whether the image in Graphics Buffer 2 is complete. If so, this updated image is displayed. If not, the image from Graphics Buffer 1 is displayed again. Motion is thus approximated by a series of discrete position jumps, ideally at a rate sufficient to render the sampled motion perceptually indiscriminable from true motion (Watson, Ahumada, & Farrell, 1986).

Unfortunately, this motion sampling is more noticeable in CGI than television or movies because, within each frame of the CGI animation, the moving object's image is at one discrete position (i.e., the object edges are sharply defined). In television and movies, object images are captured optically; hence, moving objects blur. This motion blurring provides a time sampling of the object's position, which more closely approximates true motion. There are techniques available to impose time-sample blurring on CGI (Korein & Badler, 1983; Watson, Ahumada, & Farrell,

1986), but these are difficult to implement in real time. Thus, most real-time CGI animation is left with this discrete sampling artifact, often called "motion aliasing."

How large an impact does motion aliasing have on pilots' performance in simulators? As with any other simulation artifact, its effect will depend on the task the pilot is asked to perform and the severity of the artifact in the particular simulation facility employed. Obviously, motion aliasing will have the largest performance consequences for pilotage tasks that depend strongly on peripheral motion cues (e.g., judging sink rate and attitude during landing) and in those simulators with relatively low (12–20 Hz) update rates.

These smooth, visual flow cues play a large role in pilots' sense of self-motion. Patterns of optical flow can inform pilots of their flight vector (Gibson, Olum, & Rosenblatt, 1955). In fact, visual flow stimuli (especially in the periphery) can produce a sense of motion even in the absence of corresponding vestibular motion cues. This phenomenon, termed *vection,* is indicative of what a powerful role visual cues play in motion perception and how critical it is for the graphics displays in simulators to provide adequate visual motion cues.

Visual Field. There are two important parameters used to describe pilots' visual fields: field-of-view (FOV) and field-of-regard (FOR). FOV is the area of space the pilot can see at a particular instant. FOR refers to the entire visual field that is available for sampling. Cockpits are designed with a defined eye reference point (ERP); this is the point in the cockpit where the pilot's eyes (or, more exactly, a point midway between the two eyes) is located when he or she is seated at the proper seat height and distance from the control panel. The Federal Aviation Administration (FAA) has established what minimal exterior FOVs are required for various classes of aircraft when the pilot is at ERP. Simulators are then designed with out-the-window displays that provide the same FOVs at ERP.

However, pilots do not stay situated with their eyes at ERP. Their heads are free to move within a fairly large viewing box, even when wearing safety harnesses. When they deviate from ERP, new areas of the visual scene are visible through the window apertures. Thus, the pilots have a functional FOR that is far greater than their immediate FOV (Johnson & Kaiser, 1996). Pilots utilize this expanded FOR, especially for localizing traffic and conducting taxi operations.

Until recently, the FORs of simulators' out-the-window displays were only slightly larger than the displays' FOVs. There were two major reasons for this limitation. First, the collimation optics used to focus the displays at optical infinity had extremely limited eye boxes (i.e., the viewing volume in which the display can be viewed without noticeable distortion). Moving more than a few inches from the ERP resulted in a highly warped image; a few more inches resulted in the image being completely unviewable (as the psychologist at the beginning of the chapter discovered). Second, the creation of viewable regions beyond the ERP FOV requires the graphics computer to display a larger area, which is then subsampled (usually by inserting a physical window frame aperture). Given the limitations of graphics systems discussed earlier, designers are constrained to a limited degree of oversampling.

This artificially constrained FOR can cause pilots to alter their visual strategies or create the need to implement entirely new procedures (e.g., depending on the right-seat pilot to perform all visual scans in that region). As with the high-resolution insets, forcing pilots to alter visual strategies to accommodate artifacts of the simulator's displays is highly undesirable. Thus, advancements in the visual system technology, such as those demonstrated by FlightSafety's Vital VIIIi system, are most welcome. In the Vital VIIIi, the visual system's display is designed with a single eye box for both the pilot and copilot; the eye box encompasses most of the cockpit. With this system, both pilots have nearly complete look-around capability (as can be seen in Fig. 12.4).

Another technique to give pilots look-around capability is to track their head position and dynamically redefine the graphic system's viewing frustum accord-

FIG. 12.4. Flight Safety International's VITAL VIIIi visual system, installed in the Advanced Concepts Flight Simulator at NASA Ames.

ingly (Johnson & Kaiser, 1996). However, this system is better suited for single-pilot cockpit simulation, because the resulting image can be optically appropriate for only a single eye point. Further, extreme care must be taken to ensure that the dynamic eye-point transformations are done quickly and accurately, otherwise the desired "window effect" is not achieved.

In examining the requirements for simulator visual systems, we've seen several themes emerge. First, we note that current technology is not sufficient to fully recreate the richness and complexity of the visual world; thus, the designer's task is to determine, within the limits of available hardware and software, how to best match the capability of the system to the visual requirements of the user (Proffitt & Kaiser, 1996). Second, we find that the various visual tasks pilots perform often create conflicting demands on the system capabilities; for example, rendering the fine levels of detail needed for reading airport signage and markings can compromise the graphic engine's ability to maintain the frame rate required for simulating critical motion cues. Finally, we recognize that designers simulate the visual flight environment by utilizing a combination of physical and virtual objects; the "art" of simulation design is determining how to optimize this mix to best create a compelling, high-fidelity simulation of the pilot's visual world. Vision science enhances this art by defining what information must be presented to support the pilot's critical perceptual processes.

Later, we will discuss what emerging technologies might influence this balance, as well as what new capabilities of simulation might result. But first, we examine another critical perceptual aspect of flight that simulators must support: the sense of motion.

MOTION SYSTEMS

Although the human visual system can create the illusion of self-motion, a complementary way of potentially enhancing that illusion is to move the pilot physically. Although vibration and G-seats can provide extremely limited physical movement, this section will focus on conventional motion systems, in which the entire simulator platform rotates, translates, or both. First, a brief discussion of the relevant ways pilots can sense motion, other than visually, is given. A brief discussion of how motion platforms take advantage of these sensations is given next. Then, the age-old arguments of motion platform effectiveness is covered, followed by a reconciliation and suggestions for further research.

Nonvisual Motion Sensory Capabilities

The reader should see chapter 3 for a more detailed treatment of how humans sense nonvisual motion; only a summary is provided here. Physical motion supplements the motion perceived by the eyes, and it is sensed by the vestibular organs, the skin, and the joint and muscle receptors. Humans can integrate these complementary motion cues with those from the visual system to perceive their orientation and position.

Of the nonvisual sensors, the vestibular system has received the most attention. The semicircular canals effectively result in the perception of angular velocity, and the frequency range of their peak sensitivity is between 0.1 and 10 rad/sec (Peters, 1969). This range is adequate for providing a pilot with the necessary angular cues for closed-loop control of an aircraft. The fact that the canals have a threshold value near 2 deg/sec allows for slow platform tilt to fool a pilot into believing the aircraft is undergoing a translational acceleration, as discussed later.

The otoliths in the inner ear sense translational acceleration, and their peak sensitivity range, about 0.2 to 1.5 rad/sec, is less than that of the semicircular canals. So, these organs act like a low-bandwidth accelerometer. Higher frequency accelerations, such as those induced by turbulence, are sensed by the skin, joint, and muscle receptors and must play an important role in acceleration sensing as well. Models for the muscles holding the head in place during accelerations and for body pressure dynamics are summarized by Gum (1973). Body pressure dynamics are useful out to 34 rad/sec, which extends beyond the frequency range for piloted closed-loop use.

A problem with these models is that differences exist among individuals in motion perception dynamics, thresholds, and the specific environment in which they are used. In particular, the thresholds can vary by an order of magnitude depending whether or not a subject has to perform a task, such as fly a simulator (Samji & Reid, 1992). Another unknown concerning the models is that significant uncertainty exists on the relative contributions of each sensing path. No complete picture exists on how humans integrate the multiple motion sensations to form an estimate of their true motion.

Yet, the general reasoning for providing these nonvisual motion cues is that they allow a pilot access to what will happen to the vehicle more readily than provided by the available visual cues. Accelerations are difficult to sense visually, especially when your visual scene is degraded in resolution, brightness, and richness compared to the real world. These same accelerations are perceived immediately with platform motion, and if vehicle control is dependent on feeding back that acceleration, performance and workload will likely suffer if motion is missing. Issues involved in these closures are addressed by Hess (1990; see also chap. 8).

How Motion Systems Work

Types of Platforms. Different engineering solutions can provide a simulator platform motion capability. The fundamental goal is to enable the simulator's flight deck movement with three translations (X, Y, and Z) and three rotations (pitch, roll, and yaw) in a manner that transmits forces to the pilot's vestibular and somatosensory systems similar to those experienced in actual flight. We consider the two platform designs commonly employed. The value and description of nudge bases and G-seats may be found elsewhere, such as in Lee and Bussolori (1989) and Ashworth, McKissick, and Parrish (1984).

The six-legged hexapod platform (Stewart, 1965) is the most prevalently used design today. Because all six degrees of freedom are driven by the same six actuators, the available displacement in any particular axis depends on the displacement in the other axes. The advantages of this clever hardware design are its compactness and relatively small number of moving parts, but the interaxis displacement dependency complicates its commanded movements.

When large translational motions are desired, independent drives are often used. Fig. 12.5 shows the Vertical Motion Simulator at NASA Ames, which has the world's largest displacement capability. Large-motion platforms offer significantly larger translations compared to a hexapod, and they are often used for research and development purposes at government research facilities (Dusterberry & White, 1979; Anderson, 1996). Translational displacements in these large devices can be as high as ± 30 ft (Danek, 1993).

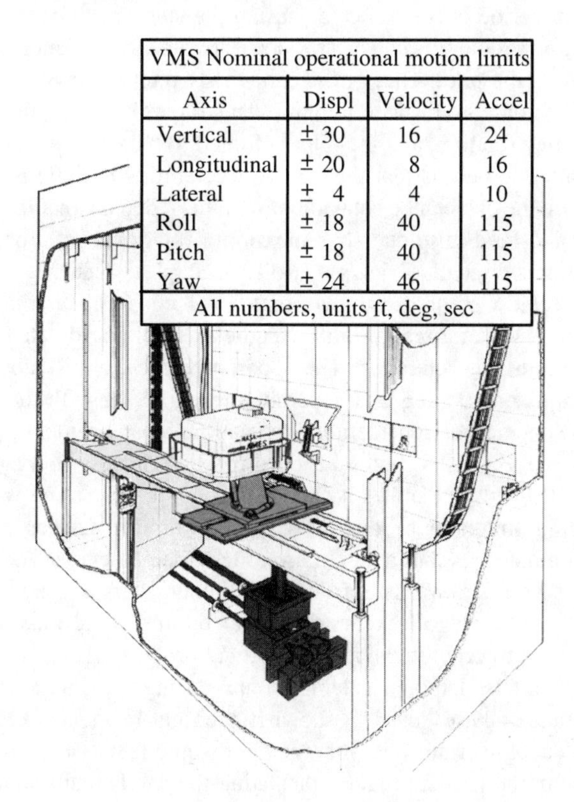

VMS Nominal operational motion limits			
Axis	Displ	Velocity	Accel
Vertical	± 30	16	24
Longitudinal	± 20	8	16
Lateral	± 4	4	10
Roll	± 18	40	115
Pitch	± 18	40	115
Yaw	± 24	46	115
All numbers, units ft, deg, sec			

FIG. 12.5. NASA Ames Vertical Motion Simulator.

How Platforms Move. Fortunately, complete vehicle motion does not have to be provided in the simulator to evoke pilot responses similar to those in flight. As acceleration frequencies go below about 0.1 rad/sec, the body's cueing receptors lose their sensitivity. So, steady state accelerations do not need to be represented. This is fortunate, because representing low-frequency accelerations requires platform displacements proportional to the inverse square of the frequency. For instance, a 0.1g vertical acceleration sinusoid with a frequency of 0.1 rad/sec requires 322 ft of vertical platform motion.

Two principal ways exist to reduce the amount of platform motion required with little-to-no objection from the pilot. The first way is to move the platform only a percentage of the full motion. The second way is to move the platform only at frequencies above a set value (with so-called washout filters). This latter way attenuates the low-frequency accelerations that quickly consume the available simulator displacement. At some point, these reductions become noticeable, and sometimes objectionable, to a pilot, as discussed later.

In some situations, a simulator's attitude degree of freedom can be used to reduce the simulator translational motion required. If an acceleration is sustained for a significant period of time, such as during the takeoff roll or the braking that occurs on landing rollout, the cab can be slowly rotated to reorient the direction of gravity relative to the pilot. That is, the cab slowly pitches down during braking, which causes a pilot to go forward into the seat straps. Of course, the visual scene does not show any rotation, and the pilot is fooled to believe the plane is decelerating. The limitation here is that the rate of acceleration buildup is governed by the rate at which the cab can be rotated before a pilot detects that the cab is rotating, which would be a false cue. The maximum rotational rate for creating the intended illusion is typically 2 deg/sec (Rolfe & Staples, 1986).

Rapid cab rotations for creating orientational cues can produce unwanted coordinated cues. When a real aircraft executes a level coordinated turn, a pilot does not feel a lateral acceleration. The apparent vertical is still along the pilot's spine. In a simulator, rotating this apparent vertical along with the pilot can be achieved at the expense of considerable translational displacement.

As an example, Fig. 12.6 shows a typical roll time history of a large transport aircraft. The left-hand plot shows it capturing a 20-deg bank and what a typical hexapod training simulator produces. While the aircraft reaches a bank of 20 degrees, the simulator reaches a maximum of less than 6 degrees and then returns to zero attitude. This action results from the washout filters that are applied in the roll axis, and this small roll excursion would require more than the available translational travel to orient the apparent vertical properly. Thus, a pilot will feel "the leans" (i.e., a false lateral acceleration cue) during the simulated maneuver.

The magnitude of these "leans" is shown in the right-hand plot of Fig. 12.6. The trace labeled "uncompensated" is what the pilot would feel as a result of the simulator roll angle in the left-hand plot. Translating the cab laterally in an attempt to alleviate this miscue achieves little, given the displacement limits of most hexa-

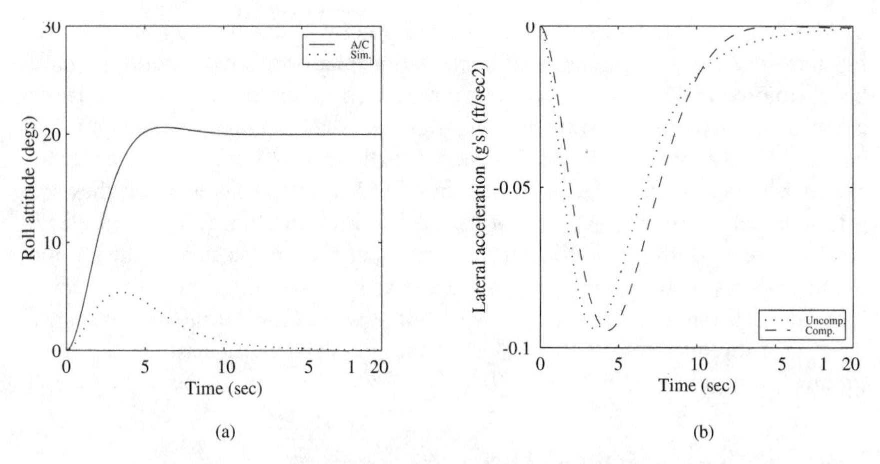

FIG. 12.6 (a). A comparison of simulator versus aircraft roll motion.
FIG. 12.6 (b). The effect of adding lateral displacement during roll motion.

pods. The trace labeled "compensated" is all that a typical hexapod can accomplish, and this compensation requires 5 ft of cab translational travel. Large research and development simulators can coordinate these maneuvers (Bray, 1972).

The pilot also feels another miscue in this maneuver. While the simulator reaches its peak roll attitude and returns to level, its roll rate produces a feeling of first rolling to the right and then of rolling to the left. So, whereas the aircraft is always rolling to the right, the simulator rolls to the right and then back to the left. This miscue is often felt. A common misperception concerning flight simulation is that some logic exists to bring the cockpit back to center at a rate below a human's sensing threshold. That is not the case. The return to zero is usually milder than the desired sensations of the maneuver, but the miscue is often still above the human sensing threshold.

The consequences of these miscues have been evaluated in several studies. For instance, Bray (1972) found that a pilot's ability to stabilize Dutch Roll motion, which is a side-to-side yaw combined with a rolling motion, degraded as the miscue increased. Jex, Magdaleno, and Junker (1978) showed that the spurious leaning cue is used by a pilot, which resulted in small improvements in tracking performance. As a result, tracking performance in a simulator could be better than that found in flight. A subsequent study evaluated different amounts and quality of lateral motion to reduce the leans and found that objective performance was not affected by the varieties of lateral motion (Jex, Jewell, Magdaleno, & Junker, 1979). However, subjective opinion revealed disconcerting and delayed side forces without proper translational platform motion. Schroeder, Chung, and Laforce (1997) found that pilot control activity increased as the leans increased, and they developed a specification on the required roll and lateral motion for satisfactory motion fidelity.

Most motion simulators provide miscues along with the primary motion cueing they are trying to render. Different motion platform drive algorithms have been explored to examine if one offers a reduction in miscues over others. These algorithms are divided into three categories: classical, optimal, and adaptive. A classical algorithm typically has fixed-filter coefficients that are selected based on experience. An optimal algorithm also has fixed-filter coefficients, but they are selected using optimal control theory in which a miscue function is minimized. An adaptive algorithm's coefficients change depending on the state of the motion platform—less motion cueing is provided near the edge of the platform's envelope and more motion cueing is provided otherwise. Few definitive advantages have been revealed among these algorithmic possibilities, at least for transport training purposes (Reid & Nahon, 1988).

The Platform Motion Debate

Now that we have covered the fundamentals of how simulators move, we address a more essential question—should they move? The question of how much motion is required, if any at all, has historically been contentious and examined by many authors (Cardullo, 1991; Boldovici, 1992; Burki-Cohen, Soja, & Longridge, 1998; Gebman, et al., 1986; Gundry, 1976; Lintern, 1987). In fact, Ed Link remembered being summoned hastily to Wright Field in the 1940s to justify the necessity of motion in his trainers (Cardullo, 1991).

Many researchers are interested in whether motion is valuable from the standpoint of pilot training. After all, that was the principal reason simulators were developed, and the vast majority of them are used today for training. However, simulators are also used for the building new aircraft and aircraft systems, as well as for accident investigations. For the former, a manufacturer's solvency often depends on the success of a new vehicle. Errors not caught in simulation, but instead revealed for the first time in flight testing, are costly and potentially hazardous. Supporting this vehicle development view, a recent study recommended that "validating simulation details, protocols, and tasks and collecting and correlating them with flight test results should be given high priority" (National Research Council, 1997). Previous human factors research has also cautioned systems development conclusions related to the issue of platform motion (Ince, Williges, & Roscoe, 1975). As such, the arguments for and against platform motion are reviewed from both a training and systems development point of view.

Arguments Against Motion. In a transfer of training study, Jacobs and Roscoe (1975) trained pilots for maneuvering motion (where the pilot creates the motion during a tracking task) with three motion conditions: no motion, washed-out motion, and random motion. The simulator was a 3-DOF General Aviation Trainer-2 motion platform, and the subjects subsequently transferred into a Piper Cherokee Arrow. No differences were measured among the motion configurations; however, all conditions (motion and no motion) improved performance in the aircraft compared to a control group that had no simulator training.

Gray and Fuller (1977) evaluated the training value of six-degree-of-freedom motion versus no motion in air-to-surface weapons delivery in an F-5B aircraft. Although simulator training improved weapon delivery performance, the presence of platform motion was not a contributing factor.

In contrast to the Jacobs and Roscoe maneuvering motion study (1975), Ryan, Scott, and Browning (1978) examined disturbancelike motion comprised of engine failures in the terminal area. In this study, the simulator was a P-3 trainer with and without six-degree-of-freedom platform motion, and the subsequent aircraft evaluations were in S-2s and T-44s. The results were the same as for Jacobs and Roscoe's study in that no performance improvements were measured in the aircraft, regardless of simulator motion presence.

Koonce (1979) also used the General Aviation Trainer-2 motion platform, but in this case, 90 pilots transferred into a Piper Aztec. Three motion configurations were examined: no motion, purely attenuated motion, and washed-out motion. Although pilots performed better in the simulator when any motion was present, no differences were measured in the aircraft. Roscoe (1980) points out that any differences that might exist between using motion or not using motion during simulator training are so small that they are difficult to measure.

Lee and Bussolari (1989) compared several motion conditions using a Level C 727 simulator (FAA, 1993). They compared six-degree-of-freedom motion versus two-degree-of-freedom motion (vertical and sway) versus almost no motion (special effects motion using approximately 1/4 of an inch travel). They found no significant differences among in performance or pilot ratings among these different motion conditions.

Finally, the standard view of the U.S. Air Force is that platform motion is not required for centerline thrust aircraft (Cardullo, 1991). For fighters performing extremely large maneuvers, the motion errors between simulation and flight are considerable. So, motion simulators are not practical for that application.

Arguments for Motion. Many studies have shown that the addition of platform motion affects pilot-vehicle performance and pilot opinion in simulators. The magnitude of these effects depends on the dynamic characteristics of the other elements in the system that the pilot is trying to control.

Young (1967) emphasized this point in a summary of several manual tracking experiments. He stressed that, as vehicles become more difficult to control, motion cues become more helpful. If the vehicle is easy to control, the additional cues supplied by motion are not required. He reviewed several experiments to support this finding. In one experiment, a pilot controlled an inverted pendulum of different lengths. As an inverted pendulum becomes shorter, it is more difficult to keep it upright. With motion, pilots could control a shorter pendulum than without motion. Sinacori (1973) also presents confirming results for an inverted pendulum. Another experiment required a pilot to recover from failures, and without motion the recovery interval increased by 100%. A third experiment showed how motion can make performance worse, but likely closer to the real-world case. For large rocket boosters, their structural flexibility can result in oscillations that affect a pilot's ability to execute precise control and even read the instruments. With motion, pilots were more gentle with the booster, and the lower pilot-vehicle performance that resulted was a more valid indicator of the real world. Without motion, a misleading estimate of pilot-vehicle performance occurred.

Hall (1989) echoes many of these results from experiments conducted on Harriers and other vehicles. He contends that for primarily open-loop, low pilot-vehicle gain, low-workload maneuvers with strong visual cues, that nonvisual cues are of little importance. If the pilot-vehicle gain rises (such as in an emergency), or the vehicle stability degrades, then the motion cues become more important.

The results of Shirley and Young (1968) add specificity to these statements. In their experiment, a pilot had to keep the wings level during a rolling disturbance. In the frequency domain, they measured the pilot's output versus roll error input. They found that the addition of motion allowed a pilot to provide lead (or predictive) compensation above 3 rad/sec. This addition of lead allowed a pilot to more tightly close the loop without a loss in system stability.

In the two studies discussed earlier on roll tracking (Jex et al., 1978; Jex et al., 1979) platform motion improved performance and reduced workload. Guidance for configuring a motion system within its constraints to provide the most useful roll and lateral acceleration cues was given. These data have been folded into a specification discussed later.

Several experiments have examined how the fundamental nature of the task affects motion requirements (Bray, 1985; Jex et al., 1978; Jex et al., 1979; Schroeder, 1999; Stapleford, Peters, & Alex, 1969; van der Vaart & Hosman, 1987). The two fundamental tasks in compensatory tracking are target following and disturbance rejection (see also chap. 8). The Fig. 12.7 block diagram may be used to explain the distinction. In compensatory tracking, the error between the target and the aircraft can develop from two sources: target motion and aircraft motion. If just the target moves, the pilot voluntarily generates motion to follow the target. This action is termed as "target following." The other source of tracking error is the external disturbance environment, which involuntarily causes an error between the target and the aircraft. In this instance, the pilot tries to mitigate, or reject, this external disturbance. The general results of these studies have shown that motion can be valuable in both instances, but in different ways.

For target following, the initial speed with which the pilot is able to close the tracking loop is not affected by motion. This makes sense, as the pilot sees, instead of feels, the initial tracking error. However, platform motion will make the error time history during the loop closure less oscillatory if the dynamic characteristics of the aircraft are poor. This is also sensible, because for these poorly fly-

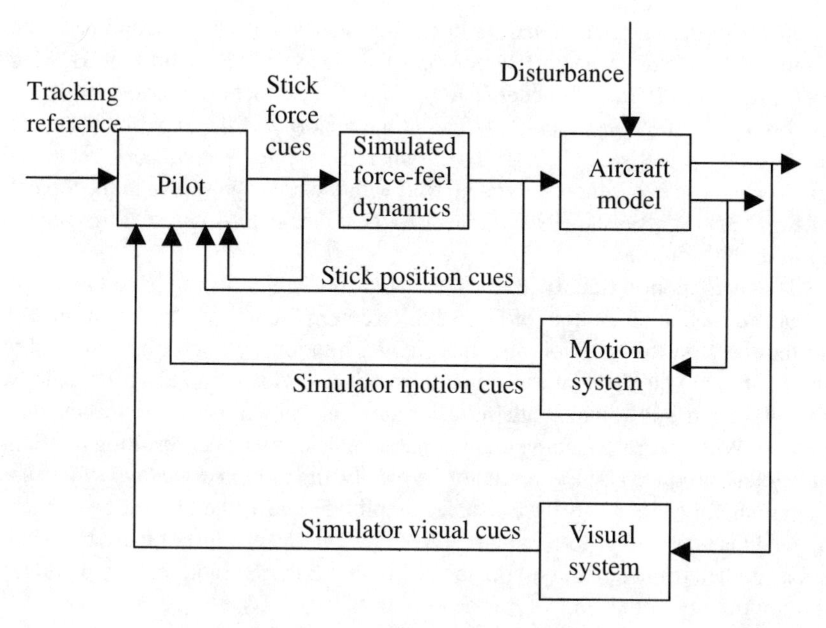

FIG. 12.7. The pilot-vehicle system in flight simulation.

ing aircraft, the pilot can productively use the motion for lead compensation in closing the loop on the target.

In contrast, for disturbance rejection, the initial speed with which the pilot can close the loop is increased with motion. Here, the pilot feels the error before seeing the error, so motion alerts to the action. Motion can also improve the error time history during the loop closure in similar ways to the target following situation. In both tasks, it is important to note that the quality of the motion, that is, its magnitude and its time shift relative to the visual scene, has a strong influence on the magnitude of these effects. So, it is not as simple as saying "motion does this, and no motion does that." The motion has to be quantified first.

Extensive data have been collected in an attempt to quantify good motion from bad motion. Metrics such as total allowable time delay, 100 msec (FAA, 1993) and allowable asynchronous motion and visual cue differences, 40 msec, have been quantified (Chung & Schroeder, 1997). For a given experiment, allowable delay is certainly dependent on the task and vehicle dynamics; however, these delays are intended to be a likely sufficient, rather than necessary, condition.

If no time-delayed or asynchronous motion cues occur, then motion quality can be specified by how its magnitude and phase shift characteristics vary with frequency. Many experiments have been conducted to evaluate the performance and workload effects of these frequency dependent characteristics, and a successful set of criteria have been developed. As described next, one criterion applies to rotational cues, and the other applies to translational cues.

Rotational Requirements. Many empirical studies have examined rotational motion requirements (Bergeron, 1970; Bray, 1972; Cooper & Howlett, 1973; Jex et al., 1978; Schroeder & Johnson, 1996; Shirachi & Shirley, 1981; Stapleford et al., 1969; van Gool, 1978), and the majority of the experiments evaluated roll-axis motion. The ranges of what constituted satisfactory rotational motion in these experiments correlate well with the criterion shown in the top half of Fig. 12.8 (Schroeder, 1999). Sinacori (1977) originally suggested the general form of this criterion.

The three motion fidelity levels in the criterion (high, medium, and low fidelity) are defined at the bottom of Fig. 12.8. To determine where a particular motion system falls on the criterion, the math model rotational velocity is compared to the rotational velocity that the simulator produces. This comparison is made by considering a sinusoidal math model rotational velocity at a frequency of 1 rad/sec. With this rotational velocity input, one determines the resulting rotational velocity produced by the simulator. Typically, the motion drive laws will cause the simulator velocity to have smaller amplitude and to be shifted by a phase angle. Once the output-to-input amplitude ratio and phase shift of these two sinusoids are determined, this point is located on the criterion plot, and the predicted motion fidelity of the simulator is determined.

Translational Requirements. Compared to rotational requirements, fewer experiments have examined translational requirements (Bray, 1973, 1985; Cooper & Howlett, 1973; Jex et al, 1979; Schroeder, 1999). The combined results of these studies correlate well with the requirements at the bottom half of Fig. 12.8. For most tasks, a large translational motion facility is required in order to be in the high-fidelity region. Thus, simulator users should be aware that most platforms will significantly attenuate translational motion cues and should use caution when extrapolating simulator findings to flight.

Now, many readers may say, this is all fine, but these results apply to simulator performance and workload comparisons without addressing the motion effects on training when flying the real aircraft. Performance, workload, and training value are not synonymous, and it is possible that the demonstrated value of motion in these situations has little training value. The only instance where quality motion has shown a training value was in a quasi-transfer experiment.

In a roll-axis disturbance rejection task, Levison (1981) examined the effects of platform motion delay in a quasi-transfer-of-training experiment. It is termed "quasi-transfer" because the value of the simulator training was measured during a separate simulator condition, instead of in an aircraft. Levison had separate groups of pilots train on a simulator in which the platform motion lagged the visual scene by 80, 200, and 300 msec. In addition, he trained in a visual-only condition. Then, all groups transferred to a simulator condition with synchronous motion and visual cues. The results showed that only the group with the shortest delay (80 msec) usefully transferred their training to the condition with

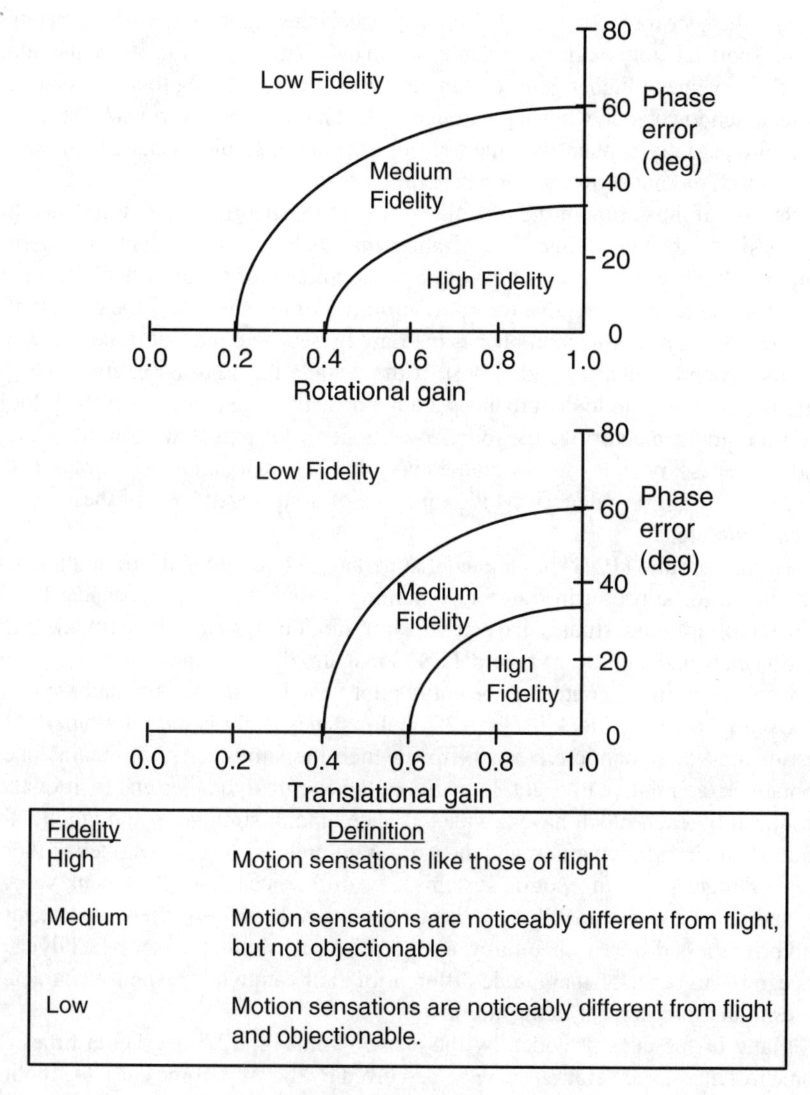

Fidelity	Definition
High	Motion sensations like those of flight
Medium	Motion sensations are noticeably different from flight, but not objectionable
Low	Motion sensations are noticeably different from flight and objectionable.

FIG. 12.8. Modified Sinacori motion fidelity criteria.

synchronous motion. A valuable aspect of this controlled experiment was that only the motion cues varied, which has never been the case in transfer experiments to an aircraft.

Reconciliation of the Motion Argument. Several factors may explain sometimes conflicted results of the motion argument (Hall, 1989). First, if the vehicle is stable and easily controlled by a pilot operating at low gain, it is certainly possible to learn the necessary flying skills from the visual cues alone. In normal operations, all transport category aircraft are stable, and in only a minority of tasks are the pilots asked to operate at higher gains (e.g., failure recovery and trying to land an aircraft in the touchdown zone when a go around should have been performed). These factors may partially explain why the transfer-of-training studies using general aviation aircraft did not show a motion benefit.

Second, if the nature of the task allows a pilot to perform it open loop, then the flying skills are knowledge based, rather than behavior dependent on external stimuli (Gregory, 1976). This type of open-loop behavior occurs when pilots perform routine landings in large transport aircraft. For instance, near the ground, the pilot of a large aircraft accomplishes the flare by setting a pitch attitude and waiting for ground contact (Field, 1985). If the vehicle has reasonable dynamics, if significant atmospheric disturbances are not affecting the path, and if the pilot is on a nominal trajectory to the touchdown zone, pilot inputs are low frequency, and the necessary cues can be obtained visually. The vast majority of air transport landings fall in this category. Motion may be of assistance if one of these conditions is violated.

Third, Cardullo (1991) has argued that perhaps the reason platform motion was not shown to be useful in transfer of training is that, by industry standards, the transfer of training studies have used poor motion systems. In particular, the motion cues had significant time delays. So, again, the usefulness of motion must be viewed within the context of the entire pilot-vehicle and pilot-simulator system.

Usually, the only block in Fig. 12.7 that matches flight is the Pilot block. The rest of the key dynamic elements, from which the simulator pilot obtains cues, contain errors relative to flight. In addition, the cueing dynamics and their quality produced by each block have a wide variance among simulators and among aircraft. Real aircraft dynamics, and the various complexity levels in modeling them, cover a large spectrum. Motion systems have differences in displacement, velocity, and acceleration capabilities. Visual systems have field-of-view, update rate, and brightness differences. Finally, and surprisingly, simulated force-feel dynamics can be an order-of-magnitude different in their frequency response characteristics than those present in the aircraft.

Many of the cues provided by the elements in Fig. 12.7 overlap at times. In some instances, accelerations may be perceived by the stick-force cues, the motion cues, and the visual cues. Depending on the task and the aircraft characteristics, sometimes these cues assist in vehicle control, and sometimes they do not.

In the context of all these possible differences and cueing overlaps, to expect a simple answer to the simple question "Is flight simulator motion necessary?" is just short of preposterous. The answer is "it depends." The value of high-quality motion has been shown to strongly depend on two factors: the task and the vehicle dynamics. These two dependencies are consistent with the control theoretic view of the value of motion. That is, a pilot does not have to generate considerable lead when controlling a vehicle with well-behaved dynamics in a task that can be accomplished almost precognitively. One might rightly argue that many of the air carrier operations fall in this category, and that most of the time, it does not seem platform motion is warranted. For some emergency procedures where violent cockpit motions might be of demonstrative value, platform motion may be warranted.

On the other hand, if considerable compensation is needed on the pilot's part to stabilize the vehicle or tightly track a target in the presence of significant disturbances, platform motion is a clear aid in improving performance and decreasing the workload. In addition, for systems design, motion can reveal problems, such as an overly sensitive aircraft, pilot-induced oscillations, and biodynamic feedback. Fixing these problems prior to first flight can save money and possibly lives.

However, whether or not the training objectives can be met without motion for vehicles that are difficult to fly or for challenging tasks is still an open question. If a pilot can learn to fly a vehicle without cues that might assist in stabilization (i.e., the motion cues), then perhaps the pilot can easily adapt to the in-flight case where these useful cues are present. It is a challenge and perhaps a safety risk to test this hypothesis. Caro (1979) points out that a better question than "Is motion needed for training?" is "For what training is motion needed?" The answer may be "none"; however, knowing what we do about the variety of dynamics that occur in aeronautical systems, it is unlikely. That variety has not been tested sufficiently.

Along similar lines, Dusterberry and White (1979) say it best with "It appears in retrospect that much of the conflict stems from a tendency of the two camps to argue their position in terms of broad generalities; that is, large motion cues *are* or *are not* necessary for pilot motion simulators. It is becoming more evident, as experience is gained, that such dogmatic generalizations are inappropriate" (p. 4).

Recommendations for Further Work

Because so many independent variables affect the value of motion, it is difficult to characterize the complete functional relationship. Lintern (1987) suggests more quasi-transfer experiments emphasizing motion be done. In a quasi-transfer experiment, a real airplane is not used to the test how effective the training is. Instead, a different simulator or configuration is used to test the transfer effectiveness. These experiments can be valuable, because the experimental conditions can be controlled precisely. Unfortunately, a valid counterargument is that flight and simulation are still going to be different, and that is where the training difference matters. Some have pointed out that it may be too dangerous and costly to conduct a transfer-of-training study for the situations where motion has shown a clear benefit (Boldovici, 1992). However, simulation is just as different from flight (if not more so) in that case as well. Regardless, it appears clear that we need to conduct more controlled experiments to achieve more definitive results. If you want to determine the effect of motion on training, then only motion should vary. A complete answer may take a long time.

FUTURE FANCIES

Having examined how technological advancements, both evolutionary and revolutionary in nature, fostered the development of flight simulators with highly realistic visual and motion qualities, we now turn to the future. What additional capabilities and systems enhancements will emerge from today's technical innovations? Will tomorrow's flight simulators outpace today's state of the art as completely as a modern system outpaces the Link Blue Box? What role will simulators play in the future of aviation?

Mark Twain sagely advised, "Never make predictions—especially about the future." That caveat aside, we would suggest several trends that are likely to continue (or emerge) in simulator systems.

1. Continuation of the improvement in simulator's cost/performance ratio. This is especially true for those component technologies (e.g., computer graphics and processor chips) that have uses across broader consumer markets. Other simulator-specific technologies (e.g., customized optics systems and motion platforms) are less likely to show rapid improvements in performance or reductions in component costs. PC-based systems are likely to play an increasing role as cost-effective training devices and midfidelity simulators (see chap. 11).
2. An emergence of networked simulator facilities, which allow pilots (and air-traffic controllers) to engage in shared simulation environments. These networks should become increasingly tolerant of hosting a variety of simulator types and platforms. The military is already conducting networked simulations, but this technology will extend to civil transport aircraft and involve modeling large segments of the national (and, ultimately, international) airspace. Such networked simulations will enable researchers to study airspace operation systems (e.g., multiple aircraft operating in a busy terminal area) as well as individual flight-crew performance.
3. A greater use of the virtual to replace the physical. As virtual environment technologies become more refined and less costly, it will become practical to depend on virtual displays (visual, auditory, haptic) to replace physical devices currently used in flight simulators (Wickens & Baker, 1995). For example, auditory processors that enable the delivery of spatially localized sounds over headphones could displace the suite of strategically located loudspeakers currently installed in the simulator cockpit.

These three trends will enable simulators to become increasingly affordable and flexible. As they do, we expect flight simulation to play a larger role in pilot training (both initial and recurrent) and flight system development. It is quite conceivable that, one day, a pilot's first flight in a new class of aircraft will be his or her check ride; simulation training will have proven that much more safe and cost effective, and the transfer of skills will have been more fully validated. In a similar vein, we foresee a time when most aviation system development, including "fly offs" between competing technologies (wherein two or more candidate solutions are compared in flight), will be completed in simulation. Flight test will be viewed as a final, validating procedure.

The military is likely to more fully embrace simulation technology to conduct virtual "war games." The development of hybrid network systems will permit the coordinated interplay of air- ground- and ocean-based forces, under a wide variety of scenarios and conditions. The flexibility of these networked simulations will permit exercises of far greater scope and complexity than actual exercises permit.

Finally, simulators are likely to play an increased role in aviation psychology research. As greater emphasis is placed on maintaining and improving safety, there is an increasing need for systematic, controlled studies of flight-crew performance. Such studies are often difficult, if not impossible, to conduct during actual aircraft operations. As networked simulators come to create a realistic emulation of the National Airspace System, their utility as a test bed for safety protocols and interventions will become unparalleled.

The relationship between flight simulation and aviation psychology will continue to be synergistic. Basic research in aviation psychology will inform simulator designers, specifying how to tune system interfaces (i.e., displays and controls) to best meet the perceptual and cognitive requirements of the users. The simulator, in turn, will remain a major tool for research in aviation psychology. As an experimental discipline, aviation psychology requires the degree of scenario control and performance assessment provided by simulators. Our flights of fancy become journeys of inquiry, exploring limits of performance within the safe confines of simulation to find ways to prevent tragic accidents in the less-forgiving environs of actual flight.

REFERENCES

Adams, J. A. (1979). On the evaluation of training devices. *Human Factors, 21,* 711–720.

Anderson, S. B. (1996). Historical review of piloted simulation at NASA Ames. In *Flight simulation—Where are the challenges?* (pp. 1–1–1–13). Neuilly-sur-Seine, France: NATO Advisory Group for Aerospace Research and Development.

Ashworth, B. R., McKissick, B. T., & Parrish, R. V. (1984). *Effects of motion base and g-seat cueing on simulator pilot performance* (Tech. Paper 2247). Hampton, VA: NASA Langley Research Center.

Bairstow, L. (1920). *Applied aerodynamics.* London: Longman, Green.

Bergeron, H. P. (1970). The effects of motion cues on compensatory tasks. In *AIAA Visual and Motion Simulation Technologies Conference* (Paper 352, pp. 1–5). New York: American Institute of Aeronautics and Astronautics.

Blaiwes, A. S., Puig, J. A., & Regan, J. J. (1973). Transfer of training and the measurement of training effectiveness. *Human Factors, 15,* 523–533.

Boldovici, J. A. (1992). *Simulator motion* (Tech. Report 961). Alexandria, VA: U.S. Army Research Institute for the Behavioral and Social Sciences.

Bray, R. S. (1972). *Initial operating experience with an aircraft simulator having extensive lateral motion* (Tech. Memo X-62155). Moffett Field, CA: NASA Ames Research Center.

Bray, R. S. (1973). A study of vertical motion requirements for landing simulation. *Human Factors, 15,* 561–568.

Bray, R. S. (1985). *Visual and motion cueing in helicopter simulation* (Tech. Memo 86818). Moffett Field, CA: NASA Ames Research Center.

Bridgeman, B., Lewis, S., Heit, G., & Nagle, M. (1979). Relation between cognitive and motor-oriented systems of visual position perception. *Journal of Experimental Psychology: Human Perception and Performance, 5,* 692–700.

Bruce, V., Green, P. R., & Georgeson, M. A. (1996). *Visual perception: Physiology, psychology, and ecology* (3rd ed.). Exeter, England: Psychology Press.

Burki-Cohen, J., Soja, N. N., & Longridge, T. (1998). Simulator platform motion—The need revisited. *International Journal of Aviation Psychology, 8,* 293–317.

Cardullo, F. M. (1991). An assessment of the importance of motion cuing based on the relationship between simulated aircraft dynamics and pilot performance: A review of the literature. In *AIAA Flight Simulation Technologies Conference* (pp. 436–447). New York: American Institute of Aeronautics and Astronautics.

Caro, P. W. (1979). The relationship between flight simulator motion and training requirements. *Human Factors, 21,* 493–501.

Chung, W. W., & Schroeder, J. A. (1997). Visual and roll-lateral motion cueing synchronization requirements for motion-based flight simulations. In *American Helicopter Society's 53rd Annual Forum* (pp. 994–1006). Alexandria, VA: American Helicopter Society.

Cooper, D. E., & Howlett, J. J. (1973). Ground based helicopter simulation. In *American Helicopter Society Symposium on Status of Testing and Model Techniques for V/STOL Aircraft* (pp. 1–18). Alexandria, VA: American Helicopter Society.

Cooper, R. E., & Harper, R. P., Jr. (1969). *The use of pilot rating in the evaluation of aircraft handling qualities* (TN-D-5153). Moffett Field, CA: NASA Ames Research Center.

Creem, S. H., & Proffitt, D. R. (1998). Two memories for geographical slant. Separation and interdependence of action and awareness. *Psychonomic Bulletin and Review, 5,* 22–36.

Danek, G. L. (1993). *Vertical motion simulator familiarization guide* (Tech. Memo 103923). Moffett Field, CA: NASA Ames Research Center.

Dusterberry, J. C., & White, M. D. (1979). The development and use of large-motion simulator systems in aeronautical research and development. In *Fifty Years of Flight Simulation.* London: Royal Aeronautical Society.

Epstein, W., & Rogers, S. J. (Eds.). (1995). *Perception of space and motion: Handbook of perception and cognition* (2nd ed.). San Diego: Academic Press.

Federal Aviation Administration. (1993). *Airplane simulator qualification* (AC 120–40B). Washington, DC: U.S. Department of Transportation.

Federal Aviation Administration. (1994). *Helicopter Simulator Qualification* (AC 120–63). Washington, DC: U.S. Department of Transportation.

Field, E. J. (1995). *Flying qualities of transport aircraft: precognitive or compensatory?* Unpublished doctoral dissertation, Cranfield University, Cranfield, England.

Flach, J. M., & Warren, R. (1995). Low-altitude flight. In P. Hancock, J. Flach, J. Caird, & K. Vicente (Eds.), *Local applications of the ecological approach to human-machine systems* (pp. 65–103). Hillsdale, NJ: Lawrence Erlbaum Associates.

Gebman, J. R., Stanley, E. L., Barbour, A. A., Berg, R. T., Kaplan, R. J., Kirkwood, T. F., & Batten, C. L. (1986). *Assessing the benefits and costs of motion for C-17 flight simulators* (Report No. R-3276-AF). Santa Monica, CA: The Rand Corporation.

Gibson, J. J., Olum, P., & Rosenblatt, F. (1955). Parallax and perspective during aircraft landings. *American Journal of Psychology, 68,* 372–385.

Gopher, D., & Sanders, A. F. (1984). S-Oh-R: Oh stages! Oh resources! In W. Prinz & A. F. Sanders (Eds.), *Cognition and motor processes* (pp. 231–253). Heidelberg, Germany: Springer.

Gray, T. H., & Fuller, R. R. (1977). *Effects of simulator training and platform motion on air-to-surface weapons delivery training* (Tech. Rep. No. AFHRL-TR-77-29). Wright-Patterson Air Force Base, OH: Air Force Research Laboratory.

Gregory, R. L. (1976). The display problem of flight simulation. In *Proceedings of the Third Flight Simulation Symposium* (pp. 3–1–3–32). London: Royal Aeronautical Society.

Gum, D. R. (1973). *Modeling of the human force and motion-sensing mechanisms* (Tech. Rep. No. AFHRL-TR-72–54). Wright-Patterson Air Force Base OH: Air Force Research Laboratory.

Gundry, J. (1976). Man and motion cues. In *Proceedings of the Third Flight Simulation Symposium* (pp. 2-1–2-19). London: Royal Aeronautical Society.

Haber, R. N. (1986). Flight simulation. *Scientific American, 255,* 96–103.

Hall, J. R. (1989). *The need for platform motion in modern piloted flight training simulators* (Technical Memorandum FM35). Bedford, England: Royal Aerospace Establishment.

Hennessy, R. T. (1999). Thoughts on using trainers to improve rotorcraft safety (pp. 225–226). In *Near-Term Gains in Rotorcraft Safety—Strategies for Investment.* Monterey, CA: Monterey Technologies.

Hess, R. A. (1990). Human use of motion cues in vehicular control. *Journal of Guidance, Control, and Dynamics, 13,* 476–482.

Hooven, F. J. (1978). The Wright brothers' flight-control system. *Scientific American, 239*(5), 132–140.

Hopkins, C. O. (1975). How much should you pay for that box? *Human Factors, 17,* 533–541.

Hubel, D. H., & Wiesel, T. N. (1968). Receptive fields and functional architecture of monkey stiate cortes. *Journal of Physiology, 195,* 215–243.

Ince, F., Williges, R. C., & Roscoe, S. N. (1975). Aircraft simulator motion and the order of merit of flight attitude and steering guidance displays. *Human Factors, 17,* 388–400.

Jacobs, R. S., & Roscoe, S. N. (1975). Simulator cockpit motion and the transfer of initial flight training. In *Proceedings of the Human Factors Society 19th Annual Meeting* (pp. 218–226). Santa Monica, CA: Human Factors Society.

Jex, H. R., Magdaleno, R. E., & Junker, A. M. (1978). Roll tracking effects of g-vector tilt and various types of motion washout. In *NASA Conference Proceedings 2060* (pp. 463–502). Moffett Field, CA: NASA Ames Research Center.

Jex, H. R., Jewell, W. F., Magdaleno, R. E., & Junker, A. M. (1979). *Effects of Various Lateral-Beam Washouts on Pilot Tracking and Opinion in the Lamar Simulator* (Tech. Rep. No. AFFDL-TR-79-3134). Wright-Patterson Air Force Base, OH: Air Force Research Laboratory.

Johnson, W. W., & Kaiser, M. K. (1996). Virtual windows for vehicle simulation. In *Proceedings of the 1996 IMAGE Conference.* Phoenix, AZ: The Image Society.

Kleiss, K. M. (1995). Visual scene properties relevant for simulating low-altitude flight: A multidimensional scaling approach. *Human Factors, 37,* 711–734.

Koonce, J. M. (1979). Predictive validity of flight simulators as a function of simulator motion. *Human Factors, 21,* 215–223.

Korein, J., & Badler, N. (1983). Temporal anti-aliasing in computer generated animation. *Computer Graphics* (SIGGRAPH '83 Proceedings), *17,* 377–388.

Lee, A. T., & Bussolari, S. R. (1989). Flight simulator platform motion and air transport pilot training. *Aviation, Space, and Environmental Medicine, 60,* 136–140.

Lender, M., & Heidelberg, P. (1917). *Improvements relating to apparatus for training aviators.* British Patent Specification 158,522.

Lender, M., & Heidelberg, P. (1918). *Safe apparatus for instruction in the management of military aeroplanes.* British Patent Specification 127,820.

Levison, W. H. (1981). *Effects of whole-body motion simulation of flight skill development* (Report No. AFOSR-TR-82-006). Washington, DC: Office of Scientific Research.

Liebowitz, H. W. (1982). The two modes of processing concept and some implications. In J. Beck (Ed.), *Organization and representation in perception* (pp. 343–363). Hillsdale, NJ: Lawrence Erlbaum Associates.

Lintern, G. (1987). Flight simulation motion systems revisited. *Human Factors Society Bulletin, 30,* 1–3.

McCarthy, R. (1993). Assembling routines and addressing representations: An alternative conceptualization of "what" and "where" in the human brain. In N. Eilan, R. McCarthy, & B. Brewers (Eds.), *Spatial representation* (pp. 373–399). Oxford, England: Blackwell Press.

Milner, A. D., & Goodale, M. A. (1996). *The visual brain in action.* Oxford, England: Oxford University Press.

Montegut, M. J. (1995). *The emergence of visual detail.* Unpublished doctoral dissertation, University of California, Santa Cruz.

Mudd, S. (1968). Assessment of the fidelity of dynamic simulators. *Human Factors, 10,* 351–358.

National Research Council Committee on the Effects of Aircraft-Pilot Coupling on Flight Safety. (1997). *Aviation safety and pilot control—understanding and preventing unfavorable pilot-vehicle interactions.* Washington, DC: National Academy Press.

Nemire, K., Jacoby, R. H., & Ellis, S. R. (1994). Simulator fidelity of a virtual environment display. *Human Factors, 36,* 79–93.

Parrish, L. (1969). *Space-flight simulation technology.* Indianapolis, IN: Howard W. Sams.

Peters, R. A. (1969). *Dynamics of the vestibular system and their relation to motion perception, spatial disorientation, and illusions* (Contractor Report 1309). Moffett Field, CA: NASA Ames Research Center.

Proffitt, D. R., & Kaiser, M. K. (1996). Hi-lo stereo fusion. *Visual Proceedings of SIGGRAPH'96,* 146.

Reid, L. D., & Nahon, M. (1988). Response of airline pilots to variations in flight simulator motion algorithms. *Journal of Aircraft, 25,* 639–648.

Rolfe, J. M., & Staples, K. J. (1986). *Flight simulation.* Cambridge, England: Cambridge University Press.

Roscoe, S. N. (1980). *Aviation psychology.* Ames: Iowa State University Press.

Ryan, L. E., Scott, P. G., & Browning, R. F. (1978). *The effects of simulator landing practice and the contribution of motion simulation to P-3 training* (Report No. 63). Orlando, FL: Navy Training Analysis and Evaluation Group.

Samji, A., & Reid, L. D. (1992). The detection of low-amplitude yawing motion transients in a flight simulator. *IEEE Transactions on Systems, Man, and Cybernetics, 22,* 300–306.

Sanders, A. F. (1991). Simulation as a tool in the measurement of human performance. *Ergonomics, 34,* 995–1025.

Schell, J. (1998). Creating VR entertainment: Suggestions for developing a successful show. *SIGGRAPH '98 Course Notes 38, Immersive Environments: Research Applications, and Magic.*

Schroeder, J. A., & Johnson, W. W. (1996). Yaw motion cues in helicopter simulation. In *Flight simulation—where are the challenges?* (pp. 5-1–5-15). Neuilly-sur-Seine, France: NATO Advisory Group for Aerospace Research and Development.

Schroeder, J. A., Chung, W. W., & Laforce, S. (1997). Effects of roll and lateral flight simulation motion gains on a sidestep task. In *Proceedings of the American Helicopter Society's 53rd Annual Forum* (pp. 1007–1015). Alexandria, VA: American Helicopter Society.

Schroeder, J. A. (1999). *Helicopter flight simulation motion platform requirements* (Tech. Paper 1999–208766). Moffett Field, CA: NASA Ames Research Center.

Shirachi, D. K., & Shirley, R. S. (1981). *Visual/motion cue mismatch in a coordinated roll maneuver* (Contractor Report 166259). Moffett Field, CA: NASA Ames Research Center.

Shirley, R. S., & Young, L. R. (1968). Motion cues in man-vehicle control. *IEEE Transactions on Man-Machine Systems, 9,* 121–128.

Sinacori, J. B. (1973). A practical approach to motion simulation. In *Proceedings of the AIAA Visual and Motion Simulation Conference* (Paper 931, pp. 1–14). New York: American Institute of Aeronautics and Astronautics.

Sinacori, J. B. (1977). *The determination of some requirements for a helicopter flight research simulation facility* (Contractor Report 152066). Moffett Field, CA: NASA Ames Research Center.

Stapleford, R. J., Peters, R. A., & Alex, F. R. (1969). *Experiments and a model for pilot dynamics with visual and motion inputs* (Contractor Report 1325). Moffett Field, CA: NASA Ames Research Center.

Stewart, D. (1965). A platform with six-degrees-of-freedom. *Proceedings of Mechanical Engineers, 180,* 371–386.

van der Vaart, J. C., & Hosman, R. J. A. W. (1987). *Compensatory tracking in disturbance tasks and target following tasks: The influence of cockpit motion on performance and control behaviour* (Report LR-511). Delft, The Netherlands: Delft University of Technology.

van Gool, M. F. C. (1978). Influence of motion washout filters on pilot tracking performance. In *Piloted Aircraft Environment Simulation Techniques* (pp. 19–1–19–5). Neuilly-sur-Seine, France: NATO Advisory Group for Aerospace Research and Development.

Watson, A. B., Ahumada, A. J., Jr., & Farrell, J. (1986). Window of visibility: Psychophysical theory of fidelity in time-sampled visual motion displays. *Journal of the Optical Society of America, 3,* 300–307.

Wickens, C. D., & Baker, P. (1995). Cognitive issues in VR. In W. Barfield & T. A. Furness III (Eds.), *Virtual environments and advanced interface design* (pp. 514–541). New York: Oxford University Press.

Young, L. R. (1967). Some effects of motion cues on manual tracking. *Journal of Spacecraft, 4,* 1300–1303.

13

Applying Crew Resource Management Theory and Methods to the Operational Environment

Paul J. Sherman
The University of Texas at Dallas

Typically, when writing about crew resource management (CRM), it is incumbent on the author to trot out a "dark and stormy approach" anecdote. Rather than relate one of these classic CRM case studies (excellent examples of which can be found in Ginnett, 1993; O'Hare & Roscoe, 1990; Wiener, Kanki, & Helmreich, 1993), I will describe one of my own experiences as a flight deck observer and use the events of this flight to illustrate how CRM theory and methods drive safety-oriented goals and practices in aviation. To this end, I will discuss the evolution of CRM and provide some detail on CRM training and assessment. Also, I will briefly describe some of the effects that increased flight deck automation can have on team performance on the flight deck. Finally, I will provide a perspective on the future of CRM.

ONE FLIGHT SEGMENT, SEVERAL INCIDENTS

The flight was scheduled overnight service from Miami to a large South American city, on a highly automated widebody (i.e., capable of both lateral and vertical navigation via a flight management system, or FMS) operated by a large U.S. carrier. I met the crew, a senior captain a few years from retirement, a primary first officer (FO) in his 40s, and a thirty-something relief FO, at the gate at about 9:45 P.M. This was the crew members' first trip together.

We pushed back on schedule. During taxi, the captain held a preflight briefing, taking particular care to review engine-out-return-to-field procedures and duties. He added that this was the first time he had flown this particular trip in a long time (he was a reserve captain and had most recently been flying to Europe). He also advised the crew that he had spent the last few hours commuting and might be fatigued later in the flight. As he was Pilot Flying, he asked that they keep an eye on his performance in case he became fatigued. The other crew members volunteered that they had flown this route recently and were relatively well rested, although the relief FO had also spent the earlier part of the day commuting.

Takeoff and climb were routine. The crew members were careful to verbalize any relevant actions and observations. The captain took particular care to state changes to automated system status during climb out (e.g., "Autothrottles coming on now"; "Autopilot coming on now"). After leaving U.S. airspace, the crew made sure to verify Latin American air traffic controllers' (ATC) transmissions by checking with each other and requesting repeats when necessary. The crew took routine rest periods. All three crew members were back in the cockpit by about 6:00 A.M., well before the start of descent. The crew briefed the approach in a fairly comprehensive manner, covering missed approach procedures for the destination airport.

Approach control was late in providing an initial descent clearance; in addition, confusion arose regarding whether the controller had cleared them to FL270 (flight level 270 or 27,000 feet) or FL200 (20,000 feet). Three callbacks were required to determine they were cleared to FL200. As a result, they were somewhat hurried to comply with the clearance. During these interchanges, the controller did not convey any terminal weather information. The captain began the descent without engaging the weather radar on his map display. The primary FO's weather radar was engaged.

The aircraft entered clouds at about FL230. Immediately, the plane was subject to continuous moderate and occasional severe turbulence. The two FOs immediately began stowing loose items and took several moments to tighten their lap and shoulder restraints. At this point the captain told the relief FO to ring the cabin for a damage and injury report. The cabin reported that several carts had been bounced around, one flight attendant was vomiting and incapacitated and several passengers were ill as well, but no one was injured. At about FL200 the approach controller provided a heading and cleared us for a visual approach; the controller provided no other information. Continual moderate and occasional severe turbulence occurred throughout the descent, necessitating two more cabin calls. The cabin crew reported that the cabin was secure, but not at full crew complement due to the incapacitated attendant.

We broke out of the clouds around 2500 feet AGL (above ground level) in moderate rain. The airport was to the left of the aircraft, at about 9 o'clock. It quickly became apparent that we were too high and too close to the airport to continue a stabilized descent. The captain called for gear down and requested an immediate 270-degree right turn to rejoin the localizer, which was approved by the approach controller.

The captain began final approach, straining to get the aircraft down onto the glideslope. After a few moments, the relief FO called for a go-around, saying "Boss, I don't think we want to do this, do you?" to which the captain immediately replied "No, we don't, tell 'em we're going around." The captain began a climb and called for gear up/flaps and slats retract; however, the primary FO had already put the gear up and moved the flaps up one notch (i.e., uncommanded) on hearing the captain concur with the relief FO's go-around call. At this time, the crew lost several seconds clearing up the confusion regarding flap settings.

At about 1000 ft AGL, the crew began discussing the upcoming approach and checked fuel status. The relief FO suggested a visual go-around, requesting vectors and altitude to bring us back around to the localizer, rather than flying the published missed approach, which would put the aircraft back into turbulence. The captain agreed with this plan, so the primary FO requested vectors. Communication problems ensued. Speaking at a measured pace, the FO used standard phraseology to request vectors and a visual approach, but the controller simply radioed back "yes, Flight [xxx], uh, go-around." The FO attempted the request again, with the same result. At this point, and after confirming with one another that there was only one other aircraft approaching the terminal, the crew decided to dispense with confirming their plan with ATC and proceed.

At the beginning of the downwind leg, the captain turned on the autopilot and made an announcement to the passengers regarding the go-around, and the relief FO called the cabin crew for a final status report. The crew quickly ran the approach checklist and configured the airplane for landing. It was obvious that heavier rain, with possible microburst conditions, were approaching the area. If this landing attempt failed, another attempt could not be made until the heavy rain left the airfield. In order to avoid the approaching weather, the captain turned to join the localizer quite early, leaving very little time to stabilize the airplane and meet the glideslope. After quickly jockeying the airplane into position, the captain landed the aircraft in a slight tailwind.

It is not my intention to malign the crew in relating this flight vignette. On the whole, they rose to the challenges that this flight presented them and performed admirably in a difficult environment. Certain actions of the crew were excellent demonstrations of crew resource management skills. Certain of their actions had significant and negative effects and are instructive examples of poor management of crew resources. I will return to this vignette later in the chapter and use both its good and bad aspects to illustrate some points about CRM.

CREW RESOURCE MANAGEMENT—ITS THEORETICAL AND APPLIED GENESIS

The acronym *CRM* has been accorded many definitions since its original formulation 20 years ago as a means to denote training that sought to reduce flight crews' operational errors (Cooper, White, & Lauber, 1980; Helmreich, Merritt, & Wilhelm, 1999). Before defining CRM in exhaustive detail, it is instructive to briefly relate some relevant developments in academia and aviation that preceded the development of CRM.

The Paradigm Shift

During the late 1950s and 1960s, many studies of group performance focused on minimal group situations (*group* is defined here as a set of interacting individuals who share common goals, are somehow interdependent, and perceive that they are part of a group [Fiedler, 1967]). Studies of this type were mainly contrived (as opposed to naturalistic) and highly controlled, with a minimum number of dependent and independent variables. The group situations constructed by researchers were often artificial, of short duration, and not very compelling to group members, so they did not always possess the demanding, cohesive characteristics of real-world group situations (Steiner, 1972).

In the years spanning the late 1960s through the late 1970s, the focus of group performance studies broadened to include the examination of real-world groups, functioning in real-world environments. Particular attention was paid to the study of team processes and performance in task-oriented, risky, and demanding situations (Bakeman & Helmreich, 1975; Barnhart et al., 1975; Radloff & Helmreich, 1968; Ruffell-Smith, 1979; see also Cooper et al., 1980). In this context, teams are distinct from groups in that team members tend to hold stakes in common situational goals and have a higher degree of interdependencies on other team members (Hughes, Ginnett, & Curphy, 1999). Despite the reduced experimental control accompanying these types of studies, this approach held the promise of yielding rich, multifaceted findings with substantial ecological validity. They allowed examination of how teams maintain or fail to maintain performance in high-risk environments and the antecedents, processes, and consequences of in situ group performance.

During the mid-1970s, several aviation accidents occurred in which flight crews flew functioning aircraft into the ground (known as controlled flight into terrain, or CFIT, accidents; National Transportation Safety Board [NTSB], 1979). Shortly thereafter, studies were undertaken to uncover possible commonalities across these accidents and other critical incidents. Researchers used several methods of inquiry, including structured interviews of pilots, studies of accident investigation records, and an examination of reports in the Aviation Safety Reporting System (ASRS; Cooper et al., 1980; Helmreich & Foushee, 1993; Murphy, 1980).

The ASRS is an ongoing project under National Aeronautics and Space Administration (NASA) sponsorship. It provides pilots with the means to submit anonymous incident reports. The database is available for public inspection.

These investigations revealed that in these accidents, a number of the errors committed involved interpersonal rather than technical deficiencies (Billings & Reynards, 1984; Cooper et al., 1980; Foushee & Manos, 1981; Murphy, 1980; Wiener, 1993a). Cooper et al. categorized many of these deficiencies as preoccupation with minor technical problems; inadequate leadership; failure to delegate tasks, assign responsibilities, and set priorities; inadequate monitoring and use of available information; and failure to communicate intent and plans. Examples of some of these decrements can be observed in the flight vignette described, such as when the captain failed to engage his weather radar display (inadequate use of available information), when the primary FO failed to communicate the descent weather to the captain (also inadequate use of available information), and when the FO set the flaps without advising the others (a failure to communicate intent and plans). To better understand how these interpersonal deficiencies are seen to affect team performance, it is helpful to conceptualize performance from a systems viewpoint.

THE SYSTEMS PERSPECTIVE: MODELING GROUP INPUTS, PROCESSES, AND OUTPUTS

Team performance, assessed as the degree of effectiveness a team demonstrates in achieving its goals, can be viewed from a systems perspective (Hughes et al., 1999). This perspective is predicated on a model generically known as the *Input-Process-Outcome* (IPO) model (Helmreich & Foushee, 1993; Maurino, 1993; see also Tannenbaum, Beard, & Salas, 1992). Table 13.1 contains a generalized list of inputs, processes, and outcomes.

TABLE 13.1
Crew performance input factors, process factors, and outcomes

Input Factors	*Process Factors*	*Outcomes*
Individual characteristics	Communication/decision tasks	Performance of the mission
Physical condition	Team formation/mgmt tasks	Performance of the crew
Crew composition	Situation awareness tasks	Individual morale, attitudes
Organizational factors	Workload management tasks	
Regulatory factors	Aircraft control tasks	
Operating environment	Procedural tasks	
Pilots' professional culture		
National culture		

Note: Adapted from Helmreich and Foushee (1993).

The IPO model holds that performance in aviation is group based—outcomes are dependent not only on individual performance, but also on how team members coordinate their activities and influence each other's process and outcome factors. This view of team performance led to better understanding of the CFIT accidents described, as well as accidents that were attributed to team failures in interpersonal communications, task management, and group decision making (for details, see Kayten, 1993).

The Model as Applied to Aviation: Characteristics and Model Dynamics

Input Factors. As Table 13.1 indicates, the input factors in aviation are categorized into individual, team, organizational, regulatory, and environmental factors, as well as professional and national culture influences. Individual input factors are defined as the individual contributions of members—their knowledge, skills, abilities, personalities, attitudes, values, and motivation, as well as their physical condition (i.e., their overall and current state of health and their current levels of fatigue and alertness). For example, pilots' experience (both in terms of overall flight hours and experience with the aircraft they currently fly) largely determines their knowledge. Team input factors reflect the combination of the attributes of the individuals who form the team—the characteristics of the group that arise from the interaction of each member's personality, intelligence, knowledge, training background, and emotional state.

Organizational, regulatory, environmental, and cultural factors are more distal inputs. They affect the operational milieu only indirectly. However, these inputs can have a pervasive effect on proximal inputs that originate with team members and their interactions. For example, pilots' knowledge, skills, and abilities are greatly affected by the organization and regulatory environment, because organizational, regulatory, environmental, and cultural factors determine, in large part, the minimum acceptable level of training and proficiency each individual is held to.

Group climate and morale can also be greatly affected by organizational variables such as the policies regarding scheduling and equipment type bidding, the nature of general worker-management relations, or the financial stability of the organization. In commercial aviation, the regulatory environment, with its precise rules for duty/rest hours and minimum proficiency levels necessary for performance of certain flight deck tasks, can also affect the nature of flight deck team interactions. For example, in the flight described, the regulatory environment permitted a reserve captain to fly international trips to both Europe and South America in the same duty month. The airline's scheduling and reserve policies also permitted this (airlines can and do disallow certain trip combinations that are perfectly valid from the regulatory standpoint, but may reduce safety margins).

The airline also permitted the crew to engage in long commutes before beginning their duty day. This too may have influenced the team's overall effectiveness.

Finally, national culture can affect the nature and quality of team interactions. National culture can be conceptualized as the overarching value system overlaying the organization's primary nation of origin. As such, it also tends to have a strong influence on superior-subordinate relations.

The input factors can be arrayed in concentric layers around the team (much like the layers of an onion), with the individual and team interaction factors comprising the innermost layers and the organizational and cultural factors comprising the outermost layers. This conceptualization is often referred to as the "organizational shell" (Hughes et al., 1999). This heuristic arrangement is seen as a useful conceptualization of the proximal and distal environment in which teams function (Hughes et al., 1999).

Process Factors. Process factors in aviation include cognitive and interpersonal functions such as communication and decision tasks (e.g., verbalizing one's own actions and inquiring into other members' actions when necessary) and team formation and management tasks (e.g., filling leadership/subordinate roles, maintaining a positive team climate). Process factors are also comprised of more physical functions such as flight control and operations of systems.

Research into communication among flight crew team members has shown that increased amounts of certain types of communication were associated with lower error rates (Foushee & Manos, 1981; Kanki & Palmer, 1993). Foushee and Manos studied the cockpit voice recordings from another researcher's (Ruffell-Smith, 1979) earlier study in which 18 three-person crews flew a simulated two-leg flight in a Boeing 747. Foushee and Manos tabulated every utterance and classified each one into categories, such as observations of operational status, inquiries, acknowledgements, and commands. The amount of speech had a significant effect. Those crews who communicated more were also those crews rated as higher performing in Ruffell-Smith's study. These crews also showed greater verbalization of flight status information, in addition to higher levels of overall communication.

It has also been demonstrated that communication patterns are associated with differing levels of crew performance. Kanki and Palmer (1993) reported on a reanalysis of flight simulation voice recording data collected by Foushee, Lauber, Baetge, and Acomb (1986). Communications from 18 of 20 total crews who flew a full-mission simulation (i.e., multiple legs, with substantial cruise phases) were subject to content analysis, similar to the technique used by Foushee and Manos. The higher performing crews, indicated by observer ratings and objective flight data, showed greater similarity of communication patterns among crew members. That is, captains and first officers from the higher performing crews uttered roughly the same proportion of commands, questions, and acknowledgements; captains and first officers from the low-performing crews showed no such correspondence.

Orasanu (1990), in correlational research examining decision-making processes among the flight crews originally studied in Foushee et al. (1986), found that crews rated as high performing exhibited higher levels of metacognitive activity (i.e., explicit definition and verbalization of plans and strategies and preparation for contingencies). Orasanu and colleagues have also found that decision making among high-performing crews in full-mission simulation studies is also characterized by the development of shared, explicitly verbalized mental models. Orasanu and Fisher (1992) coded the utterances of two- and three-person commercial airline crews performing full-mission simulations and found that higher performing crews (as judged by expert raters) talked about the problems faced in the flight, discussed how the problems should be solved, and discussed task assignments for solving problems, more than did lower performing crews. In other words, high-performing crews defined the problems they faced, discussed them to ensure that all crew members shared an understanding of them, and made explicit plans for how to solve them. More detailed descriptions of the experimental work leading to these conclusions are available in Orasanu (1993).

Finally, inquiry into team formation by Ginnett (1987, 1993) showed that captains rated as effective performers were more likely to hold briefings that (a) established their authority while simultaneously inviting crew members' active participation, (b) established explicit expectations for performance of group tasks without going into minute detail concerning how to perform tasks, and (c) acknowledged that others in the aviation environment, such as cabin crew and/or maintenance, could aid in accomplishing the flight-deck team's goals. This work will be elaborated on further later in the chapter.

Outcome Factors. The outcome factors in the IPO model can include mission and crew performance results such as the overall safety and efficiency of the flight (i.e., the degree to which the flight was conducted in a safe manner and with adherence to company and regulatory guidelines), and individual and organizational outcomes such as satisfaction, motivation, and morale (Helmreich & Foushee, 1993; McGrath, 1964). In many domains where this model applies, the ultimate outcome measure is the accident rate; however, the base rate of accidents in commercial aviation is very low, which makes it difficult to assess the effects of the various input and process variables. Other measures, such as critical incidents, expert observer ratings, error records, and self-report measures of attitudes, values, and job satisfaction, provide alternate assessments of group outcomes (Helmreich & Foushee, 1993; Reason, 1997). The utility of these various surrogate measures will be discussed in greater detail next.

Recursiveness: A Feature of the IPO Model. In the IPO model, group inputs generally determine the nature of the processes and outcomes. An important feature of the model, however, is its recursive nature: outcomes can change the nature of group input factors, thereby changing group processes. For example, exemplary completion of a flight, with large safety margins and compliance to organizational and regulatory requirements, can lead to positive morale, improved attitudes, and a favorable team climate. These outcomes can then contribute to the input factors of future operations. This recursiveness can also work in a negative manner—that is, negative outcomes (for example, poor team performance or intentional violations of procedures or policies) may lead to reduced morale and motivation, setting of norms for flight-deck behavior that compromise safety, and other input factors.

Applying the IPO Model to Flight-Deck Team Training

As the results of the research programs described were disseminated through the operational community, several commercial flight training departments endeavored to impart the information to pilots in hopes that it might help them to avoid experiencing similar performance decrements in subsequent flights. This intervention was broadly termed Cockpit (later Crew) Resource Management. Lauber (1984, p. 20) defined CRM as "using all available resources—information, equipment, and people—to achieve safe and efficient flight operations." Regulatory bodies have shown broad support for this definition. The Federal Aviation Administration's Advisory Circular 120–51c (FAA, 1998) states that "CRM refers to the effective use of all available resources. . . . [it] is one way of addressing the challenge of optimizing the human/machine interface and accompanying interpersonal activities." The Joint Aviation Administration (JAA) defines CRM as "the effective utilisation of all available resources (e.g. crew members, aeroplane systems, and supporting facilities) to achieve safe and efficient operation" (JAA, 1998). This definition of CRM is adopted for this chapter.

A means of increasing the probability that team members adequately utilize the resources available to them would be to provide members with techniques to better manage these resources. As Salas and colleagues elaborate (Salas et al., 1999):

> CRM training [is] a family of instructional strategies designed to improve teamwork in the cockpit by applying well-tested training tools (e.g., performance measures, exercises, feedback mechanisms) and appropriate training methods (e.g., simulators, lectures, videos) targeted at specific content (i.e., teamwork knowledge, skills, and attitudes). (p. 163)

This definition of CRM training complements what Helmreich and colleagues see as the primary goal of CRM training, which is to provide crews with tools and techniques to prevent, mitigate, and manage errors. This point will be further explored later in the chapter. In general, though, it can be said that one way to improve teamwork would be to provide an intervention or series of interventions that lead team members to better utilize their own and other members' knowledge, skills, and abilities.

To further illustrate this point, consider how CRM training might address the events of the South American flight described at the beginning of the chapter. One potentially effective training intervention might be one that helped the crew to avoid underutilizing available information such as the weather situation. Another intervention might attempt to increase crews' awareness of (and compensation for) differences in pilot–air traffic control interaction in the Latin American versus U.S. operational environments. Still another intervention might aim to increase the tendency for pilots to consistently verbalize actions they take that can have a critical effect on other's tasks and the flight as a whole. Finally, training may seek to impart techniques so crews could plan for rare but perilous contingencies such as severe turbulence during approach or uncommunicative controllers.

Helmreich and Foushee (1993) describe the theoretical mechanism for how training interventions accomplish these goals. They state that communication to pilots of information concerning effective and ineffective team performance, in concert with efforts to change and/or reinforce attitudes regarding appropriate management of flight-deck tasks, should lead to improvements in team processes and team performance.

Others have further elaborated on the theoretical mechanism underlying CRM training, as well as the primary goal and overall approach to providing training. Hendy and colleagues (Hendy & Ho, 1998; Hendy & Lichacz, 1999), drawing on a model that posits humans as goal-directed, error-correcting systems, hold that the objective of CRM training is to provide participants with methods to better manage the finite resources of time, knowledge, and attention. In IPO model terms, this view is process oriented. It suggests that decision making, because it determines what courses of action team members take in the operational environment, is primary. Therefore, training should be focused on pilots' decisions. That is, training should do the following: (a) reinforce good decision making, (b) reduce or eliminate decisions that lead to nonoptimal use of time (e.g., task loading oneself during a busy approach and leaving the other pilot idle or vice versa), (c) lead to explicit elicitation of knowledge from others when appropriate (e.g., a captain new to a particular operational theater failing to elicit a veteran first officer's vast operational experience), and (d) ensure that the crew members regulate their attention so it does not wander from critical tasks (e.g., avoiding situations where the entire crew becoming engrossed in a minor technical problem, without clearly defining who has responsibility for the flight controls).

This position is considered in concert with a view of learning that describes training efficacy as a function of whether training leads participants to apply lesson concepts into knowledge, knowledge into skills, and skills into everyday practice (Hendy & Ho, 1998). Thus, the favored approach to influencing participants' decision making is to (a) provide participants with the theoretical underpinnings of decision making, information processing, and team performance concepts; (b) have participants apply this information by assessing "canned" situations such as flight vignettes, accident reports, and case studies; and (c) allow participants to practice assessing situations and acting on their assessments, in role-playing activities and high-fidelity simulations.

The approaches described (Helmreich & Foushee, 1993; Hendy & Ho, 1998; Hendy & Lichacz, 1999) were devised years after the first CRM programs were instituted and thus have the benefit of hindsight. Early training interventions did not have the advantage of rigorously defined theoretical contexts. In the hope of fostering greater understanding of the evolution of CRM theory and practice, the next section describes how CRM training interventions were initially formulated by the aviation training community and provides some detail on how pilots were instructed.

EARLY APPROACHES TO IMPROVING TEAM PERFORMANCE ON THE FLIGHT DECK

Initial implementations of CRM training during the early 1980s sought to bring about behavioral change by describing to participants the negative effects of subordinate crew members failing to speak up and imparting to participants an awareness of their own "managerial style," in seminar-type classes based on Blake and Mouton's (1964) work on managerial effectiveness. Training was focused on encouraging pilots to adapt their behavior to augment or compensate for their purported "style of management." For example, a captain who was seen as excessively authoritarian would be encouraged to solicit more information and team participation from junior officers. A first officer or flight engineer who was characterized as lacking assertiveness would be encouraged to speak up during flight operations and be more assertive in relation to senior officers (Helmreich et al., 1999b).

Although pilots generally regarded the concepts introduced in these early training attempts as worthy of attention, this approach was judged by some researchers, practitioners, and pilots to be too vague to bring about change in pilots' flight deck behavior (Helmreich et al., 1999b). As Helmreich et al. state, early CRM courses "advocated general strategies of interpersonal behavior without providing clear definitions of appropriate behavior in the cockpit. Many employed games and exercises unrelated to aviation to illustrate concepts." (1999b, p. 19). These early interventions were seen as inadequate because, although they introduced CRM concepts, they often did not strongly link the information imparted in the classroom to practices in the operational environment.

Refinement of CRM Training

By the late 1980s, CRM training was shifting from generalized, "psychological-" and management-based approaches to more aviation-relevant operationally focused methods. One impetus for this was the desire by airlines to evaluate and improve their approaches to training. Another was the effort by the research community to provide the aviation industry with improved metrics for measuring training efficacy (Helmreich et al., 1999b; Orlady & Foushee, 1987).

Many airlines began to implement training programs focused on bringing about distinct, observable behaviors in specific phases of flight operations. For example, the focus of some CRM training modules from this period was driven by team formation and leadership research such as Ginnett's work (1987, 1993). Ginnett, under NASA sponsorship, studied 20 three-person 727–200 crews as they came together for the first time to fly an actual (not simulated) 2- to 5-day series of flights. Ginnett's selection procedures ensured that the majority of the three-person teams were composed of members who had never worked together in previous flights. Furthermore, each of these crews was led by a captain previously identified by the airline's training pilots as either particularly skilled or particularly unskilled at creating effective, high-performing teams. Observers who were blind to the captains' ratings observed the crews from the moment of first meeting to the end of the line trip.

Ginnett found that captains previously identified by their management pilots as particularly skilled at creating high-performing teams were found to have held crew briefings and debriefings that (a) established their competence as leaders while concurrently engaging the other crew members to be active participants in the operation, (b) established explicit expectations (i.e., norms) for performance and reinforced them with their own behavior, and (c) expanded the definition of the flight-deck crew to include others (such as cabin crew or dispatch) who may be in a position to help accomplish the team's goals.

Drawing on this type of scholarly work, training programs began to stress development of effective briefing procedures (such as those just described) as one way to foster group cohesion and the growth of shared expectations for the flight. However, these techniques and strategies, for all their increased specificity, were often taught in seminar-type situations (rather than in more operationally relevant settings) and sometimes employed terminology and examples that were not relevant to the aviation domain. This may have reduced pilots' training "buy-in" (i.e., acceptance of training). Some pilots had difficulty applying the concepts to their operating environment, and some tended to reject the training (Helmreich & Wilhelm, 1989). Additionally, it became evident that as pilots became more familiar with and practiced CRM techniques, others who strongly influenced the operational environment (i.e., check airmen, instructors, corporate policymakers) but had no exposure to CRM training, could inadvertently undermine the effects of CRM training (Chidester, 1993).

This contributed to the conjecture that merely providing a single incidence of CRM training to pilots was insufficient to bring about pervasive, lasting change. In order for training to be effective, others within the organization had to be aware of and support the training intervention, and the training had to be supported with recurrent training. Some support for this claim is provided by Helmreich et al. (1999b).

Two airlines that began providing CRM training in the late 1980s were examined. The measure of CRM concepts used was the Cockpit Management Attitudes Questionnaire, a paper-and-pencil instrument that measures pilots' endorsement or rejection of statements representing specific domains of interpersonal performance on the flight deck. Helmreich and colleagues (1999b) documented a decrease in acceptance of basic CRM concepts over a period of several years at one organization. They attributed this finding to the fact that the airline did not provide CRM training to instructor and check pilots and provided no recurrent CRM training to pilots. The airline that showed no decrease in acceptance provided both initial and recurrent CRM training.

It should be noted that this study did not (and could not) control for other organizational effects, such as the financial health of the measured organizations or the nature of pilot-management relations, and it contrasted two different training programs. Although Helmreich et al.'s (1999b) findings are promising, further research is certainly warranted.

Partly as an outgrowth of this supposition and the suggestive empirical results, CRM training began to be extended to those who had responsibility for training and of air crews—instructors and check pilots. Concomitant with the refinement of CRM training, metrics to gauge CRM training efficacy were being developed and refined.

MEASURING CRM TRAINING EFFICACY

Because CRM training is intended to effect both specific and broadly drawn behavioral changes on the flight deck, there are no objective "pass/fail" indicators of successful CRM training. It was (and in many ways, still is) difficult to ascertain the efficacy of CRM training. Any intervention strategy, however, that seeks to change people's behavior must establish its efficacy—that is, the intervention must be shown to accomplish the effects it purports to bring about.

True to CRM's roots in psychological inquiry, evaluation tools generally focus on the standard dependent variables of attitude self-reports and behavioral assessments by others. The most widely utilized measures of CRM training effectiveness are pre- and posttraining measurement of crews' attitudes toward CRM-related concepts using survey methods and evaluations of crews' simulator and flight-deck performance.

Attitude Measurement and Its Use in CRM Training

In aviation, survey research methods are seen by researchers and practitioners to be useful both in determining whether training has resulted in desired attitude change and in identifying potential threats to safety that are addressable in training. Although some researchers have demonstrated that in certain situations survey research is a poor substitute for behavioral measurement (Wicker, 1969,

1971; Wise, 1996), when well designed and used in the right context, surveys can provide clear indications of probable behavior.

Investigators have demonstrated attitude and behavior linkages when (a) individuals' perceptions of the situation are congruent with the attitudes, (b) subjective norms (the implicit rules for behavior in place in a particular situation) influence behavior in the direction implied by the attitude, (c) the attitude is based on direct experience with the attitude object, and (d) there is a strong association between attitude expression and behavior (Fazio, 1986; Fazio & Zanna, 1981). Complementing this view, Ajzen and Fishbein (1977) state that when an attitude is general and the behavior is circumscribed, there should not be a close attitude-behavior correspondence. If a pilot endorses the attitude item "Crew members should work with each other to improve the safety of flight," it should not be expected that this would predict specific behaviors during flight operations. However, when both the attitude and the behavior are specific and closely related, there is likely to be close attitude-behavior correspondence. A pilot who endorses the item "I let other crew members know when my workload is becoming (or about to become) excessive," however, is seen as more likely to include behaviors reflective of this item in their repertoire of actions on the flight deck.

The use of attitude surveys in aviation will be elaborated on in a later section of this chapter; what follows is a more in-depth discussion of the ways researchers define and operationalize the term *attitude* and attitude-behavior correspondence (i.e., how well a measured attitude predicts behavior).

Attitudes as Indicators of Behavior. The study of attitudes is one of the more venerable traditions of social psychology. McGuire (1985) reports that J. B. Watson and sociologists Thomas and Znaniecki regarded social psychology as the study of attitudes. Theorists have differed in their definition of the attitude construct. The most broadly drawn definitions of an attitude describe it as any categorization of an object along an evaluative dimension (Fishbein & Ajzen, 1975). McGuire (1985) more precisely defines attitude as "a unifying mediational construct that provides an economical and provocative depiction of the interrelations between a set of antecedent conditions and consequent responses" (p. 240) or, more broadly, as generally covert "responses that locate objects of thought on dimensions of judgment" (p. 238). As such, attitudes are distinguishable from constructs such as knowledge, opinions, values, habits, motives, traits, emotions, and interests. A more precise distinction between attitudes and values can be drawn: values are broader in scope than attitudes, encompassing many "objects" in a single value judgment (Sherman, 1997).

Seminal work in the attitude domain was done with the implicit assumption that there was a one-to-one correspondence between attitudes and behavior; this was later seen as an overly simplistic representation of the attitude-behavior linkage (McGuire, 1985). By the 1970s, researchers were questioning the nature of this linkage and challenging the usefulness of the construct, given its low predictive utility. Wicker (1969, 1971) represents the high-water mark of attitude criticism, arguing for abandonment of attitude studies on the grounds that they did not adequately predict behavior.

Fazio (1986) countered these objections by claiming that behavior is a function of the person's "perceptions in the immediate situation in which the attitude object is encountered" (p. 207). This formulation assumes that behavior is to a large extent driven by perceptions. These perceptions of the attitude object in the immediate situation, as well as perceptions of the situation as a whole (that is, the environment in which the individual encounters the attitude object), help generate an individual's definition of the event. The individuals' definition of the event in turn is mediated by their attitudes. Attitudes guide information processing in the social situation, thus guiding an individual's perceptions of the situation and, consequently, the individual's behavior in relation to the situation.

In Fazio's (1986) view, subjective norms (i.e., individuals' expectations for behavior in a particular situation) affect attitude-behavior linkage. This has important implications; it suggests that when normative guidelines are in opposition to an individual's attitudes, the attitude-behavior consistency will be reduced. Also, attitude-behavior consistency can be reduced if the attitude is not retrieved from memory when the individual encounters the attitude object. Evidence for this was provided by Snyder and Kendzierski (1982), who found greater attitude-behavior consistency among subjects whose positive attitudes toward a particular behavior were "activated" by a confederate in an experimental situation.

Attitudes in Aviation: Training Design and Evaluation Using Survey Methods. One survey tool for design and evaluation of CRM training that came into widespread use during the early and mid-1980s was the Cockpit Management Attitudes Questionnaire (CMAQ; Gregorich, Helmreich, & Wilhelm, 1990; Helmreich, 1984; Helmreich, Wilhelm, & Gregorich, 1988). The CMAQ items comprise three scales that measure attitudes toward specific domains of interpersonal performance on the flight deck. The Communication and Coordination scale encompasses "communication of intent and plans, delegation of tasks and assignment of responsibilities, and the monitoring of crew members" (Gregorich et al., 1990, p. 689). The Command Responsibility dimension incorporates the notions of "appropriate leadership and its implications for the delegation of tasks and responsibilities" (p. 689). The third scale, Recognition of Stressor Effects, measures the respondent's "consideration of—and possible compensation for—stressors" (p. 689).

These scales have been used to investigate relationships between pilots' attitudes toward management of flight and crew interactions. The predictive validity of these constructs has been established by demonstrating their ability to discriminate between expert observer evaluations of effective and ineffective behavior in line operations (Helmreich, Foushee, Benson, & Russini, 1986). In this validation study, a group of expert observers (i.e., supervisory pilots) in a single airline rated a group of 114 pilots with whom they had direct operational experience on overall flight-deck management skill. Pilots were sorted into high-rated and low-rated groups. Discriminant analysis was used to contrast the CMAQ scale responses of pilots judged to be above average and those judged to be below average in overall flight deck management skill. Discriminant analysis is a technique whereby subjects' responses on a set of stimuli are used to generate a function that is employed to classify subjects into a (usually small) number of categories (Klecka, 1980; Lachenbruch, 1975). The meanings of the categories are subject to interpretation by the researcher. In this study, it was found that the discriminant function correctly classified 95.7% of pilots into their rated group (i.e., above average or below average in flight-deck management skill).

The Flight Management Attitudes Questionnaire (FMAQ; Helmreich, Merritt, Sherman, Gregorich, & Wiener, 1993), an expanded version of the CMAQ, contains additional items probing attitudes toward command and leadership, crew communication and coordination, stress recognition, work values, and team behaviors. These items appear in Table 13.2.

TABLE 13.2
Items measuring flight management attitudes, from the FMAQ

1. The captain should take physical control and fly the aircraft in emergency and nonstandard situations.
2. Even when fatigued, I perform effectively during critical times in a flight.
3. The airline's rules should not be broken—even when the employee thinks it is in the airline's best interests.
4. Senior staff deserve extra benefits and privileges.
5. I let other crew members know when my workload is becoming (or about to become) excessive.
6. Debriefing the flight can be a valuable learning experience.
7. My decision-making ability is as good in emergencies as in routine flying conditions.
8. I am more likely to make judgment errors in an emergency.
9. Successful flight-deck management is primarily a function of the captain's flying proficiency.
10. I am ashamed when I make a mistake in front of my other crew members.
11. Crew members should not question actions of the captain except when they threaten the safety of the flight.
12. I am less effective when stressed or fatigued.
13. An essential captain duty is training first officers.
14. My performance is not adversely affected by working with an inexperienced or less-capable crew member.
15. Crew members should monitor each other for signs of stress or fatigue.
16. Personal problems can adversely affect my performance.
17. A truly professional crew member can leave personal problems behind when flying.
18. Except for total incapacitation of the captain, the first officer should never assume command of the aircraft.
19. Written procedures are necessary for all in-flight situations.
20. A true professional does not make mistakes.
21. Crew members should mention their stress or physical problems to other crew before or during a flight.

The FMAQ also includes items designed to assess group-level differences on dimensions of culture described by Hofstede (1980, 1991) and dimensions described by Merritt (1996) and Merritt and Helmreich (1996). These dimensions of culture represent broad characterizations of national cultural groups' level of individualism, deference to authority, achievement orientation, and ability to deal with ambiguity. Subsequent psychometric analyses of the expanded instrument have established its psychometric properties (Merritt, 1997), although further work to reestablish the relationship between these updated attitude measures and pilot performance would be appropriate. In general though, the attitude items of the CMAQ/FMAQ questionnaires are circumscribed and closely reflect operational situations, and they have served the CRM community well as a tool with which to assess baseline CRM attitudes and to measure training efficacy. The next section provides information on how crews' CRM skills can be assessed more directly.

Performance Evaluation: Expert Ratings in the Simulator and on the Line

In contrast to indirect assessment of crews' CRM skills through survey research, trained raters can be used to evaluate the actual performance of airline crews in simulations and in actual operations (Fowlkes, Lane, Salas, Franz, & Oser, 1994; Wilhelm, 1991). This technique, as with survey research, is empirically based in earlier psychological research.

Performance Ratings by Trained Raters. There is a considerable body of evidence showing the effectiveness of trained raters on performance rating tasks (Law & Sherman, 1995). Researchers have shown that trained observer ratings are less lenient on average than are peer ratings (Fiske & Cox, 1960; Freeberg, 1969; Rothaus, Morton, & Hanson, 1965). Furthermore, observers who have been trained to avoid typical rating troubles demonstrate reduced leniency (where leniency refers to a pervasive shift in mean ratings from the scale midpoint to the upper (i.e., better) end of the scale [Law & Sherman, 1995]) and less halo bias (that is, ratings that show a lack of discrimination of a ratee's performance on distinct rating dimensions [Law & Sherman, 1995; Ivancevich, 1979]), and increased accuracy (Borman, 1975). Training raters in the use of a particular rating instrument, however, affords even greater reduction in rating errors (Bernardin & Walter, 1977; Gordon, 1972).

During the mid- and late 1980s, an effort was focused on developing a rating tool that would allow trained raters to assess specific aspects of crew performance, both in line-oriented flight training (LOFT) and in regular operations. LOFT is a simulator-based training session for aircrews that provides instruction in both technical and CRM skills in an environment that seeks to recreate as close as possible the conditions found in actual flight operations (Butler, 1993; Lauber

& Foushee, 1981). The result of this work was the LLC, for "Line and Line-oriented simulation Checklist." (Helmreich, Butler, Taggart, & Wilhelm, 1994; Helmreich, Klinect, & Wilhelm, 1999a; Wilhelm & Helmreich, 1996)

The LLC prompts for ratings of crews' performance of specific behaviors that are indicative of effective CRM. The item statements were derived from an analysis of behaviors that were present in a number of incidents and accidents where crews demonstrated effective teamwork (Helmreich et al., 1995). The items can be grouped into several overarching categories generally representing team formation and maintenance, leadership/followership, and communication and coordination of intent and plans. An illustrative team formation and maintenance item would be "Briefings are operationally thorough, and address crew coordination and planning for potential problems. Expectations are set for how possible deviations from normal operations are to be handled, e.g., rejected takeoff, engine failure after rotation, go-around at destination" (p. 2). The instrument also contains two global ratings, one for overall crew technical proficiency and one for overall crew effectiveness.

The instrument provides a four-choice behaviorally anchored rating scale for each item. This means that the ratings assigned to crews for each item are based on the degree to which desired behavior is present (or absent). A "Poor" rating on an item is indicated by the following description: "Observed performance (on this item) is significantly below expectations. This includes instances where necessary behavior was not present, and examples of inappropriate behavior that was detrimental to mission effectiveness." A "Minimum Expectations" rating is seen as one where "Observed performance meets minimum requirements but there is ample room for improvement. This level of performance is less than desired for effective crew operations." A rating of "Standard" denotes that "The demonstrated behavior promotes and maintains crew effectiveness. This is the level of performance that should be normally occurring in flight operations." Finally, an "Outstanding" rating means that a crew's performance ". . . represents exceptional skill in the application of specific behaviors, and serves as model for noteworthy and effective teamwork."

The LLC was utilized by raters (usually pilots, instructors, or researchers) who were typically trained to calibration by watching, evaluating, and then discussing videotaped simulator-based flight vignettes that had been previously judged by other expert raters as representative of high-, average-, and poor-performing crews. After raters had been adequately calibrated, they would observe real crews in line or simulator operations and provide overall ratings of the crews' performance on the specific behavioral markers and on the overall technical and CRM performance items. The LLC also includes a set of behavioral markers prompting for ratings of the crews' performance with regard to operation of automated systems. This is seen as important because crews' use of automation can affect how they communicate and coordinate with each other during flight. These automation-related items inquire as to whether crews communicated changes to automated system parameters to others, remained vigilant for automation failures, and used different automation capabilities at appropriate times, across different phases of flight (i.e., predeparture, takeoff/climb, cruise, and approach/landing). The rating items (including the automation-related items) are presented in Table 13.3 (Wilhelm & Helmreich, 1996).

Recently, the LLC has been expanded to capture information from raters regarding crew errors and their error avoidance strategies, as well whether threats to safety originating from the operational environment (e.g., severe weather, mechanical failures, or poorly performing air traffic controllers) occurred during the rated flight segment (Helmreich et al., 1999a).

A psychometric study of the LLC provides some evidence that the LLC is a reliable measure of crew performance. A 1995 examination of 34 pilots training for CRM evaluator positions in a major U.S. airline demonstrated that one group of raters viewing a crew previously judged as demonstrating average to above-average performance in a full-mission simulation produced interrater agreement coefficients of .82 for overall crew effectiveness ratings and .94 for overall technical proficiency ratings. Another group of raters viewing a below-average crew yielded agreement coefficients of 1.0 for overall crew effectiveness and .78 for technical proficiency ratings (Law & Sherman, 1995).

Although these are encouraging findings, this study has some limitations. It was a post hoc examination of archival data, and the study authors were not privy to the details of the rater training sessions. A more rigorous psychometric evaluation of the LLC would require control and observation of the training sessions, as well as gathering of rater demographic data to ensure equivalence across the rater groups. Also, this study did not establish agreement coefficients for the other,

TABLE 13.3
Behavioral checklist items from the LLC

1. Captain coordinates flight-deck activities to establish proper balance between command authority and crew member participation and acts decisively when the situation requires.
2. Team concept and environment for open communications established and/or maintained (e.g., crew members listen with patience, do not interrupt or "talk over," do not rush through the briefing, make eye contact as appropriate).
3. Briefings are operationally thorough, interesting, and address crew coordination and planning for potential problems. Expectations are set for how possible deviations from normal operations are to be handled (e.g., rejected takeoff, engine failure after liftoff, go-around at destination).
4. Operational decisions are clearly stated to other crew members and acknowledged and include cabin crew and others when appropriate (e.g., good cross-talk between pilots, everyone "on same page").
5. Crew members demonstrate high levels of vigilance in both high and low workload conditions (e.g., active monitoring, scanning, cross-checking, attending to radio calls, switch settings, altitude callouts, crossing restrictions).
6. Crew prepares for expected or contingency situations including approaches, weather, and so forth (e.g., stays "ahead of curve").
7. Automated systems used at appropriate levels (i.e., when programming demands reduce situational awareness and create overloads, the level of automation is reduced or disengaged). Automation is effectively used to reduce workload.
8. Crew members verbalize and acknowledge entries and changes to automated systems parameters.

more specific LLC items. This too would be done in a more rigorous psychometric evaluation of the LLC.

The LLC, although still in need of psychometric fleshing out, has proved to be a useful tool in evaluating crews' performance. It allows crew-based, rather than just individual-based, evaluations. It also allows real-time observation, either in actual operations or in simulated missions. Finally, it focuses on behaviors whose absences were shown to be implicated in several commercial aviation accidents (Helmreich et al., 1995), and thus the instrument possesses a high degree of ecological validity.

FURTHER REFINEMENT OF CRM CONCEPTS

As CRM programs evolved during the late 1980s and early 1990s, it was perceived by some inside the aviation field that the focus and message of CRM had become diffuse. This was evidenced by pilot perceptions of CRM—many saw it simply as a training module intended to foster flight deck harmony. Even worse, some saw it as an attempt to take authority away from the captain. CRM's raison d'être as a set of tools to manage error was not (and to this day is still not) always recognized by those who received CRM training (Merritt & Helmreich, 1997).

In response to the tendency for some to view (and sometimes teach) CRM as synonymous with almost any type of interpersonal consciousness-raising seminar (Chidester, 1993), more empirically minded practitioners and researchers sought to reclaim it by repatriating it with its original goals. Most recently, researchers have drawn on Wiener's (1993b) and Reason's (1990, 1997) conceptualizations of error management to further refine CRM training, placing renewed emphasis onto the prevention, recognition, and mitigation of crew error and other threats to safety (Helmreich, Klinect, & Wilhelm, 1999b; Helmreich & Merritt, 1998; Jones & Tesmer, 1999).

Merritt and Helmreich, (1997; see also Helmreich & Merritt, 1998), also drawing on Reason's and Wiener's work, note that commission of error is unavoidable in any human endeavor and call for a "normalization" of error in both training and operational environments. In this view, acknowledgement and acceptance of the inevitability of error paves the way for the design and implementation of safeguards and countermeasures against error. The goals of training and operational practices, then, become (a) reducing the likelihood of error, (b) "trapping" errors before they have an operational effect, and (c) mitigating the consequences of error (Merritt & Helmreich, 1997; Wiener, 1993b).

As a result of the work described, some designers of current CRM training programs are beginning to emphasize specific behavioral techniques intended to foster recognition, avoidance, and amelioration of error and threats to safety. An example of this is related by Jones and Tesmer (1999), who report that, based on data collected during safety audits at a major U.S. airline, in a number of cases crew acceptance of suboptimal ATC commands led to a number of error conditions during flight operations. Examples of suboptimal commands from ATC include situations such as when a controller requests that a crew follow a flight path that it is not possible to follow on a particular aircraft or when a controller issues a command that would place severe time pressure on the crew. Another example of suboptimal ATC commands would include the approach controller in the flight vignette related earlier stating "yes, Flight [xxx], uh, go-around," instead of issuing vector instructions.

This finding led the airline to begin training crews to avoid acquiescence to suboptimal ATC commands as a means of avoiding error. This is in keeping with the recommendations of Salas and his collaborators (Salas et al., 1999), who state that effective CRM training programs should focus on training pilots in "what to do [i.e., avoid commission of error, remain vigilant for errors, and mitigate their effects when they do occur] rather than how to feel" (p. 164; also see Hendy & Ho, 1998; Hendy & Lichacz, 1999).

Recent evidence establishes that CRM training of this type is in fact successful. Salas and colleagues (Salas et al., 1999) recently evaluated a number of military crews undergoing CRM training against crews serving as control groups (who did not undergo CRM training). Measuring crews' behavior on the flight deck as well as pilots' reactions to training, Salas et al. established that crews who underwent training performed better than crews who did not. In some cases, crews who were trained in CRM showed 8% to 20% more "teamwork behaviors" (p. 166) on the flight deck. Although this validation of CRM training is of course limited to this particular training program, it is significant because it shows that CRM *can* be validated. It remains to be seen, of course, whether other, different programs are as efficacious as this one. It is, nonetheless, a very encouraging finding.

This (re)conceptualization of CRM as management of error and other threats to safety also seems to be well received by industry and regulatory bodies. Worldwide, several organizations that promulgate recommendations and/or requirements for aircrew human factors training now cite error avoidance, trapping, and mitigation as the main goals of CRM-type training (FAA, 1995, 1998; Joint Aviation Administration [JAA], 1998).

Accompanying this shift in (or more accurately, a return to) the stated purpose of CRM training, CRM training evaluation methods have broadened to incorporate a more complete, systemic view of the environment in which the crew performs.

The Growth of Systemic Evaluation Methods

In the early years, it was thought that an airline could simply "do" CRM in a one-time application, and crews would forevermore practice CRM skills on the line. However, the evidence presented suggesting decay in CRM acceptance over time (Helmreich et al., 1999b) argues that successful application of CRM training requires continual, broad, organizationwide support.

In recognition of this, Helmreich et al. (1999b) have called for a more comprehensive approach to the construction and evaluation of CRM training. They advocate using five main sources of data on flight operations to continually revise and improve CRM training: (a) training and checking evaluations of pilots (by those who have themselves undergone CRM training), (b) incident reports, (c) flight data from Flight Operations Quality Assurance (FOQA) programs, (d) attitude surveys of aircrews, and (e) observations of line operations by trained observers. The data sources are seen as reflective of the "big picture," as they encompass the flight deck environment and the organizational milieu within which flight deck operations take place.

This last piece of data, it should be mentioned, should always be gathered under conditions of anonymity and nonjeopardy, in order to reduce any demand characteristics that can accompany evaluation of CRM skills in a jeopardy environment. Taken together, these sources of information can help organizations pinpoint the most pertinent threats to safety in their particular operational environment by using these convergent sources of data to distinguish between real and spurious threats. In turn, training can then be adapted to address the most salient threats faced by crews.

In theory, the training adaptation process would be continuous. In practice, however, airlines tend to return periodically to their CRM training programs, in some cases after an incident or accident has pointed to the existence of safety threats that may not be adequately addressed by current training.

The flight vignette described at the beginning of this chapter took place within such a context. The airline I observed for was interested in identifying whether the operational environment posed risk factors that were not being adequately addressed in training. They requested that an independent team of researchers and flight-deck observers evaluate their Latin American operations using the techniques described by Helmreich et al. (1999b). As a flight-deck operations observer for this airline, my task was to rate aircrews' performance using the LLC and to administer the FMAQ to crews, for completion after the flight. However, the airline was also particularly interested in gaining outsiders' perspectives of their operations. They encouraged observers to write up a verbose description of any particularly noteworthy flights and, if possible, to provide an analysis of the flight, with particular attention to identifying pervasive threats to safety. What follows is a (deidentified) version of my original description, updated to reflect the threat and error management concepts described.

Applying Current Thinking to the Flight Observation

The flight segment described at the beginning of this chapter is an instructive example of the systems-oriented nature of commercial flight operations. It also demonstrates how flight crew errors can be committed and compounded during a flight, and also how effective group performance can mitigate the consequences of errors.

From the outset of this flight, the stage was set for potential problems. Both the captain and relief FO had commuted for several hours before this long overnight flight, which added to the fatigue normally experienced on an 8-hour overnight flight. This illustrates the effect that individual situational factors can have on an operation. In addition, the captain was relatively unaccustomed to flying this particular route in particular, and into Latin America in general. Over the past few months he had mainly flown trips to Europe, where the quality of ATC service is generally higher. These factors implicate the importance of organizational factors (i.e., the captain being eligible for this trip), regulatory factors (the availability of this particular trip to this particular crew), and cross-cultural factors (the captain having expectations for ATC service that were based on U.S. and European air traffic environments).

In the crew's defense, they performed several actions that in all likelihood helped them to avoid getting into even more difficulties than they did. During the pretakeoff briefing, the captain did well to advise the FOs of his fatigue and request that they monitor him. This set the tone for open communications and a nonpunitive approach to correcting one another's errors. Additionally, the FOs' statements that they were familiar with the trip made it more likely that the captain would feel comfortable relying on their knowledge of the particulars of the operation.

The crew made numerous errors during the descent; several of these errors were the direct consequence of preceding errors (i.e., several sequences of errors could be characterized as error chains, where the commission of one error precipitated the occurrence of other errors). The captain degraded situation awareness by failing to engage the weather radar at the top of descent, and the primary FO erred by not mentioning or expressing concern about the weather picture that appeared on his map display.

Although it is impossible to ascertain exactly why the captain failed to use the weather radar, it can be conjectured that, given his recent flying experience in the United States and Europe, he assumed weather was not a factor because the controller did not mention it and did not provide vectors for deviating around it. In general, controllers in the United States and Europe take care to pass along weather information derived from pilot reports and are proactive in providing vectors to deviate around trouble spots. It is likely that the captain, having been used to U.S. and European controllers' service, was lulled into a false sense of security by the lack of weather information from the arrival airport's approach controller. However, having flown into Latin America in the past, he should have been aware of differences between North American and Latin American controllers' facility with English and taken steps to ensure he wasn't being sent into severe conditions without warning. He could have done this by employing a number of strategies, such as further questioning the ATC, eliciting opinions from the FOs as to their assessment of the situation, or (preferably) both. This is an example of an environment-specific threat to safety that is addressable in CRM training.

The FOs also share some responsibility for the incident, as they had been flying Latin American trips lately and presumably were familiar with the operational vagaries of the region. More importantly, the primary FO had a weather picture on his map display. Regardless of whether he failed to notice the red blotches on the radar, or if he noticed them but did not discuss his concerns with the captain, he was remiss. In general, the captain's failure to use the available hardware resources adequately, the probable overestimation of the level of service provided by this particular approach controller, and the failure of the FO to call attention to the weather situation that appeared on his radar proved to be the catalyst for many of the subsequent series of errors and consequent threats to safety.

Once the bouncing started, the crew managed cabin issues well. The relief FO gathered and forwarded cabin reports quickly and efficiently, showing good communication and followership skills. In this situation, the presence of a third crew member to manage cabin calls and deal with possible damage or injuries was invaluable. Had this been a shorter flight, a relief FO would not have been present, and the effort to manage both the descent tasks and the cabin calls most likely would have overwhelmed a two-person crew. In this case, the requirement to fly the flight with an augmented crew increased safety margins on this flight, which again demonstrates the pervasive effect of the regulatory environment.

When we broke out of the clouds and began the approach the captain quickly recognized that he was not lined up correctly and should be commended for immediately requesting a turn back onto the glide slope. Unfortunately, the execution of this plan was flawed; either the turn was too tight, the descent was too gradual, or both. The relief FO deserves praise for quickly and unequivocally calling for a go-around. Although it is my inclination to commend the primary FO for his quick reactions during the moments after calling the go-around, I am not completely familiar with this airline's procedures and duties for the FO in a missed approach situation. The primary FO in fact (as I suspect) may have contributed to the ensuing confusion by moving the flaps handle uncommanded. However, the primary FO performed well by unequivocally advocating a sound plan for the go-around (i.e., staying low and attempting a visual approach). This was an appropriate reaction to two threats to safety—the turbulence and the uncomprehending controller. Although it could be claimed that, as regards the approach controller, the safest course of action would have been to fly the default (i.e., the published) missed approach profile, the crew's unspoken but unanimous assessment seemed to be that, given the continuing difficulty with approach control, it was best to attempt the most expeditious route to the airfield rather than to await "resequencing" while in a holding pattern.

In summary, this flight revealed several sources of safety threats specific to this type of operational environment that are addressable in training as well as in revision of policies and/or procedures. Among these are the possible fatigue induced by commuting, the difficulty encountered by the captain in adjusting his expectations to the Latin American ATC environment, and the possibility that regulatory and organizational rules for flying certain trips while on reserve may contribute to lowered safety margins in certain cases. This flight also showed how effective teamwork can mitigate the consequences of errors and other negative effects. This crew allowed themselves to be led down a dangerous path and, through effective group functioning, removed themselves from further danger.

It should be noted that in the van on the way to the crew hotel, at the captain's initiative, the crew engaged in a lengthy debriefing and self-critique of their performance during the flight. The captain especially was very honest in his assessment of his errors; at one point he stated, "Today I made three mistakes—I trusted ATC when I shouldn't have, I didn't turn on my radar when I should have, and I never should've tried to line up for that first approach."

AUTOMATION ON THE FLIGHT DECK—ANOTHER POTENTIAL SOURCE OF SAFETY THREATS

Another potential source of error and threats to safety has its roots in technology-induced changes in team performance (i.e., the interface between groups and automated systems). This section will briefly relate some of the effects automated systems can have on group performance and will describe how automation can affect aircrew performance in particular.

Automation and Aircrews

Automation can be broadly defined as the replacement of a human being with a machine function. (See chap. 9 for a more in-depth discussion.) The application of automation in aviation dates back to the later efforts of the Wright brothers. From simple mechanical systems responsible for automatic adjustments of aircraft pitch to today's computerized integrated flight systems, flight-deck automation has progressed to the point where pilots can now delegate almost all flight-deck tasks to an array of different systems (Curran, 1992; Wickens, 1992).

Since the widespread application of automation in aviation beginning in the 1970s, the overall accident rate has decreased. However, this overall decrease masks the existence of a trend in incidents and accidents that are attributable in whole or in part to pilots' interaction with automated systems (Billings, 1997).

In several instances, automated systems have behaved in accordance with their design specifications, but at inappropriate times (FAA, 1996). This inappropriate performance is akin to what Reason (1990) terms a "latent" error (and the conditions that bring it about a "latent pathogen") that harmlessly resides in a system until the specific conditions calling it into play are met. In other cases, crews have failed to detect or the automation has failed to inform crews of system malfunctions (Billings, 1997). Despite the best intentions of aircraft and software designers, as

Wiener (1993b) observes, automation cannot completely eliminate human error and may in some cases exacerbate it. It is likely that this is because automation was introduced to aviation with what some feel were unrealistically high expectations and without training that adequately prepared crews for its best use (Sherman, Hines, & Helmreich, 1997).

Initially, the widely held view that human limitations were seen as potential barriers to achieving greater flight safety was one of the main motives for increased deployment of automation in aviation. As the technology to control more aspects of flight became available, designers expected that pilots would adopt a more passive approach to tasks, using automation as often as possible. However, this approach proved to be untenable for a variety of reasons. As Wiener (1993b) and Billings (1997) note, humans are still expected to use the automation—far from removing the possibility of error from an operation, automation in some cases merely relocates the opportunity for human error. Recently, a refined view of automation has developed that accounts for both the intended and unintended effects of automation deployment.

In the view advanced by Billings (1997) and others, automation is characterized by (a) complexity (i.e., it is difficult for the operator to understand), (b) tight coupling (i.e., it creates dependencies among subsystems that may not be obvious to the operator), (c) autonomy (i.e., it has the ability to initiate changes to flight parameters such as speed, pitch, and power), and (d) inadequate feedback (i.e., it doesn't always notify the operator of changes it is making) (Perrow, 1984; Woods, 1996).

Group Communication and Coordination Effects

Automation can affect group processes by affecting the quality and quantity of human-to-human interaction in team endeavors. Danaher (1980) points out that there are intangible benefits from the interaction of humans in a complex system stemming from the opportunity for richer, more flexible information transfer, and that these benefits are reduced when the interaction is limited to the human-computer interface. This is implicitly recognized by at least some FAA airline inspectors, who recently reported that pilots at one major U.S. airline do not practice adequate verification and monitoring of one another's automated system inputs (Dornheim, 1996). Generally speaking, if crews do not verbalize and confirm their inputs to an automated system, then setup errors can turn into performance errors.

Hulin (1989) observes that when automated systems fail, the operators frequently need to increase their communication and workload. This is because automation failures in group-operated systems lead to situations where the information required to diagnose and correct failures may be dispersed throughout the crew, and thus intracrew communication is critical.

These observations suggest that operators will frequently need to increase their levels of communication and coordination when utilizing automation, in order to ensure safe accomplishment of tasks. This assertion has found support in survey research and laboratory and simulator-based studies.

Using high-fidelity simulators and scripted flight scenarios, Veinott and Irwin (1993) found in a comparison of crews flying DC9 and MD88 aircraft (a nonautomated and automated version of the same airplane, respectively) that crews in the latter airplane communicated more, showing higher frequencies of commands, observations, questions, replies, and acknowledgments. Bowers, Deaton, Oser, Prince, and Kolb (1993), in a study of student pilots who flew a simulated flight scenario, found an increase in overall communications for participants operating highly automated versus nonautomated aircraft simulations. It is thought that the need to apprise the other crew member of the automation's status accounts for this higher rate of communication.

Complementing these findings, McClumpha, James, Green, and Belyavin (1991) conducted a study of attitudes toward flight deck automation among pilots in the United Kingdom, using a survey that assessed attitudes in several content areas, including workload, crew interaction, feedback, training, specific aspects of flying skills, and general evaluations of automation. One of the findings was that pilots of automated aircraft expressed frustration with the task of keeping informed regarding other crew members' use of automated systems.

Automation as Crew Member

Due to the highly interrelated nature of crew-automation performance on the flight deck, many researchers and industry groups view the FMS as part of the aircrew (Air Line Pilots Association [ALPA], 1996; Billings, 1997; Woods, 1996). That is, because the FMS performs strategic tasks (sometimes with a high degree of autonomy), it is seen as an "electronic crew member" that carries out group process functions and interacts with live crew members (Helmreich, 1987, p. 70). Additionally, on highly automated flight decks, many actions that are manually performed by pilots can be carried out solely by interacting with the FMS, via a data entry and readout device generically known as the control display unit (CDU). Thus, considering the FMS as a crew member (albeit a silent and inflexible one) has a considerable degree of utility.

Overall, this understanding of automation as a crew member and the data cited strongly suggest that the manner in which crew members should work with their "electronic peer" is similar to the manner in which crews should work with each other (Helmreich, Chidester, Foushee, Gregorich, & Wilhelm, 1990, p. 13). That is, for optimal performance, aircrews should be trained to inquire into and verify the performance of automation and ensure that it is not exceeding design limitations.

This realization has received greater attention in CRM programs during the last decade. Initially, as airlines first took delivery of highly automated aircraft, pilots transitioning to these aircraft were taught to use automation as much as possible, presumably because of the belief that this would prevent error and increase operating efficiency (Hopkins, 1993; Laming, 1993). Soon, however, it became apparent to pilots and training departments that complacency, communication and coordination effects, degraded situation awareness, loss of proficiency, and other of the above-mentioned effects could be observed in actual line performance (Helmreich, Hines, & Wilhelm, 1996; see also chap. 9.)

Incorporating Automation Issues Into CRM Training

Attempting to counteract these deleterious effects, training departments at a few airlines formulated and disseminated guidelines (or "philosophies") for automation management (Degani & Wiener, 1994). Generally, these automation use philosophies stressed the aircrew's responsibility to remain proficient and comfortable with all levels of automation. Some training departments also began to integrate these philosophies of automation management into existing CRM training. In the early 1990s Delta Airlines introduced an "Introduction to Aircraft Automation" (IA^2) training course as an augmentation to regular CRM training. This course encouraged aircrews to remain proficient in the use of all levels of automation by explicitly stating that they were responsible for maintaining their hand-flying (i.e., nonautomated flying) skills. It also stressed the need to exercise judgment in utilizing the degree of automation most appropriate for a given situation (Byrnes & Black, 1993).

Although these efforts to prevent negative consequences of automaton use are admirable, it has been observed that without pervasive organizationwide support, initiatives like these often meet with limited success (Degani & Wiener, 1994). This assertion is based on the systems view described earlier. Degani and Wiener take the position that, because multiple factors affect flight crew performance, an intervention can only succeed if it is reinforced by other people and policies within the organization. These automation-specific initiatives have been accompanied by an FAA-led transition of fleet training programs from traditional, time-based training requirements to more individualized, proficiency-based training requirements. The new requirements are referred to as the Advanced Qualification Program ([AQP]; FAA, 1991).

In summary, it is fair to say that the philosophies and practices of CRM training with regard to automation management are in a state of transformation. The FAA Human Factors Team (1996), in its report "The Interfaces Between Flight Crews and Modern Flight Deck Systems," recommends that all automated aircraft flight training programs focus on imparting the types of lessons taught in IA^2, using a variety of training methods. In particular, they recommend using as teaching tools incidents and occurrences involving automation that have been experienced by pilots, as well as descriptions of pilots' "mental models" (p. 94) of automation. They also recommend that airlines support these initiatives throughout the organizational structure, not just in the training department. Because the availability and abilities of automation will only increase in the future, integrating it more effectively into the aircrew team and recognizing when it increases and decreases risk are the great challenges to the CRM community.

CONCLUDING REMARKS

It is fair to say that the aviation industry has over the past two decades seen an almost total integration of CRM principles into aviation training. In most cases, this has probably been beneficial. However, as many have pointed out (Gregorich & Wilhelm, 1993; Salas et al., 1999; Wise, 1996), CRM training, to be truly successful, must be evaluated.

It should be remembered that there is no such beast as a successful "generic CRM course." Organizations, and the cultures they create around them, differ significantly. Even within a single organization, fleets can differ, both in their microcultures and in the operational vagaries they face. It also goes without saying that instructors differ in their skills and effectiveness. Each instantiation of CRM training—in each new environment—is applied in different conditions, and its effects are mitigated by these different conditions. Although many programs may, in fact, transfer successfully from fleet to fleet, from one airline to another, or from one culture to another, many programs may not.

This state of affairs demands that training efficacy be established in each new situation. Even if the principles underlying a CRM training program are sound and based in empirical observation and the justification for training is undeniable, the focus of CRM training, the execution of training, and the training's integration into the operational environment will always be in need of scrutiny.

This last point is crucial. As the threat-and-error model postulates (and the continuing occurrence of accidents and incidents indicates), there is a seemingly inexhaustible variety of external threats to safety and internal sources of crew error. For CRM to remain an effective countermeasure, training must continually be assessed, so that the risks present in a particular operational environment are being addressed adequately by training. Put another way (and speaking prescriptively), the bulk of the work to be done in CRM research and training should be focused on ensuring that those who fly aircraft are always "trained to the threat."

REFERENCES

Air Line Pilots Association. (1996, September). *Automated cockpits and pilot expectations: A guide for manufacturers and operators.* Herndon, VA: Author.

Ajzen, I., & Fishbein, M. (1977). Attitude-behavior relations: A theoretical analysis and review of empirical research. *Psychological Bulletin, 84,* 888–918.

Bakeman, R., & Helmreich, R. (1975). Cohesiveness and performance: Covariation and causality in an undersea environment. *Journal of Experimental Social Psychology, 11,* 478–489.

Barnhart, W., Billings, C., Cooper, G., Gilstrap, R., Lauber, J., Orlady, H., Puskas, B., & Stephens, W. (1975). *A method for the study of human factors in aircraft operations* (NASA Technical Memorandum TM X-62, 472). Moffett Field, CA: NASA Ames Research Center.

Bernardin, H. J., & Walter, C. S. (1977). Effects of rater training and diary-keeping on psychometric error in ratings. *Journal of Applied Psychology, 62,* 64–69.

Billings, C. E. (1997). *Aviation automation: The search for a human-centered approach.* Mahwah, NJ: Lawrence Erlbaum Associates.

Billings, C. E., & Reynards, W. D. (1984). Human factors in aircraft incidents: Results of a 7-year study. *Aviation, Space, and Environmental Medicine, 55,* 960–965.

Blake, R. R., & Mouton, J. S. (1964). *The managerial grid.* Houston: Gulf Press.

Borman, W. C. (1975). Effects of instruction to avoid halo error on reliability and validity of performance evaluation ratings. *Journal of Applied Psychology, 60,* 556–560.

Bowers, C., Deaton, J., Oser, R., Prince, C., & Kolb, M. (1993). The impact of automation on crew coordination and performance. In R. S. Jensen & D. Neumeister (Eds.), *Proceedings of the Seventh International Symposium on Aviation Psychology* (pp. 573–577). Columbus: The Ohio State University.

Butler, R. E. (1993). LOFT in CRM training. In E. L. Wiener, B. G. Kanki, & R. L. Helmreich (Eds.), *Cockpit resource management* (pp. 231–259). San Diego, CA: Academic Press.

Byrnes, R. E., & Black, R. (1993). Developing and implementing CRM programs: The Delta experience. In E. L. Wiener, B. G. Kanki, & R. L. Helmreich (Eds.), *Cockpit resource management* (pp. 421–443). San Diego, CA: Academic Press.

Chidester, T. R. (1993). Critical issues for CRM. In E. L. Wiener, B. G. Kanki, & R. L. Helmreich (Eds.), *Cockpit resource management* (pp. 315–336). San Diego, CA: Academic Press.

Cooper, G. E., White, M. D., & Lauber, J. K. (1980). *Resource management on the flightdeck: Proceedings of a NASA/industry workshop held at San Francisco, California, June 26–28, 1979* (NASA Conference Publication 2120). Moffett Field, CA: NASA Ames Research Center.

Curran, J. (1992). *Trends in advanced avionics.* Ames: Iowa State University Press.

Danaher, J. W. (1980). Human error in ATC systems operations. *Human Factors, 22,* 535–545.

Degani, A., & Wiener, E. L. (1994). *On the design of flightdeck procedures* (NASA Contractor Report No. 177642). Moffett Field, CA: NASA Ames Research Center.

Dornheim, M. A. (1996). America West undergoes first post-Valujet inspection. *Aviation Week and Space Technology, 145*(9), 22–23.

Fazio, R. H. (1986). How do attitudes guide behavior? In R. M. Sorrentino & E. T. Higgins (Eds.), *Handbook of motivation and cognition* (pp. 205–243). New York: Guilford.

Fazio, R. H., & Zanna, M. P. (1981). Direct experience and attitude-behavior consistency. In L. Berkowitz (Ed.), *Advances in experimental social psychology* (Vol. 14 pp. 161–202). New York: Academic Press.

Federal Aviation Administration. (1991). *Advanced qualification program* (Advisory Circular 120–54). Washington, DC: Author.

Federal Aviation Administration. (1995). *National plan for civil aviation human factors: An initiative for research and application* (draft version). Washington, DC: Author.

Federal Aviation Administration. (1996). *FAA Human Factors Team report on the interfaces between flight crews and modern flight deck systems.* Washington, DC: Author.

Federal Aviation Administration. (1998). *Crew resource management training* (Advisory Circular 120–51c). Washington, DC: Department of Transportation.

Fiedler, F. E. (1967). *A theory of leadership effectiveness.* New York: McGraw-Hill.

Fishbein M., & Ajzen, I. (1975). *Belief, attitude, intention and behavior: An introduction to theory and research.* Reading, MA: Addison-Wesley.

Fiske, D. W., & Cox. J. A., Jr. (1960). The consistency of ratings by peers. *Journal of Applied psychology, 44,* 11–17.

Foushee, H. C., Lauber, J. K., Baetge, M. M., & Acomb, D. B. (1986). *Crew factors in flight operations III: The operational significance of exposure to short-haul air transport operations.* (NASA Technical Memorandum 88322). Moffett Field, CA: NASA Ames Research Center.

Foushee, H. C., & Manos, K. L. (1981). *Within-cockpit communication patterns and flight crew performance* (NASA Technical Paper 1875). Moffett Field, CA: NASA Ames Research Center.

Fowlkes, J. E., Lane, N. E., Salas, E., Franz, T., & Oser, R. L. (1994). Improving the measurement of team performance: The TARGETs methodology. *Military Psychology, 6,* 47–61.

Freeberg, N. E. (1969). Relevance of rater-ratee acquaintance in the validity and reliability of ratings. *Journal of Applied Psychology, 53,* 518–524.

Ginnett, R. G. (1987). *First encounters of the close kind: The first meetings of airline flight crews.* Unpublished doctoral dissertation. Yale University, New Haven, CT.

Ginnett, R. (1993). Crews as groups: Their formation and their leadership. In E. L. Wiener, B. G. Kanki, & R. L. Helmreich (Eds.), *Cockpit resource management* (pp. 71–98). San Diego, CA: Academic Press.

Gordon, M. E. (1972). An examination of the relationship between the accuracy and favorability of ratings. *Journal of Applied Psychology, 56,* 49–53.

Gregorich, S. E., Helmreich, R. L., & Wilhelm, J. A. (1990). The structure of cockpit management attitudes. *Journal of Applied Psychology, 75,* 682–690.

Gregorich, S. E., & Wilhelm, J. A. (1993). Crew resource management. In E. L. Wiener, B. G. Kanki, & R. L. Helmreich (Eds.), *Cockpit resource management* (pp. 173–198). San Diego, CA: Academic.

Helmreich, R. L. (1984). Cockpit management attitudes. *Human Factors, 26,* 583–589.

Helmreich, R. L. (1987). Flight crew behavior. *Social Behaviour, 2,* 63–72.

Helmreich, R. L., Butler, R. E., Taggart, W. R., & Wilhelm, J. A. (1994). *The NASA/University of Texas/FAA line/LOS checklist: A behavioral marker-based checklist for CRM skills assessment* (University of Texas Aerospace Crew Research Project Technical Report 94–02). Austin: University of Texas at Austin.

Helmreich, R. L., Butler, R. E., Taggart, W. R., & Wilhelm, J. A. (1995). *Behavioral markers in accidents and incidents: Reference list* (University of Texas Aerospace Crew Research Project Technical Report 95–01). Austin: University of Texas at Austin.

Helmreich, R. L., Chidester, T. R., Foushee, H. C., Gregorich, S., & Wilhelm, J. A. (1990). How effective is cockpit resource management training? *Flight Safety Digest, 9*(5), 1–17.

Helmreich, R. L., & Foushee, H. C. (1993). Why crew resource management? Empirical and theoretical bases of human factors training in aviation. In E. L. Wiener, B. G. Kanki, & R. L. Helmreich (Eds.), *Cockpit resource management* (pp. 3–45). San Diego, CA: Academic Press.

Helmreich, R. L., Foushee, H. C., Benson, R., & Russini, W. (1986). Cockpit resource management: Exploring the attitude-performance linkage. *Aviation, Space, and Environmental Medicine, 57,* 1198–1200.

Helmreich, R. L., Hines, W. E., & Wilhelm, J. A. (1996). *Crew performance in advanced technology aircraft: Observations in four airlines* (University of Texas Aerospace Crew Research Project Technical Report 96–8). Austin: University of Texas at Austin.

Helmreich, R. L., Klinect, J. R., & Wilhelm, J. A. (1999a). *The line/LOS checklist, version 6.0: A checklist for human factors skills assessment, a log for external threats, and worksheet for flight crew error management* (University of Texas Team Research Project Technical Report 99–01). Austin: University of Texas at Austin.

Helmreich, R. L., Klinect, J. R., & Wilhelm, J. A. (1999b). Models of threat, error, and CRM in flight operations. In R. S. Jensen, B. Cox, J. D. Callister, & R. Lavis (Eds.), *Proceedings of the Tenth International Symposium on Aviation Psychology* (pp. 677–682). Columbus: The Ohio State University.

Helmreich, R. L., & Merritt, A. C. (1998). *Error and error management* (University of Texas Aerospace Crew Research Project Technical Report 98–03). Austin: University of Texas at Austin.

Helmreich, R. L., Merritt, A. C., Sherman, P. J., Gregorich, S. E., & Wiener, E. L. (1993). *NASA/UT/FAA Flight management attitudes questionnaire (FMAQ): A new questionnaire containing items from the original CMAQ, cross-cultural items to tap national differences, and an attitudes toward automation scale* (NASA/University of Texas/FAA Technical Report 93–4). Austin: University of Texas at Austin.

Helmreich, R. L., Merritt, A. C., & Wilhelm, J. A. (1999). The evolution of crew resource management training in commercial aviation. *International Journal of Aviation Psychology, 9,* 19–32.

Helmreich, R. L., & Wilhelm, J. A. (1989). When training boomerangs: Negative outcomes associated with cockpit resource management programs. In R. S. Jensen (Ed.), *Proceedings of the Fifth International Symposium on Aviation Psychology* (pp. 692–697). Columbus: The Ohio State University.

Helmreich, R. L., Wilhelm, J. A., & Gregorich, S. E. (1988). *Revised version of the cockpit management attitudes questionnaire (CMAQ) and CRM seminar evaluation form* (NASA/UT Technical Report 88–3). Austin: University of Texas at Austin.

Hendy, K. C., & Ho, G. (1998). *Human factors of CC-130 operations: Vol. 5. Human factors in decision making* (DCIEM No. 98-R-18). Toronto, Canada: Defence and Civil Institute of Environmental Medicine.

Hendy, K. C., & Lichacz, F. (1999). Controlling error in the cockpit. In R. S. Jensen, B. Cox, J. D. Callister, & R. Lavis (Eds.), *Proceedings of the Tenth International Symposium on Aviation Psychology* (pp. 658–663). Columbus: The Ohio State University.

Hofstede, G. (1980). *Culture's consequences: International differences in work-related values.* Beverly Hills, CA: Sage.

Hofstede, G. (1991). *Cultures and organizations: Software of the mind.* Maidenhead, England: McGraw-Hill.

Hopkins, R. (1993, March 31–April 6). Backing up of approaches [Letter to the editor]. *Flight International,* 40.

Hughes, R. L., Ginnett, R. C., & Curphy, G. J. (1999). *Leadership: Enhancing the lessons of experience.* Boston: Irwin McGraw/Hill.

Hulin, C. L. (1989). The role of communications in group operations of automated systems. In *Proceedings of the Human Factors Society 33rd Annual Meeting* (pp. 775–777). Santa Monica, CA: Human Factors Society.

Ivancevich, J. M. (1979). Longitudinal study of the effects of rater training on psychometric error in ratings. *Journal of Applied Psychology, 64,* 502–508.

Joint Aviation Administration. (1998). *Leaflet No. 5: Crew resource management—flight crew.* JAA Administrative & Guidance Material, Section Four: Operations, Part Three: Temporary Guidance Leaflets (JAR-OPS), 01.02.98.

Jones, S. G., & Tesmer, B. (1999). A new tool for investigating and tracking human factors issues in incidents. In R. S. Jensen, B. Cox, J. D. Callister, & R. Lavis (Eds.), *Proceedings of the Tenth International Symposium on Aviation Psychology* (pp. 696–701). Columbus: The Ohio State University.

Kanki, B. G., & Palmer, M. T. (1993). Communication and crew resource management. In E. L. Wiener, B. G. Kanki, & R. L. Helmreich (Eds.), *Cockpit resource management* (pp. 99–136). San Diego, CA: Academic Press.

Kayten, P. (1993). The accident investigator's perspective. In E. L. Wiener, B. G. Kanki, & R. L. Helmreich (Eds.), *Cockpit resource management* (pp. 283–314). San Diego, CA: Academic Press.

Klecka, W. R. (1980). *Discriminant analysis: Quantitative applications in the social sciences series, no. 19.* Thousand Oaks, CA: Sage.

Lachenbruch, P. A. (1975). *Discriminant analysis.* NewYork: Hafner.

Laming, J. (1993, June 9–15). Who controls the aircraft? [Letter to the editor]. *Flight International,* 140.

Lauber, J. K. (1984). Resource management in the cockpit. *Air Line Pilot, 53,* 20–33.

Lauber, J., & Foushee, H. C. (1981). *Guidelines for line oriented flight training (Vol. 1).* (NASA Conference Publication 2184). Moffett Field, CA: NASA Ames Research Center.

Law, J. R., & Sherman, P. J. (1995). Do raters agree? Assessing inter-rater agreement in the evaluation of air crew resource management skills In R. S. Jensen & L. A. Rakovan (Eds.), *Proceedings of the*

Eighth International Symposium on Aviation Psychology (pp. 608–612). Columbus: The Ohio State University.

Maurino, D. (1993). Cross-cultural perspectives in human factors training: The lessons from the ICAO human factors programme. In R. S. Jensen & D. Neumeister (Eds.), *Proceedings of the Seventh International Symposium on Aviation Psychology* (pp. 606–611). Columbus: The Ohio State University.

McClumpha, A. J., James, M., Green, R. G., & Belyavin, A. J. (1991). Pilots' attitudes to cockpit automation. In *Proceedings of the Human Factors Society 35th Annual Meeting* (pp. 107–111). Santa Monica, CA: Human Factors Society.

McGrath, J. E. (1964). *Social psychology: A brief introduction.* New York: Holt, Rinehart, and Winston.

McGuire, W. J. (1985). Attitudes and attitude change. In G. Lindzey & E. Aronson (Eds.), *The handbook of social psychology* (Vol. 1, pp. 233–341). New York: Random House.

Merritt, A. C. (1996). *National culture and work attitudes in commercial aviation: A cross-cultural investigation.* Unpublished doctoral dissertation. University of Texas at Austin.

Merritt, A. C. (1997). Replicating Hofstede: A study of pilots in eighteen countries. In R. S. Jensen & L. A. Rakovan (Eds.), *Proceedings of the Ninth International Symposium on Aviation Psychology* (pp. 667–672). Columbus: The Ohio State University.

Merritt A. C., & Helmreich, R. L. (1996). Human factors on the flightdeck: The influence of national culture. *Journal of Cross-Cultural Psychology, 27*(1), 5–24.

Merritt, A. C., & Helmreich, R. L. (1997). CRM: I hate it, what is it? (Error, stress, culture). In *Proceedings of the Orient Airlines Association Air Safety Seminar* (pp. 123–134). Jakarta, Indonesia, April 23, 1996.

Murphy, M. R. (1980). Analysis of eighty-four commercial aviation incidents: Implications for a resource management approach to crew training. In *Proceedings of the Annual Reliability and Maintainability Symposium* (pp. 298–306). Piscataway, NJ: IEEE.

National Transportation Safety Board. (1979). *Aircraft accident report: United Airlines, Inc., McDonnell-Douglas, DC-8–61, N8083U, Portland, Oregon, December 28, 1978* (NTSB-AAR-79–7). Washington, DC: Author.

O'Hare, D., & Roscoe, S. N. (1990). *Flightdeck performance: The human factor.* Ames: Iowa State University Press.

Orasanu, J. (1990). *Shared mental models and crew decision making* (Technical Report No. 46). Princeton, NJ: Princeton University Cognitive Science Laboratory.

Orasanu, J. (1993). Decision-making in the cockpit. In E. L. Wiener, B. G. Kanki, & R. L. Helmreich (Eds.), *Cockpit resource management* (pp. 137–172). San Diego, CA: Academic Press.

Orasanu, J., & Fisher, U. (1992). Distributed cognition in the cockpit: Linguistic control of shared problem solving. In *Proceedings of the 14th Annual Conference of the Cognitive Science Society* (pp. 189–194). Hillsdale, NJ: Lawrence Erlbaum Associates.

Orlady, H. W., & Foushee, H. C. (1987). *Cockpit resource management training* (NASA CP-2455). Moffett Field, CA: NASA Ames Research Center.

Perrow, C. (1984). *Normal accidents.* New York: Basic Books.

Radloff, R., & Helmreich, R. (1968). *Groups under stress: Psychological research in SEALAB II.* New York: Appleton-Century-Crofts.

Reason, J. (1990). *Human error.* New York: Cambridge University Press.

Reason, J. (1997). *Managing the risks of organizational accidents.* Aldershot, England: Ashgate.

Rothaus, P., Morton, R. B., & Hanson, P. G. (1965). Performance appraisal and psychological distance. *Journal of Applied Psychology, 49,* 48–54.

Ruffell-Smith, H. P. (1979). *A simulator study of the interaction of pilot workload with errors, vigilance, and decisions* (NASA Technical Memorandum 78482). Moffett Field, CA: NASA Ames Research Center.

Salas, E., Prince, C., Bowers, C. A., Stout, R. J., Oser, R. L., & Cannon-Bowers, J. A. (1999). A methodology for enhancing crew resource management training. *Human Factors, 41,* 161–172.

Sherman, P. J. (1997). *Aircrews' evaluations of flight deck automation training and use: Measuring and ameliorating threats to safety.* Unpublished doctoral dissertation. University of Texas at Austin.

Sherman, P. J., Hines, W. E., & Helmreich, R. L. (1997). The risks of automation: Some lessons from aviation and implications for training and design. In C. Johnson (Ed.), *Proceedings of the Workshop on Human Error and Systems Development* (pp. 133–138). Glasgow, Scotland: University of Glasgow.

Snyder, M., & Kendzierski, D. (1982). Action on one's attitudes: Procedures for linking attitudes and actions. *Journal of Experimental Social Psychology, 18,* 165–183.

Steiner, I. D. (1972). *Group processes and productivity.* New York: Academic Press.

Tannenbaum, S. I., Beard, R. L., & Salas, E. (1992). Team building and its influence on team effectiveness: An examination of conceptual and empirical developments. In K. Kelley (Ed.), *Issue, theory, and research in industrial/organizational psychology* (pp. 117–153). Amsterdam: Elsevier.

Veinott, E. S., & Irwin, C. M. (1993). Analysis of communication in the standard vs. automated aircraft. In R. S. Jensen & D. Neumeister (Eds.), *Proceedings of the Seventh International Symposium on Aviation Psychology* (pp. 584–588). Columbus: The Ohio State University.

Wickens, C. D. (1992). *Engineering psychology and human performance* (2nd ed.). New York: HarperCollins.

Wicker, A. W. (1969). Attitudes versus actions: The relationship of verbal and overt behavioral responses to attitude objects. *Journal of Social Issues, 25,* 41–78.

Wicker, A. W. (1971). An examination of the "other variables" explanation of attitude behavior inconsistency. *Journal of Personality and Social Psychology, 19,* 18–30.

Wiener, E. L. (1993a). Crew coordination and training in the advanced technology cockpit. In E. L. Wiener, B. G. Kanki, & R. L. Helmreich (Eds.), *Cockpit resource management* (pp. 199–229). San Diego, CA: Academic Press.

Wiener, E. L. (1993b). *Intervention strategies for the management of human error* (NASA Contractor Report 4547). Moffett Field, CA: NASA Ames Research Center.

Wiener, E. L., & Curry, R. E. (1980). Flight deck automation: Promises and problems. *Ergonomics, 23,* 995–1011.

Wiener, E. L., Kanki, B. G., & Helmreich, R. L. (1993). *Cockpit resource management.* San Diego, CA: Academic Press.

Wilhelm, J. A. (1991). Crew member and instructor evaluations of Line Oriented Flight Training. In R. L. Jensen (Ed.), *Proceedings of the Sixth International Symposium on Aviation Psychology* (pp. 362–367). Columbus: The Ohio State University.

Wilhelm, J. A., & Helmreich, R. L. (1996). *The Line/LOS Checklist for check pilots: A short form for evaluation of crew human factors skills in line flight settings* (University of Texas Aerospace Crew Research Project Technical Report 96–6). Austin: University of Texas at Austin.

Wise, J. A. (1996, April). *CRM and the emperor's new clothes.* Paper presented to the Third Global Flight Safety and Human Factors Symposium, Auckland, New Zealand.

Woods, D. D. (1996). Decomposing automation: Apparent simplicity, real complexity. In R. Parasuraman, & M. Mouloua (Eds.), *Automation and Human Performance* (pp. 3–18). Mahwah, NJ: Lawrence Erlbaum Associates.

ACKNOWLEDGMENTS

The author wishes to thank Robert C. Ginnett, Keith Hendy, and the volume editors for their comments on this work. The author would also like to thank Susan L. Hura for her insightful comments on earlier drafts of this work.

14

Assessing Cognitive Aging in Piloting

Pamela S. Tsang
Wright State University

The U.S. Census Bureau (1999) projected that between 1998 and 2025, the average age of the population will become progressively older in every nation. In a report titled *The Human Capital Initiative* (National Behavioral Science Research Agenda Committee, 1992), the psychological science community identified several major problem areas in need of basic and applied research to help strengthen our human capital. Maximizing the potential of older adults to maintain vitality and sustain productivity was deemed vital (Vitality for Life Committee, 1993; Committee on the Changing Nature of Work, 1993). At the same time, the nature of work is changing, driven largely by rapid technological advances. The impact of these worldwide trends will be particularly acute in the domain of aviation, which faces constant technological changes.

There are both theoretical and practical motivations for studying cognitive aging in piloting. Piloting is an exact, demanding, highly trained skill that is difficult to reproduce in laboratory. It involves not only multiple, individual cognitive functions but also their intricate interplay. This skill could serve as a rich test bed for studying the higher level cognitive functioning that would not be in operation with simple laboratory tasks. Further, the results would add to an empirical database that could provide an objective basis for debates pertaining to aging and complex job performances in general and to the Age 60 Rule in particular.

ABOUT THE AGE 60 RULE

Since 1959, the Age 60 Rule (a Federal Aviation Regulation [FAR]) prohibits any air carrier from using the services of any person as a pilot, and prohibits any person from serving as a pilot, on an airliner if that person has reached his or her 60th birthday. In 1995, the Federal Aviation Administration (FAA) extended the Age 60 Rule to commuter and air taxi operations that include 10 to 30 seat aircraft. In contrast, in 1997, the European Joint Aviation Authority (JAA) revised their regulation to allow pilots of ages 60 to 64 to operate in multipilot crews, provided only one pilot in that crew is age 60 or above. By 1999, United Kingdom, Denmark, The Netherlands, and Iceland operated under this JAA policy. The opposite directions taken by the United States FAA and the European JAA is but one illustration of the controversy that has surrounded the Age 60 Rule for over four decades. It is beyond the scope of the present chapter to delineate all the arguments put forth by both sides. But interested readers should consult Birren and Fisher (1995b), FAA (1995), and Mohler (1981) for representative viewpoints. Here, only summary arguments from each view are presented.

The FAA has argued that maintaining the Age 60 Rule is necessary to maintain the current level of safety for airline travel (FAA, 1995). The FAA holds that significant medical and psychological defects occur at an increasing rate as age increases. Sudden incapacitation due to medical defects such as heart attacks and strokes becomes more frequent after age 60. The FAA argues that neither the medical defects of concern nor cognitive declines could be predicted reliably by any available tests or criteria. In addition, performance checks only verify the state of a pilot's performance at the time of the checks. The FAA also contends that several studies found an increasing risk of accidents with increasing age (e.g., Golaszewski, 1983; Office of Technology Assessment, 1990). Therefore, allowing a crew member older than age 60 would compromise the safety designed into the redundancy of having a multimember crew.

There are three major counterarguments. One, in-flight incapacitation in airline operations is rare with an FAA estimate of one accident per 8,307,082,800 flying hours. In fact, there has never been a passenger fatality in commercial airline operations due to pilot incapacitation because the second-in-command pilot has been able to take control (Odenheimer, 1999). Two, many members of the medical community believe that medical testing is now sufficiently advanced that could discriminate those individuals who can continue to fly safely after age 60 and those who cannot (e.g., Mohler, 198l; Odenheimer, 1999; Simons, Valk, Krol, & Holewijin, 1996; Stuck, van Gorp, Josephson, Morgenstern, & Beck, 1992). Proponents of this view further point out that if the currently available tests are considered to be sufficiently reliable for testing younger individuals and for the use of granting special medical exemptions, they should be reliable enough for evaluating older individuals. Three, recent analyses of the accident data (see following review) have failed to reveal a clear and significant relationship between age and accident rate (Stuck et al., 1992).

The present chapter begins with an examination of the higher level cognitive demands of modern cockpits on pilots. The chapter then focuses on the paramount role that experience plays in age-related cognitive changes. Methodological issues with assessing age effects on pilot performance are discussed. Age-related changes in simulator performance and accident rate are then examined. Last, an empirical approach to a better understanding of potential cognitive aging effects on piloting is proposed.

COGNITIVE DEMANDS OF PILOTING AND AGING

Technological advances have not only changed how airplanes fly, but also have expanded the conditions in which airplanes can fly (see chap. 2). Cockpits evolved from the early days with practically no instruments to filling up an expansive cockpit panel in the front, on the sides, and overhead. With the introduction of glass cockpits in the 1980s, several information channels could be consolidated onto one multifunction display. Accompanying all the technological changes is the changing role of the pilot from primarily a manual controller (see chap. 8) to primarily a supervisory controller (see chap. 9). With properly instrumented airplanes and airports, landing now can be accomplished at near zero visibility. The sweeping ramifications of the technological changes is evidenced by their treatment throughout this book and by the account of a veteran airline pilot, Captain Buck (1995), who had professionally experienced the Douglas DC-2 through the Boeing 747. This chapter focuses on the ramifications of the changes in the cognitive demands placed on the pilot and how these demands might be met as the pilot ages.

Some cognitive functions judged to be essential for piloting are: perceptual processing (e.g., instrument and out-the-window monitoring), memory (e.g., domain knowledge in long-term memory and air traffic controller's instructions in working memory), problem solving (e.g., fault diagnosis), decision making (e.g., whether to carry out a missed approach procedure), psychomotor coordination (e.g., flight control), and time-sharing or juggling multiple tasks (e.g., instrument monitoring while communicating with ATC; maintaining aircraft stability while navigating). Excellent overviews of the age effects on these cognitive functions are readily available in the literature (e.g., Birren & Schaie, 1996; Craik & Salthouse, 2000; Hardy & Parasuraman, 1997; Park & Schwarz, 2000; Perfect & Maylor, 2000; Rabbitt, 1997; Stern & Carstensen, 2000; Tsang, 1992, 1997b). Undoubtedly, much can, and should, be learned from the cognitive aging literature. But much of the ensuing discussion in this chapter pertains to the need for caution in extrapolating from the general population to pilots.

This section focuses on the *central executive control*—one of the most important higher level cognitive functioning of the modern pilot due to the multitude of functions that need to be performed and the increasing supervisory role of the pilot (e.g., Adams, Tenny, & Pew, 1991). A pilot in command can be likened to a CEO (chief executive officer) of a business. Executive decisions are made and carried out based on the CEO's corporate knowledge, perception of the state of affairs, memory of resources, and experience of strategic and coordinated deployment of resources under prevailing constraints. Recent research shows that executive control is an attentional skill that develops with training, is subject to voluntary strategic control, and has its inherent limitations (e.g., Gopher, 1996). The central executive control certainly involves a number of elementary cognitive processes, but it is more than a mere collection of them (e.g., Cepeda, Kramer, & Gonzalez de Sather, 2001; Gopher, 1996; Kramer, Hahn, & Gopher, 1999). Two components of the executive control will be examined in detail here: attention switching and attention sharing.

Attention Switching

The prevailing experimental paradigm used in studying attention switching is the *task switching paradigm.* In this paradigm, participants are asked to alternate attention between tasks when responding to sequences of stimuli. The primary measure is the *switch cost,* estimated by taking the difference between the response times obtained from a trial that does not require a switch and one that does. The switch is hypothesized to entail several subprocesses: the inhibiting of responses to the previous task, reconfiguring the appropriate processing algorithm, and the selecting and preparing of the next response (e.g., Kramer, Hahn, & Gopher, 1999).

Several common observations emerge from recent investigations. One, switch costs have been observed across several task domains (e.g., verbal, numerical, spatial), suggesting a higher level function that spans a variety of tasks. Two, the switching processes appear to be distinct from other processes that support performance such as perceptual speed and working memory (e.g., Cepeda et al., 2001; Gopher, Armony, & Greenshpan, 2000; Kimberg, Aguirre, & D'Esposito, 2000; Kramer, Hahn, & Gopher, 1999; Rogers & Monsell, 1995; Rubinstein, Meyer, & Evans, 2001; but see Salthouse, Fristoe, McGuthry, & Hambrick, 1998). Of interest here is whether the switch cost is disproportionately large for older adults. Two sample studies will be reviewed for illustrative purposes.

Kramer, Hahn, and Gopher (1999) studied the switch cost of younger (aged 30 and younger) and older adults (aged 60 to 75) in a series of experiments. Rows of digits (e.g., "3 3 3") were presented, and participants indicated whether the number of the digits (Task A) or the value of the digits (Task B) were greater than or less than five. In homogeneous blocks, the same task was performed in all the trials within a block. In heterogeneous blocks, a signal to switch to the other task was presented at unpredictable intervals within a block. A single, linear function did not fit well to the response times of the older and younger participants, suggesting differential aging effects on the switching performance. Most interestingly, an Age by Practice interaction showed that although the older participants exhibited greater switch cost than the younger participants early in practice did, no difference was observed late in practice. Further, the older participants maintained the age-equivalent switch cost over a 2-month retention period. Older participants also benefited just as much as younger participants from additional preparation time (lengthened response-stimulus interval [RSI]). However, when the memory load of the task was increased by requiring participants to remember to switch every five trials without additional external cues, the older participants had significantly greater switch cost.

In another study, Kray and Lindenberger (2000) studied participants between the ages of 20 and 80 years in three task domains. In all cases, the attributes of both tasks were simultaneously presented by the same stimulus. For example, in the figural task domain, depending on the instruction, participants determined either the shape or the color of a figure. The switching instructions were also given ahead of time such that the participants would have to remember when to switch without external cues. These two manipulations were purposely designed to induce executive control by providing inherent task uncertainty and minimizing external cueing from the environment. Substantial switch cost was observed, but no age effect was detected when contrasting switch and nonswitch trials in a heterogeneous block. However, larger switch costs between homogeneous and heterogeneous blocks were observed for the older participants. As observed by Kramer, Hahn, and Gopher (1999), age-related switch cost was detected only when the memory load was increased by having to remember when to switch within a heterogeneous block. In Kray and Lindenberger's study, substantial practice reduced the age difference, but did not eliminate it. The authors noted that performance might not have reached asymptote. As reported by several other researchers, longer preparation time (RSI) reduced switch costs, but did not interact with age.

So far, few studies have applied this paradigm to the pilot population. Gopher's (1982) study is an exception. In this study, Gopher incorporated a dichotic listening task into the pilot selection test battery of the Israeli Air Force. Two independent streams of information were simultaneously presented, one to each ear. A group of 2,000 flight cadets was required to detect digit names in the ear designated as the relevant channel. Midstream, a tone indicated whether the same ear remained, or the other ear became, the relevant channel. Performance of the dichotic listening task was found to improve the predictive validity of the selection battery. Switching errors were particularly predictive of flight training success. Although this study did not address age effects, the results suggested that the central executive control involved in the switching task is the kind of higher level cognitive functioning that would be particularly revealing about an individual's piloting capability.

To summarize, switch costs were observed across a variety of task domains. Attention switching accounts for performance variance beyond that of the individual component tasks. Age-related deficit in switch cost also appears to be distinct from that of the component tasks. Reduced memory load and extended practice can reduce or eliminate the age difference. In addition, older participants are just as able as younger participants to maintain the acquired attention-switching skill and to make use of longer preparation time to reduce switch cost (see also Cepeda et al., 2001; Meiran, Gotler, & Perlman, 2001).

Attention Sharing

> The pilot is like a juggler who starts with two balls and then has a third added. It's necessary to keep all the balls in motion, and as we'll show, additional balls are frequently thrown into the pilot's juggling job, so there are many more than three in the air all the time. (Buck, 1995, p. 10)

Whereas attention switching concerns the selective aspect of attention, attention sharing or *time-sharing* concerns the dividing of processing resources among concurrent activities. That time-sharing is a critical element of flight performance has been demonstrated by a modest predictive value of dual- task performance to flight training success (e.g., Damos, 1978; Damos & Lintern, 1981; Gopher, 1993, Griffin & McBride, 1986; Jorna, 1989) and by superior pilot dual- or multiple-task performance (e.g., Bellenkes, Wickens, & Kramer, 1997; Damos, 1993; Tham & Kramer, 1994).

There are two major limitations to time-sharing: limited resources (e.g., Kahneman, 1973) and skill (e.g., Gopher, 1996). All controlled processes (as opposed to automatic processes) require attentional resources. The greater the task demand, the more resources are needed. Two or more tasks can be time-shared (performed simultaneously) without performance degradation from the single task level if the total demand of the tasks and their management do not exceed the resources available. Further, according to the multiple resource model (Wickens, 1984, 1987), which hypothesizes specialized attentional resources for specific processing (see chaps. 4 & 7), the less similar the resource demands of the component time-shared tasks, the less resource competition between them, and the higher the level of achievable performance. However, this model also predicts that resource allocation between tasks can be facilitated when the component tasks require similar resources. This is because dissimilar resources are not exchangeable, just like additional hydraulic fluid cannot help an airplane that is running out of fuel. A growing body of literature indicates that the managing of limited resources or allocating resources among multiple tasks can be improved by training or by the use of more effective strategies (e.g., Gopher, 1993, 1996; Kramer, Larish, & Strayer, 1995; Tsang & Wickens, 1988), both of which are hallmarks of skills.

A common laboratory paradigm for studying time-sharing performance is the *dual-task paradigm.* Variations of the paradigm include the secondary task technique (see Guttentag, 1989; see also chap. 4), the variable priority method (see Gopher, Weil, & Siegel, 1989; Kramer et al., 1995), and the optimum-maximum method (see Navon, 1985; Tsang & Shaner, 1998). An important feature of these methods is the manipulation of task priorities to examine the central executive or allocation control. One task would be designated the high-priority task, and participants are instructed to protect its performance even though this might lower the performance of the other task(s). Two aspects of time-sharing performance are of interest. *Time-sharing efficiency* refers to the level of performance achieved and is commonly inferred from decrement scores obtained by taking the difference between the single- and dual-task performances. *Allocation control* refers to the resource management and is commonly assessed by the extent to which the participant could achieve performance levels commensurate with externally imposed task importance.

Like attention switching, research shows that time-sharing accounts for performance variance beyond that associated with the component tasks. For example, Damos and Wickens (1980) found time-sharing improvement with practice distinct from any improvement of single-task performance. Also, Damos and Smist (1982) observed significant time-sharing performance differences across groups of participants employing different dual-task strategies, despite their comparable single-task performance. In other words, dual tasks are not just more complex single tasks (e.g., Kramer et al., 1995). Age-related declines in time-sharing also have been shown to be distinct from those of single-task performance. For example, Salthouse, Fristoe, Lineweaver, and Coon (1995) found that age-related variance attributable to dual-task performance remained after variance attributable to single-task performance was taken into account. In reviewing the literature, Kramer et al. (1995) noted age-related deficits specific to dual-task performance and disproportionate age-related divided attention costs.

For example, Kramer et al. (1995) had participants time-share a monitoring task and an alphabet-arithmetic task. Two priority training strategies were used. For the fixed priority (FP) group, the two tasks were always emphasized equally. For the variable priority (VP) group, the two tasks were emphasized equally and unequally in different blocks. Practice benefited both the younger (18 to 29 years old) and older (60 to 74 years old) participants but did not eliminate the age-related dual-task deficit entirely. Variable priority training was particularly effective in improving dual-task performance, whereas FP training improved mostly single-task performance. In a later session, the VP group learned a new task combination faster and achieved a higher level of performance than the FP group. The successful transfer to a new task combination reinforces the idea that time-sharing is a higher level cognitive function that can span across tasks.

In a follow-up study, Kramer, Larish, Weber, and Bardell (1999) doubled the amount of practice from the previous study with new task combinations. The initial task combination involved a target-canceling task and a pursuit-tracking task. The previously observed VP advantages were replicated. In addition, VP, but not FP, training substantially reduced the age-related dual-task performance decrement. A moment-by-moment analysis of the tracking performance indicated that the VP training benefit could be attributed to both reduced response interference and more rapid allocation of attention between the two tasks. The VP group again demonstrated more successful and rapid transfer to a novel set of tasks. Finally, the VP participants showed significant retention of the original training tasks over 45 to 60 days, whereas the FP participants showed a decline.

The next two studies examine age effects specifically on pilot's time-sharing performance; both used the optimum-maximum method to exercise the executive control. Tsang and Shaner (1998) examined 90 participants across a broad age range (20 to 80 years old). A group of active pilots, presumed to have expertise in time-sharing, was contrasted to a group of nonpilots. Demanding and flight-relevant laboratory tasks (i.e., previously demonstrated to correlate with simulator or flight performance) were used. Participants time-shared a continuous acceleration-controlled tracking task with either a discrete memory or spatial orientation processing task.

The results indicated an age-related deficit in time-sharing efficiency above and beyond that at the single-task level. Also, pilots time-shared more efficiently than nonpilots did. Further, the more similar the resource demand of the component time-shared tasks, the larger the age-related dual-task decrement and the greater the difference between pilots and nonpilots (see Fig. 14.1). This is consistent with: (a) the multiple-resource prediction of heightened resource competition between tasks demanding similar resources, (b) the notion that age-related dual-task deficits are related to reduced processing efficiency, and (c) the notion that pilots having expertise in time-sharing could perform at higher efficiency.

Younger participants also exhibited better resource management than participants aged 60 and above. And pilots had better resource management than nonpilots. Pilots were better able to sacrifice the low-priority performance in order to achieve a higher level of performance for the high priority task (see Fig. 14.1). Most interesting, pilot expertise appeared to reduce some of the age-related decrement in time-sharing efficiency and resource management. However, although the beneficial effects of long-term extended training were evident, even pilots with a fairly high mean total flight hours of 5,547 were not completely immune to age effects.

In a second study, Tsang (1997c) had 76 participants between the ages of 30 and 69 perform tasks similar to the previous study. Thirty-six participants were active pilots with a mean total flight hours of 4,801. Moment-by-moment analysis was performed on the continuous tracking task when it was time-shared with a spatial or a verbal discrete task that either required a manual or a speech response. Results showed that discrete manual responses clearly interfered with the tracking performance (see Fig. 14.2), but speech did not. At the moment that the discrete response was made, tracking error increased temporarily with a corresponding decreased level of control activity (control speed). This pattern was observed across all ages, although the older groups (ages 50 and above) generally had higher tracking error and less control activity. However, whereas the older pilots emitted comparable amounts of control activity as the older nonpilots, the older pilots had significantly lower tracking error. This strongly suggested that the observed dual-task interference was not entirely motoric in nature but had a more central

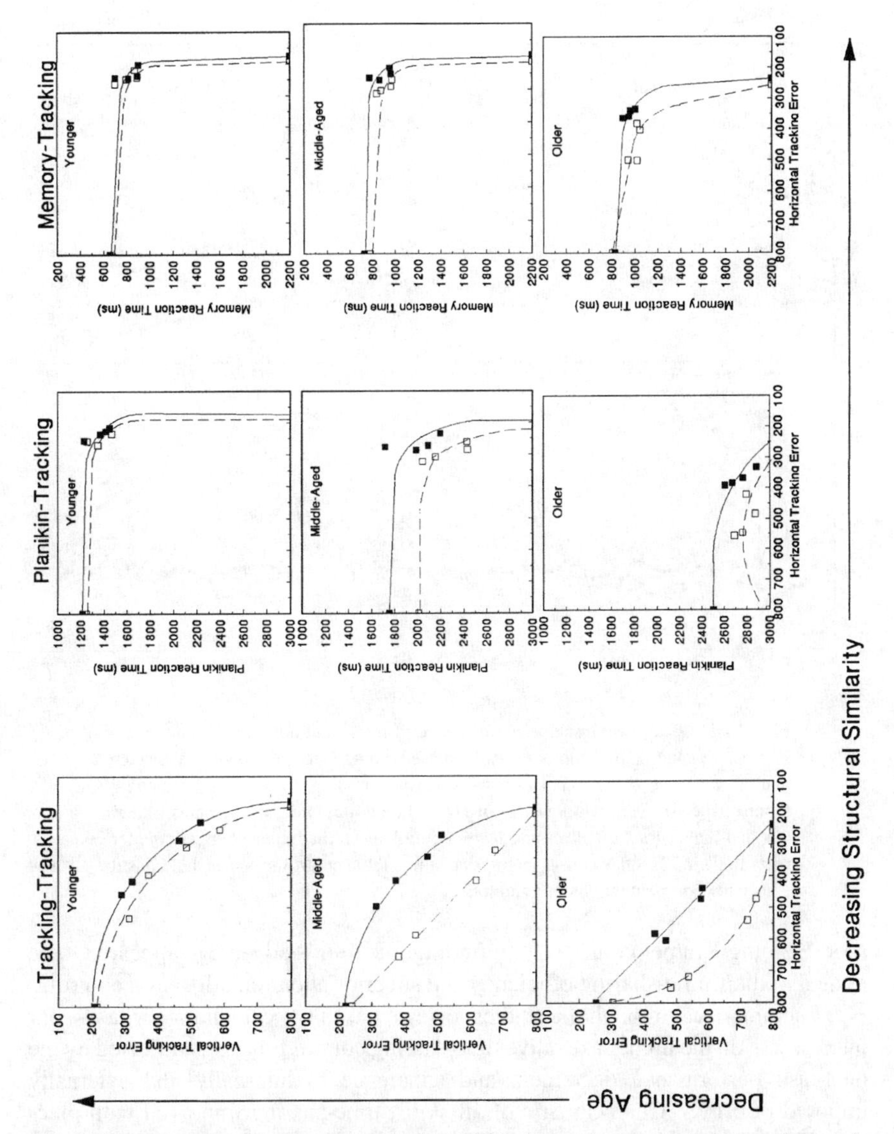

FIG. 14.1. Performance operating characteristics (POCs) plotting the joint performance of two tasks obtained under different priority conditions. The leftmost task pair had two identical tracking tasks competing for the same processing resources. The middle task pair had two spatial tasks with differential response processing demands. The rightmost task pair was the most dissimilar, with a spatial tracking task and a verbal memory task responded to by speech. In the upper leftmost panel, the younger participants exhibited graded performance trade-offs (as performance of one task improved, performance of the other degraded) as task priorities varied. With decreasing similarity of the resource demand between the two tasks (from left to right), performance trade-offs diminished, but time-sharing efficiency improved with the POCs departing from the origin (poor performance). Performance trade-offs became more erratic with increased age (from top to bottom), more so for nonpilots (dashed lines) than for pilots (solid lines). The POCs also moved increasingly closer to the origin, with increased age showing a reduction in time-sharing efficiency. Effects of age and experience were particularly acute with increased structural similarity.

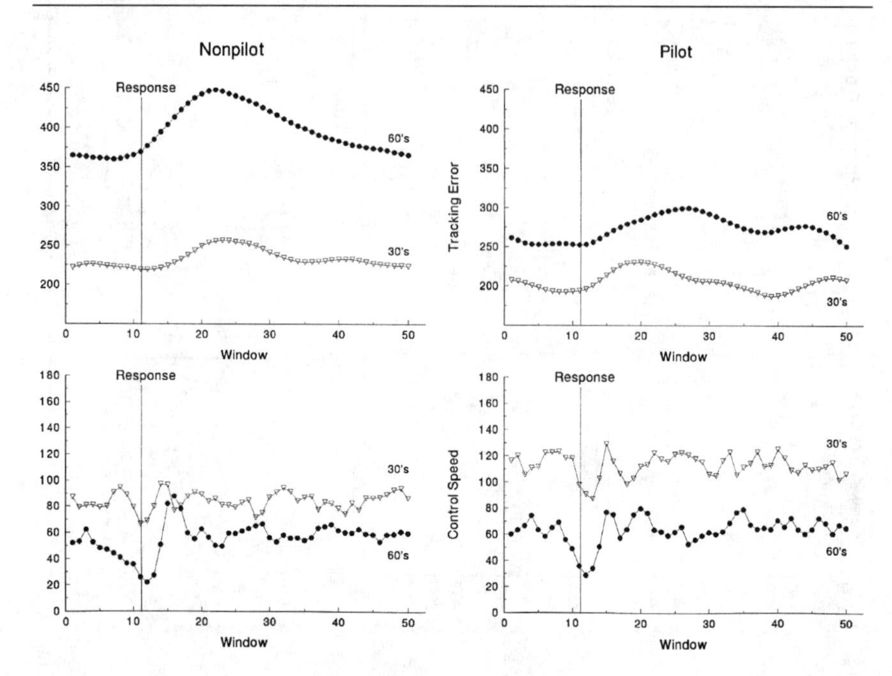

FIG. 14.2. Dual-task tracking performance of pilots and nonpilots from two age groups. Tracking error temporarily increased immediately following a discrete manual response to the other task with a corresponding drop in control activity (control speed). Each window was of 100 ms duration. Older participants in their 60s had higher tracking error and lower control speed than their younger counterparts in their 30s. Older pilots (right) had comparable control speed as, but considerably lower error than, older nonpilots.

root. Echoing Kramer et al.'s (1999) finding, the pilot's advantage appeared to be related to their time-sharing efficiency and superior attention allocation control.

To summarize, although the studies reviewed used typical laboratory tasks, our interest was on the higher executive time-sharing control functions reflected by the dual-task performance decrement and adherence to internally and externally imposed priorities. Characteristic of all skills, time-sharing improved with practice. The VP training designed specifically to exercise the executive control was particularly effective in improving and retaining time-sharing skill, facilitating the transfer of the acquired skill to new task combinations, and reducing age-related dual-task performance decrements. Age-related dual-task deficit was observed in the studies reviewed here and in others (e.g., Crossley & Hiscock, 1992; McDowd, 1986; McDowd & Shaw, 2000; Nestor, Parasuraman, & Haxby, 1989; Ponds, Brouwer, & van Wolffelaar, 1988; Salthouse, Rogan, & Prill, 1984), but it is not always observed in the literature (e.g., Braune & Wickens, 1985; Somberg & Salthouse, 1982; Wickens, Braune, & Stokes, 1987). Possible accounts for the inconsistencies are that age-related dual-task deficits tend not to be significant until age 60 or beyond and only for intensively demanding tasks. Composition (and thereby similarity of resource demands) of the time-shared tasks and relevant time-sharing experience also affected time-sharing performance. Both practice in the laboratory setting and actual flight experience appeared to moderate age effects. In the next section, we explore further the extent to which extensive experience might moderate age-related deficits in tasks specific to the domain of expertise.

THE EXPERTISE ADVANTAGE

One of the most distinguished aviators, General Chuck Yeager, wrote:

> The question I'm asked most often and which always annoys me is whether I think I've got "the right stuff." . . . The question annoys me because it implies that a guy who has "the right stuff" was born that way. . . . All I know is I worked my tail off to learn how to fly, and worked hard at it all the way. And in the end, the one big reason why I was better than average as a pilot was because I flew more than anybody else. If there is such a thing as "the right stuff" in piloting, then it is experience. (Yeager & Janos, 1985, p. 319)

General Yeager's sentiment is entirely consistent with the extant view that expertise is largely acquired rather than innate. The development and attainment of expertise has gained much attention in the cognitive aging literature because of the complex interactions of the two. Whereas many cognitive functions tend to decline with age, experience and skilled performance tend to increase. In this section, we are particularly interested in the potential mitigating effects of pilot expertise on age-related declines. General findings pertaining to the nature of expertise will first be presented. It is proposed that expertise could mitigate certain age-related declines by circumventing certain intrinsic processing limitations. Empirical research on the interactive effects of aging and expertise on both laboratory task and job performances are then examined.

Characterizing Expertise

First, we rarely accord the status of expertise to performance that does not involve fairly complex skills. Expertise is mostly learned, acquired through many hours of deliberate practice (e.g., Adams & Ericsson, 1992; Chi, Glaser, & Farr, 1988; Druckman & Bjork, 1991; Ericsson, 1996; Glaser, 1987; Kliegl & Baltes, 1987). The practice is generally intensive and involves self-monitoring and error correction with an intent to improve. Deliberate practice is distinguished from playful interactions, which include routine work and public performances that are motivated by external social and monetary rewards (Ericsson & Charness, 1997). Once performance has reached a certain acceptable level, amateurs and employees rarely continue to expend time and effort on further improvement. Consequently, there is often only a weak relation between the amount of experience (i.e., time on the job) and the level of expertise (e.g., Ericsson & Charness, 1997; Horn & Masunaga, 2000; Salthouse, 1991).

A fundamental difference between novices and experts is the amount of acquired *domain-specific knowledge.* In addition to having acquired declarative knowledge (facts), experts have a large body of procedural (how-to) knowledge. With practice over time, many procedural rules (productions) become concatenated into larger rules that can produce action sequences efficiently (Druckman & Bjork, 1991).

The expertise advantage goes beyond a quantitative difference. The organization of knowledge is fundamentally different between experts and novices. An expert's knowledge is well structured so that retrieving information is much facilitated. Experts' problem representations are organized around principles that may not be apparent in the surface of the problem, whereas novices' problem representations are organized around the literal objects and explicit events (Glaser, 1987). The large body of organized knowledge enables experts to readily see meaningful patterns, make inferences from partial information, continuously update their perception of the current situation, and anticipate future conditions (Glaser, 1987; see also chap. 4). An accurate account of the current situation allows the experienced pilot to rapidly retrieve the appropriate course of action directly from memory. The large number of procedural rules that experts possess enable them to perform a task automatically as soon as they recognize the conditions. But expert problem solving is more than just quick retrieval of stored solutions to old problems. Expertise is also associated with effective application of a large amount of knowledge in reasoning to cope with novel problems (Charness, 1989; Horn & Masunaga, 2000). In contrast, due to their incomplete and unorganized knowledge, novices must engage in slow, deliberate, sequential search for specific information and possible solutions.

Additionally, experts show metacognitive capabilities that are not present in novices (Druckman & Bjork, 1991). These capabilities include knowing what one does and does not know, planning ahead, efficiently apportioning one's time and attentional resources, and monitoring and editing one's efforts to solve a problem (Glaser, 1987). Once attained, expertise has to be actively maintained (e.g., Krampe & Ericsson, 1996). As Yacovone, Borosky, Bason, and Alcov (1992) put it, "Flying an airplane is not like riding a bike. If you haven't done it recently, you might not remember how" (p. 60).

The Paradoxical Relationship Between Age and Expertise

Although basic cognitive performance assessed by laboratory tasks or psychometric tests generally reveal significant age differences, older experts are often found to perform comparably to younger experts in their domain of expertise. It is therefore proposed that in the same manner that experts could circumvent many of the intrinsic processing limitations, experts may also circumvent many of the performance declines associated with the natural process of aging. For example, Hunt and Hertzog (1981) pointed out that older experienced individuals already have their "frames of knowledge" that can be used readily to solve problems. In contrast, even though younger individuals may be faster, their frames have yet to be developed.

Empirical support for this view abounds in the literature. Murrell and Humphries (1978) had four groups of participants: older experienced professional simultaneous interpreters (with an average of 16 years of experience), younger professionals (7 months of experience), and older and younger naive participants. In a laboratory speech shadowing task (similar to one used for actual training), the professionals had significantly lower error and larger ear-voice span than the naive participants. Age effects were observed only among the naive participants. In a study of skilled typists, Salthouse (1984) used a typing task and a choice reaction time task; both requiring keystroke responses to rapidly presented alphanumeric characters. Age-related decrements were found only for the reaction time task. Salthouse proposed that the older typists used a compensatory-anticipatory-processing strategy to overcome their slower perceptual motor processes. Dollinger and Hoyer (1996) had younger and older medical laboratory technologists and matched novices tested on a domain-specific and a domain-general visual inspection task. The domain-specific stimuli consisted of high-quality color slide reproductions of grain stains containing bacterial morphology typically encountered in a clinical pathology laboratory. The domain-general stimuli consisted of geometric objects that varied along several dimensions. Age differences were found for both the technologists and novices in single- and dual-task domain-general conditions. But, no age difference in the dual-task performance was found among the technologists in the domain-specific condition. Similarly, when Krampe and Ericsson (1996) studied the cognitive-motor skills in expert and accomplished amateur pianists, the older pianists had slower processing speed in general, but only the older amateurs showed age-related declines in a music-related task. The older expert pianists' reproduction of polyrthythms that demanded precise intermanual coordination was comparable to that of their younger counterparts. (See also Bosman, 1994; Charness, 1981; Lindenberger, Kliegl, & Bates, 1992.)

In the aviation context, Szafran (1968) found older commercial pilots extracted relevant information from a noisy background less efficiently than younger pilots but used more optimal strategies to overcame the reduced efficiency. Taylor et al. (1994) and Yesavage, Dolhert, and Taylor (1994) found that despite slower processing speed as assessed by psychometric test (Digit Symbol), older pilots were just as proficient as younger pilots in their simulator performance. Tsang (1997c) and Tsang and Shaner (1998) found that pilots had superior time-sharing performance to nonpilots, even though they did not differ in their memory span, perceptual-motor speed, or intelligence.

The findings just presented are mostly from laboratory studies, albeit using data from domain-related tasks or simulator performance. Charness (2000), Park (1994), and Salthouse and Maurer (1996) surveyed the literature to determine if similar phenomena are observed in actual job performances. Frequently cited are two meta-analyses (McEvoy & Cascio, 1989; Waldman & Avolio, 1986) that revealed no relationship between age and job performance even though negative relations between age and cognitive abilities (e.g., Salthouse, 1991) and positive relationships between cognitive abilities and job performance (e.g., Hunter & Hunter, 1984) were found. One potential disadvantage of the meta-analysis approach is that some of the real effects could be masked. For example, a null effect could be nothing more than positive relationships in some contexts canceling out negative relationships in other contexts. However, specific examples of the paradox in job performances also can be provided.

Giniger, Dispernzieri, and Eisenberg (1983) examined the relation of age (25 to over 65 years) and experience to garment industry workers' performance. Jobs that required speed were distinguished from those that demanded skill. Older workers surpassed the younger ones in both job categories; experience rather than age determined performance. Murrell and colleagues observed the paradoxical relationships in several industrial contexts. Murrell and Edwards (1963) observed that the cutting time in the tool room of two older workers (58 and 61 years) was substantially less than two younger workers (21 years old), although the older workers took longer in comprehending the instructions and preparing the work. Murrell, Powesland, and Forsaith (1962) examined the speed and accuracy of pillar drilling of experienced drillers and inexperienced participants. The free-drilling task required aiming the drill at a punch mark with zero error. The older experienced drillers (44 to 60 years) took about the same time as the younger experienced drillers (24 to 33 years). The young inexperienced participants (20 to 21 years) were faster than the older inexperienced (50 to 58 years), but the young inexperienced were less accurate than all the rest. Murrell and Griew (1965) pointed out that these observations were especially noteworthy because the tasks examined demanded perceptual-motor speed that was known to be particularly age sensitive.

Circumventing Intrinsic Processing Limitations

Although there appears to be a number of ways in which experts bypass intrinsic processing limitations, the present discussion focuses on two mechanisms that have been proposed to overcome the severe limits of short-term memory and the slowing that accompanies aging.

Long-Term Working Memory (LTWM). Long-term working memory is a form of memory that Ericsson and Kintsch (1995) proposed to emerge as expertise develops and is a defining feature of advanced level of skill (Ericsson &

Delaney, 1998). It is functionally independent of working memory or short-term memory (STM). Whereas STM capacity is thought to be seven plus or minus two chunks that would be forgotten within seconds unless rehearsed, LTWM is described to have a larger capacity that persists for over a period of minutes (or even hours). The critical aspect of experts' working memory is not the amount of information stored per se, but rather how the information is stored and indexed in long-term memory (LTM). With a meaningful system for organizing information that already would have been built in LTM through expertise development, even very briefly seen, seemingly random, information might be organized. Retrieval cues can then be devised and used to access information in LTM quickly. Ericsson and Charness (1997) pointed out that experts acquire memory skill to meet specific demands of encoding and accessibility in specific activities. For this reason, their skill is unlikely to transfer from one domain to another.

For example, Chase and Ericsson (1982) and Staszewski (1988) reported three individuals who, after extensive practice, developed a digit span in the neighborhood of a 100 numbers. Being avid runners, the participants related the random digits to facts related to running that were already in their LTM (e.g., date of the Boston Marathon, best indoor mile time). These individuals had a normal short-term memory span (seven plus or minus two) when the studies began and after practice would again demonstrate the normal span when materials other than digits were tested. To further illustrate the workings of LTWM, Ericsson and Kintsch (1995) described the medical diagnosis process that requires one to store numerous individual facts in working memory. Medical experts would have already developed a retrieval structure in LTM that would foster accurate encoding of patient information and effective reasoning. Indeed medical experts were found to be better able to recall important information at a higher conceptual level that subsumed specific facts and to produce more effective diagnosis. The main function of LTWM appeared then to be providing working memory support for reasoning about and evaluating diagnostic alternatives (Norman, Brooks, & Allen, 1989; Patel & Groen, 1991).

Processing Speed. Although increased slowing is one of the most reliable performance effects associated with advancing age, its cause is still hotly debated (see for example, Birren & Fisher, 1995a; Fisher, Duffy, & Katsikopoulos, 2000; Salthouse, 1996). But as the examples illustrate, older experts are often found to perform domain-specific tasks at a comparable level as younger experts. It's been put forth that slower processing can be compensated to a large extent by an increased knowledge base. The large body of structured knowledge allows experts to perceive and apprehend the situation, to frame the problem, and to anticipate future events including eventual retrieval conditions. Further, experts would have available a multitude of productions such that responses can be swiftly executed. Reduction in processing speed could be offset by replacing relatively slow, effortful searches and computations with relatively fast retrievals and executions. Charness (2000) points out that younger adults are rarely going to be equal to the older adults in knowledge in most domains because expert knowledge systems are acquired only through extensive deliberate practice.

In summary, older, experienced individuals do not necessarily perform more poorly than their younger counterparts in tasks specific to their domain of expertise. In fact, we are beginning to learn about the mechanisms by which experts can maintain a higher level of performance despite advancing age. Ultimately, because the basic difference between experts and novices lies not in their basic cognitive abilities but on the body of accessible and usable knowledge that experts possess and the means by which experts circumvent intrinsic processing limitations, simple laboratory task or psychometric performances are unlikely to reveal age-related changes in job performances.

METHODOLOGICAL ISSUES

This section covers selected methodological topics that are particularly germane to the domain of cognitive aging in piloting. These topics include the metrics of pilot performance and experience, the merits and shortcomings of cross-sectional and longitudinal experimental designs, the confounds of age and experience effects, and the difficulty with generalizing from laboratory findings to pilot performance. Readers should consult Hertzog (1996), Kausler (1990), Rabbitt (2000), and Salthouse (2000) for a more general and comprehensive coverage of the methodology for studying cognitive aging.

Metrics of Pilot Performance and Experience

Because piloting is highly complex and multifaceted, its assessment has escaped simple solutions (e.g., Damos, 1996). Major approaches to assessing pilot performance include evaluating in-flight job performance or its approximations such as simulator or flight-related laboratory task performance. Accident rate and experience also have been treated as a proxy to pilot performance. The strengths and drawbacks of each of these approaches will be discussed.

The most common form of pilot performance evaluation is a pass/fail in-flight evaluation by an instructor or a check pilot. Like other subjective measures, this form of evaluation potentially could be influenced by personal biases and is difficult to compare and analyze quantitatively. Objective in-flight performance also could be recorded and analyzed quantitatively off-line. However, determining the appropriate variables to examine and according meaning to the different facets of pilot performance is far from straight forward.

The large number of flight parameters and performance variables are equally problematic for high-fidelity simulator studies. In practice, proficiency in high-fidelity simulators is evaluated subjectively as is done in flight. Low-fidelity simulator performance and complex laboratory task performance afford more focused evaluation. Typical measures include degree deviation from a flight course, tracking error, and response time. Importantly, simulators offer the flexibility of presenting specific scenarios designed for retraining or testing certain skills. The performance recorded can be examined analytically with systematic manipulation of the variables of interest (e.g., weather condition, level of automation, training phase). Although these measures offer the most diagnostic information under controlled conditions, they could be criticized as possessing only limited generality and validity.

Accident rate has been argued to be the ultimate measure of safety, which presumably is related to pilot skill. However, there are several major difficulties associated with interpreting accident data. First, the definition of accident is equivocal. Accidents tend to have multiple causes (Perrow, 1999; Reason, 1990). Exogenous factors that the pilot cannot control include weather conditions, mechanical problems, and traffic congestion near airports (e.g., Li, Baker, Grabowski, & Rebok, 2001). Some investigators include all accidents in their evaluation; others include only those for which the pilot is implicated. Even when the pilot is implicated, there is no simple way to determine whether age is a causative factor. Further, different investigators use different algorithms to compute accident rate (e.g., number of accidents per total annual flight hours, per 1000 active airmen, per number of takeoffs and landings), making it difficult to draw conclusions across studies. To complicate matters, it is often necessary to extract the information needed to compute the accident rate from multiple databases such as the National Transportation Safety Board, the FAA Airmen Certification database, and the FAA Medical History database, each with its own reporting system. For accident rate to be more useful, there needs to be a more systematic, consolidated database of the information needed.

Second, aviation accidents are too rare to allow rigorous statistical analysis (Kay et al., 1994). The varying degrees of statistical sophistication used in the different investigations render reconciling inconsistencies reported across investigations difficult. Comparisons across studies are difficult also because different investigations cover time spans ranging from 1 year (e.g., Mortimer, 1991) to over a decade (e.g., Rebok et al., 1999), and age groups are demarcated at different points. In any case, accident rate is a metric of trends and does not reveal the skill level of any individual.

Last, one might expect pilot performance to correlate positively with pilot experience, but no simple, single index of pilot experience exists. Lifetime accumulation of total flight hours and recent flight hours (e.g., hours flown in last 12 months) are common benchmarks. But the expertise literature (reviewed earlier) shows that although a large amount of time is needed to develop skill, the number of hours by itself is not be an accurate index of expertise. In fact, the relationship between pilot performance and experience tends to be weak and nonlinear (e.g.,

Li et al., 2001). Several investigations reported no significant relationship between total flight hours and performance or accident rate at all (e.g., Kay et al., 1994; Tsang, 1997a; Yesavage, Taylor, Mumenthaler, Noda, & O'Hara, 1999).

Some investigators use the pilot's class of medical as a crude index of experience because airline operations have the strictest medical standard (Class I) and private pilots the lowest (Class III). Other investigators use pilot certificates and ratings (e.g., single-engine, instrument, commercial, and air transport pilot (ATP)) as a demarcation of experience. However, the difference in the skills required for two disparate aircraft, such as a Boeing 777 airliner and a small, single-engine Cessna Skyhawk, is difficult to quantify. In any case, certificates and ratings do not distinguish the quality of performance within a category. Yet another index is the various FAR's operations under Part 91 (general aviation), Part 121 (airline), and Part 131 (commercial and air taxi) operations. Note that these various categories of experience are not mutually exclusive. For example, although all airline pilots must have a Class I medical certificate and an ATP rating, not all holders of Class I medical have an ATP rating or are an airline pilot. Because it is unlikely that a single index can adequately represent pilot performance and experience, Birren and Fisher (1995b) propose adopting as many different approaches as is feasible.

Cross-Sectional and Longitudinal Methods for Assessing Age Effects

There are two major experimental designs for assessing age effects. In a cross-sectional design, people from different age brackets are included in different age groups, all assessed at essentially the same time. In a longitudinal design, the same group of participants serve at all levels of the age variable, assessed at different time periods that can be widely separated.

A major criticism of the cross-sectional design is that the differences observed across age groups do not necessarily reflect a true age difference in skill. For example, a difference in the amount of time required to reprogram the on-board computer between older and younger pilots could be due to a cohort effect of the younger pilots having more computer experience in general. To ascertain that the observed difference is due to age change, it is necessary to rule out other potential contributors to the difference such as educational, health, and cohort effect. One methodological solution is to balance the potentially confounding variables across age groups. But this could lead to reduced generalizability of the results to all pilots. For example, because more younger pilots tend to have a college education (see Hansen & Oster, 1997), only older pilots with a college education might be included in the study. The age difference then is more likely to be an underestimate of the true difference from more representative groups of younger and older pilots that vary in educational level as well as age level. Although the inclusion of all age and educational levels in a cross-sectional design would seem ideal, the study would become unwieldy quickly if all levels of all suspected confounds are manipulated factorially. To alleviate this problem, systematic research on identifying the critical variables that account for a substantial amount of performance variance is needed. Meanwhile, effort to control for commonly known suspected influences (e.g., sensory acuity and training) is advisable.

In a longitudinal design, each participant is his or her perfectly matched counterpart at every contrasted age level. The obvious difficulty with the longitudinal design is the impracticality of conducting a study that covers an extended period. Kausler (1990) pointed out three additional difficulties. One is selective attrition whereby not all members of the original group could serve as participants on all following evaluations. Causes for attrition vary from death to unwillingness or inability to return. A confound is produced if attrition covaries with original performance. For example, individuals of lower skill level who did poorly initially might be less inclined to return for more "difficult" testing. One solution is to restrict the analyses to the scores earned throughout the study by the nondropouts. But this could create the problem of not knowing the extent to which the age change manifested longitudinally by superior individuals (who remained in the study) can be generalized to those of average and below-average ability (who dropped out of the study).

Two, it is practically impossible to avoid practice effects given that the same participants are tested repeatedly. The extent to which the performance at a later testing reflects a true level of performance or that of practice would be difficult to distinguish. The third difficulty concerns time-of-measurement effects. In addition to age, much could have changed during the intervening periods of measurement. For example, some participants could have obtained more advanced training whereas others decreased their level of flight-related activities. The investigator and the experimental/operational environment/equipment also could be different from one time period to another (Buck, 1995; Kay et al., 1994).

Indeed, Kay et al. (1994) observed that cross-sectional and longitudinal studies did not always suggest the same age effects, even when using data from the same population over the same time period. Although neither method is categorically superior, the two experimental designs contribute differently to the understanding of age effects, and both should be used.

Confounds of Age and Experience

As is common in most professions, piloting experience tends to accumulate with age. The difference between age groups in a cross-sectional design and the difference between periods of measurement in a longitudinal design therefore would not be just age, but also experience. The confounds between age and experience is practically unavoidable. The true age effect is likely to be underestimated due to the protective effects of experience. This would be particularly problematic if the investigative interest is the pure effects of age and the population of interest is the total population. However, the population of interest here is not the general population but the pilot population. Piloting experience should not be treated as an extraneous variable, but taken into account by methodological and statistical means. For instance, hierarchical regression analysis (see Pedhazur, 1997) could be used to examine the relative contribution of age and experience. Alternatively, experience levels could be crossed with age groups in a factorial experiment for evaluating the interactive effects of age and experience.

There Are No Old Airline Pilots. A rather unique problem exists if one's interest is not just experience but expertise in piloting. The group of civilian pilots considered to be most rigorously trained and most experienced is air carrier pilots, who are required to retire by age 60 in the United States. However, cognitive aging effects are generally not easily detectable until above age 60. Establishing the extent to which age affects expert pilots is therefore methodologically challenging.

One solution that has been suggested is to have general aviation (GA) pilots age 60 and above as surrogates for older air carrier pilots (e.g., Kay et al., 1994). The obvious objection is that the two groups simply are not the same. But the proponent's expectation is that because the GA surrogates, as a group, tend to have fewer total flight hours and less-rigorous monitoring of health and skills, conclusions drawn from this group are most probably going to be conservative (i.e., an overestimate of the true accident rate). Implications from these conclusions on the Age 60 Rule, if in error, would err on the safe side. Though not perfect, the proposed solution is an alternative to not having any data at all on older pilots.

Variability in Individual Differences. Rabbitt (1993c) states that "cognitive gerontology is not the study of decline in performance with age, but rather of the enormously *increased variance, both between and within individuals,* in older populations" (p. 201; see also Birren & Schroots, 1996; Czaja, 1990; Schaie, 1990). Many factors have been proposed to account for the increased variability. Different physiological, sensory, and cognitive systems change at different rates. Verbal ability and crystallized intelligence tend to increase with age and then were maintained into old age. Spatial ability, fluid intelligence, and response speediness tend to show decline with increased age (e.g., Horn & Masunaga, 2000; Rabbitt, 1993b). Life experiences also become increasingly divergent with increased age. Some skills are maintained by practice, whereas others are lost through disuse.

Rabbitt (1997) described an early study conducted by Shooter, Schonfield, King, and Welford (1956) to contrast training and aging effects. They examined the amount of retraining needed for ex-tram drivers between the ages of 20 and 70 to become bus drivers. The percentage of drivers who passed the retraining in 3 weeks declined with age. However, with some additional training provided in 6 weeks or less, over 75% of the older drivers (ages 61 to 67) successfully retrained for a complex demanding skill. In another study, Rabbitt (1993a) examined the practice effects on the performance of a two-choice visual search task. Both the "young" (ages 50 to 64) and "old" (ages 65 to 80) groups continued to improve for 36 sessions. The average age difference was 94 ms, compared to an average practice effect of 232 ms. That is, the practiced older participants were faster than the unpracticed younger participants. Both of these studies showed how a relatively small amount of practice could overshadow the age effects.

Tsang and Voss (1996) examined the response distributions from a group of pilots and a group of nonpilots between the ages of 20 and 79. Performance included single- and dual-task tracking error, spatial processing response times, and verbal memory reaction times. Not surprisingly, the range of performance broadened with increased age. The degree of overlap of the response distributions between two groups (the 40 to 59 and 60 to 79 age groups) was estimated by the probability that a randomly selected individual from one group would perform better than a randomly selected individual from another. Overall, there was an estimated probability of .2 to .3 that an older individual would time-share more efficiently or have better attention allocation control than a middle-aged individual. In other words, 2 to 3 people over 60 years old out of 10 are likely to time-share better than a randomly selected person between the ages of 40 and 59. A closer examination revealed that the obtained estimates varied along several dimensions: the specific performance measure (tracking error versus response time), pilot versus nonpilot, and single- versus dual-task performance. The probability that a randomly selected pilot could time-share better than a nonpilot also increased with age.

Because the increased variability could be attributed to a multitude of factors that change at different rates and exert variable influences on performance, the amount of variances accounted for by chronological age itself is generally small. For example, the University of Manchester longitudinal studies (Rabbitt, 1993c) revealed that correlations between performance indices and chronological age, although significant, were all modest (seldom exceeding $r = .35$, or some 10% of population variance). One implication is that the usefulness of age as predictor of performance decreases with age. Consequently, Birren and Schroots (1996) and Rabbitt (1993b), among others, recommend that in addition to examining group means, variability measures should be included as much as possible. Note also that the larger the variance present, the larger the sample size is required to attain the same level of statistical power.

Generalizing Laboratory Findings to Real-World Operations

External validity refers to the extent to which local results from a specific laboratory study generalize to the population of interest in the relevant context. This is the goal of all research, but external validity is of particular concern in the process of understanding aging effects and pilot performance. Ideally, the group of research participants should resemble the population of interest on all relevant dimensions. But most of what we understand about the aging process comes from data obtained from nonpilots. And pilots might belong to a more select group that excludes individuals of lower ability or individuals more susceptible to age decrements (e.g., Braune & Wickens, 1984b; Institute of Medicine, 1981). There simply have not been many large-scale and systematic investigations on the pilot population. Second, piloting is a complex skill that requires considerable training to acquire and regular practice to sustain. Such expertise has shown to be able to mitigate some typical age effects (see McFarland, Graybiel, Liljencrantz, & Tuttle, 1939). Researchers therefore have urged for caution in extrapolating simple laboratory task findings to real-world competence (e.g., Birren & Fisher, 1995b; Hunt & Hertzog, 1981; Schaie, 1988).

However, this does not mean that laboratory studies with nonpilot participants cannot produce useful information. On the contrary, laboratory studies are among the most efficient means for identifying causal relationships between the suspected variables and specified measures. But it is important that the variables to be studied, both independent and dependent, be demonstrated to be related to piloting. This could be achieved by task analysis, correlations with flight or simulator performance, consistency with existing data, and agreement among subject matter experts.

For example, Gopher (1993) demonstrated that physical fidelity is not essential for positive transfer from laboratory task training to actual pilot performance. Gopher studied three groups of flight cadets in the Israeli Air Force flight school that received differential attentional training. The training task was a computer game called the Space Fortress, which was developed specifically for psychological experimentation. The task contained multiple dynamic components identified to require high-level attentional management skills similar to those required of a pilot in the cockpit. One group received specific task strategies training (e.g., how to control the ownship). Another group received more general attentional control or resource management strategies training. The third group served as a control and did not receive any strategy training. All three groups practiced for 10 hours on the task. The control group received the lowest score, and the specific task strategy group received the highest score on the Space Fortress game. After completing his or her jet training, each cadet's flight performance was evaluated by a flight instructor. Here, the two groups that received strategy training did not differ from each other, and both outperformed the control group. This suggested that the strategy training was beneficial to flight performance even though the training task bore no physical resemblance to the operational task. Thus, well-conceived and principled laboratory studies could be highly informative, but generalization from laboratory studies to operational performance cannot be automatically assumed.

Summary

Given the complexity of assessing age effects on pilot performance, one methodological strategy that is necessary is the use of converging operations (see Garner, Hake, & Eriksen, 1956). That is, use multiple approaches to examine various behavioral manifestations of the same underlying process and seek convergence or agreement among them. The pattern of results borne out of the various metrics examined should be meaningful. Neither cross-sectional nor longitudinal studies are categorically superior. Each yields different information and is subject to different deficiencies. Because of the considerable influence of experience, aging effects cannot be assessed with no regard to the experience level, and domain-specific tasks should be the basis for evaluation. The increased variability with advancing age has two important implications: (a) a sufficiently large sample is required to provide reliable results, and (b) chronological age is unlikely to serve as a useful indicator of an individual's performance.

MANIFESTATION OF AGING IN SIMULATOR PERFORMANCE AND ACCIDENT DATA

So far, we have surveyed primarily laboratory task and nonpiloting job performances. This section focuses directly on pilots' simulator performance and accident data.

Flight Simulator Performance

With advanced PC (personal computer) technology, low-cost simulators are becoming more versatile and prevalent (see chap. 11). But even simulators of the highest degree of fidelity achievable today cannot duplicate actual flight (see chap. 12). Caution, therefore, is needed when applying simulator results to actual flight performance.

Braune and Wickens (1984b) administered five laboratory tasks (tracking, Sternberg memory, maze tracking, hidden figures, and dichotic listening) to instrument-rated pilots between the ages of 20 and 60. These tasks had previously been shown to be age sensitive for nonpilots (Braune & Wickens, 1984a). Performance on these tasks was used to predict performance of a series of maneuvers on a GAT-2 twin engine simulator and a simulated air traffic control (ATC) communication task. Results showed that the laboratory task performance generally correlated with the maneuvering and communication performance, the memory performance being the best predictor. Whereas age effects were previously observed in all five tasks with nonpilots, age effects were detected in only two of the tasks (hidden figures and dichotic listening) with pilots. The authors suggested that pilots, as a group, might belong to a more select group that was less susceptible to age effects. Because the tasks were deemed flight relevant (as demonstrated by the correlation between task and simulator performances), pilot experience could also have provided some protection against certain age-related declines.

In an extensive investigation, Hyland, Kay, and Deimler (1994) studied a group of more experienced B727-rated male pilots with a minimum of 5,000 total flight hours who ranged in age from 41 to 71. A team of simulator experts, instructors, check pilots, and aviation psychologists were involved in creating three simulator scenarios. The low workload scenario entailed routine maneuvers such as takeoff and landing. The moderate workload scenario entailed more challenging maneuvers such as steep turn and missed approach. The high workload scenario entailed emergency/abnormal maneuvers such as holding pattern with an engine fire. For each maneuver, the team defined the component actions required and the desired level of performance. In addition to a subjective evaluation of pilot performance, deviations from error-free simulator performance were recorded. Age did not correlate significantly with the objective deviation scores, but older pilots received lower subjective evaluation ratings in all three scenarios.

The pilots in this study also performed three PC-based task batteries. Task battery performance was then validated against simulator performance. The first task battery, CogScreen, was developed by Horst and Kay (1991) and consisted of cognitive tests of memory, selective attention, time-sharing, visual-spatial, verbal-sequential, and psychomotor performance. CogScreen was designed to provide a computer-based cognitive function screening test for medical certification of commercial pilots. The second battery, WOMBAT (Roscoe, 1997), was initially conceived of as a pilot-selection device and measured vigilance and the ability to simultaneously perform several tasks. The set of tasks included pursuit tracking, pattern recognition, mental rotation, and working memory. The third battery, Flitescript (Stokes, Belger, & Zhang, 1990), required pilots to listen to an ATC communication sequence and either recall the sequence or select the correct graphic depiction of the described situation from a set of alternatives. Whereas the first two batteries were meant to assess pilots' domain-general cognitive skills, the third was meant to assess domain-specific skills.

Test scores from all three batteries correlated with pilot age. This was expected for CogScreen and WOMBAT, because they included tasks that prior literature suggested were likely to be age sensitive. Significant correlation was found between CogScreen and Wombat performances, suggesting that these two batteries assessed similar underlying cognitive abilities. In contrast, Flitescript scores did not correlate with either WOMBAT or CogScreen scores. Further, the three task batteries did not correlate with simulator performance with only one exception. A CogScreen composite score correlated with the subjective evaluation of pilots' emergency/abnormal maneuvers performance. Of note, the task battery professed to be domain specific, Flitescript, did not produce any score that correlated with simulator performance. It is also the one battery that did not include any higher level cognitive functioning such as the requirement of performing multiple tasks. Importantly, even though age correlated with the performance of all three task batteries, age did not correlate with simulator performance.

In another study, Taylor et al. (1994) examined a group of younger (21 to 34 years) and a group of older (51 to 74 years) licensed male, mostly instrumented-rated, pilots with a median of 1500 total flight hours. They flew in a Frasca 141 simulator, which emulated a single-engine airplane similar to a Cessna 172 with "through-the-window" display. During a simulated flight, the pilots needed to avoid other aircraft and monitor the instrument panel for engine malfunctions. The readback and execution of 16 ATC messages presented during a simulated flight were evaluated. Speech rate, message length, and age significantly affected performance. Contrary to a common finding of a disproportionately large age effect with increased difficulty, no significant interaction between age and either speech rate or message length was observed. The researchers proposed two possible explanations. One, the relatively small sample size could have produced insufficient power to detect real effects. Two, error rate was generally high for the longer ATC messages; the older group's mean error rate was around 90% and might suggest a floor effect. Yet another possible explanation for the lack of an age-by-difficulty interaction was that experience and familiarity with the standard ATC phraseology could have provided some modest protection against age effects for the older pilots.

In the next two studies, Yesavage et al. (1999) and Taylor, O'Hara, Mumenthaler, and Yesavage (2000) studied 100 active aviators ages 50 to 69 with 300 to 15,000 hours of total flight experience. Most of the pilots were instrument rated and had a Class I medical certificate (the lower the class number, the more rigorous the medical requirement), but no air carrier pilots were included. Pilots were tested on a Frasca 141 flight simulator with "through-the-window" display. Three different emergency situations were interjected during a simulated flight: carburetor icing, drop of engine oil pressure, and suddenly approaching air traffic. Twenty-three simulator performance measures (deviations from ideal or assigned positions) were recorded. A summary simulator score was calculated by obtaining the mean of five composite simulator scores (deviation from flight course, accuracy of dialing in communication frequencies, traffic avoidance, instrument monitoring, execution of the approach). The pilots also performed the CogScreen-AE (Aeromedical Edition) task battery. The CogScreen scores were reduced to five composite scores: (a) speed/working memory, (b) visual associate memory, (c) motor coordination, (d) tracking, and (e) attribute identification.

Increased age was significantly associated with a lower simulator score even after controlling for flight experience. In fact, flight experience did not correlate with the summary simulator score. Regression analyses indicated a curvilinear relationship between age and the summary simulator score, with the youngest and oldest having the lowest performance. The range of performance from the younger and older pilots overlapped considerably. Pilot age explained 22% or less of the performance variance on the various individual tasks and 18% of variance in the summary simulator score. Four CogScreen scores were found jointly to explain 45% of the variance in the summary simulator score. The speed/working memory scores had the highest correlation with the summary simulator score, tracking the lowest, and attribute identification none. Age significantly improved the prediction of the summary simulator score, but accounted for only an additional 6% variance. This suggested that age played a much smaller role than cognitive performance in predicting simulator performance. Given that the various CogScreen scores contributed independently to improve the prediction of simulator performance, Taylor et al. (2000) pointed out that it is unlikely that any single cognitive measure can adequately predict pilot performance.

In summary, the simulator studies reviewed collectively covered an age range that spanned from 20 to 74 years old, experience level from that typical of GA pilots to that of air carrier pilots, and simulators of single-engine airplane to jet liner. Consistent age effects were readily observed in domain-general, laboratory cognitive task performances, suggesting that pilots were not immune to aging effects. But only inconsistent and small age effects were observed in simulator performance, suggesting that experience could be mitigating some of the age-related declines in domain-relevant performance.

Accident Data

Because of the important role that experience plays, the present review focuses on sample investigations that have statistically examined age and experience concomitantly. Investigations of military aviation are not included because military aviators typically retire relatively young.

A report known as the Hilton Age 60 Study covered accidents operating under Parts 91, 121, and 135, with pilots holding Class I, II, or III medical certificates between 1976 and 1988 (Kay et al., 1994). Analysis of variance (ANOVA) was used to analyze the accident rate, and chi-square statistic was used to analyze frequency data. The major conclusions can be summarized as follows. For Class I pilots between the ages of 30 and 59 with more than 2,000 hours total flight time (minimum requirement for Part 121 pilots), recent flight time, but not total flight time (2000 to 10,000 hrs.), interacted with pilot age. Accident rate (number of accidents per annual flight hours) decreased with increased age (from 30s to mid 40s) then leveled off for pilots with more than 250 hours recent flight time. No reliable age effect was detected for those with ATP ratings who were employed by major airlines and who had more than 700 recent flight hours. For Class II pilots between the ages of 30 and 69, accident rate decreased with age, leveling off for older pilots. Accident rate for the 60 to 65 group did not differ from that of the 55 to 59 group, but was lower than that of the 65 to 69 group. For Class III pilots between the ages of 30 and 69 with more than 500 total flight hours and more than 50 recent flight hours, the accident rate decreased with age, leveling off for older pilots. A slight increase in accident rate for ages 63 through 69 was detected. In short, a slight age-related increase in accident rate was detected only for pilots with considerably fewer total and recent flight hours than a typical airline pilot.

Li et al. (2001) used a multivariate logistic regression modeling approach (see Hosmer & Lemeshow, 1989) to describe the relationship between pilot error involved in a crash and a set of independent variables that included endogenous pilot characteristics (age, gender, total flight time, and certificate rating) and exogenous crash circumstances (time and location of crash, weather condition, type of aircraft, and crash severity). Airplane and helicopter crashes that occurred during 1983 to 1996 under regulations 91, 121, or 135 were analyzed separately. Among the crash circumstances examined, instrument meteorological conditions (adverse weather conditions) were most predictive of pilot error. Crashes at on-airport locations where takeoffs and landings took place were also more likely than crashes away from airports.

Neither pilot age nor gender was independently associated with pilot error in major airline or commuter/air taxi crashes. For GA pilots, those under age 20 had a significantly higher accident rate than the rest (including those over age 60). The odds of pilot error decreased as total flight time increased and certificate rating advanced. Pilot error was more likely to be attributed to pilots with student/private pilot licenses than other GA pilots were. For commuter/air taxi crashes, pilots who held airline transport certificates had a lower prevalence of pilot error than commercial pilots did. A nonlinear relation between total flight hours and accident rate was observed: the safety benefit of flight experience was greater in the early stage (GA) and diminished as total flight time increased (commuter/air taxi and major airlines). However, advanced certificate rating continued to have a protective effect against pilot error for pilots with high hours. These results showed that the true benefit of flight experience would not be captured entirely by the total accumulated flight time alone because the type of flying also played a significant role. Li et al. (2001) concluded that although certain pilot characteristics such as flight experience contributed to pilot error rate, these endogenous factors in general were less important than environmental factors such as weather and crash locations.

McFadden (1997) also used a logistic regression technique to model the influence of pilot's age, flying experience (total flight hours), risk exposure (flight hours in the last 6 months), and employer (major/nonmajor airline) on incidents identified by the FAA to be related to pilot error in the period of 1986 to 1992. Examples of incidents were near midair collisions, landing at the wrong airport, and landing with the gear up. Only U.S. airline pilots were included in this analysis. Youth, inexperience, and nonmajor airline employer (with operating revenues under 1 billion dollars) independently contributed to increased risk of pilot-error incidents. After controlling for the variables listed, no gender difference was detected. McFadden (1996) observed similar results when pilot-error accidents (that involved death, serious injury, or substantial aircraft damage) were analyzed.

Broach (2000a) focused on pilots characteristic of the population to which the Age 60 Rule applied (airline, commuter/air taxi pilots). Accident rate (number of accidents per annual hours flown occurring under Parts 121 and 135) of professional pilots holding Class I medical and ATP certificates for the period of 1988 through 1997 were examined. A U-shaped function relating age and accident rate indicated that the youngest (up to 30 years of age) and oldest groups (60 to 63) had the highest and most variable accident rate. The mean accident rate for the 60 to 63 age group did not differ significantly from that of the 55 to 59 age group.

In a later analysis, Broach (2000b) broadened the database to include commercial pilots and pilots holding Class II medical certificates. The U-shaped relationship between age and accident rate was again observed. The 60 to 63-year-old pilots had a statistically higher accident rate than the 55 to 59-year-old pilots. Broach concluded that, consistent with findings from an earlier investigation (Golaszewski, 1983), accident rate was related to pilot age. However several caveats to Broach's conclusion should be considered.

First, Broach (2000a, 2000b) did not take experience (e.g., total flight time, recent flight time) into account when examining the relationship between age and accident rate. This would be problematic because it is known that experience could moderate age effects. However, considering both sets of data from Broach (2000a, 2000b) collectively could provide a glimpse of the effects of pilot ratings and class of medical. Inclusion of Class II medical and commercial pilots in the second investigation led to a significant difference in accident rate between the 55 to 59 and 60 to 63 age groups that was not observed when only Class I medical and ATP pilots were examined in the first. This strongly suggested that age was not the sole driver of accident rates because the same age span was examined in both. Note that Golaszewski's (1983) data also suggested that age effect was evident only in relatively inexperienced pilots with low total flight hours (less than 1,000) and low recent flight hours (less than 50).

Second, Broach (2000a, 2000b) noted that the U-shaped function relating age and accident rate indicated that accident rate tended to increase with age. Several details of this function are informative. One, the age 60 and above group did not include any pilots that were airline pilots, as did the younger groups. Because airline operations have the lowest accident rate, the accident rate for the 60 to 63 age group might be inflated. Two, no significant difference was observed between the 55 to 59 and 60 to 63 groups when only the typically more experienced ATP and Class I medical pilots were considered. Three, the accident rate for the ATP and Class I medical pilots in the 60 to 63 age group was .25 compared to .24 to .26 for their youngest counterparts. This was hardly a large difference. When the commercial and Class II medical pilots were included, the youngest groups had an even higher accident rate than the 60 to 63 group.

Summary. Because accidents have multiple causes that include exogenous and endogenous factors, establishing a causal relationship between accident rate and age could be procedurally complicated. Recent analytical efforts have not revealed any overwhelming evidence to indicate that increased age is a significant contributor to increased accident rate. When an age-related increase was observed, the increase tended to be small (e.g., Broach, 2000a), and the accident rate associated with the oldest group tended to be comparable to or better than that of the youngest groups (Broach, 2000b). Total flight time did not appear to be a significant factor beyond the early stage of training that Li et al. (2001) referred to as the experience-building stage. However, the type of pilot rating continued to have a significant effect beyond the initial stage. Consistently observed is the relationship between recent flight time and accident rate, implicating the importance of continual practice for maintaining a complex skill such as piloting. Negative age effects tended to be found only for pilots with low total and recent flight time and less-advanced rating (e.g., low-hour private pilot).

TOWARD A BETTER UNDERSTANDING OF THE INTERPLAY BETWEEN AGING AND PILOTING

Does Age Affect Piloting?

Charness (2000) observes that "the weight of evidence favors the conclusion that today's older adults will show inferior performance to young ones on almost any laboratory task that taps unpracticed abilities" (p. 109). If age affects basic abilities that support higher level performance, the more complex performance necessary in professional life also should be impaired. Except that complex operational/professional performance, as reviewed earlier, does not necessarily exhibit declines with age. In fact, many actual job performances can be maintained at a high level into the sixth or seventh decade and beyond. Cognitive slowing commences in the 20s, but one certainly would not conclude that therefore 30-year-olds must fly more poorly than 20-year-olds. By their mid to late 60s, many individuals experience some memory declines. If this necessarily disrupts job performance, then decisions made by the U.S. Supreme Court (with a mean age of the justices being 67 in 2001) should be suspect. Similarly, the presence of age effects does not mean that pilot performance will be compromised necessarily (see McFarland, 1954).

It is also clear from the literature that due to large individual variability, there is not one age when all cognitive capabilities degrade for everybody (e.g., Rabbitt, 1993b; Schaie, 1993). The prevailing data strongly suggest that the quest for determining a specific cutoff age that could be applied generally is likely to be futile (e.g., Czaja, 1990; McFarland, 1955). Clearly, other avenues need to be sought if the goal is to determine the skilled performance of individuals in order to maximize the potential of older adults and to strengthen the human capital. The substantial influence of individual differences and experience on age effects have important implications on the focus of the investigation and how it should proceed.

The Paramount Role of Experience/Expertise

It is now clear that age effects cannot be assessed independently of one's experience level. Therefore, care must be taken in how piloting experience is indexed. The number of years of experience in a domain generally is only weakly related to the performance. This is because it is the amount of deliberate practice that determines expertise and because expertise needs to be actively maintained. For example, the number of chess competitions attended and the number of baseball games played in major leagues have not been found to accurately predict performance (Ericsson, Krampe, & Tesch-Roemer, 1993). Total flight hours, by itself, also has not proven to be particularly useful (e.g., Wiggins & O'Hare, 1995). Advanced ratings and certificates appear to be more reliable, but only crude, indicators of skill. Additional research for establishing finer indices of flight experience therefore would be particularly useful.

The expertise literature reveals that experts do not posses extraordinary intelligence nor extraordinary basic cognitive abilities. They are subject to the same basic limitations of cognitive processing such as that of attention and working memory. Therefore, psychometric tests like memory span for arbitrary lists or laboratory tasks that involve simple stimuli and tasks that minimize any effects of previously acquired knowledge and skill will not inform about the level of expertise. To capture the extent to which experienced pilots could maintain their piloting skills at any age, it is essential that domain-specific tasks be used. Gopher's (1993) study (see earlier) demonstrated that high physical fidelity is nonessential for relevant skills to develop. The successful transfer of training from a complex computer game with multiple, dynamic components to actual flight performance demonstrates the criticality of cognitive fidelity.

Although many simplistic laboratory tasks with little relevance to flight continue to be used to assess pilot performance, the present knowledge points to the need for considerable updating. Due to the multitude of functions required and the increased supervisory role of today's pilot, the type of cognitive functions critical to flying are, and therefore the focus of research should be, higher level functions such as those involving executive control and resource management (see chap. 7), problem solving and decision making (see chap. 6), and reaction to epochs of heavy workload and emergency situations (see chaps. 4 & 9).

An Empirical Approach

The approach proposed here follows that advanced by Kliegl and Baltes (1987) and Ericsson and Charness (1997). Kliegl and Baltes advocated that "growth and decline can be better understood if studied at limits of performance and under controlled laboratory conditions simulating real-life settings of expertise acquisition and development" (p. 114). Ericsson and Charness reasoned that if experts have acquired their superior performance by circumventing specific constraints in their domains, then representative tasks that incorporate these constraints need to be used so that the natural performance of experts can be reproduced under controlled conditions in the laboratory. They pointed out that superior performance by experts is exhibited reliably only under conditions that capture the essence of expert performance, such as the conditions of competition for athletes or difficult cases in medical diagnosis for medical experts.

It should be emphasized again that in identifying representative tasks for the present purposes, the objective is not physical realism. Even the most powerful computer cannot duplicate the world, and the process of validating simulation fidelity is equivocal. The solution offered by Adams (1979) and Gopher and Kimchi (1989) is to rely on sound psychological principles for determining whether the critical elements have been captured.

Several methods are available for selecting domain-representative tasks. One is through a job analysis. However, pilot performance is complex, and there has not been any contemporary, comprehensive job analysis in the literature (see Damos, 1996). But the aviation psychology literature is replete with information concerning flight-relevant component tasks. Flight relevance is indicated when pilot performance is superior to nonpilot performance, reliable correlation with actual or simulator performance is obtained, or task performance is predictive of flight training success. In areas where there is a lack of established information, new empirical work is needed. Though much more effortful, this is necessary for expanding and updating the database as the conceptualization of the piloting task itself changes with new understanding and technological advances.

The proposed approach can be further illustrated by an empirical example. As described earlier, Tsang and Shaner (1998) assessed the multiple-task performance of pilots and nonpilots as a function of age. Age-sensitive and flight-relevant tasks were purposely designed to be demanding and to require exact resource management control. Pilots' time-sharing performance was found to be superior to nonpilots', despite a lack of difference in their basic perceptual and cognitive abilities. But the expertise mitigation of the age-related dual-task decrements was modest. Possibly, even though the individual tasks were flight-relevant, the conditions needed for testing the higher level cognitive component may not have been sufficiently so. Executive control in the cockpit comprises more than performing two unrelated tasks that just happened to be present in close temporal and spatial proximity.

To more accurately ascertain the extent to which expertise could mitigate age-related decrements in time-sharing, the flight meaningfulness of the time-sharing will need to be manipulated. Flight relevance could be manipulated along several dimensions. One, a more integrated time-sharing environment could be designed. Instead of presenting three unrelated laboratory tasks, a flight environment incorporating all three previously represented cognitive components could be developed. The more integrated task might require flight control (tracking) while detecting and making spatial judgments about aircraft in the area (spatial orientation) or responding to air traffic control queries (memory). Two, the individual tasks could be made more or less domain specific. For example, a three-dimensional flight control task could be used instead of line tracking; the control dynamics of the flight control task and the relevance of the memory task (e.g., ATC instructions vs. nonsense syllables) could be manipulated. Three, one essence of time-sharing is attention allocation control among demanding, dynamic tasks. The flight meaningfulness of the time-sharing condition could be manipulated by using a prioritization scheme that is more natural to flight. For example, in contrast to arbitrarily assigning higher priority to any task, the fuel level could be manipulated to impose various degrees of urgency for maintaining the flight path. Maintaining altitude (or not going below certain altitude) is another naturally high-priority task for pilots. Note that priority manipulations do not alter the task structurally. It is primarily a high-level function manipulation.

One advantage of this simulation approach is that the question of interest can be addressed in a controlled environment. Though much more complex and closer to the operational tasks than typical laboratory tasks, the task parameters still can be systematically varied and relatively easy to modify. Performance measures are well defined, and their interpretation can be guided by previous research. Obviously, this simulation is of low physical fidelity. What guarantee is there that this extension of the research will be operationally significant? For certain, no one experiment by itself will be able to establish operational significance. The approach is to methodically incorporate elements deemed critical in approximating operational conditions. The extent to which expertise would mitigate age effects is expected to change systematically as the degree of flight relevance is systematically varied. This is the prediction of the principle that expertise is domain specific, a principle that has been examined extensively in the literature and a principle that has received support from a variety of domains. Although far from a guarantee, the trend obtained with systematic manipulation is the best information available.

CONCLUSION

Recent research efforts have better identified the relevant variables pertaining to the intricate relationship between aging and piloting. Although age effects appear to be pervasive, they are often not large, are associated with large individual variability, and under some circumstances, can be greatly reduced by training. The expertise literature reveals that the differences between experts and novices are complex, and the mechanisms by which experts could circumvent intrinsic limitations are sophisticated. Importantly, skilled performance, including its protection against age-related declines, is highly domain specific. Given the multiplicity and the dynamic nature of the demands in the cockpit, the higher level cognitive functions of the pilots deserve particular attention.

The cognitive aging literature reveals great individual variability in aging. This basic finding and observations of actual job performances show that age accounts little for performance. The aviation literature reveals the difficulty of relying on accident data for understanding age effects. Performance-based evaluations, therefore, appear to be the most promising approach for evaluating age effects. However, simple laboratory tasks and psychometric tests are unlikely to capture the full capability of experts. Following the recommendations of Kliegl and Baltes (1987) and Ericsson and Charness (1997), sufficiently demanding and domain-specific conditions need to be used. Here, the key is cognitive fidelity and not surface similarity.

The investigation of the complex issues of aging and piloting are not without practical obstacles. Notably, aging effects are variable, pilot performance is hard to measure, and because of current ruling there are no pilots beyond the early 60s that would be considered to be most expert in the industry (air carrier pilots). Direct testing is therefore not possible. But this difficult scenario is not unique in scientific investigations. A systematic approach bounded by known principles and tried methodologies is the keystone to any scientific investigations and would not be an exception here (see Kantowitz, 1992; Gopher & Kimchi, 1989).

It is therefore not a question of whether pilots experience age effects. Age effects in basic cognitive processing such as processing speed do not always manifest in professional or expert performance. More instructive questions include the following: What are some of the most pertinent cognitive functions essential to flight performance? How might the development of an affordable, practical, objective performance test battery be expedited by psychologists and medical experts? What performance support in terms of task redesign, interface enhancement, and computer aid could be provided? Last, a question asked only too seldom, What gains accompany aging? In addition to on-the-job knowledge and experience, caution and wisdom are generally thought to increase with age. These are attributes important to flight, and they are not easy to come by.

> The thing that is highly significant here is that certain of the more purely psychological processes step in to save the situation when the physiological service has become impaired. When this occurs we have a glimpse of man at his highest level of adjustment rectifying and correcting lower by higher mental process, psychologically organizing himself more and more intricately to compensate for neuromuscular losses. Society should do its utmost to relieve the players in the supreme game from all possible inhibiting weights. (Miles, 1933, p. 120)

Perhaps not all is lost when Saint-Exupéry (1943) said, "I have had to grow old." (p. 19)

REFERENCES

Adams, J. A. (1979). On the evaluation of training devices. *Human Factors, 21,* 711–720.

Adams, M. J., Tenny, Y. J., & Pew, R. W. (1991). *Strategic workload and the cognitive management of advanced multi-task systems.* Wright-Patterson Air Force Base, OH: Crew Systems Ergonomics Information Analysis Center.

Adams, R. J., & Ericsson, K. A. (1992). *Introduction to cognitive processes of expert pilots* (DOT/FAA/RD-92/12). Washington, DC: U.S. Department of Transportation, Federal Aviation Administration.

Bellenkes, A. H., Wickens, C. D., & Kramer, A. F. (1997). Visual scanning and pilot expertise: The role of attentional flexibility and mental model development. *Aviation, Space, and Environmental Medicine, 68,* 569–579.

Birren, J. E., & Fisher, L. M. (1995a). Aging and speed of behavior: Possible consequences for psychological functioning. *Annual Review of Psychology, 46,* 329–353.

Birren, J. E., & Fisher, L. M. (1995b). Rules and reason in the forced retirement of commercial airline pilots at age 60. *Ergonomics, 38,* 518–525.

Birren, J. E., & Schaie, K. W. (1996). *Handbook of the psychology of aging* (4th ed.). San Diego, CA: Academic.

Birren, J. E., & Schroots, J. J. F. (1996). History, concepts, and theory in the psychology of aging. In J. E. Birren & K. W. Schaie (Eds.), *Handbook of the psychology of aging* (4th ed., pp. 3–23). San Diego, CA: Academic.

Bosman, E. A. (1994). Age and skill differences in typing related and unrelated reaction time tasks. *Aging and Cognition, 1,* 310–322.

Braune, R., & Wickens, C. D. (1984a). *Individual differences and age-related performance assessment in aviators. Part 1: Battery development and assessment* (Final Tech. Rep. EPL-83-4/NAMRL-83-1). Urbana–Champaign: University of Illinois, Engineering Psychology Laboratory.

Braune, R., & Wickens, C. D. (1984b). *Individual differences and age-related performance assessment in aviators. Part 2: Initial battery validation* (Final Tech. Rep. EPL-83-7/NAMRL-83-2). Urbana-Champaign: University of Illinois, Engineering Psychology Laboratory.

Braune, R., & Wickens, C. D. (1985). The functional age profile: An objective decision criterion for the assessment of pilot performance capacities and capabilities. *Human Factors, 27,* 681–693.

Broach, D. (2000a). *Pilot age and accident rates Report 3: An analysis of professional air transport pilot accident rates by age* (unpublished Civil Aeromedical Institute report prepared for Congress). Oklahoma City, OK: FAA Civil Aeromedical Institute Human Resources Research Division.

Broach, D. (2000b). *Pilot age and accident rates Report 4: An analysis of professional ATP and commercial pilot accident rates by age* (unpublished Civil Aeromedical Institute report prepared for Congress). Oklahoma City, OK: FAA Civil Aeromedical Institute Human Resources Research Division.

Buck, R. N. (1995). *The pilot's burden: Flight safety and the roots of pilot error.* Ames: Iowa State University Press.

Cepeda, N. J., Kramer, A. F., & Gonzalez de Sather, J. C. M. (2001). Changes in executive control across the life span: Examination of task-switching performance. *Developmental Psychology, 37,* 715–730.

Charness, N. (1981). Aging and skilled problem solving. *Journal of Experimental Psychology: General, 110,* 21–38.

Charness, N. (1989). Age and expertise: Responding to Talland's challenge. In L. W. Poon, D. C. Rubin, & B. A. Wilson (Eds.), *Everyday cognition in adulthood and late life* (pp. 437–456). Cambridge, England: Cambridge University Press.

Charness, N. (2000). Can acquired knowledge compensate for age-related declines in cognitive efficiency? In S. H. Qualls & N. Abeles (Eds.), *Psychology and the aging revolution: How we adapt to longer life* (pp. 99–117). Washington, DC: American Psychological Association.

Chase, W. G., & Ericsson, K. A. (1982). Skill and working memory. In G. H. Bower (Ed.), *The psychology of learning and motivation* (Vol. 16, pp. 1–59). New York: Academic Press.

Chi, R., Glaser, M. T. H., & Farr, M. J. (Eds.) (1988). *The nature of expertise.* Hillsdale, NJ: Lawrence Erlbaum Associates.

Committee on the Changing Nature of Work (1993, October). Human Capital Initiative Report 1—The changing nature of work. *American Psychological Society Observer, Special Issue.* Washington, DC: American Psychological Society.

Craik, F. I. M., & Salthouse, T. A. (2000). *The handbook of aging and cognition* (2nd ed.). Mahwah, NJ: Lawrence Erlbaum Associates.

Crossley, M., & Hiscock, M. (1992). Age-related differences in concurrent-task performance of normal adults: Evidence for a decline in processing resources. *Psychology and Aging, 7,* 499–506.

Czaja, S. (Ed.) (1990). *Human factors research needs for an aging population.* Washington, DC: National Academy.

Damos, D. L. (1978). Residual attention as a predictor of pilot performance. *Human Factors, 20,* 435–440.

Damos, D. L. (1993). Meta-analysis to compare the predictive validity of single- and multiple-task measures to flight performance. *Human Factors, 35,* 615–628.

Damos, D. L. (1996). Pilot selection batteries: Shortcomings and perspectives. *International Journal of Aviation Psychology, 6,* 199–209.

Damos, D. L., & Lintern, G. (1981). A comparison of single- and dual-task measures to predict simulator performance of beginning student pilots. *Ergonomics, 24,* 673–684.

Damos, D. L., & Smist, T. E. (1982). Individual differences in multi-task response strategies. *Aviation, Space, and Environmental Medicine, 53,* 1177–1181.

Damos, D. L., & Wickens, C. D. (1980). The identification and transfer of timesharing skills. *Acta Psychologica, 46,* 15–39.

Dollinger, S. M. C., & Hoyer, W. J. (1996). Age and skill differences in the processing demands of visual inspection. *Applied Cognitive Psychology, 10,* 225–239.

Druckman, D., & Bjork, R. A. (Eds.). (1991). *In the mind's eye: Enhancing human performance* (chap. 4). Washington, DC: National Academy.

Ericsson, K. A. (1996). The acquisition of expert performance: An introduction to some of the issues. In K. A. Ericsson (Ed.), *The road to excellence* (pp. 1–50). Mahwah, NJ: Lawrence Erlbaum Associates.

Ericsson, K. A., & Charness, N. (1997). Cognitive and developmental factors in expert performance. In P. J. Feltovich, K. M. Ford, & R. R. Hoffman (Eds.), *Expertise in context* (pp. 3–41). Cambridge, MA: MIT Press.

Ericsson, K. A., & Delaney, P. F. (1998). Working memory and expert performance. In R. H. Logie & K. J. Gilhooly (Eds.), *Working memory and thinking* (pp. 93–114). Hillsdale, NJ: Lawrence Erlbaum Associates.

Ericsson, K. A., & Kintsch, W. (1995). Long-term working memory. *Psychological Review, 105,* 211–245.

Ericsson, K. A., Krampe, R. Th., & Tesch-Roemer, C. (1993). The role of deliberate practice in the acquisition of expert performance. *Psychological Review, 100,* 363–406.

Federal Aviation Administration. (1995). *The Age 60 Rule* (Docket No. 27264). Washington, DC: Author.

Fisher, D. L., Duffy, S.A., & Katsikopoulos, K. V. (2000). Cognitive slowing among older adults: What kind and how much? In T. J. Perfect & E. A. Maylor (Eds.), *Models of cognitive aging* (pp. 87–124). New York: Oxford University Press.

Garner, W. R., Hake, H. W., & Eriksen, C. W. (1956). Operationalism and the concept of perception. *Psychological Review, 63,* 149–159.

Giniger, S., Dispenzieri, A., & Eisenberg, J. (1983). Age, experience, and performance on speed and skill jobs in an applied setting. *Journal of Applied Psychology, 68,* 469–475.

Glaser, R. (1987). Thoughts on expertise. In C. Schooler & K. W. Schaie (Eds), *Cognitive functioning and social structure over the life course* (pp. 81–94). Norwood, NJ: Ablex.

Golaszewski, R. (1983). *The influence of total flight time, recent flight time and age on pilot accident rates* (Final Report DTRS57-83-P-80750). Bethesda, MD: Acumenics Research and Technology.

Gopher, D. (1982). A selective attention test as a predictor of success in flight training. *Human Factors, 24,* 173–184.

Gopher, D. (1993). The skill of attention control: Acquisition and execution of attention strategies. In D. Meyer & S. Kornblum (Eds.), *Attention and performance XIV* (pp. 299–322). Hillsdale, NJ: Lawrence Erlbaum Associates.

Gopher, D. (1996). Attention control: Explorations of the work of an executive controller. *Cognitive Brain Research, 5,* 23–38.

Gopher, D., Armony, L., & Greenshpan, Y. (2000). Switching tasks and attention policies. *Journal of Experimental Psychology: General, 129,* 308–339.

Gopher, D., & Kimchi, R. (1989). Engineering psychology. *Annual Review of Psychology, 40,* 431–455.

Gopher, D., Weil, M., & Siegel, D. (1989). Practice under changing priorities: An approach to the training of complex skills. *Acta Psychologica, 71,* 147–177.

Griffin, G. R., & McBride, D. K. (1986). *Multitask performance: Predicting success in naval aviation primary flight training* (NAMRL 1316). Pensacola, FL: U.S. Naval Air Station, Naval Aerospace Medical Research Laboratory.

Guttentag, R. E. (1989). Age differences in dual-task performance: Procedures, assumptions, and results. *Developmental Review, 9,* 146–170.

Hansen, J. S., & Oster, C. V., Jr. (Eds.). (1997). *Taking flight.* Washington, DC: National Academy.

Hardy, D. J., & Parasuraman, R. (1997). Cognition and flight performance in older pilots. *Journal of Experimental Psychology: Applied, 3,* 313–348.

Hertzog, C. (1996). Research design in studies of aging and cognition. In J. E. Birren & K. W. Schaie (Eds.), *Handbook of the psychology of aging* (4th ed., pp. 24–37). San Diego, CA: Academic.

Horn, J. L., & Masunaga, H. (2000). New directions for research into aging and intelligence: The development of expertise. In T. J. Perfect & E. A. Maylor (Eds.), *Models of cognitive aging* (pp. 125–159). New York: Oxford University Press.

Horst, R. L., & Kay, G. G. (1991). COGSCREEN: Personal computer-based tests of cognitive function for occupational medical certification. In *Proceedings of the Sixth International Symposium on Aviation Psychology* (pp. 733–739). Columbus: The Ohio State University.

Hosmer, D. W., & Lemeshow, S. (1989). *Applied logistic regression.* New York: Wiley.

Hunt, E., & Hertzog, C. (1981), *Age related changes in cognition during the working years* (Final Report). Seattle, WA: University of Washington, Department of Psychology.

Hunter, J. E., & Hunter, R. F. (1984). Validity and utility of alternative predictors of job performance. *Psychological Bulletin, 96,* 72–98.

Hyland, D. T., Kay, E. J., & Deimler, J. D. (1994). *Age 60 Rule research, Part IV: Experimental evaluation of pilot performance* (Tech. Rep. No. DOT/FAA/AM-94–23). Washington, DC: Federal Aviation Administration, Office of Aviation Medicine.

Institute of Medicine. (1981). *Airline pilot age, health and performance.* Washington, DC: National Academy.

Jorna, P. G. A. M. (1989). Prediction of success in flight training by single- and dual-task performance. In *AGARD Conference Proceedings, AGARD-CP-458* (21–1-21–10). Neuilly-sur-Seine, France: Advisory Group for Aerospace Research and Development.

Kahneman, D. (1973). *Attention and effort.* Englewood Cliffs, NJ: Prentice Hall.

Kantowitz, B. H. (1992). Selecting measures for human factors research. *Human Factors, 34,* 387–398.

Kausler, D. H. (1990). *Experimental psychology, cognition, and human aging.* New York: Springer-Verlag.

Kay, E. J., Hillman, D. J., Hyland, D. T., Voros, R. S., Harris, R. M., & Deimler, J. D. (1994). *Age 60 Rule research, Part III: Consolidated data base experiments final report* (Tech. Rep. No.

DOT/FAA/AM-92–22). Washington, DC: Federal Aviation Administration, Office of Aviation Medicine.

Kimberg, D. Y., Aguirre, G. K., & D'Esposito, M. (2000). Modulation of task-related neural activity in task-switching: An fMRI study. *Cognitive Brain Research, 10,* 189–196.

Kliegl, R., & Baltes, P. B. (1987). Theory-guided analysis of mechanisms of development and aging through testing-the-limits and research on expertise. In C. Schooler & K. W. Schaie (Eds.), *Cognitive functioning and social structure over the life course* (pp. 95–119). Norwood, NJ: Ablex.

Kramer, A. F., Hahn, S., & Gopher, D. (1999). Task coordination and aging: Explorations of executive control process in the task switching paradigm. *Acta Psychologica, 101,* 339–378.

Kramer, A. F., Larish, J. F., & Strayer, D. L. (1995). Training for attentional control in dual task settings: A comparison of young and old adults. *Journal of Experimental Psychology: Applied, 1,* 50–76.

Kramer, A. F., Larish, J. F., Weber, T. A., & Bardell, L. (1999). Training for executive control: Task coordination strategies and aging. In D. Gopher & A. Koriat (Eds.), *Attention and performance XVII Cognitive regulation of performance: Interaction of theory and application* (pp. 617–652). Cambridge, MA: MIT Press.

Krampe, R. Th., & Ericsson, K. A. (1996). Maintaining excellence: Deliberate practice and elite performance in young and older pianists. *Journal of Experimental Psychology: General, 125,* 331–359.

Kray, J., & Lindenberger, U. (2000). Adult age differences in task switching. *Psychology and Aging, 15,* 126–147.

Li, G., Baker, S. P., Grabowski, J. G., & Rebok, G. W. (2001). Factors associated with pilot error in aviation crashes. *Aviation, Space, and Environmental Medicine, 72,* 52–58.

Lindenberger, U., Kliegl, R., & Bates, P. B. (1992). Professional expertise does not eliminate age differences in imagery-based memory performance during adulthood. *Psychology and Aging, 7,* 585–593.

McDowd, J. M. (1986). The effects of age and extended practice on divided attention performance. *Journal of Gerontology, 41,* 764–769.

McDowd, J. M., & Shaw, R. J. (2000). Attention and aging: A functional Perspective. In F. I. M. Craik & T. A. Salthouse (Eds.), *The handbook of aging and cognition* (pp. 221–292). Mahwah, NJ: Lawrence Erlbaum Associates.

McEvoy, G. M., & Cascio, W. F. (1989). Cumulative evidence of the relationship between employee age and job performance. *Journal of Applied Psychology, 74,* 11–17.

McFadden, K. L. (1996). Comparing pilot-error accident rates of male and female airline pilots. *Omega, International Journal of Management Science, 24,* 443–450.

McFadden, K. L. (1997). Predicting pilot-error incidents of US airline pilots using logistic regression. *Applied Ergonomics, 28,* 209–212.

McFarland, R. A. (1954). Psycho-physiological problems of aging in air transport pilots. *Journal of Aviation Medicine, 25,* 210–220.

McFarland, R. A. (1955, January 17). Aging is airman's greatest foe. *Aviation Week,* 108.

McFarland, R. A., Graybiel, A., Liljencrantz, E., & Tuttle, A. D. (1939). An analysis of the physiological and psychological characteristics of 200 civil air line pilots. *Journal of Aviation Medicine, 10,* 160–210.

Meiran, N., Gotler, A., & Perlman, A. (2001). Old age is associated with a pattern of relatively intact and relatively impaired task-set switching abilities. *Journal of Gerontology: Psychological Sciences, 56B,* P88–P102.

Miles, W. R. (1933). Age and human ability. *Psychological Review, 40,* 99–123.

Mohler, S. R. (1981). Reasons for eliminating the "Age 60" regulation for airline pilots. *Aviation, Space, and Environmental Medicine, 52,* 445–454.

Mortimer, R. G. (1991). Some factors associated with pilot age in general aviation crashes. In *Proceedings of the Sixth International Symposium on Aviation Psychology* (pp. 770–775). Columbus: The Ohio State University.

Murrell, H., & Edwards, E. (1963). Field studies of an indicator of machine tool travel with special reference to the ageing worker. *Occupational Psychology, 37,* 267–275.

Murrell, K. F. H., & Griew, S. (1965). Age, experience and speed of response. In A. T. Welford & J. E. Birren (Eds.), *Behavior, aging, and the nervous system* (pp. 60–66). Springfield, IL: Charles C. Thomas.

Murrell, K. F. H., & Humphries, S. (1978). Age, experience, and short-term memory. In M. M. Gruneberg, P. E. Morris, & R. N. Sykes (Eds.), *Practical aspects of memory* (pp. 363–365). London: Academic Press.

Murrell, K. F. H., Powesland, P. F., & Forsaith, B. (1962). A study of pillar-drilling in relation to age. *Occupational Psychology, 36,* 45–52.

National Behavioral Science Research Agenda Committee. (1992, February). Human Capital Initiative. *American Psychological Society Observer, Special Issue.* Washington, DC: American Psychological Society.

Navon, D. (1985). Attention division or attention sharing? In M. I. Posner & O. S. M. Marin (Eds.), *Attention and performance XI* (pp. 133–146). Hillsdale, NJ: Lawrence Erlbaum Associates.

Nestor, P. G., Parasuraman, R., & Haxby, J. V. (1989). Attentional costs of mental operations in young and old adults. *Developmental Neuropsychology 5,* 141–158.

Norman, G. R., Brooks, L. R., & Allen, S. W. (1989). Recall by expert medical practitioners and novices as a record of processing attention. *Journal of Experimental Psychology: Learning, Memory, and Cognition, 15,* 1166–1174.

Odenheimer, G. (1999). Function, flying, and the Age 60 Rule. *Journal of American Geriatrics Society, 47,* 910–911.

Office of Technology Assessment. (1990). *Medical risk assessment and the Age 60 Rule for airline pilots.* Washington, DC: U.S. House of Representatives, Committee on Public Works and Transportation, Subcommittee on Investigations and Oversight.

Park, D. (1994). Aging, cognition, and work. *Human Performance, 7,* 181–205.

Park, D., & Schwarz, N. (Eds.). (2000). *Cognitive aging: A primer.* Philadelphia: Psychology Press.

Patel, V. L., & Groen, G. J. (1991). The general and specific nature of medical expertise: A critical look. In K. A. Ericsson & J. Smith (Eds.), *Towards a general theory of expertise: Prospects and limits* (pp. 93–125). Cambridge, England: Cambridge University Press.

Pedhazur, E. J. (1997). *Multiple regression in behavioral research: explanation and prediction.* Orlando, FL: Harcourt Brace.

Perfect, T. J., & Maylor, E. A. (Eds.). (2000). *Models of cognitive aging.* New York: Oxford University.

Perrow, C. (1999). *Normal accidents: Living with high-risk technologies.* New York: Basic Books.

Ponds, R. W. H. M., Brouwer, W. H., & van Wolffelaar, P. C. (1988). Age differences in divided attention in a simulated driving task. *Journal of Gerontology, 43,* 151–156.

Rabbitt, P. M. A. (1993a). Crystal quest: An examination of the concepts of "fluid" and "crystallised" intelligence as explanations for cognitive changes in old age. In A. D. Baddeley & L. Weiskrantz (Eds.), *Attention, selection, awareness and control* (pp. 188–230). New York: Clarendon Press/Oxford University Press.

Rabbitt, P. M. A. (1993b). Does it all go together when it goes? The nineteenth Bartlett Memorial lecture. *Quarterly Journal of Experimental Psychology (40A),* 385–434.

Rabbitt, P. M. A. (1993c). Methodological and theoretical lessons from the University of Manchester longitudinal studies of cognitive changes in normal old age. In J. J. F. Schroots (Ed.), *Aging, health and competence: The next generation of longitudinal research* (pp. 199–219). New York: Academic Press.

Rabbitt, P. M. A. (1997). Ageing and human skill: A 40th anniversary. *Ergonomics, 40,* 962–981

Rabbitt, P. M. A. (2000). Measurement indices, functional characteristics, and psychometric constructs in cognitive aging. In T. J. Perfect & E. A. Maylor (Eds.), *Models of cognitive aging* (pp. 160–187). New York: Oxford University Press.

Reason, J. (1990). The contribution of latent human failures to the breakdown of complex systems. *Royal Society of London Philosophical Transactions (327),* 475–484.

Rebok, G. W., Grabowski, J. G., Baker, S. P., Lamb, M. W., Willoughby, S., & Li, G. (1999). *Pilot age and performance as factors in aviation crashes.* Poster presented at the Annual Meeting of the American Psychological Association, Boston.

Rogers, R., & Monsell, S. (1995). Costs of a predictable switch between simple cognitive tasks. *Journal of Experimental Psychology: General, 124,* 207–231.

Roscoe, S. N. (1997). Predicting and enhancing flightdeck performance. In R. Telfer & P. Moore (Eds.), *Aviation training, pilot, instructor, and organisation* (pp. 195–208). Aldershot, England: Averbury.

Rubinstein, J. S., Meyer, D. E., & Evans, J. E. (2001). Executive control of cognitive process in task switching. *Journal of Experimental Psychology: Human Perception and Performance, 27,* 763–797.

Saint-Exupéry, A. (1943). *The little prince* (K. Woods, Trans.). New York: Harcourt, Brace & World.

Salthouse, T. A. (1984). Effects of age and skill in typing. *Journal of Experimental Psychology: General, 113,* 345–371.

Salthouse, T. A. (1991). Expertise as the circumvention of human processing limitations. In K. A. Ericsson & J. Smith (Eds.), *Toward a general theory of expertise* (pp. 286–300). Cambridge, England: Cambridge University Press.

Salthouse, T. A. (1996). The processing-speed theory of adult age differences in cognition. *Psychological Review, 103,* 403–428.

Salthouse, T. A. (2000). Methodological assumptions in cognitive aging research. In F. I. M. Craik & T. A. Salthouse (Eds.), *The handbook of aging and cognition* (2nd ed., pp. 467–498). Mahwah, NJ: Lawrence Erlbaum Associates.

Salthouse, T. A., Fristoe, N. M., Lineweaver, T. T., & Coon, V. E. (1995). Aging of attention: Does the ability to divide decline? *Memory & Cognition, 23,* 59–71.

Salthouse, T. A., Fristoe, N., McGuthry, K. E., & Hambrick, D. Z. (1998). Relation of task switching to speed, age, and fluid intelligence. *Psychology and Aging, 13,* 445–461.

Salthouse, T. A., & Maurer, T. J. (1996). Aging, job performance, and career development. In J. E. Birren & K. W. Schaie (Eds.), *Handbook of the psychology of aging* (pp. 353–364). San Diego, CA: Academic.

Salthouse, T. A., Rogan, J. D., & Prill, K. (1984). Division of attention: Age differences on a visually presented memory task. *Memory & Cognition, 12,* 613–620.

Schaie, K. W. (1988). The impact of research methodology on theory building in the developmental sciences. In J. E. Birren & V. L. Bengtson (Eds.), *Emergent theories of aging* (pp. 43–57). New York: Springer.

Schaie, K. W. (1990). The optimization of cognitive functioning in old age: Predictions based on cohort-sequential and longitudinal data. In P. B. Baltes & M. M. Baltes (Eds.), *Successful aging: Perspectives from the behavioral sciences* (pp. 94–117). New York: Cambridge University Press.

Schaie, K. W. (1993). The course of adult intellectual development. *American Psychologist, 49,* 304–313.

Shooter, A. M. N., Schonfield, A. E. D., King, H. F., & Welford, A. T. (1956). Some field data on the training of older people. *Occupational Psychology, 30,* 204–215.

Simons, M., Valk, P. J. L., Krol, J. R., & Holewijn, M. (1996). Consequences of raising the maximum age limit for airline pilots. In *Proceedings of the Third ICAO Global Flight Safety and Human Factors Symposium* (pp. 248–255). (Human Factors Digest No. 13). No. CIRC 266-AN/158.

Somberg, B. L., & Salthouse, T. A. (1982). Divided attention abilities in young and old adults. *Journal of Experimental Psychology, 8,* 651–663.

Staszewski, J. J. (1988). Skilled memory and expert mental calculation. In M. T. H. Chi & R. Glaser (Eds.), *The nature of expertise* (pp. 71–128). Hillsdale, NJ: Lawrence Erlbaum Associates.

Stern, P. C., & Carstensen, L. L. (Eds.). (2000). *The aging mind: Opportunities in cognitive research.* Washington, DC: National Academy.

Stokes, A. F., Belger, A., & Zhang, K. (1990). *Investigation of factors comprising a model of pilot decision making: Part II. Anxiety and cognitive strategies in expert and novice aviators* (Final Technical Report ART-90-8/SCEEE-90–2). Wright-Patterson Air Force Base, OH: AMRL, Human Engineering Division.

Stuck, A. E., van Gorp, W. G., Josephson, K. R., Morgenstern, H., & Beck, J. C. (1992). Multidimensional risk assessment versus age as criterion for retirement of airline pilots. *Journal of the American Geriatics Society, 40,* 536–532.

Szafran, J. (1968). Psychophysiological studies of aging in pilots. In G. A. Talland (Ed.), *Human aging and behavior* (pp. 37–74). New York: Academic Press.

Taylor, J. L., O'Hara, R., Mumenthaler, M. S., & Yesavage, J. A. (2000). Relationship of CogScreen-AE to flight simulator performance and pilot age. *Aviation, Space, and Environmental Medicine, 71,* 373–380.

Taylor, J. L., Yesavage, J. A., Morrow, D. G., Dolhert, N., Brooks, J. O., III, & Poon, L. W. (1994). The effects of information load and speech rate on younger and older aircraft pilots' ability to execute simulated air-traffic controller instructions. *Journal of Gerontology, 49,* P191-P200.

Tham, M., & Kramer, A. (1994). Attentional control and piloting experience. In *Proceedings of the Human Factors and Ergonomics Society 38th Annual Meeting* (pp. 31–35). Santa Monica, CA: Human Factors and Ergonomics Society.

Tsang, P. S. (1992). A reappraisal of aging and pilot performance. *International Journal of Aviation Psychology, 2,* 193–212.

Tsang, P. S. (1997a). Age, flight experience, time-sharing performance, and perceived workload. In *Proceedings of the Ninth International Symposium on Aviation Psychology* (pp. 42–46). Columbus: The Ohio State University and Association of Aviation Psychologists.

Tsang, P. S. (1997b). Age and pilot performance. In R. Telfer & P. Moore (Eds.), *Aviation training: Pilot, instructor and organization* (pp. 21–40). Brookview, VT: Averbury.

Tsang, P. S. (1997c). A microanalysis of age and pilot time-sharing performance. In D. Harris (Ed.), *Engineering psychology and cognitive ergonomics: Vol. 1. Transportation systems* (pp. 245–251). Brookfield, VT: Ashgate.

Tsang, P. S., & Shaner, T. L. (1998). Age, attention, expertise, and time-sharing performance. *Psychology and Aging, 13,* 323–347.

Tsang, P. S., & Voss, D. T. (1996). Boundaries of cognitive performance as a function of age and piloting experience. *International Journal of Aviation Psychology, 6,* 359–377.

Tsang, P. S., & Wickens, C. D. (1988). The structural constraints and the strategic control of resource allocation. *Human Performance, 1,* 45–72.

U.S. Census Bureau. (1999). *World population profile: 1998* (Report WP/98). Washington, DC: U.S. Government Printing Office.

Vitality for Life Committee. (1993, December). Human Capital Initiative Report 2—Vitality for life: Psychological research for productive aging. *American Psychological Society Observer, Special Issue.* Washington, DC: American Psychological Society.

Waldman, D. A., & Avolio, B. J. (1986). A meta-analysis of age differences in job performance. *Journal of Applied Psychology, 71,* 33–38.

Wickens, C. D. (1984). Processing resources in attention. In R. Parasuraman & D. R. Davies (Eds.), *Varieties of attention* (pp. 63–98). New York: Academic Press.

Wickens, C. D. (1987). Attention in aviation. In *Proceedings of the Fourth International Symposium on Aviation Psychology* (pp. 602–608). Columbus: The Ohio State University.

Wickens, C. D., Braune, R., & Stokes, A. (1987). Age differences in the speed and capacity of information processing: 1. A dual-task approach. *Psychology and Aging, 2,* 70–78.

Wiggins, M., & O'Hare, D. (1995). Expertise in aeronautical weather-related decision making: A cross-sectional analysis of general aviation pilots. *Journal of Experimental Psychology: Applied, 1,* 305–320.

Yacovone, D. W., Borosky, M. S., Bason, R., & Alcov, R. A. (1992). Flight experience and the likelihood of U.S. Navy mishaps. *Aviation, Space, and Environmental Medicine, 63,* 72–74.

Yeager, C., & Janos, L. (1985). *Yeager: An autobiography.* New York: Bantam.

Yesavage, J. A., Dolhert, N., & Taylor, J. L. (1994). Flight simulator performance of younger and older aircraft pilots. Effects of age and alcohol. *Journal of the American Geriatric Society, 42,* 577–582.

Yesavage, J. A., Taylor, J. L., Mumenthaler, M. S., Noda, A., & O'Hara, R. (1999). Relationship of age and simulated flight performance. *Journal of the American Geriatrics Society, 47,* 819–823.

Contributors

Evan A. Byrne is currently a Supervisory Human Performance Investigator and chief of the Human Performance Division in the Office of Aviation Safety at the National Transportation Safety Board (NTSB). He has been at the NTSB since 1996 and has been involved as a human performance specialist in a number of major aviation accidents. Before joining the NTSB he was a postdoctoral research associate at the Cognitive Science Laboratory at The Catholic University of America. He is an instrument-rated private pilot.

Thomas R. Carretta received his Ph.D. in psychology in 1983 from the University of Pittsburgh. Currently, he is a research psychologist in the Crew Systems Development Branch of the Human Effectiveness Directorate of the Air Force Research Laboratory at Wright-Patterson Air Force Base, Ohio. Prior to his current position, he spent over 10 years in the Manpower and Personnel Research Division of the Air Force Research Laboratory in San Antonio, Texas, working on air crew selection and classification issues. His professional interests include personnel measurement, selection, and individual and group differences.

Ralph Norman Haber and **Lyn Haber** married in 1972 and have worked together professionally ever since. They formed their firm, Human Factors Consultants, in 1988, specializing in providing quantitative answers to complex questions. Their military and business consulting has included work on assessing product fit, productivity, safety, and proficiency; developing training programs; and improving product design and product fit. For example, they have consulted with the U.S. Air Force and the U.S. Department of Transportation. Their legal consulting work has included problems of perception and memory, eyewitness testimony, and recovered memory, as well as human factors analyses of accidents. The Habers have in common a professional lifetime focused on experimental research, but their areas of individual expertise differ: Ralph first specialized in perception and memory as an experimental psychologist, then in human factors and artificial intelligence. Lyn first specialized in experimental research in language analyses and language disorders, then in evaluation and training procedures. Today, they are both adjunct professors of psychology at the University of California at Riverside and research associates in psychology at the University of California at Santa Cruz. Previously, they were both affiliated with the University of Illinois at Chicago from 1979 to 1994; the University of Rochester from 1964 to 1979 (1973 to 1979 for Lyn); Temple University from 1968 to 1973 (Lyn); and Yale University from 1958 to 1964 (Ralph). Ralph was chairman of the Department of Psychology at the University of Rochester from 1967 to 1971. Ralph received his Ph.D. from Stanford University in psychology in 1957. Lyn received her Ph.D. from the University of California at Berkeley in linguistics in 1970. Between them, they have published nine books and 275 scientific articles, and given 150 conference and convention presentations. They live in a remote part of the Sierra Nevada Mountains at 6,000 feet, surrounded by peaks, deer, and incredible beauty.

Ronald A. Hess received B.S., M.S., and Ph.D. degrees in Aerospace Engineering from the University of Cincinnati. He is a professor and vice chairman of the Department of Mechanical and Aeronautical Engineering at the University of California–Davis. Prior to joining the faculty at the University of California, he was a research scientist at NASA Ames Research Center. He is an associate Fellow of the American Institute of Aeronautics and Astronautics (AIAA) and a senior member of the Institute of Electrical and Electronic Engineers (IEEE). He is an associate editor of the *AIAA Journal of Aircraft,* the *IEEE Transactions on Systems, Man, and Cybernetics,* and the *Proceedings of the Institution of Mechanical Engineers, Part G: Journal of Aerospace Engineering.*

Mary K. Kaiser is a research psychologist in the Human Factors Research and Technology Division at NASA Ames Research Center. She received her Ph.D. in psychology from the University of Virginia, and was a postdoctoral fellow at the University of Michigan in applied experimental psychology before joining Ames in 1985. From 1996 through 1999, she was an associate editor of the *Journal of Experimental Psychology: Human Perception and Performance.*

David O'Hare is a senior lecturer in psychology at the University of Otago, Dunedin, New Zealand. He obtained his undergraduate and Ph.D. degrees at the University of Exeter (England) and was formerly a lecturer at the University of Lancaster (England) before moving to New Zealand in 1982. He is the co-author (with Stan Roscoe) of *Flightdeck Performance: The Human Factor* (Iowa State

University Press) and has published an edited book on *Human Performance in General Aviation* (Ashgate). He is the associate editor (Pacific-Rim) for *The International Journal of Aviation Psychology* (Lawrence Erlbaum Associates). He has been a keen sailplane and licenced private pilot.

Raja Parasuraman is director of the Cognitive Science Laboratory and professor of psychology at The Catholic University of America in Washington, D.C. He received a B.Sc. in electrical engineering from Imperial College, University of London (1972) and a M.Sc. in applied psychology (1973) and a Ph.D. in psychology from the University of Aston, Birmingham, England (1976). He has carried out research on attention, aging, aviation, event-related brain potentials, functional brain imaging, signal detection, vigilance, and workload. His research in these areas has been supported by the National Aeronautics and Space Administration (NASA), the National Institutes of Health (NIH), the U.S. Navy, as well as by private foundations. His books include *The Psychology of Vigilance* (1982), *Varieties of Attention* (1984), *Event-Related Brain Potentials* (1990), *Automation and Human Performance* (1996), and *The Attentive Brain* (1998). He was chair of the Human Performance and Cognition Study Section for the NASA Neurolab Mission in 1995, a member of the Human Development and Aging Study Section of NIH from 1992 to 1995, and a member of the National Research Council's Panel on Human Factors in Air Traffic Control Automation from 1994 to 1998. He is currently chair of the National Research Council Panel on Human Factors. He was elected a Fellow of the American Association for the Advancement of Science (1994), the American Psychological Association (1991), the American Psychological Society (1991), and the Human Factors and Ergonomics Society (1994).

John Patrick is at the University of Cardiff, Wales and is an experienced researcher in the cognitive aspects of training, including problem solving in industrial contexts. He has published various books and journal articles.

Malcolm James Ree is Associate Professor at the Center for Leadership Studies at Our Lady of the Lake University, San Antonio, Texas. Formerly, he was the senior scientist in the Space Warfighter Training Branch of the Human Effectiveness Directorate of the Air Force Research Laboratory, San Antonio, Texas. He received his Ph.D. in psychometrics in 1976 from the University of Pennsylvania. His professional interests include human abilities, individual differences, the intelligence–job performance nexus, and statistics.

Jeffery A. Schroeder received a B.S. and M.S. from Purdue University and a Ph.D. from Stanford University in aeronautics and astronautics. He has been at NASA Ames Research Center for 13 years working in the areas of flight controls, displays, and flight simulation cueing. He is currently the deputy chief of the Flight Control and Cockpit Integration Branch, and also serves as an adjunct faculty member at San Jose State University.

Paul J. Sherman teaches Human-Computer Interaction at the University of Texas at Dallas and manages Intuit's User-Centered Design group based in Plano, Texas. Previously, Paul was an independent consultant and usability engineer with

Austin Usability. Before relocating to Texas, he was at Lucent Technologies in New Jersey, where he supervised the user interface design of several telecommunications management applications and led efforts to develop cross-product user interface standards. Paul received his Ph.D. in 1997 from the University of Texas at Austin. His research focused on how pilots' use of automated systems on the flight deck affected their individual and team performance. As part of this work, Paul logged over 145 flights observing pilots on the flight deck during commercial airline operations.

Pamela S. Tsang is an associate professor at Wright State University in Dayton, Ohio. Previously, she was a postdoctoral research associate at NASA Ames Research Center. She received her A.B. from Mount Holyoke College and her Ph.D. from the University of Illinois at Urbana–Champaign in experimental/engineering psychology. Her current research interests are aviation psychology, attention and time-sharing performance, mental workload, and cognitive aging.

Michael A. Vidulich received his B.A. from the State University College of New York at Potsdam in 1977, M.A. from the Ohio State University in 1979, and Ph.D. from the University of Illinois in 1983, where Christopher Wickens was his graduate adviser. In the mid-1980s he worked in the Human Performance and Workload group at NASA Ames Research Center. As part of that group, he helped to develop the NASA TLX and SWORD subjective workload measurement tools. Since 1987, he has worked at Wright-Patterson Air Force Base on the development of mental workload and situation awareness metrics and their application to Air Force systems. He was on the editorial board of *Human Factors* from 1990 to 2000. He is also a member of the adjunct faculty at Wright State University.

Christopher D. Wickens is currently a professor of experimental psychology, head of the Aviation Research Laboratory, and associate director of the Institute of Aviation at the University of Illinois at Urbana–Champaign. He also holds an appointment in the Department of Mechanical and Industrial Engineering and the Beckman Institute of Science and Technology. He received his A.B. degree from Harvard College in 1967 and his Ph.D. from the University of Michigan in 1974 and served as a commissioned officer in the U.S. Navy from 1969 to 1972. He is a Vietnam veteran. His research interests involve the application of the principles of human attention, perception, and cognition to modeling operator performance in complex environments, and to designing displays to support that performance. Particular interest has focused on aviation and air traffic control. Dr. Wickens is a member and Fellow of the Human Factors and Ergonomics Society and received the society's Jerome H. Ely Award in 1981 for the best article in the *Human Factors* journal, and the society's Paul M. Fitts Award in 1985 for outstanding contributions to the education and training of human factors specialists. He was elected to the Society of Experimental Psychologists. He was also elected Fellow of the American Psychological Association, and in 1983 he received the Franklin Taylor Award for Outstanding Contributions to Engineering Psychology from Division 21 of that association. In 2000 he received the Henry L. Taylor Founder's Award from the Aerospace Human Factors Association. He served as a Distinguished Visiting Professor at the Department of Behavioral Sciences and Leadership, U.S. Air Force Academy in 1983–1984, 1991–1992, and 1999–2000. From 1995 to 1998 he chaired a National Research Council Panel to Examine the Human Factors of Automation of the Air Traffic Control System. In 1995 he was a Luckman Award finalist for Excellence in Undergraduate Education at the University of Illinois. In 2001 he received the Federal Aviation Administration's Excellence in Aviation Award for his continued contributions in aviation research and education. He is the author or editor of numerous books, including *Engineering Psychology and Human Performance* and *Flight to the Future: Human Factors in Air Traffic Control,* and has published over 145 book chapters or articles in peer-refereed publications. He is an avid mountain climber.

Laurence R. Young, Sc.D., is the Apollo Program Professor of Astronautics at the Massachusetts Institute of Technology (MIT), and the director of the National Space Biomedical Research Institute. In 1962, he joined the MIT faculty and co-founded the Man-Vehicle Laboratory, which does research on the visual and vestibular systems, visual-vestibular interactions, space motion sickness, flight simulation, and manual control and displays. In 1991 Prof. Young was selected as a Payload Specialist for Spacelab Life Sciences 2. He spent two years training at Johnson Space Center and served as Alternate Payload Specialist during the October 1993 mission. He has been principal investigator on five Spacelab experiments. His contributions to the field of human factors research have been recognized by the Paul Hansen Award of the Aerospace Human Factors Association, the Dryden Lectureship in Research, the Jeffries Medical Research Award of the American Institute of Aeronautics and Astronautics (AIAA), and the Franklin Taylor Award of the Institute of Electrical and Electronic Engineers (IEEE). Most recently he was awarded the prestigious Koetser Foundation prize for his contributions to neuroscience. A member of the National Academy of Engineering, the Institute of Medicine, and the International Academy of Astronautics, he is the author of more than 250 journal articles, largely in the areas of space physiology and human factors, including review chapters in *Fundamentals of Aerospace Medicine* and the *Handbook of Physiology.* His translation of Ernst Mach's *Fundamentals of the Theory of Movement Perception* was published in December 2001.

Author Index

All page numbers in italic type refer to citations in the references of each chapter.

B

C

D

I

J

K

N

Q

R

Subject Index

A

D

I

J